Biodiversity, Ecology and Conservation

Biodiversity, Ecology and Conservation

Editor: Isaac Hughes

www.callistoreference.com

Callisto Reference,
118-35 Queens Blvd., Suite 400,
Forest Hills, NY 11375, USA

Visit us on the World Wide Web at:
www.callistoreference.com

ISBN: 978-1-63239-936-6 (Hardback)

Printed in China.

Cataloging-in-Publication Data

Biodiversity, ecology and conservation / edited by Isaac Hughes.
p. cm.
Includes bibliographical references and index.
ISBN 978-1-63239-936-6
1. Biodiversity. 2. Ecology. 3. Biodiversity conservation. I. Hughes, Isaac.
QH541.15.B56 B56 2018
333.95--dc23

Table of Contents

Preface IX

Chapter 1 **Computer-Assisted Photo Identification Outperforms Visible Implant Elastomers in an Endangered Salamander, *Eurycea tonkawae*** 1
Nathan F. Bendik, Thomas A. Morrison, Andrew G. Gluesenkamp, Mark S. Sanders, Lisa J. O'Donnell

Chapter 2 **Assessing Environmental DNA Detection in Controlled Lentic Systems** 9
Gregory R. Moyer, Edgardo Díaz-Ferguson, Jeffrey E. Hill, Colin Shea

Chapter 3 **Threatened and Endangered Subspecies with Vulnerable Ecological Traits also have High Susceptibility to Sea Level Rise and Habitat Fragmentation** 18
Allison M. Benscoter, Joshua S. Reece, Reed F. Noss, Laura A. Brandt, Frank J. Mazzotti, Stephanie S. Romañach, James I. Watling

Chapter 4 ***Conus:* First Comprehensive Conservation Red List Assessment of a Marine Gastropod Mollusc Genus** 28
Howard Peters, Bethan C. O'Leary, Julie P. Hawkins, Kent E. Carpenter, Callum M. Roberts

Chapter 5 **Tourism Revenue as a Conservation Tool for Threatened Birds in Protected Areas** 40
Rochelle Steven, J. Guy Castley, Ralf Buckley

Chapter 6 **Optimized Spatial Priorities for Biodiversity Conservation in China: A Systematic Conservation Planning Perspective** 48
Ruidong Wu, Yongcheng Long, George P. Malanson, Paul A. Garber, Shuang Zhang, Diqiang Li, Peng Zhao, Longzhu Wang, Hairui Duo

Chapter 7 **Conservation Investment for Rare Plants in Urban Environments** 59
Mark W. Schwartz, Lacy M. Smith, Zachary L. Steel

Chapter 8 **Short-Term Dispersal Response of an Endangered Australian Lizard Varies with Time of Year** 68
Mehregan Ebrahimi, C. Michael Bull

Chapter 9 **The Implications of Habitat Management on the Population Viability of the Endangered Ohlone Tiger Beetle (*Cicindela ohlone*) Metapopulation** 75
Tara M. Cornelisse, Michelle K. Bennett, Deborah K. Letourneau

Chapter 10 **Predicting Environmental Suitability for a Rare and Threatened Species (Lao Newt, *Laotriton laoensis*) using Validated Species Distribution Models** 83
Amanda J. Chunco, Somphouthone Phimmachak, Niane Sivongxay, Bryan L. Stuart

Chapter 11 **The Genetic Integrity of the *Ex Situ* Population of the European Wildcat (*Felis silvestris silvestris*) is Seriously Threatened by Introgression from Domestic Cats (*Felis silvestris catus*)**.....96
Kathrin A Witzenberger, Axel Hochkirch

Chapter 12 **DNA Barcode Identification of Podocarpaceae—The Second Largest Conifer Family**.....108
Damon P. Little, Patrick Knopf, Christian Schulz

Chapter 13 **Mammalian Collection on Noah's Ark: The Effects of Beauty, Brain and Body Size**.....119
Daniel Frynta, Olga Šimková, Silvie Lišková, Eva Landová

Chapter 14 **The Push and Pull of Land use Policy: Reconstructing 150 Years of Development and Conservation Land Acquisition**.....131
Maria João Santos, Terry Watt, Stephanie Pincetl

Chapter 15 **Managing Conservation Reliant Species: Hawai'i's Endangered Endemic Waterbirds**.....140
Jared G. Underwood, Mike Silbernagle, Mike Nishimoto, Kim Uyehara

Chapter 16 **Growth Dynamics of the Threatened Caribbean Staghorn Coral *Acropora cervicornis*: Influence of Host Genotype, Symbiont Identity, Colony Size, and Environmental Setting**.....149
Diego Lirman, Stephanie Schopmeyer, Victor Galvan, Crawford Drury, Andrew C. Baker, Iliana B. Baums

Chapter 17 **A Longitudinal Study of Steller Sea Lion Natality Rates in the Gulf of Alaska with Comparisons to Census Data**.....158
John M. Maniscalco, Alan M. Springer, Pamela Parker, Milo D. Adkison

Chapter 18 **Genetic Diversity and Structure of *Sinopodophyllum hexandrum* (Royle) Ying in the Qinling Mountains, China**.....167
Wei Liu, Dongxue Yin, Jianjun Liu, Na Li

Chapter 19 **Great Apes and Biodiversity Offset Projects in Africa: The Case for National Offset Strategies**.....181
Rebecca Kormos, Cyril F. Kormos, Tatyana Humle, Annette Lanjouw, Helga Rainer, Ray Victurine, Russell A. Mittermeier, Mamadou S. Diallo, Anthony B. Rylands, Elizabeth A. Williamson

Chapter 20 **Fast Growing, Healthy and Resident Green Turtles (*Chelonia mydas*) at Two Neritic Sites in the Central and Northern Coast of Peru: Implications for Conservation**.....195
Ximena Velez-Zuazo, Javier Quiñones, Aldo S. Pacheco, Luciana Klinge, Evelyn Paredes, Sixto Quispe, Shaleyla Kelez

Chapter 21 **Genetic Variability and Population Structure of *Disanthus cercidifolius* subsp. *Longipes* (Hamamelidaceae) based on AFLP Analysis**.....207
Yi Yu, Qiang Fan, Rujiang Shen, Wei Guo, Jianhua Jin, Dafang Cui, Wenbo Liao

Chapter 22 **Sperm Preservation by Freeze-Drying for the Conservation of Wild Animals** **220**
Takehito Kaneko, Hideyuki Ito, Hidefusa Sakamoto, Manabu Onuma, Miho Inoue-Murayama

Chapter 23 **Trends in Extinction Risk for Imperiled Species in Canada** **224**
Brett Favaro, Danielle C. Claar, Caroline H. Fox, Cameron Freshwater, Jessica J. Holden, Allan Roberts, UVic Research Derby

Permissions

List of Contributors

Index

Preface

Over the recent decade, advancements and applications have progressed exponentially. This has led to the increased interest in this field and projects are being conducted to enhance knowledge. The main objective of this book is to present some of the critical challenges and provide insights into possible solutions. This book will answer the varied questions that arise in the field and also provide an increased scope for furthering studies.

Biodiversity studies the variety of life on Earth. It is vital for the survival of humans. The interaction of the organisms among themselves and with their environment is referred to as ecology. Their conservation is of utmost importance as it is impossible to sustain life without ecology or biodiversity. This book studies, analyses and upholds the pillars of biodiversity, ecology and conservation, and its utmost significance in modern times. It includes some of the vital pieces of work being conducted across the world, on various topics related to these areas of study. In this book, using case studies and examples, constant effort has been made to make the understanding of concepts of conservation as easy and informative as possible, for the readers.

I hope that this book, with its visionary approach, will be a valuable addition and will promote interest among readers. Each of the authors has provided their extraordinary competence in their specific fields by providing different perspectives as they come from diverse nations and regions. I thank them for their contributions.

Editor

1

Computer-Assisted Photo Identification Outperforms Visible Implant Elastomers in an Endangered Salamander, *Eurycea tonkawae*

Nathan F. Bendik[1]*, Thomas A. Morrison[2], Andrew G. Gluesenkamp[3], Mark S. Sanders[4], Lisa J. O'Donnell[4]

1 Watershed Protection Department, City of Austin, Austin, Texas, United States of America, **2** Wyoming Cooperative Fish and Wildlife Research Unit, Department of Zoology and Physiology, University of Wyoming, Laramie, Wyoming, United States of America, **3** Texas Parks and Wildlife Department, Austin, Texas, United States of America, **4** Balcones Canyonlands Preserve, City of Austin, Austin, Texas, United States of America

Abstract

Despite recognition that nearly one-third of the 6300 amphibian species are threatened with extinction, our understanding of the general ecology and population status of many amphibians is relatively poor. A widely-used method for monitoring amphibians involves injecting captured individuals with unique combinations of colored visible implant elastomer (VIE). We compared VIE identification to a less-invasive method – computer-assisted photographic identification (photoID) – in endangered Jollyville Plateau salamanders (*Eurycea tonkawae*), a species with a known range limited to eight stream drainages in central Texas. We based photoID on the unique pigmentation patterns on the dorsal head region of 1215 individual salamanders using identification software Wild-ID. We compared the performance of photoID methods to VIEs using both 'high-quality' and 'low-quality' images, which were taken using two different camera types and technologies. For high-quality images, the photoID method had a false rejection rate of 0.76% compared to 1.90% for VIEs. Using a comparable dataset of lower-quality images, the false rejection rate was much higher (15.9%). Photo matching scores were negatively correlated with time between captures, suggesting that evolving natural marks could increase misidentification rates in longer term capture-recapture studies. Our study demonstrates the utility of large-scale capture-recapture using photo identification methods for *Eurycea* and other species with stable natural marks that can be reliably photographed.

Editor: Jan M. Hemmi, University of Western Australia, Australia

Funding: Work was primarily funded by the City of Austin, through staff time and purchase of materials. This includes the City of Austin Balcones Canyonland Preserve and Watershed Protection Department. The Texas Parks and Wildlife Department also provided staff assistance with this project. The funding agencies had no role in study design, data collection and analysis, decision to publish, or preparation of the manuscript (with the exception of the authors and acknowledged individuals).

Competing Interests: The authors have declared that no competing interests exist.

* E-mail: nathan.bendik@austintexas.gov

Introduction

Global declines in amphibian abundance and diversity have received considerable attention in recent years because of the alarming rates of loss, particularly relative to other major taxonomic groups [1]. In some cases, entire populations or even species have disappeared in the course of a few years [2]. The swiftness of these declines, and our limited understanding of the general ecology and population status of many species (22.5% of amphibians are considered 'data deficient' in the IUCN species status classification [1]) provide strong motivation to develop more rapid, reliable and non-invasive methods for monitoring populations in the hopes of identifying major threats and developing conservation remedies [3].

Many of the most common and powerful methods for collecting demographic and movement data (e.g. capture-recapture) involve physically capturing, handling and marking study organisms and re-identifying them during subsequent surveys [4]. However, amphibians are often challenging to mark due to their darkly pigmented, sensitive skin and their small size. Common methods for tagging amphibians include tattooing, branding, toe-clipping, passive integrated transponder tagging and implanting colored elastomers under the skin of captured animals [5]. This latter method – visual implant elastomers (VIE) – has become popular because it can be used to identify both larval and adult stages (and metamorphosis between the two), though observer identification ability and mark retention are significant issues [6].

Photographic identification (photoID) is an increasingly popular technique used in capture-recapture studies. Advances in digital image analysis tools and pattern recognition algorithms have accelerated the application of photoID to a wide range of species with natural marking patterns, including amphibians [7,8,9], reptiles [10], terrestrial mammals [11] and fishes [12], among others (for review, see Table 1 in [13]). While computer-assisted photoID is only useful in species with variable natural markings, it has the advantage of being relatively inexpensive (entailing only a digital camera and computer), requiring only basic technical expertise beyond the development of image analysis tools, and allowing large numbers of individuals to be re-identified. In the context of monitoring populations of endangered or threatened species, developing survey methods that minimize the potential

Table 1. Summary of datasets used in false rejection rate (FRR) and false acceptance rate (FAR) error calculations.

Photo dataset	Years	#Indiv.	#Captures	#Matching photo pairs (recaptures)	Error Rates					
					VIE FRR	VIE FAR	VIE error sample size[a]	Photo FRR	Photo FAR	PhotoID error sample size[a]
Low-quality	2007	473	965	896	0.0190 (0.0003)	0.0178 (0.0002)	554	0.1591 (0.0001)	0.0132 (0.0004)	264
High-quality	2008–2010	742	1367	1090				0.0076 (0.0002)	0.0000 (0.0000)	356

Standard errors are included in parentheses.
[a]per iteration.

handling effects and welfare concerns related to animal capture is paramount [14].

Regardless of identification technique, if markings change over time or are relatively invariable across individuals, misidentification errors are likely. These errors are problematic in capture-recapture studies because they violate the assumption that individuals are correctly identified; when ignored, errors can bias demographic parameter estimates [15,16]. Often, studies do not assess or report probability of making misidentification errors [13], which makes it difficult to contrast the relative performance of one method against another (though see [5,8]).

Jollyville Plateau salamanders (*Eurycea tonkawae*) are endemic to central Texas and one of four closely-related species that are currently proposed for protection under the US Endangered Species Act [17]. Here, we test the performance of a computer-assisted photoID system and a VIE system in a wild population of *E. tonkawae* using available data on individuals captured over a four-year period. Our goals are to (1) illustrate the advantages and limitations of using either of these methods, with the bottom line being the ability to accurately identify individual salamanders; (2) determine the accuracy of automated photo matching using both 'high' and 'low-quality' photographs; and (3) provide tips and example code for determining whether photoID can be applied to a novel organism and/or research setting.

Materials and Methods

Study Organism and Study Site

Eurycea tonkawae is a small (total length typically <75 mm) neotenic salamander endemic to springs, spring-fed headwater streams, and wet caves in northwest Austin, Texas and surrounding areas, with a total range of <110 km^2 [18,19]. Our study site includes a small, occasionally intermittent spring (Lanier Spring) and adjacent stream bed (Bull Creek) within the City of Austin's Balcones Canyonland Preserve, Travis County, Texas. Salamanders were captured with aquarium nets using a drive survey technique that exposed all available surface cover (rocks, leaves, algae) within the study area (64 m^2). Sampling consisted of 42 total capture-recapture surveys consisting of 14 primary periods with consecutive 3-day secondary periods between 2007 and 2010.

Visible Implant Elastomers

Salamanders >16 mm snout-vent length were anaesthetized in a solution of 0.25 g Tricaine S (MS-222)/L of naturally-buffered spring water and then marked using VIE tags (Northwest Marine Technology Inc., Shaw Island, Washington). Sterile 28-gauge syringes were used to inject small amounts (2–20 μL) of elastomer just underneath the skin to form a bead. Each salamander was given three to four unique VIE tags using a combination of seven different colors in five locations on the body. Captured and recaptured salamanders were photographed (details below) initially so that their natural marks could serve as a secondary mark to check our VIE identifications. Using this photographic database, each recaptured salamander was compared to a previous capture of the same individual (based on VIE identification) by manually (i.e., 'by-eye') comparing their natural marks (melanophore and iridophore pigmentation patterns). This process allowed us to confirm or correct most VIE identifications, although a small number (<2%) of salamanders with missing or unidentifiable VIE tags were impossible to match manually. When referring to the identification techniques used throughout this paper, we use the phrase VIE as shorthand to refer to the entire VIE-based identification system, which includes this manual photo-matching validation step. Thus, any errors reported as 'VIE' errors occurred after both visually checking VIE tags and manually checking photos for confirmation of a match.

Animal Welfare

The anaesthetization procedure was reviewed and approved by a veterinary professional (Chris Sanders, DVM). The VIE tagging procedures we used were approved by the IACUC of the University of Texas at Austin for a separate study designed to evaluate the effects of marking on individual growth and performance of two closely related species, *Eurycea sosorum* and *E. nana*. (IACUC protocol #07072701). The co-author (AGG) of that study also conducted and oversaw all VIE tagging for the present study. Additionally, care was taken to ensure animals being held were maintained at the ambient water temperature by keeping them in flow-through mesh boxes within their habitat during processing or in containers with frequent water changes. The data we present here were not collected primarily for this study, but capture-recapture (and associated VIE tagging) was carried out by the City of Austin Watershed Protection Department for the purpose of understanding the population dynamics of *E. tonkawae*. Field collections were conducted under Texas Parks and Wildlife Scientific Permit SPR-1005-1515.

Computer-assisted PhotoID

The second identification technique involved computer-assisted salamander identification using digital photographs of pigmentation patterns in the dorsal head regions of individuals. Salamanders were photographed in a shallow water-filled tray with a white background initially using a Nikon Coolpix E995 'point-and-shoot' camera (for 24 surveys in 2007) but then were later photographed using an advanced DSLR system: Nikon D80, 90 mm macro lens and two close-up flashes (for 18 surveys from 2008–2010). Photos were taken handheld, with an effort to ensure that the focal plane and salamander dorsum were parallel. DSLR camera settings included an aperture of f/16 or smaller, a shutter speed of 1/160 s or faster, and high-quality JPEG settings.

After cropping each photograph to include only the head, all photographs in our database were compared and matched using open-source pattern identification software called 'Wild-ID' (ver. 1.0.1 [13]; http://www.dartmouth.edu/~envs/faculty/bolger.html). To date, Wild-ID has been used to accurately identify wildebeest (*Connochaetes taurinus* [20]) and giraffe (*Giraffa camelopardis* [13]). This software uses the SIFT (Scale Invariant Feature Transform) algorithm to characterize variable patterns within photographs and compare all combinations of photographs in a database [21]. SIFT is a convenient pattern-matching algorithm because keypoints are robust to variation in scale and rotation of the photograph [21]. SIFT localizes keypoints within each photo based on local information gradients (Figure 1). Image pairs are then scored and ranked based on the similarity of their keypoint maps. Similarity scores range from 0.0 to 1.0 and provide a standardized measure of pattern resemblance between each image pair. Similarity scores are determined by iteratively comparing geometrically self-consistent subsets of keypoints within pairs of images. During the scoring, each image pair is compared in only one direction (i.e., image A vs. image B, but not B vs. A) and matches are compiled sequentially in the order they were collected (for full details, see [13]).

Initial tests using Wild-ID to match photographs of *E. tonkawae* indicated that high scores were exceptionally predictive of correct matches, based on visual confirmation and VIE identification. Thus, we developed a two-step photo matching process to rapidly identify photo matches using Wild-ID (Figure 2). The majority of photo pairs could be accepted or rejected as matches simply based on similarity scores [13]. Examination of the distribution of similarity scores (Figure 3) suggested that scores above 0.1 likely indicated true matches in the high-quality dataset. Thus, we used this threshold as an automated cut-off: images above this threshold were deemed matches, while those below were considered potential matches. Below this threshold score of 0.1 we visually checked the top 100 highest scoring rank-1 image pairs. We chose 100 images for this criterion because it provided an acceptable balance among effort, speed and coverage of likely matches. Based on these two criteria, we generated a capture-history of photos for each unique individual. R scripts [22] used to generate capture histories are included in the supporting information (Files S1 and S2). We used the program 'Wdump' to compile scores of all pairwise image comparisons generated by Wild-ID (distributed with Wild-ID, ver. 1.0.1).

Since photograph quality can affect matching success [23], images for matching were segregated into two groups based on camera type: a 'low-quality' group (photos taken in 2007 with the point-and-shoot camera) and a 'high-quality' group (photos taken between 2008 and 2010 with the DSLR). Images from the two groups were of the same population and included many of the same individuals, but matching occurred only within (and not between) groups. The cameras used to acquire the images differed in resolution (3.14 vs. 10.2 megapixels), format (point-and-shoot vs. DSLR), and spanned several years of technological advances (e.g. release years of 2001 and 2006), resulting in perceptible differences in image quality. Overall, the low-quality dataset suffered from the combined effects of reduced resolution from poorer focus and a more variable subject angle relative to the high-quality dataset. These problems were exacerbated by the use of a lower quality camera, but can also be caused by the photographer's skill level and factors outside the researcher's control, such as the subject's size, speed, and behavior. The incidental use of these two different camera types/technologies in this dataset allowed us to examine the consequences of photo quality on computer-assisted photographic identification.

Performance of Identification Methods

To assess the accuracy of identification techniques, we calculated two error metrics common to biometric recognition studies: false rejection rates (FRR) and false acceptance rates (FAR) [24]. We define FRR as the frequency of failing to match two captures (either photos or VIE tags) of the same individual:

$$FRR = \frac{(\#\,\text{false rejections})}{(\#\,\text{identification attempts})}.$$

FAR is the frequency of falsely matching two captures (either photos or VIE tags) of different individuals and is calculated as.

$$FAR = \frac{(\#\,\text{false acceptances})}{(\#\,\text{identification attempts})}.$$

To detect errors in our datasets, we compared capture histories between the two identification methods and manually identified where errors occurred. We first identified VIE errors based on mismatches with the photoID dataset. Importantly, we assumed that visual confirmation of pigmentation patterns in photos provided the baseline for establishing 'truth' between two potential matches. For example, if VIEs indicated an incorrect ID for a high-scoring photoID match, we would examine the photos in question

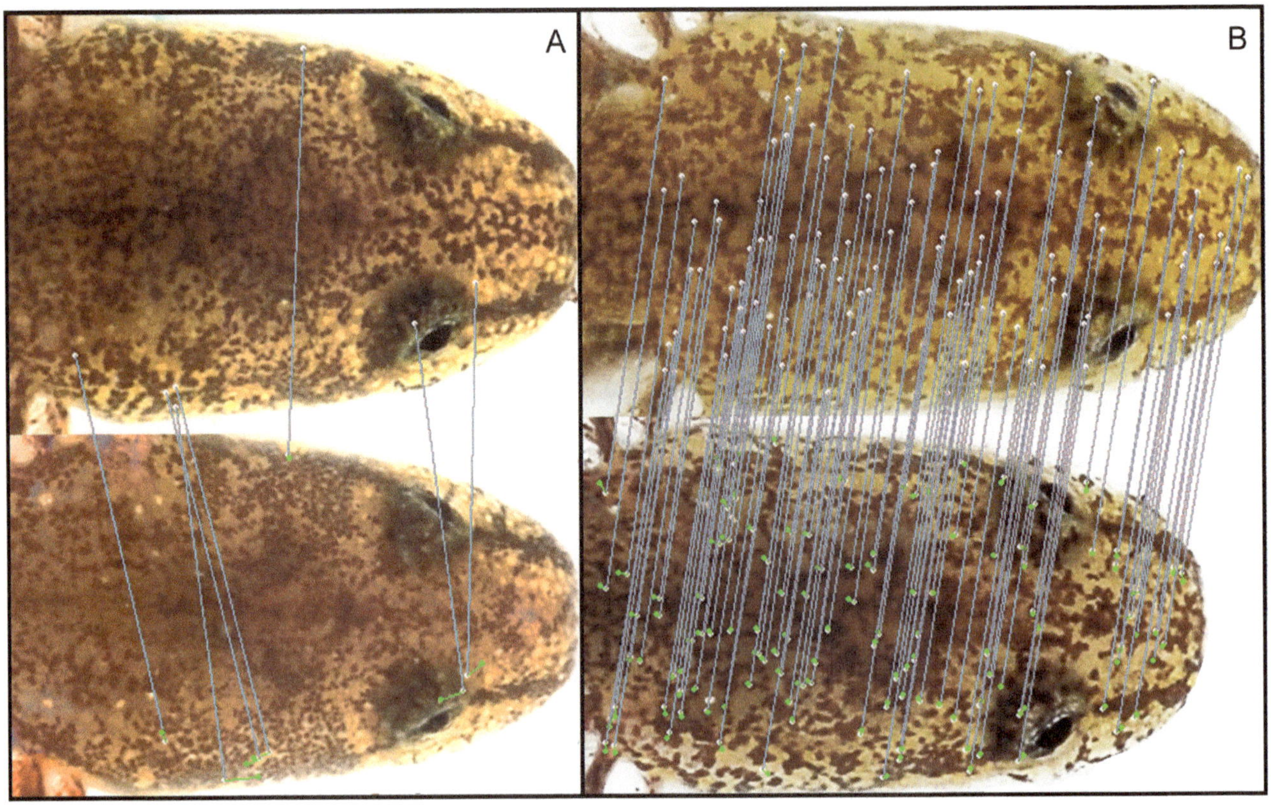

Figure 1. Head pattern recognition in *Eurycea tonkawae*. Pair of images from (A) two different individuals and (B) the same individual one year apart. Lines connect matching SIFT features.

to determine whether a photoID false acceptance or a VIE false rejection had occurred. Similarly, if photoID failed to match a VIE pair, we would visually check the photos from the two captures to determine whether the non-match was a VIE false acceptance or a photoID false rejection. We acknowledge the possibility that some errors could have been missed by the combined VIE and photoID identification methods. However, double false acceptance errors should be very infrequent given that we visually confirmed all true matches and pigmentation patterns have high variability between individuals (Figure 1). Further, the double false rejection rate (i.e., $FRR_{VIE} \times FRR_{photoID}$) and double false acceptance rate (i.e., $FAR_{VIE} \times FAR_{photoID}$) were extremely low in this study (see Results).

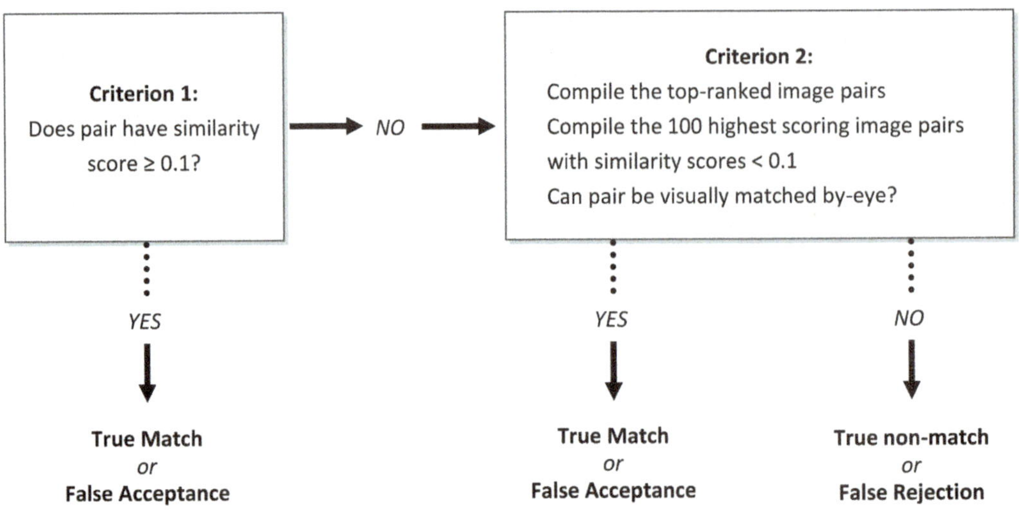

Figure 2. Diagram of the two-step image-matching process. Image pairs above the score threshold of 0.1 were considered matches (criterion 1). The top 100 first-ranked image pairs below 0.1 were compared by eye.

One difficultly we encountered in calculating error rates was that individuals recaptured more than once had multiple, non-independent recapture pairs. Pairs of captures from individuals with large numbers of captures would have been over-represented relative to those with only a few captures. Therefore, we randomly selected a single pair of captures for each unique individual and used this subset of data as the basis for calculating expected error rates. We repeated this subsampling procedure 1000 times and reported the mean error rates across all iterations.

Performance Over Time

Some natural marking patterns are known to change over time [25], a process that reduces the probability of correctly matching photographs. Therefore, we tested whether the time interval between captures influenced similarity scores in each photo dataset. Again, because some individuals were recaptured on more than one occasion, we used a subsampling procedure that selected a single photo pair per individual. Since many recaptures occur over short time intervals, we made the probability of selecting a particular photo pair proportional to the frequency with which 100-day time intervals were represented in the dataset in order to ensure longer periods were included in subsamples. Given the subsample, we tested whether similarity scores were inversely related to the time interval between captures. We used a Spearman rank test because of heteroscedasticity and lack of normality in similarity scores over time. We repeated the subsampling procedure 1000 times and reported the proportion of iterations in which similarity scores declined significantly over time, where our criterion was $P(\rho=0) < 0.01$. Additionally, each photo incorrectly rejected by photoID was visually compared to true matches to quantify the mismatches due to either changing natural marks or photo quality. All statistical computations were performed using R [22].

Results

During 2007, a total of 473 salamanders were marked with VIE, photographed, released, and recaptured (mean no. recaptures = 0.9) over eight primary sampling occasions (24 total surveys), resulting in 965 low-quality photos. Six sampling occasions (18 total surveys) from 2008–2010 resulted in 742 VIE-marked individuals and 1367 high-quality photos (mean no. recaptures = 0.8). None of the 1213 marked salamanders had lost all of their VIE tags, although three individuals were double-marked (i.e. inadvertently marked on two separate occasions) due to VIE misidentification errors. We did not observe any cases where natural marks conflicted with VIE tags when examining image pairs post-hoc to evaluate type and source of error.

VIE misidentification errors were detected by visually examining photoID mismatched image pairs. VIE tags were 2.5-fold more likely to generate false rejections than the high-quality photoID method, and were more likely to generate false acceptances than both photoID datasets (Table 1). FAR and FRR were similar for VIE tags due to the nature of the VIE error-generating process (Table 1). For all VIE false rejections, the mean time to the first misidentification was 10.2 months, or an average of 3.7 primary periods and 1.6 recaptures before an individual's first VIE misidentification error.

Computer-assisted photoID produced high matching success, particularly in the high-quality dataset. Based on bootstrapped pairwise comparisons, photoID of high-quality photos produced an FRR 20 times lower than the low-quality dataset (Table 1). The FAR was also much higher for the low-quality dataset compared to an FAR of zero for high-quality photos (Table 1). Visual examination of the falsely-rejected photo pairs (57 individuals) revealed that only 5% of photo errors were the result of evolving natural marks, and 11% were the result of both evolving marks and poor photo quality (e.g. blurry or low resolution photos). The remaining errors (84%) were due solely to poor photo quality.

Similarity scores declined over time in both low-quality ($\rho = -0.31$; $P<0.01$ for 100% of iterations) and high-quality ($\rho = -0.71$; $P<0.01$ for 100% of iterations) datasets, suggesting that matching performance decreases as the time interval between two matching photographs increases (Figure 3).

Discussion

One of the most powerful and flexible methods for monitoring animal populations is capture-recapture, a technique that requires accurate identification of individuals over time. Our computer-assisted identification scheme using Wild-ID software exhibited extremely high success for identifying individual salamanders over time with greater accuracy than nearly all other computer-assisted identification studies where error rates have been clearly reported [13]. Scores of rank-1 matches only overlapped rank-1 non-matches marginally (Figure 3) in our high-photo-quality dataset, highlighting the exceptional discriminatory power of score-based identification for *E. tonkawae*. However, false rejection and false acceptance rates increase dramatically when using lower-quality photographs and score-based image matching. Thus, the advantage of semi-automation via score-sorting diminishes when photo quality is poor, requiring visual inspection for more image pairs. Our results suggest that automated photo matching with Wild-ID in which correct matches can be distinguished based on score alone may be suitable for large data sets where photo quality is high. Despite the enormous time-savings compared to manual photo matching [11], computer-assisted identification can still be a time-consuming process for studies that include thousands of images. We suspect the approach using score-based filtering will be most useful to researchers who obtain high-resolution photographs of animals that can be posed with consistent lighting. This technique could be particularly powerful for obtaining rapid (several days of surveying) estimates of population abundance, as it does not require specialized equipment or extensive training.

The predictive ability of Wild-ID scores to correctly identify individuals allowed us to easily cross-check the accuracy of VIE-based matches. Several studies have noted tag loss and migration of VIE tags in amphibians [6,26], which can violate the capture-recapture assumption that marks do not change over time. Although we did not track retention of individual marks, overall VIE mark retention was generally high, which is consistent with observations from other salamander studies [5,27,28]. However, even with high mark retention, VIE read errors and data recording errors can contribute to misidentifications. To reduce identification error, we manually compared photographs of VIE captures and recaptures during data entry to confirm VIE identifications. Despite this, we still encountered instances of inadvertent VIE double-marking and loss/misidentification of more than one VIE mark. These types of errors were extremely difficult to correct during the VIE data validation step (i.e., manual photo-match confirmation) due to the large number of possible matches in our data set. However, these VIE errors were easily identified by computer-assisted photoID. Our analysis revealed that VIE identification had an FRR 2.5 times higher than for photoID. Overall, photoID of *E. tonkawae* resulted in a false rejection rate that was considerably lower than other photoID studies with comparably-sized databases (for review, see [13]). FAR was higher in the VIE dataset than both photoID datasets

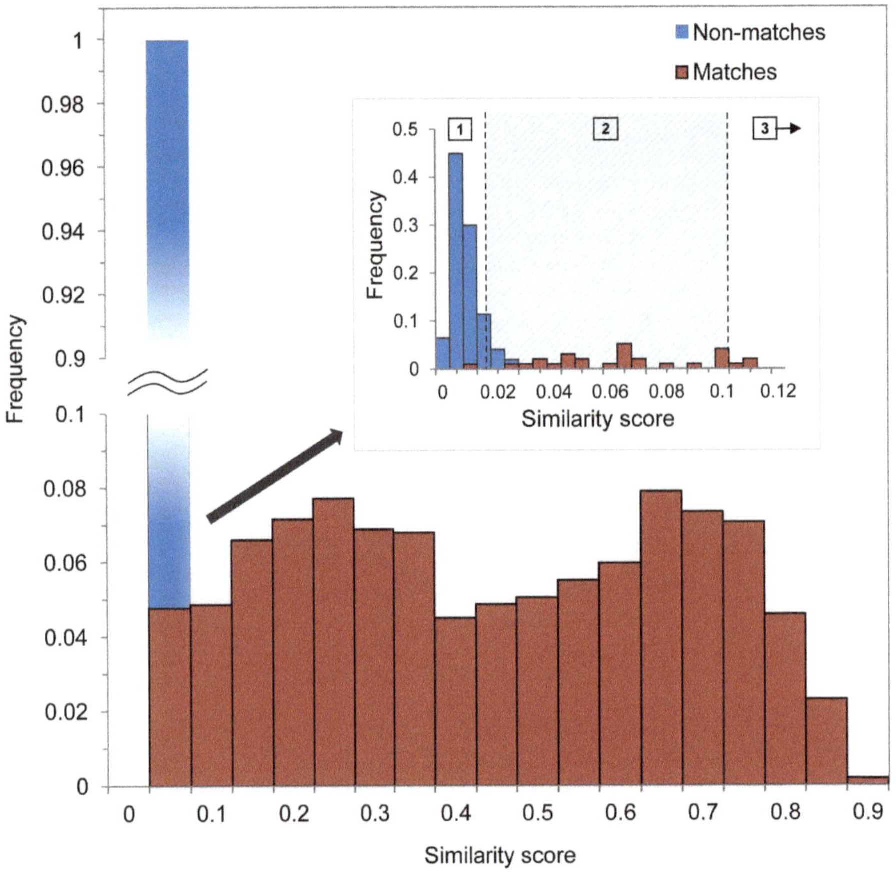

Figure 3. Similarity scores and matching success. Frequency of similarity scores for image pairs from different (black) and the same individuals (grey) from the high-quality dataset. Inset shows the lower range of similarity scores for top ranked image pairs only. The shaded region between dashed lines (2) indicates the range (similarity scores: 0.017–0.1) in which we visually (by-eye) compared 100 potential matching pairs. Similarity scores above this range (3) always involved photo pairs from the same individuals (according to VIE tags). Below this range (1), all but one photo pair came from non-matching individuals (according to VIE tags).

(Table 1). Unlike false rejections, false acceptance errors do not lead to new unique identifications. Thus, false rejection errors produce relatively greater bias in capture-recapture model estimates [15] and are of greater concern in capture-recapture studies.

Our experience with *Eurycea* salamanders suggests that VIE errors and photoID errors are mostly due to lost or evolving marks, respectively. It is important to note that VIE errors almost always result in false acceptances and false rejections. There is a fixed (known) number of VIE color combinations in any study, and if a novel color combination is observed, the researcher knows that an error occurred. In contrast, there are effectively infinite pigmentation patterns, so evolving patterns typically produce a false rejection but not a false acceptance [11]. Thus, some caution should be taken when comparing error rates across identification methods because their effect on the structure of capture histories, and the way in which errors would be modeled, are different.

Because of high identification accuracy of existing pigmentation patterns, the majority of errors in semi-automated photo-matching for this and similar species are more likely to occur due to evolving natural marks (if photo quality is high). The most common example of evolving natural marks we encountered was due to melanophore expansion and contraction, rendering the overall look of a salamander either lighter or darker. Rapid color changes have been observed in larval *Ambystoma* [29] and we documented individuals exhibiting dramatic changes in melanophore size within 24 hours. Patterns also change due to growth, development and gravidity [7]. For example, melanophores in young individuals expand during growth, making it difficult to recognize individuals that have considerably changed size since their last observation. Despite these potential challenges, error rates were still low when using high-quality images across all time intervals.

Although evolving natural marks did not significantly inhibit the accuracy of photoID of *E. tonkawae* during this four-year study, there was a negative relationship between similarity scores of matching pairs and the time interval between successive captures (Figure 4). Our results suggest matching success remains high, even across long intervals, as long as image quality is high. Lower quality matching image pairs have lower similarity scores over time and likely generate higher error rates. Species whose natural marks change quickly over time, either seasonally or annually, will require relatively short intervals between capture periods and high capture rates within periods to maintain sufficiently low error rates. Thus, the relationship between misidentification rate and time interval between captures depends on (1) how quickly marks evolve between successive captures, and (2) on the quality of photographs.

Additional optimization of the identification software and photo-collection methods could have conceivably improved our

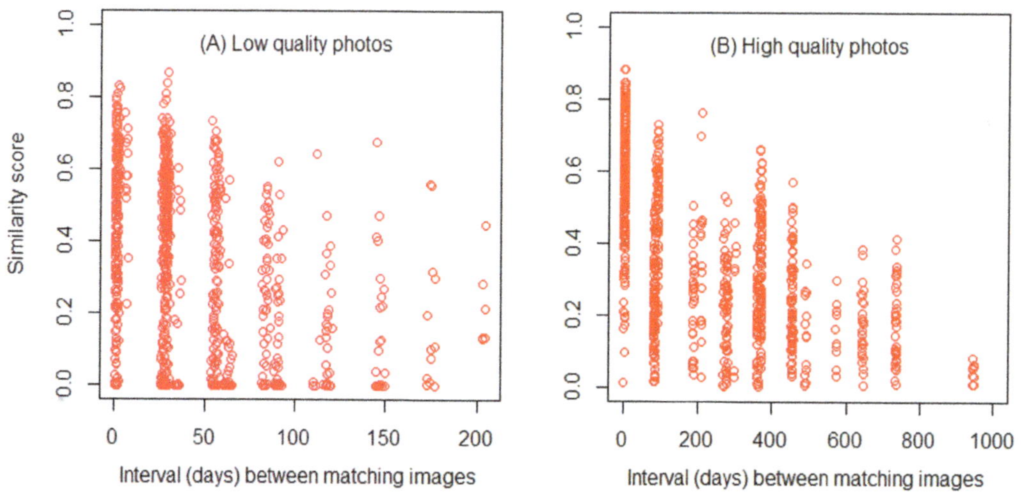

Figure 4. Effect of time on similarity score. Similarity scores versus days between photo captures of the same individual for (A) low-quality ($n = 896$) and (B) high-quality photos ($n = 1090$). Tables.

matching success. Given the low initial error rates observed with Wild-ID, we did not test other software platforms nor attempt to further optimize the Wild-ID scoring algorithm. Photo ID applications aimed at species with more challenging natural marking patterns [20] or species with evolving marks [19] may require either optimization of the Wild-ID scoring algorithm [13] or modeling of the error process. Indeed, new methods are already being developed to accommodate photoID errors in capture-recapture modeling [30] including errors involving evolving natural marks [16].

The use of photo-only based identification for *Eurycea tonkawae* has a number of advantages over VIE based approaches in addition to substantial reductions in false rejection and false acceptance rates. PhotoID capture-recapture is (1) less invasive, as it does not require injections or anaesthetization; (2) faster, both in terms of time required in the field to mark individuals and time in the office to process data; (3) less expensive, since it requires less time and overall equipment costs are lower for large studies (camera setup has high initial cost, but VIE consumables are expensive and frequently need replacement); and (4) requires less experience, since it is easier to produce a quality photograph than to inject VIE tags into a small salamander in the field. Photographic identification with Wild-ID shows considerable promise as a substitute for VIE marking in spring-dwelling *Eurycea*, and potentially other taxa as well.

Conservation Implications

The development of an inexpensive, fast, accurate and relatively non-invasive method for tracking *Eurycea* is timely. Four salamander species (*E. chisholmensis*, *E. naufragia*, *E. tonkawae* and *E. waterlooensis*) endemic to central Texas were recently proposed for protection under the Endangered Species Act of 1973, highlighting the need for a better understanding of their ecology [17]. Until recently, most published research on central Texas *Eurycea* has focused on their unique morphological variation [31–34], physiology [35,36], distribution [37] and taxonomy [18,38,39]. Despite ample scientific interest and numerous anthropogenic threats facing these species [40], surprisingly few studies have focused on understanding the population dynamics, life history or dispersal of any of the central Texas *Eurycea* salamanders (though see [41,42]). These facets of the salamanders' ecology are central to conservation efforts and largely require individual-level identification. While recent research is beginning to shed light on the ecology of spring-dwelling *E. naufragia* [43,44] and *E. tonkawae* [19,45], much remains to be learned about the ecology and population dynamics of these and other central Texas *Eurycea*. The advantages of photography-based identification over VIE marking, as demonstrated here, may facilitate an expansion of ecological knowledge about these endangered, karst-dwelling species that will ultimately help guide conservation efforts and sound management practices.

Acknowledgments

Bennet Vance, Blake Sissel, Rob Clayton, Liza Colucci, Alisha Shah, Leah Gluesenkamp, Josh O'Brien and the Texas Parks and Wildlife Department were instrumental in this research. We are grateful to the two referees and the associate editor for their constructive comments and suggestions.

Author Contributions

Provided critical review and revision of article: LJO AGG MSS. Conceived and designed the experiments: NFB TAM LJO AGG MSS. Performed the experiments: NFB TAM LJO AGG MSS. Analyzed the data: NFB TAM. Contributed reagents/materials/analysis tools: NFB AGG MSS LJO. Wrote the paper: NFB TAM.

References

1. Stuart SN, Chanson JS, Cox NA, Young BE, Rodrigues ASL, et al. (2004) Status and trends of amphibian declines and extinctions worldwide. Science 306: 1783–1786. doi:10.1126/science.1103538.
2. Wake DB, Vredenburg VT (2008) Are we in the midst of the sixth mass extinction? A view from the world of amphibians. Proc Natl Acad Sci U S A 105: 11466–11473. doi:10.1073/pnas.0801921105.

3. Lips KR, Brem F, Brenes R, Reeve JD, Alford RA, et al. (2006) Emerging infectious disease and the loss of biodiversity in a Neotropical amphibian community. Proc Natl Acad Sci U S A 103: 3165–3170.
4. Amstrup SC, McDonald TL, Manly BFJ (2010) Handbook of Capture-Recapture Analysis. Princeton: Princeton University Press. 335 p.
5. Bailey LL (2004) Evaluating elastomer marking and photo identification methods for terrestrial salamanders: Marking effects and observer bias. Herpetol Rev 35: 38–41.
6. Grant EHC (2008) Visual implant elastomer mark retention through metamorphosis in amphibian larvae. J Wildl Manage 72: 1247–1252. doi:10.2193/2007-183.
7. Church DR, Bailey LL, Wilbur HM, Kendall WL, Hines JE (2007) Iteroparity in the variable environment of the salamander *Ambystoma tigrinum*. Ecology 88: 891–903. doi: 10.1890/06-0896.
8. Gamble L, Ravela S, McGarigal K (2008) Multi-scale features for identifying individuals in large biological databases: an application of pattern recognition technology to the marbled salamander *Ambystoma opacum*. J Appl Ecol 45: 170–180. doi:10.1111/j.1365-2664.2007.01368.x.
9. Knox CD, Cree A, Seddon PJ (2012) Accurate identification of individual geckos (*Naultinus gemmeus*) through dorsal pattern differentiation. New Zealand Journal of Ecology 37: 7. Available: http://www.nzes.org.nz/nzje/new_issues/NZJEcol37_1_KnoxIP.pdf. Accessed 2012 Nov 11.
10. Sacchi R, Scali S, Pellitteri-Rosa D, Pupin F, Gentilli A, et al. (2010) Photographic identification in reptiles: a matter of scales. Amphibia-Reptilia 31: 489–502. doi:10.1163/017353710X521546.
11. Morrison TA, Yoshizaki J, Nichols JD, Bolger DT (2011) Estimating survival in photographic capture–recapture studies: overcoming misidentification error. Methods Ecol Evol 2: 454–463. doi:10.1111/j.2041-210X.2011.00106.x.
12. Arzoumanian Z, Holmberg J, Norman B (2005) An astronomical pattern-matching algorithm for computer-aided identification of whale sharks *Rhincodon typus*. J Appl Ecol 42: 999–1011. doi:10.1111/j.1365-2664.2005.01117.x.
13. Bolger DT, Morrison TA, Vance B, Lee D, Farid H (2012) A computer-assisted system for photographic mark–recapture analysis. Methods Ecol Evol 3: 813–822. doi:10.1111/j.2041-210X.2012.00212.x.
14. McMahon C, van den Hoff J, Burton H (2005) Handling intensity and the short- and long-term survival of elephant seals: Addressing and quantifying research effects on wild animals. Ambio 34: 426–429. doi:10.1639/0044-7447(2005)034[0426:HIATSA]2.0.CO;2.
15. Lukacs PM, Burnham KP (2005) Estimating population size from DNA-based closed capture–recapture data incorporating genotyping error. J Wildl Manage 69: 396–403. doi:10.2193/0022-541X(2005)069<0396:EPSFDC>2.0.CO;2.
16. Yoshizaki J, Pollock KH, Brownie C, Webster RA (2009) Modeling misidentification errors in capture-recapture studies using photographic identification of evolving marks. Ecology 90: 3–9. doi:10.1890/08-0304.1.
17. Fish and Wildlife Service (2012) Endangered and threatened wildlife and plants; Endangered status for four central Texas salamanders and designation of critical habitat; Proposed rule. Federal Register 77: 50767–50854.
18. Chippindale PT, Price AH, Wiens JJ, Hillis DM (2000) Phylogenetic relationships and systematic revision of central Texas hemidactyliine plethodontid salamanders. Herpetological Monographs 14: 1–80. doi:10.2307/1467045.
19. Bowles BD, Sanders MS, Hansen RS (2006) Ecology of the Jollyville Plateau salamander (*Eurycea tonkawae*: Plethodontidae) with an assessment of the potential effects of urbanization. Hydrobiologia 553: 111–120. doi:10.1007/s10750-005-5440-0.
20. Morrison TA, Bolger DT (2012) Wet season range fidelity in a tropical migratory ungulate. J Anim Eco 81: 543–552. doi:10.1111/j.1365-2656.2011.01941.x.
21. Lowe DG (2004) Distinctive image features from scale-invariant keypoints. Int J Comput Vis 60: 91–110. doi:10.1023/B:VISI.0000029664.99615.9422.
22. R Development Core Team (2012) R: A language and environment for statistical computing. Vienna, Austria: R Foundation for Statistical Computing. Available: http://www.R-project.org/. Accessed 2013 Feb 18.
23. Kelly MJ (2001) Computer-aided photograph matching in studies using individual identification: an example from Serengeti cheetahs. J Mammal 82: 440–449. doi: 10.1644/1545-1542(2001)082<0440:CAPMIS>2.0.CO;2.
24. Jain AK (2007) Technology: Biometric recognition. Nature 449: 38–40. doi:10.1038/449038a.
25. Hastings KK, Hiby LA, Small RJ (2008) Evaluation of a computer-assisted photograph-matching system to monitor naturally marked harbor seals at Tugidak Island, Alaska. J Mammal 89: 1201–1211. doi:10.1644/07-MAMM-A-151.1.
26. Campbell TS, Irvin P, Campbell KR, Hoffmann K, Dykes ME, et al. (2009) Evaluation of a new technique for marking anurans. Appl Herpetol 6: 247–256. doi:10.1163/157075409X420042.
27. Davis TM, Ovaska K (2001) Individual recognition of amphibians: effects of toe clipping and fluorescent tagging on the salamander *Plethodon vehiculum*. J Herpetol 35: 217–225. doi:10.2307/1566111.
28. Phillips CT, Fries JN (2009) An evaluation of visible implant elastomer for marking the federally listed fountain darter and the San Marcos salamander. N Am J Fish Manage 29: 529–532. doi:10.1577/M08-138.1.
29. Fernandez PJJ, Collins JP (1988) Effect of environment and ontogeny on color pattern variation in Arizona tiger salamanders (*Ambystoma tigrinum nebulosum* Hallowell). Copeia 1988: 928–938.
30. Link WA, Yoshizaki J, Bailey LL, Pollock KH (2010) Uncovering a latent multinomial: Analysis of mark–recapture data with misidentification. Biometrics 66: 178–185. doi:10.1111/j.1541-0420.2009.01244.x.
31. Emerson ET (1905) General anatomy of *Typhlomolge rathbuni*. Proceedings of the Boston Society of Natural History 32: 43–76.
32. Wake DB (1966) Comparative osteology and evolution of the lungless salamanders, family Plethodontidae. Memoirs of the Southern California Academy of Sciences 4: 1–111.
33. Wiens JJ, Chippindale PT, Hillis DM (2003) When are phylogenetic analyses misled by convergence? A case study in Texas cave salamanders. Syst Biol 52: 501–514. doi:10.1080/10635150390218222.
34. Clemen G, Sever DM, Gerven H (2009) Notes on the cranium of the paedomorphic *Eurycea rathbuni* (Stejneger, 1896) (Urodela: Plethodontidae) with special regard to the dentition. Vertebrate Zoology 59: 157–168.
35. Kezer J (1952) Thyroxin-induced metamorphosis of the neotenic salamanders *Eurycea tynerensis* and *Eurycea neotenes*. Copeia 1952: 234–237.
36. Sever D (1985) Sexually dimorphic glands of *Eurycea nana*, *Eurycea neotenes* and *Typhlomolge rathbuni* (Amphibia, Plethodontidae). Herpetologica 41: 71–84.
37. Sweet SS (1982) A distributional analysis of epigean populations of *Eurycea neotenes* in central Texas, with comments on the origin of troglobitic populations. Herpetologica 38: 430–444.
38. Sweet SS (1984) Secondary contact and hybridization in the Texas cave salamanders *Eurycea neotenes* and *E. tridentifera*. Copeia 1984: 428–441. doi:10.2307/1445201.
39. Hillis DM, Chamberlain DA, Wilcox TP, Chippindale PT (2001) A new species of subterranean blind salamander (Plethodontidae: Hemidactyliini: *Eurycea: Typhlomolge*) from Austin, Texas, and a systematic revision of central Texas paedomorphic salamanders. Herpetologica 57: 266–280.
40. Chippindale PT, Price AH (2005) Conservation of Texas spring and cave salamanders (*Eurycea*). In: Lannoo M, editor. Amphibian Declines: The conservation status of United States species. Berkley and Los Angeles: University of California Press. 193–197.
41. Bruce R (1976) Population structure, life history and evolution of paedogenesis in salamander *Eurycea neotenes*. Copeia 1976: 242–249.
42. Tupa D, Davis W (1976) Population dynamics of San Marcos salamander, *Eurycea nana* Bishop. Tex J Sci 27: 179–195.
43. Biagas TD, Hall AS, Ritzer AL, Pierce BA (2012) Time of day does not affect detection in visual-encounter surveys of a spring-dwelling salamander, *Eurycea naufragia*. Southwest Nat 57: 162–165. doi:10.1894/0038-4909-57.2.162.
44. Pierce BA, Christiansen JL, Ritzer AL, Jones TA (2010) Ecology of Georgetown salamanders (*Eurycea naufragia*) within the flow of a spring. Southwest Nat 55: 291–297. doi:10.1894/WL-30.1.
45. Bendik NF, Gluesenkamp AG (2013) Body length shrinkage in an endangered amphibian is associated with drought. J Zool. doi:10.1111/jzo.12009.

2

Assessing Environmental DNA Detection in Controlled Lentic Systems

Gregory R. Moyer[1]*, Edgardo Díaz-Ferguson[1,4], Jeffrey E. Hill[2], Colin Shea[3]

1 United States Fish and Wildlife Service, Conservation Genetics Laboratory, Warm Springs, Georgia, United States of America, 2 Program in Fisheries and Aquatic Sciences Tropical Aquaculture Laboratory, University of Florida, Ruskin, Florida, United States of America, 3 Department of Biology, Tennessee Technological University, Cookeville, Tennessee, United States of America, 4 Department of Fisheries and Allied Aquacultures, Auburn University, Auburn, Alabama, United States of America

Abstract

Little consideration has been given to environmental DNA (eDNA) sampling strategies for rare species. The certainty of species detection relies on understanding false positive and false negative error rates. We used artificial ponds together with logistic regression models to assess the detection of African jewelfish eDNA at varying fish densities (0, 0.32, 1.75, and 5.25 fish/m^3). Our objectives were to determine the most effective water stratum for eDNA detection, estimate true and false positive eDNA detection rates, and assess the number of water samples necessary to minimize the risk of false negatives. There were 28 eDNA detections in 324, 1-L, water samples collected from four experimental ponds. The best-approximating model indicated that the per-L-sample probability of eDNA detection was 4.86 times more likely for every 2.53 fish/m^3 (1 SD) increase in fish density and 1.67 times less likely for every 1.02 C (1 SD) increase in water temperature. The best section of the water column to detect eDNA was the surface and to a lesser extent the bottom. Although no false positives were detected, the estimated likely number of false positives in samples from ponds that contained fish averaged 3.62. At high densities of African jewelfish, 3–5 L of water provided a >95% probability for the presence/absence of its eDNA. Conversely, at moderate and low densities, the number of water samples necessary to achieve a >95% probability of eDNA detection approximated 42–73 and >100 L, respectively. Potential biases associated with incomplete detection of eDNA could be alleviated via formal estimation of eDNA detection probabilities under an occupancy modeling framework; alternatively, the filtration of hundreds of liters of water may be required to achieve a high (e.g., 95%) level of certainty that African jewelfish eDNA will be detected at low densities (i.e., <0.32 fish/m^3 or 1.75 g/m^3).

Editor: Brian Gratwicke, Smithsonian's National Zoological Park, United States of America

Funding: This work was funded by the United States Fish and Wildlife Service Region 4 Aquatic Invasive Species Program. The funders had no role in study design, data collection and analysis, decision to publish, or preparation of the manuscript.

Competing Interests: The authors have declared that no competing interests exist.

* Email: greg_moyer@fws.gov

Introduction

Assessing the distribution, abundance, and dynamics of populations or species frequently requires the collection and identification of individuals from sample locations. As such, species detection is fundamental to scientific disciplines such as phylogenetics, conservation biology, and ecology. The idea of a species being either present or absent from a collection of sites has a long history in ecology, as it provides the foundation for assessing the status and dynamics of species at local and landscape scales. Reliable species detection during sampling, however, can be difficult to achieve, especially for species that are present in low abundances such as threatened and endangered taxa and, in some cases, newly invaded species [1–3].

Recent advances in molecular and forensic methods have provided innovative tools for detecting marine and aquatic organisms that may circumvent the aforementioned limitations [4–6]. One tool that holds particular promise is environmental DNA (eDNA). Defined as short DNA fragments that an organism leaves behind in non-living components of the ecosystem (i.e., water, air or sediments) [7–8], eDNA can be used to detect the presence (or absence) of a species through cells or tissues found in the environment containing the genetic material. In aquatic systems, genetic material can be collected via water filtration through a micron screen and tested for presence of the target species using specific genetic markers via polymerase chain reaction (PCR), quantitative PCR (qPCR) or direct sequencing of the PCR product. The basic technique outlined above raises the possibility to detect and monitor target taxa, particularly rare species, in aquatic environments while eliminating extraneous noise generated by the presence of (potentially numerous) non-target taxa. Consequently, eDNA has garnered increased attention for use with endangered aquatic organisms [2,6] and aquatic invasive species [1,9,10].

Recently, there has been increased attention and scrutiny regarding eDNA detection methodologies [11–13]; yet, little consideration has been given to the utility and accuracy of eDNA presence/absence data with respect to rare or difficult-to-detect taxa [14,15]. For example, what is the certainty of a species being detected via eDNA methods (i.e., what is the false positive error rate); in contrast, if a species fails to be detected using eDNA, then is it truly absent or is it present but simply not detected (i.e., what is the false negative error rate)? The latter, which is termed Process Type II Error [16], characterizes the imperfect detection of species and is of particular concern when using presence/absence data to make inferences regarding the predominant factors influencing the status, distribution, and dynamics of species. The confounded nature of non-detection and true absence imposes a fundamental problem when using eDNA presence/absence data, and failing to explicitly account for incomplete detection in a study design or analysis could lead to biased results and potentially unreliable inferences [17].

Occupancy modeling approaches are widely used in ecological research and management because they can effectively and efficiently account for potential biases associated with imperfect detection [18] and misidentification [19] of species. Occupancy models use records of species detections and non-detections during repeated surveys at a given location, and the resulting capture histories can be used to estimate model parameters of interest (e.g., detectability or occupancy). It follows that there is great potential for applying occupancy modeling approaches to eDNA detection/non-detection data [14,17]. Incorporating such an approach into eDNA experimental designs could allow researchers to account for sources of bias such as false-negative [20] and false positive [19] measurement errors.

The African jewelfish (*Hemichromis letourneuxi*) is an aquatic invasive species that has spread throughout southern Florida since its introduction in 1965 [21]. More recently African jewelfish has spread into coastal river systems of west-central and southwest Florida and canals and wetlands along the Atlantic Coast from Cape Canaveral south [22]. Continued spread inland and northward in peninsular Florida is occurring (Hill, unpublished data). The species has the potential to compete with native [23,24] and non-native species [25] and reduces survival of native fishes in seasonal refuges of short hydroperiod wetlands of the Florida Everglades [26], making its introduction and spread a threat to native ichthyofauna of Florida. In an effort to detect and monitor the spread of African jewelfish, eDNA markers have been established for the species [27]. Aquarium experiments using African jewelfish have shown a positive correlation between eDNA detection and fish density [27], with limited detection of eDNA (1 L water sample) occurring at densities less than 13 fish/m^3. Difficulties associated with the detection of African jewelfish eDNA when densities are less than 13 fish/m^3 is concerning given that the spread or introduction of this species could transpire at densities much lower than the observed value [28,29]. Further, the issue of incomplete detection and the complexities of abiotic and biotic factors influencing eDNA detection raise more general questions regarding appropriate eDNA sampling strategies for monitoring programs focused on rare species and for understanding the spatial distribution of eDNA.

The goal of our study was to assess the detection of African jewelfish eDNA in a controlled lentic system at varying fish densities. Our specific objectives were to 1) determine the most effective water stratum for the detection of eDNA, 2) estimate true and false positive eDNA detection rates at varying fish densities, and 3) assess the number of water samples necessary to minimize the risk of false negative errors when developing eDNA sampling protocols.

Materials and Methods

Experimental Design

We used four artificial ponds as mesocosms to estimate and compare detection probabilities of the African jewelfish at differing densities. The four earthen ponds were located at the University of Florida's Tropical Aquaculture Laboratory, in Ruskin, Florida. The dimensions of each pond were approximately 18 m×7.5 m, with an average depth of 1.4 m. On June 10, 2013, 30 days prior to the introduction of fish, ponds were drained, pond bottoms and banks pressure washed, and remaining excess debris removed. Hydrated lime (CaOH) was then applied at a rate of 22.6 kg/pond to ensure that there were no remaining live fish in each pond. Over the course of 72 hrs, ponds were allowed to fill naturally from ground water. Once ponds were full, pH, dissolved oxygen, and temperature were monitored weekly. Three, 1-L eDNA water samples were taken from each pond 15 days before the introduction of the fish to check for any potential contamination. Samples were processed as outlined below, and African jewelfish DNA was absent in each pond of the study system before the fish introduction.

To begin the experiment, each of the four study ponds was stocked with a known number of fish (average TL = 69.92 mm; average wt. = 5.36 g): pond I contained 0 fish (control), pond II contained 60 fish (low density: 0.32 fish/m^3 or 1.7 g/m^3), pond III contained 330 fish (moderate density: 1.75 fish/m^3 or 9.35 g/m^3), and pond IV contained 990 fish (high density: 5.24 fish/m^3 or 28.08 g/m^3; Table 1). Note that all animal research was approved by the University of Florida, Institute of Food and Agriculture Sciences, Animal Research Committee (Approval # 002-13RUS). We stratified each pond into nine transects, three of which were located in one third of the pond (section 1), three in the middle third (section 2), and three in the remaining third (section 3). Three sample locations were then selected along each transect and randomly assigned one of three water column positions to each sample: surface, middle, or bottom. Following the first 24 hrs (i.e., day 1), 27 1-L water samples were collected from each pond at the specified locations using a Van Dorm collection bottle to ensure adequate coverage (depth and surface area). A kayak was used to move between transects and quadrants and was cleaned between ponds with Alconox detergent (1:100 dilution, Alconox, Inc) to avoid contamination. This protocol was repeated on days 5 and 10, for a total of 81 samples in each pond and 324 samples total. On each sampling day, pond temperature was recorded.

Molecular methods

Each water sample was treated with 1 mL of 3M sodium acetate (pH 5.2) and 33 mL 95% non-denature ethanol for DNA/tissue preservation and refrigerated on site until filtration. Each water sample was filtered on site and filter paper frozen until extraction date. DNA was extracted following the protocol of Díaz-Ferguson et al. [27], however; the MOBIO Power Water DNA Isolation kit was substituted for the Rapid Water Isolation kit. Final DNA templates were eluted in 45 uL of buffer provided with the kit and then an ethanol precipitation was conducted to improve quality and concentration of the yielded DNA.

Taqman qPCR assays were employed to detect the presence of African jewelfish eDNA in each water sample collected from the four experimental ponds using primers AJFq3 and AJFRq2 and probe Pr028373859 designed for the target species (Díaz-Ferguson et al. 2014). Taqman assays were optimized for 20 uL reactions using DNA normalized to a concentration of 25 ng/uL and Taqman core reagents as follows: 2.0 uL of 5× *Taq* reaction buffer (Applied Biosystems, Inc), 2.5 uL $MgCl_2$ (25 mM), 0.5 uL of each dNTP (1 mM), 1 uL of each primer (10 uM each), 0.20 uL probe Pr028373859 (10 uM), 0.5 uL AmpErase (Uracil-N-glycosylase), and 0.20 uL Amplitaq Gold *Taq* polymerase (5 U/uL, Applied Biosystems, Inc).

All assays were conducted using the following thermal profile: 60 C (1 min), initial denaturation at 95°C for 10 min., followed by

Table 1. Pond number, pond volume, fish density, number of fish stocked, average temperature, and number of African jewelfish detections by water stratum across 27 water samples conducted on days 1, 5, and 10 of the study.

Pond	Day	Volume (m^3)	Density (no. m^{-3})	Number fish	Temperature (C)	Detections by strata		
						Surface	Middle	Bottom
1	1	189	0	0	30.20	0	0	0
1	5	189	0	0	30.17	0	0	0
1	10	189	0	0	28.3	0	0	0
2	1	189	0.32	60	30.77	0	0	0
2	5	189	0.32	60	29.77	0	0	0
2	10	189	0.32	60	28.23	0	1	0
3	1	189	1.75	330	30.7	1	1	0
3	5	189	1.75	330	29.87	0	0	0
3	10	189	1.75	330	28.23	0	0	0
4	1	189	5.24	990	30.77	2	2	3
4	5	189	5.24	990	29.77	2	0	1
4	10	189	5.24	990	28.3	6	2	7

35 cycles of 95 C (15 s) and 60 C (1 min.) Detection of DNA from each sample was performed using a 7500 Fast Real Time PCR machine (Applied Biosystems, Inc.). Taqman assay quality controls consisted of repetition of all qPCR results and inclusion of two negative qPCR controls (substitution of distilled water for DNA) and a positive qPCR inhibition control for each qPCR plate. The positive control consisted of a water sample taken from each pond, spiked with 5–10 mg/uL lyophilized tissue from *H. letourneuxi*, filtered, DNA extracted, and DNA used as a positive confirmation that our qPCR reactions were working correctly in the presence of potential inhibitors. In addition, we sequenced 25% of the positive Taqman assays (following the protocol outlined by Díaz-Ferguson et al. 2014) for confirmation that the qPCR product was truly that of African jewelfish. All sequences were imported into GENEIOUS v4.8.5 alignment editor (Biomatters, available from http://www.geneious.com/), ends trimmed, aligned by eye, and compared for base pair composition and similarity with other African jewelfish sequences previously deposited in the GenBank.

Statistical analyses

Our primary interest was to assess per-sample eDNA detection rates (i.e., prevalence of false-negative errors); however, we also evaluated the prevalence of false positive errors in the sample data. False-negative errors represented instances where, for a given 1 L water sample, qPCR DNA amplification failed to detect African jewelfish when it was known to be present in a pond. In contrast, false-positive errors represented instances where the species was detected when it was known to be absent from a pond (i.e., the control pond). To estimate eDNA detection probabilities, we fitted logistic regression models [30] relating eDNA detection/non-detection data to two pond-level factors, fish density and water temperature, and one sample-level factor, position in the water column. Water temperatures varied among ponds and time periods; however, there was a general decline in temperature across all ponds over the 10-day study period owing to a cold front that moved through the region. We assumed that fish density remained constant over the course of the study. We observed no fish mortality in any pond over the 10-day period; hence, we believe that the assumption of constant fish density was valid. For both data types, the dependent variable was the detection or non-detection (binary coded) of eDNA from individual water samples. Note that we excluded control pond data from the analysis of eDNA detection because the pond did not contain fish. Conversely, only the control pond data were used to estimate false positive error rates, because any positive eDNA detections in the control pond were, by definition, false positives.

Regular logistic regression cannot account for dependence (i.e., autocorrelation) among repeated samples, and we suspected that repeated water samples taken from particular sections and transects were dependent [31]. For example, it was possible that individuals, and hence their eDNA, were concentrated in particular areas of each pond, which would tend to inflate false negative errors as eDNA would not be present in some areas and, hence, unavailable for detection. To account for dependence among samples, we fitted hierarchical logistic regression models to the DNA detection/non-detection data [32]. For our study, the log-odds of eDNA detection, η_{hijk}, was modeled as:

$$\eta_{hijk} = \beta_0 + \beta_{Qhk} X_{Qhk} + \mu_{0i} + \mu_{1j}$$

where β_0 was the intercept, β_Q was the effect of pond- (h) or sample-level (k) factors (fish density, water temperature, and water column position) X_Q on eDNA detection, and μ_{0i} and μ_{1j} were the section- (i) and transect (j)-level random effects, respectively, that were assumed to be normally distributed with mean of zero and random effect-specific variance [33]. The random components u_{0i} and u_{1j} represented unique effects associated with sections and transects, respectively, that were unexplained by pond- and sample-level covariates. Because there were no DNA detections in the control pond (see Results), it was not possible to model false positive errors as a function of covariates. Thus, we fit a single "intercept only" logistic regression model (i.e., the log-odds of false positive errors, $\eta_{hijk} = \beta_0$) to estimate the false positive error rates and estimated the likely number of false positive errors in the false negative data (i.e., data from ponds with fish) assuming a sample size of 243. We used Markov Chain Monte Carlo (MCMC) as implemented in OpenBUGS software, version 3.2.1 [34] to fit candidate hierarchical logistic regression models. All models were fit using 200,000 iterations, a 50,000 iteration burn in (i.e., the first 50,000 MCMC iterations were dropped), and diffuse priors.

We used an information-theoretic approach [35] to evaluate the relative fit of candidate models relating pond- and sample-level characteristics to eDNA detection/non-detection data. For data from the three ponds that were stocked with fish, we developed 16 models representing relations between various combinations of pond- and sample-specific predictors and eDNA detection. The pond temperature predictor represented the measured temperature at the surface, middle, and bottom of each pond during each sample day, which resulted in three temperature measurements per pond on days 1, 5, and 10 of the study; the fish density predictor represented the known density of fish in each pond (low, moderate, high); and the water column position predictors included 'middle' and 'bottom', with 'surface' samples serving as the statistical baseline. The categorical water column position predictors were binary coded as 0 (surface) or 1 (middle and bottom). To facilitate model-fitting, we standardized both continuous predictors, water temperature and fish density, with mean of zero and standard deviation of one.

Prior to fitting candidate models, we evaluated the relative-fit of four different variance structures using the global (all predictors) model by fitting models that contained several combinations of random effects for sections and transects. The four variance structures included (1) no random effects, (2) a random intercept associated with individual sections, (3) a random intercept associated with individual transects, and (4) a random intercept associated with individual sections and transects. The best approximating variance structure was identified using the Deviance Information Criterion (DIC). The DIC is a Bayesian measure of model fit or adequacy, with smaller DIC indicating a better approximating model [36]. We then evaluated the relative fit of the 16 candidate models using DIC and calculated DIC weights following Link and Barker [37], which range from 0–1 with the best approximating candidate model having the highest weight. We considered the most plausible models to be those with DIC weights within 10% of the best-approximating model, which is similar to Royall's general rule-of-thumb of 1/8 or 12% for evaluating strength of evidence [38].

We assessed the precision of parameter estimates for each model by calculating 95% Bayesian credible intervals [39], which are analogous to 95% confidence intervals. To facilitate interpretation, we also calculated odds ratios (OR) for each fixed-effect pond- and sample-level predictor variable [40]. We assessed MCMC convergence for each model in the confidence set using the diagnostics detailed by Gelman and Rubin [41]. Lastly, we assessed the adequacy of the global model (Goodness of fit) by calculating a Bayesian p-value using the discrepancy measure method [42]. Extreme Bayesian p-values (i.e., ≤0.05 or ≥0.95) indicate that a model does not adequately describe the data.

Using parameter estimates from the best-approximating model, we also calculated cumulative detection probabilities to evaluate the number of 1-L water samples required to achieve a specified level of certainty that low, moderate, and high density African jewelfish populations would be detected at least once. We also calculated per-sample detection probabilities as a function of fish density.

Results

A total of 28 detections of African jewelfish eDNA were made across 324 individual 1-L water samples collected from the four experimental ponds (Table 1). Sequence confirmation of qPCR fragments showed between 83–89% query coverage (percent of the query sequence that overlaps the subject sequence) and a percent sequence similarity between 92–96% that corresponded to *Hemichromis* (GenBank accession numbers: KJ553580.1, JQ667546.1, JN026744, GU817297.1, AY662793.1, KJ553529.1).

Excluding 81 water samples collected from the control pond (i.e., no fish present), there were 28 detections across the 243 collections taken from ponds that contained varying densities of fish, which corresponded to an overall eDNA detection rate of ~12%. The detection of eDNA was more prevalent on the surface and bottom (each with 39% of the detections) when compared to the middle of the water column. Notably, there were no false positive detections among the 81 control samples.

The best-approximating error structure for the logistic regression models relating eDNA detection/non-detection data to pond- and sample-level covariates included no random effects associated model intercepts and slopes, indicating no substantial dependence among pond transects or quadrants. The assessment of model adequacy using the discrepancy measure method indicated that the global model provided an adequate description of the data, with a Bayesian p-value of 0.58. The confidence model set consisted of four models that contained various combinations fish density, water temperature, and sample position in the water column (Table 2). The best-approximating model contained density, temperature, and middle and was 1.44, 2.44, 3.54, and 5.57 times more plausible than the next best-approximating models in the confidence set (Table 2).

Parameter estimates from best-approximating model indicated that the per-sample probability of eDNA detection was strongly and positively related to fish density and negatively related to water temperature (Table 3). Odds ratios (OR) indicated that African jewelfish were 4.86 time more likely to be detected for every 1 SD (2.53 fish/m^3) increase in fish density, whereas the species was 1.67 times less likely to be detected for every 1 SD (1.02 C) increase in pond temperature (Table 3). Parameter estimates for the remaining covariates in the confidence model set, bottom and middle, indicated that per-sample detection was highest in collections taken from the surface (i.e., detection was negatively related to middle and bottom); however, the parameter estimates were considered imprecise as the 95% credible intervals contained zero (Table 3).

Although there were no false positive errors associated with the control data in this study, it was still possible to estimate the probability and number of false positive errors under the assumption of binomially distributed data with a sample size of 81 (i.e., the number of 1-L water samples in the control pond). The parameter estimate from the logistic regression model fit to the control data indicated that false positive errors were very unlikely,

Table 2. Deviance, effective number of parameters (pd), deviance information criterion (DIC), ΔDIC, DIC weights (w_i), and Bayesian p-values (p-value) for the confidence set of logistic regression models relating African jewelfish eDNA detections to pond- and sample-level factors.

Model	Deviance	pd	DIC	ΔDIC	w_i	p-value
Intercept, middle, density, temperature	125.70	3.69	129.39	0.00	0.39	0.60
Intercept, density, temperature	127.20	2.95	130.15	0.75	0.27	0.60
Intercept, middle, bottom, density, temperature	126.50	4.63	131.13	1.74	0.16	0.58
Intercept, bottom, density, temperature	127.90	3.94	131.84	2.45	0.11	0.58
Intercept, middle, density	130.00	2.71	132.71	3.32	0.07	0.57

with an estimated per-sample probability of 0.014 (Table 4). Across the 243 1-L water samples conducted in the three stocked ponds, the per-sample false-positive error rate estimate of 0.014 suggested that the likely number of false positives averaged 3.62 and ranged from 0–12.

Using parameter estimates from the best-approximating model, the number of 1-L water samples required to achieve a specified level of certainty varied considerably depending on the density of African jewelfish aggregates (Figure 1). Per-sample detection probabilities plotted as a function of fish density demonstrated that although the per-sample probability of detection increased with fish density, the relationship was nonlinear across the range of densities used in this study (Figure 2).

Note that our best-approximating model included temperature and not time; yet, we expected a positive relationship between time and eDNA detection. Unfortunately it was difficult to differentiate the effects of temperature and time because temperature declined over the 10-day period in the entire system (Table 1). To address this, we constructed a model that included time instead of temperature. We observed a weak positive relationship (i.e., for every 1 day increase, the log-odds of detection increased by 0.10; Table 4) and based on the DIC (130.76 vs. 129.39), the time model had less support than the temperature model in explaining eDNA detection. Given our findings and that the time since invasion is rarely known, we opted to use temperature in the models because it was more relevant to our study objective.

Discussion

The ability to detect individuals at low densities in aquatic habitats is critical for successful control and management of invasive species [43] and for the conservation of threatened and endangered organisms [3,44]. Unfortunately, rarity typically presents problems when dealing with both spatial sampling and detectability [44]. This issue is not new and like traditional sampling methods designed to detect rare or elusive species, eDNA sampling methods will suffer the same biases and problems.

Table 3. Parameter estimates (Mean), standard deviations (SD), 95% credible intervals, and odds ratios (OR) from the confidence set of logistic regression models relating African jewelfish eDNA detections to pond- and sample-level factors.

Model	Parameter	Mean	SD	Lower 95%	Upper 95%	OR
Best						
	Intercept	−2.78	0.39	−3.62	−2.08	
	Middle	−0.81	0.50	−1.83	0.14	0.44
	Density	1.58	0.30	1.04	2.22	4.86
	Temperature	−0.51	0.23	−0.97	−0.07	0.60
2nd best						
	Intercept	−2.98	0.38	−3.78	−2.31	0.05
	Density	1.56	0.30	1.02	2.19	4.75
	Temperature	−0.51	0.23	−0.96	−0.08	0.60
3rd best						
	Intercept	−2.70	0.45	−3.64	−1.88	
	Middle	−0.91	0.55	−2.01	0.15	0.40
	Bottom	−0.23	0.50	−1.21	0.74	0.80
	Density	1.60	0.30	1.05	2.23	4.93
	Temperature	−0.53	0.23	−1.00	−0.08	0.59
False positive						
	Intercept	−4.20	0.78	−5.91	−2.86	

Also reported is the single parameter estimate (intercept) associated with the logistic regression model fit to the control data to estimate the probability of false positive errors.

Table 4. Parameter estimates (Mean), standard deviations (SD), and 95% credible intervals for the logistic regression model relating African jewelfish eDNA detections to pond- and sample-level factors.

Parameter	Mean	SD	Lower 95%	Upper 95%
Intercept	−3.25	0.52	−4.33	−2.29
Middle	−0.86	0.50	−1.89	0.08
Density	1.54	0.29	1.01	2.17
Day	0.10	0.06	−0.01	0.22
Deviance	126.70	2.853		
DIC	130.76			

This model is identical to the best approximating model listed in Table 3 but with time (day) substituted for temperature.

Therefore the development of methods and models that properly account for imperfect detection of eDNA should be a vital first step in designing and implementing detection and monitoring surveys for rare organisms that rely on eDNA methods [16,17,45].

While our study and numerous others have illustrated a positive and often significant relationship between organismal density and eDNA detection [14,17,27,46], our basic understanding of the biotic and abiotic factors influencing eDNA detection is still in its infancy [however, see 47], with the majority of studies focusing on type I and II errors associated with the molecular method itself [e.g.,3,11,16]. In contrast, the focus of our study was to assess the false negative error rate termed Process Type II Error by Darling and Mahon [16]. While Darling and Mahon [16] recognized that the estimation of false negative and false positive error rates is important for eDNA assay development, they acknowledged that few if any studies effectively address this issue. Our study sought to provide a quantitative approach for estimating and understanding sampling efficiency for African jewelfish.

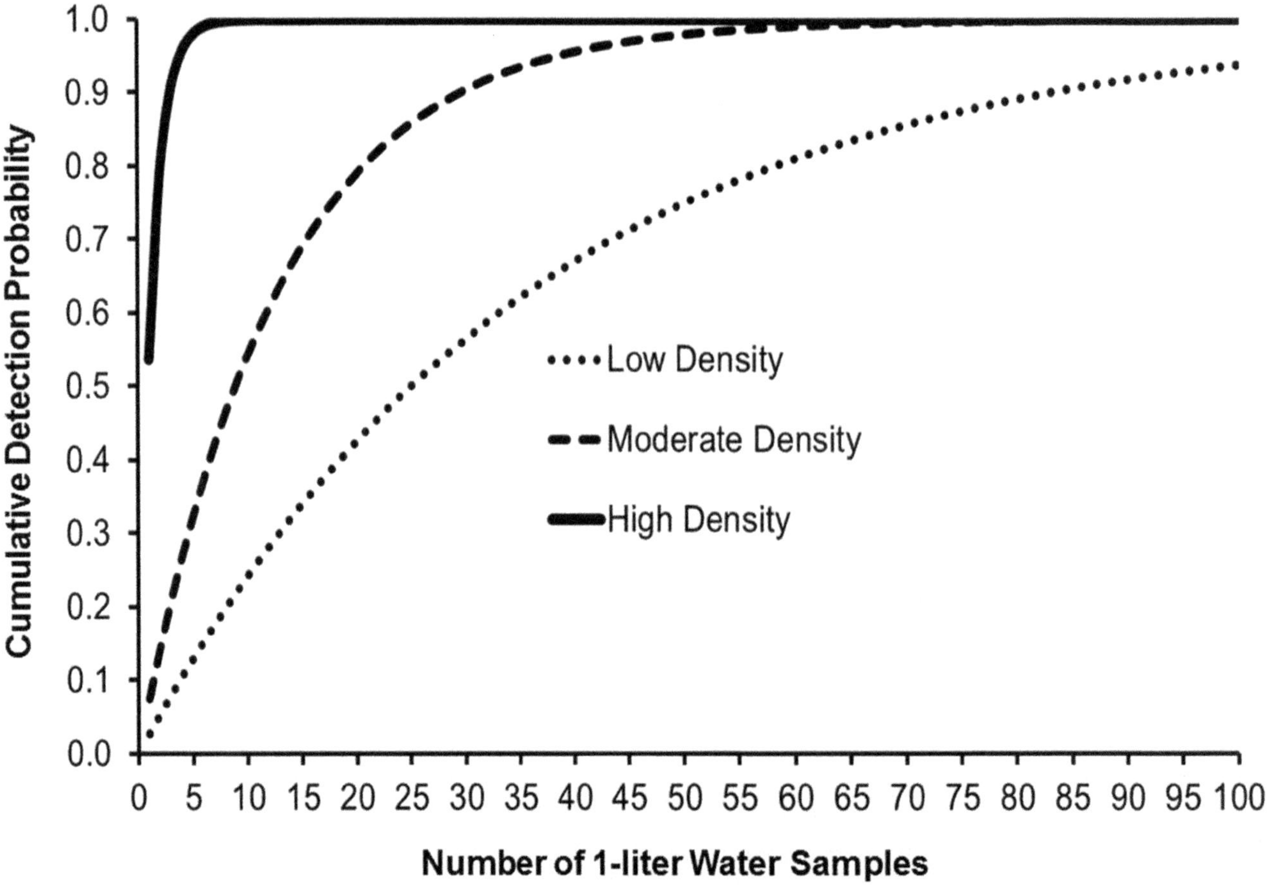

Figure 1. Predicted cumulative African jewelfish eDNA detection probability with an increasing number of 1-L water samples. Detection estimates are based on parameter estimates from the best-approximating hierarchical logistic regression models relating African jewelfish eDNA detection/non-detection data to pond- and sample-level covariates and were calculated for low density (0.32 fish/m^3), moderate density (1.75 fish/m^3), and high density (5.24 fish/m^3) populations assuming a water temperature of 28 C.

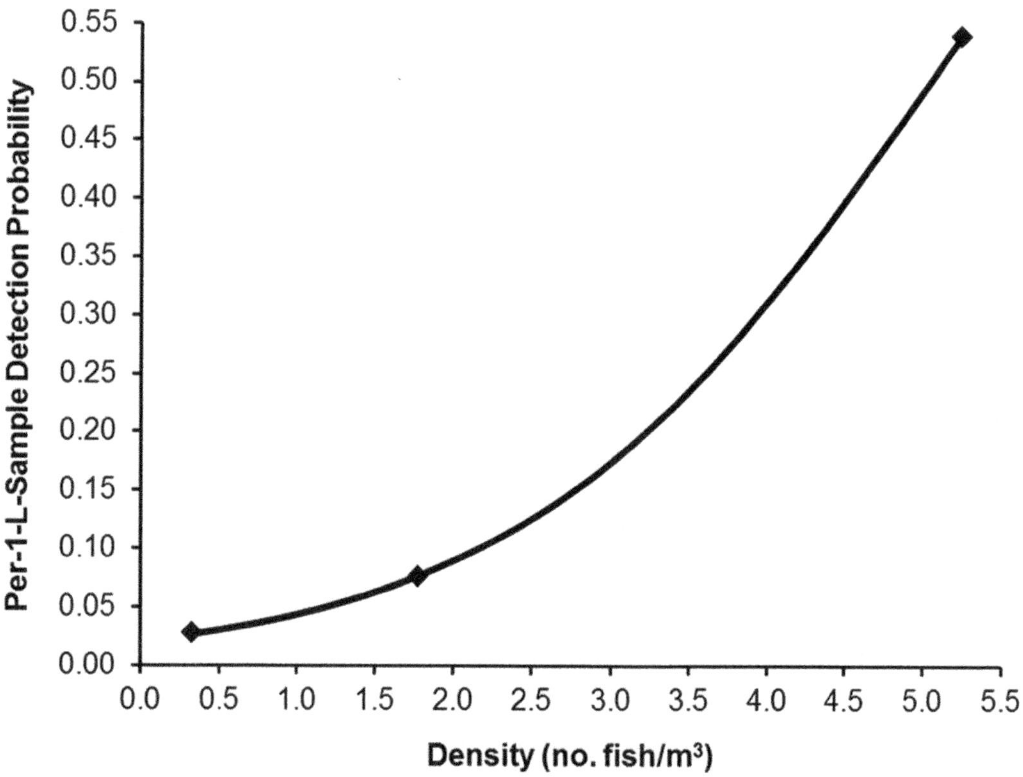

Figure 2. Predicted per-1-L water sample DNA detection probability with increasing densities of African jewelfish in experimental ponds. Detection estimates are based on parameter estimates from the best-approximating hierarchical logistic regression models relating African jewelfish eDNA detection/non-detection data to pond- and sample-level covariates. Filled diamonds represent the low (0.32 fish/m^3), moderate (1.75 fish/m^3), and high densities (5.24 fish/m^3) used in this study.

The eDNA from living macrofauna most likely originates from urine and feces, epidermal tissues, or other secretions such as reproductive fluids and reproductive cells. Most of this material is introduced into the water column as large particles (>1000 um) that remain at the surface for a limited amount of time before sinking or breaking apart [13,48]; thus, the surface provides a logical place to survey for eDNA and it is also relatively efficient to collect surface samples when compared to soil samples from the bottom of a lentic system. While eDNA studies have concentrated sampling efforts near the surface [3,49], none have justified their sampling approach. Once introduced, African jewelfish seek and remain on the bottom of an earthen pond (J. Hill personal observation); therefore, we suspected that samples taken from the bottom would show a significant increase in eDNA detection. In contrast, our results indicated that the best section of the water column to sample and detect eDNA was the surface and to a lesser extent the bottom. Our findings support the pattern that, at least in small lentic systems, eDNA remains at the surface level for a given time period before settling to the bottom or until degradation occurs.

Elevated temperature can accelerate the rate of eDNA degradation. Degradation can occur directly by denaturing the DNA or indirectly by increasing enzymatic activity and microbial metabolism [47]. Parameter estimates from our best-approximating model indicated that eDNA detection was negatively related to water temperature such that the species was 1.67 times less likely to be detected for every 1 SD (1.02 C) increase in stream temperature. A similar observation was found for preliminary eDNA persistence trials of African jewelfish held in aquaria. In these trials, African jewelfish eDNA was found to degrade between 25 and 33 C (E. Díaz-Ferguson unpublished data). While the negative effect of temperature in our study likely reflects the combined effects of lower degradation rates under lower temperatures and more eDNA in the system as time progressed, the influence of time on eDNA detection was assumed minimal – a finding supported by Díaz-Ferguson et al. [27] who found a non-significant relationship between African jewelfish eDNA detection and time in aquaria held at a constant temperature over a seven day period. Thus, we believe that temperature was a significant factor influencing eDNA detection in our study. This finding suggests that to minimize the negative influence of temperature on species detection rates, the implementation of eDNA monitoring programs in the relatively warm waters of the tropics and subtropics should be cautioned if ambient water temperatures exceed 29–30 C.

At high densities of African jewelfish (5.24 fish/m^3), the filtration of 3–5 L of water (or the filtration of 3–5, 1-L water samples) should provide a high degree of confidence (95–100% probability) to confirm the presence or absence of its eDNA. However, if only a 1-L water sample was collected from our pond containing 990 fish, then our ability to detect eDNA would be approximately 55%. For our ponds that contained 330 and 60 fish, we had a 7% and a 3% chance of detecting African jewelfish eDNA if a 1-L water sample was taken. Conversely, at moderate and low densities, the number of water samples necessary to achieve a 95–100% probability of eDNA detection would approximate 42–73 L and >100 L, respectively. Our findings highlight a well-known and important concept with hypothesis testing – statistical power [31]. Scientist and resource managers using eDNA methods must agree on the level of accepted error prior to hypothesis testing [16], but depending on the hypothesis being tested, they will often want to keep both type I and type II errors small. Clearly, then, the only way to minimize eDNA false negatives is to improve the power of the test while keeping the significance level (α) constant. To do so requires increasing the sensitivity (= power) of the test either by increasing the sample size (volume of water tested) or by increasing the sensitivity of the eDNA marker. Our eDNA marker appears sufficiently sensitive because the observed theoretical lower limit of qPCR detection using our eDNA marker was similar to other eDNA studies [27]; therefore, if minimizing type I and type II errors is a priority when using eDNA to monitor the leading edge of invasion for African jewelfish, then a large volume of water must be screened.

Our findings raise a vexing problem when designing eDNA sampling strategies for aquatic species that are rare – the density of the organism is usually unknown; hence, the amount of water necessary to detect the organism with a high level of certainty also will be unknown. Minimizing false negatives (i.e., increasing detection probability) will come at the cost of filtering more water (either in the form of more volume or more samples). The filtration of hundreds of liters of water through a small micron filter tends to be problematic (filters clog) and often expensive (e.g., filtration of 100, 1-L bottles each through its own filter); alternatively, devices that maximize water volume such as inline filters or plankton nets, may prove valuable for eDNA monitoring of rare species [13], but will be contingent on the particle size of aqueous eDNA being emitted from the organism of concern [13].

While our study was not specifically designed to address the issue of taking numerous smaller samples vs. one large sample (e.g., 50, 1-L water samples vs 1, 50-L water sample), we recommend the former, especially for rare species in lentic systems. In a lentic environment, the taxon in question may have specific (patchy) habitat requirements; thus, stratifying by habitat and conducting numerous smaller-volume samples would be preferred over taking one large-volume sample. Furthermore, more habitat and sample-specific covariates (i.e., all the interesting heterogeneity within the sampling site) are available for potential use in an occupancy model when collecting numerous smaller-volume samples vs. obtaining one large-volume sample.

Finally, our study emphasizes the difficulties of inferring detection probabilities for an organism inhabiting a natural system simply from aquarium trials. First, even when one fish is placed in an aquarium, it is difficult to simulate lower densities necessary for inferring accurate detection probabilities (i.e., there is a nonlinear relationship between eDNA detection and density, see Figure 2). For example, Díaz-Ferguson et al. [27] used one fish (5.45 g) in a 75.5 L aquarium to simulate their lowest density; however, this approximated to a density of 13 fish/m^3 (or 70.85 g/m^3) in our pond experiment and was a value greater than our highest density pond. Thus, if the species of concern is rare, then the estimation of detection probabilities should be conducted in a larger controlled system that can simulate the rarity of the organism in its natural setting. For an organism inhabiting a lotic system, the use of artificial streams [50] or raceways may be necessary. Alternatively, the estimation of detection probabilities may be unattainable for an endangered organism due to its rarity; thus, the use of a surrogate species (e.g., a close congener with similar life history attributes) may be necessary. Second, aquarium trial experiments of African jewelfish failed to detect eDNA from a 1-L water sample at a density of 13 fish/m^3 [27]. In the present study, a 1-L water sample taken at this density should always detect African jewelfish; thus, aquarium experiments appear to have underestimated the detection of African jewelfish eDNA. There are a variety of potential explanations for the discrepancy in detection probabilities including behavioral (antagonistic behavior, increased movement) and environmental (wind/wave action); regardless, our study demonstrated the complexities of extrapolating eDNA detection probabilities from a controlled to a natural environment.

Our results and those of Díaz-Ferguson et al. [27] indicate that detection probabilities for African jewelfish can be imperfect (i.e., <1) and vary spatially or temporally in response to local environmental conditions. As such, presence-absence data derived from eDNA-based methods (e.g., the proportion of sites where a species was detected) where the density of African jewelfish is low will be negatively biased and could have profound implications when determining the leading edge of invasion for this species if imperfect detection is not taken into account. Potential biases associated with incomplete detection could be alleviated by formally estimating detection probabilities under an occupancy modeling framework [51,52]; alternatively, the filtration of hundreds of liters of water may be required to detect African jewelfish at low densities (i.e., <0.32 fish/m^3 or 1.7 g/m^3) with a desired level of confidence.

Acknowledgments

We thank J. Herod and J. Stoeckel for logistical support, C. Givens for assistance with sample collection, K. Dowling for assistance with fish collection, and L. Lawson, Jr. for pond preparation. This manuscript benefited from comments by two anonymous reviewers. Use of trade names throughout the manuscript does not constitute endorsement by the United States Fish and Wildlife Service.

Author Contributions

Conceived and designed the experiments: GRM EDF JEH CS. Performed the experiments: EDF JEH. Analyzed the data: GRM CS. Contributed reagents/materials/analysis tools: JEH GRM CS. Contributed to the writing of the manuscript: JEH EDF GRM CS.

References

1. Ficetola G F, Miaud C, Pompanon F, Taberlet P (2008) Species detection using environmental DNA from water samples. Biol Lett 4: 423–425.
2. Goldberg CS, Pilliod D, Arkle R, Waits L (2011) Molecular detection of vertebrates in stream water: a demonstration using rocky mountain tailed frogs and Idaho giant salamanders. PLOS ONE 6: e22746
3. Jerde CL, Mahon AR, Chadderton WL, Lodge DM (2011) "Sight-unseen" detection of rare aquatic species using environmental DNA. Conserv Lett 4: 150–157.
4. Valentini A, Pompanon F, Taberlet P (2009) DNA barcoding for ecologists. Trends Ecol Evol 24: 110–117.
5. Lodge D, Turner C, Jerde C, Barnes M, Chadderton L, et al. (2012) Conservation in a cup of water: estimating biodiversity and population abundance from environmental DNA. Mol Ecol 11: 2555–2558.
6. Thomsen P, Kielgast J, Iversen L, Wiuf C, Rasmussen M, et al. (2012) Monitoring endangered freshwater biodiversity using environmental DNA. Mol Ecol 21: 2565–2573.
7. Foote A, Thomsen P, Sveegaard S, Wahlberg M, Kielgast J, et al. (2012) Investigating the potential use of environmental DNA (eDNA) for genetic monitoring of marine mammals PLOS ONE 7: e4178.
8. Taberlet P, Coissac E, Hajibabaei M, Rieseberg L (2012) Environmental DNA. Mol Ecol 21: 1789–1793.
9. Dejean T, Valentini A, Miquel C, Taberlet P, Bellemain E, et al. (2012) Improved detection of an alien invasive species through environmental DNA barcoding: the example of the American bullfrog *Lithobates catesbeianus*. J Appl Ecol 49: 953–959.
10. Takahara T, Minamoto T, Doi H (2013) Using environmental DNA to estimate the distribution of an invasive fish species in ponds. PLOS ONE 8: e56584.

11. Wilcox TM, McKelvey KS, Young MK, Jane SF, Lowe WH, et al. (2013) Robust detection of rare species using environmental DNA: the importance of primer specificity. PLOS ONE 8: e59520.
12. Deiner K, Altermatt F (2014) Transport distance of invertebrate environmental DNA in a natural river. PLOS ONE 9: e88786.
13. Turner C, Barnes M, Charles C, Xu Y (2014) Particle size distribution and optimal capture of macrobial eDNA. Methods Ecol Evol In press.
14. Pilliod D, Goldberg C, Arkle R, Waits L (2013) Estimating occupancy and abundance of stream amphibians using environmental DNA from filtered water samples. Can J Fish Aquat Sci 70: 1123–1130.
15. Díaz-Ferguson E, Moyer G (2014) History, applications, methodological issues and perspectives for the use of environmental DNA (eDNA) in marine and freshwater environments. Rev Biol Trop. In press.
16. Darling JA, Mahon AR (2011) From molecules to management: Adopting DNA-based methods for monitoring biological invasions in aquatic environments. Environ Res 111: 978–988.
17. Schmidt BR, Kéry M, Ursenbacher S, Hyman OJ, Collins JP (2013) Site occupancy models in the analysis of environmental DNA presence/absence surveys: a case study of an emerging amphibian pathogen. Method Ecol Evol 4: 646–653.
18. MacKenzie D, Nichols JD, Lachman GB, Droege S, Royle JA, et al. (2002) Estimating site occupancy rates when detection probabilities are less than one. Ecology 84: 2200–2207.
19. Royle JA, Link WL (2006) Generalized site occupancy models allowing for false positive and false negative errors. Ecology 87: 835–841.
20. Tyre AJ, Tenhumberg B, Field SA, Niejalke D, Parris K, et al. (2003) Improving precision and reducing bias in biological surveys: estimating false-negative error rates. Ecol Appl 13: 1790–1801.
21. Rivas LR (1965). Florida fresh water fishes and conservation. Q J Florida Acad Sci 28: 255–258.
22. Langston JN, Schofield PJ, Hill JE, Loftus WF (2010) Salinity tolerance of the African jewelfish *Hemichromis letourneuxi*, a non-native cichlid in south Florida (USA). Copeia 2010: 475–480.
23. Schofield PJ, Loftus WF, Brown ME (2007) Hypoxia tolerance of two centrarchid sunfishes and an introduced cichlid from karstic Everglades wetlands of southern Florida, USA. J Fish Biol 71(sd): 87–99.
24. Dunlop-Hayden KL, Rehage JS (2011) Antipredator behavior and cue recognition by multiple Everglades prey to a novel cichlid predator. Behavior 148: 795–823.
25. Porter-Whitaker A, Rehage JS, Liston SE, Loftus WF (2012) Multiple predator effects and native prey responses to two non-native Everglades cichlids. Ecol Freshwat Fish 21: 375–385
26. Rehage JS, Linston SE, Dunker KJ, Loftus WF (2013) Fish community responses to the combined effects of decreased hydroperiod and nonnative fish invasions in a karst wetland: are Everglades solution holes sinks for native fishes? Wetlands 1–15.
27. Díaz-Ferguson E, Herod J, Galvez J, Moyer G (2014) Development of molecular markers for eDNA detection of the invasive African jewelfish (*Hemichromis letourneuxi*): a new tool for monitoring aquatic invasive species in National Wildlife Refuges. Manag Biol Invasion 5:121–131.
28. Britton JR, Gozlan RE (2013) How many founders for a biological invasion? Predicting introduction outcomes from propagule pressure. Ecology 94: 2558–2566.
29. Woodford DJ, Hui C, Richardson DM, Weyl OL (2013) Propagule pressure drives establishment of introduced freshwater fish: quantitative evidence from an irrigation network. Ecol Appl 23: 1926–1937.
30. Agresti A (2002) Categorical data analysis. 2nd edition. Hoboken: John Wiley and Sons, Inc. 744 p.
31. Sokal RR, Rohlf FJ (1995) Biometry: the principles and practice of statistics in biological research. New York: Freeman. 885 p.
32. Snijders T, Bosker R (199) Multilevel analysis: an introduction to basic and advanced multilevel modeling. Thousand Oaks: Sage. 368 p.
33. Bryk AS, Raudenbush SW (2002) Hierarchical linear models: applications and data analysis methods. 2nd edition. Newbury Park: Sage. 512 p.
34. Lunn D, Spiegelhalter D, Thomas A, Best N (2009) The BUGS project: evolution, critique, and future directions. Stat Med 28: 3049–3067.
35. Burnham KP, Anderson DR (2002) Model selection and inference: an information-theoretic approach, 2nd edition. New York: Springer-Verlag. 518 p.
36. Spiegelhalter DA, Best NG, Carlin BP, van der Linde A (2002) Bayesian measures of model complexity and fit. J R Stat Soc Series B Stat Methodol 64: 583–639.
37. Link WA, Barker RJ (2009) Bayesian inference with ecological applications. San Diego: Academic Press. 400 p.
38. Royall RM (1997) Statistical evidence: a likelihood paradigm. New York: Chapman and Hall. 191 p.
39. Congdon P (2001) Bayesian statistical analysis. New York: Wiley. 556 p.
40. Hosmer DW Jr, Lemeshow S (2000) Applied logistic regression. New York: Wiley. 392 p.
41. Gelman A, Rubin DB (1992) Inference from iterative simulation using multiple sequences. Stat Sci 7: 457–511.
42. Gelman A, Meng XL, Stern H (1996) Posterior predictive assessment of model fitness via realized discrepancies. Stat Sinica 6: 733–759.
43. Hulme PE (2009) Trade, transport and trouble: managing invasive species pathways in the era of globalization. J Appl Ecol 46: 10–18.
44. MacKenzie DI, Nichols JD, Sutton N, Kawanishi K, Bailey LL (2005) Improving inferences in population studies of rare species that are detected imperfectly. Ecology 86: 1101–1113.
45. Yoccoz NG (2012). The future of environmental DNA in ecology. Mol Ecol 21: 2031–2038.
46. Pilliod D, Goldberg C, Arkle R, Waits L (2014) Factors influencing detection of eDNA from a stream dwelling amphibian. Mol Ecol Resour 14: 109–116.
47. Barnes M, Turner C, Jerde C, Renshaw M, Chadderton L, et al. (2014) Environmental conditions influence eDNA persistence in aquatic systems. Environ Sci Technol 48: 1819–1827.
48. Wotton RS, Malmqvist B (2001) Feces in aquatic systems. Bioscience 51: 537–544.
49. Minamoto T, Yamanaka H, Takahara T, Honjo M, Kawabata Z (2012) Surveillance of fish species composition using environmental DNA. Limnology 13: 193–197.
50. Hutson AM, Toya LA, Tave D (2012) Production of the endangered Rio Grande silvery minnow, *Hybognathus amarus*, in the conservation rearing facility at the Los Lunas Silvery Minnow Refugium. J. World Aquacult Soc 43: 84–90.
51. MacKenzie DI, Nichols JD, Lachman GB, Royle JA, Pollock KH, et al. (2006) Occupancy estimation and modeling: inferring patterns and dynamics of species occurrence. San Diego: Elsevier-Academic Press. 324 p.
52. Royle JA, Dorazio RM (2008) Hierarchical modeling and inference in ecology: The analysis of data from populations, metapopulations, and communities. San Diego: Academic Press. 464 p.

3

Threatened and Endangered Subspecies with Vulnerable Ecological Traits Also Have High Susceptibility to Sea Level Rise and Habitat Fragmentation

Allison M. Benscoter[1]*, Joshua S. Reece[2], Reed F. Noss[3], Laura A. Brandt[4], Frank J. Mazzotti[1], Stephanie S. Romañach[5], James I. Watling[1]

1 University of Florida, Fort Lauderdale Research and Education Center, Davie, Florida, United States of America, **2** Valdosta State University, Department of Biology, Valdosta, Georgia, United States of America, **3** University of Central Florida, Department of Biology, Orlando, Florida, United States of America, **4** United States Fish and Wildlife Service, Davie, Florida, United States of America, **5** United States Geological Survey, Southeast Ecological Science Center, Davie, Florida, United States of America

Abstract

The presence of multiple interacting threats to biodiversity and the increasing rate of species extinction make it critical to prioritize management efforts on species and communities that maximize conservation success. We implemented a multi-step approach that coupled vulnerability assessments evaluating threats to Florida taxa such as climate change, sea-level rise, and habitat fragmentation with in-depth literature surveys of taxon-specific ecological traits. The vulnerability, adaptive capacity, and ecological traits of 12 threatened and endangered subspecies were compared to non-listed subspecies of the same parent species. Overall, the threatened and endangered subspecies showed high vulnerability and low adaptive capacity, in particular to sea level rise and habitat fragmentation. They also exhibited larger home ranges and greater dispersal limitation compared to non-endangered subspecies, which may inhibit their ability to track changing climate in fragmented landscapes. There was evidence for lower reproductive capacity in some of the threatened or endangered taxa, but not for most. Taxa located in the Florida Keys or in other low coastal areas were most vulnerable to sea level rise, and also showed low levels of adaptive capacity, indicating they may have a lower probability of conservation success. Our analysis of at-risk subspecies and closely related non-endangered subspecies demonstrates that ecological traits help to explain observed differences in vulnerability and adaptive capacity. This study points to the importance of assessing the relative contributions of multiple threats and evaluating conservation value at the species (or subspecies) level when resources are limited and several factors affect conservation success.

Editor: Brock Fenton, University of Western Ontario, Canada

Funding: Funding was provided by United States Fish and Wildlife Service, National Park Service (Everglades and Dry Tortugas National Park through the South Florida and Caribbean Cooperative Ecosystem Studies Unit), United States Geological Survey (Greater Everglades Priority Ecosystems Science), Kresge Foundation, and Florida Fish and Wildlife Conservation Commission. The views expressed here do not necessarily represent the views of the United States Fish and Wildlife Service, nor the United States Department of the Interior. Use of trade, product, or firm names does not imply endorsement by the United States Government. The funders had no role in study design, data collection and analysis, decision to publish, or preparation of the manuscript.

Competing Interests: The authors have declared that no competing interests exist.

* E-mail: ambenscoter@gmail.com

Introduction

Assessing the vulnerability (sensitivity+exposure to threats) and adaptive capacity (ability to adjust to threats) of species helps identify the most important factors affecting species survival and prioritize limited conservation resources. In addition to pressures such as habitat fragmentation [1], invasive species [2], disease [3], and overexploitation [4], climate is a critical factor affecting the survival and distribution of species. Climate affects biodiversity at multiple spatial and taxonomic scales, and recent climate change is already linked to alterations in species phenology, survival, and distribution [5,6,7,8]. Climate change may present unique challenges to threatened and endangered taxa (hereafter referred to as endangered), because these species often exist in small populations, have limited genetic variation, and may have ecological traits that make them vulnerable to rapid environmental change. It is cost-effective to prioritize conservation efforts towards species and communities that have a relatively high probability of conservation success, and provide ecological, evolutionary, social, or economic value.

Florida is a national hotspot of endemism [9,10], making biological conservation significant at local, national, and broader levels. There are also many taxa at risk of extinction, and only three other states in the country have more federally endangered taxa than Florida (Hawaii, California, and Alabama). Of the 572 endangered animal taxa listed on the federal register in the United States, 53 occur in Florida (approximately 10% of all listed taxa). Of these 53 federally listed taxa, 23 are designated as subspecies (approximately 43% of Florida's listed taxa), and 19 of the 23 subspecies are endemic (unique) to Florida [11]. Understanding the vulnerability of endangered subspecies in this region to factors such as climate change and habitat fragmentation may help reduce extinction rates for Florida's endangered vertebrates.

Numerous pressures affect native species persistence in Florida, including climate change. Florida is particularly vulnerable to sea level rise, with about 10% of its land area less than 1 m above current sea level [12]. Most of the Florida Keys are predicted to be inundated or drastically altered under a sea level rise scenario of 0.6 m [13,14], and substantial tracts of the Everglades, other low-lying coastal areas, and barrier islands across the state will be submerged or severely modified with 1 m of sea level rise [15]. Many areas that are not inundated by increased sea level will be susceptible to storm surges, flooding, erosion, and other risks [5]. Although Florida's climate is predicted to warm less than other regions in North America [5,16], a climate inventory over the past 35 to 108 years indicated Florida is experiencing greater climate extremes, with trends of increased summer and fall maximum temperatures and decreased winter and spring minimum temperatures [17]. The intensity of tropical storms is also predicted to increase, although frequency may decrease [5,18]. Given the link between extreme climate events and the decline of local populations [13,19], an increase in frequency or intensity of extreme climate may threaten endangered species that exist in small fragmented populations.

High human population growth rates and non-native species invasions also threaten native fauna of this region. Florida's human population growth rate was the third highest in the United States between 2000 and 2010 [20]. Florida also has a high incidence of non-native plant and animal species [21,22]. These pressures coupled with high biodiversity and limited conservation resources make setting conservation priorities extremely important.

Vulnerability assessments are a valuable tool that aid in prioritizing conservation efforts through the evaluation of threats and their impacts on species and communities. The Standardized Index for Vulnerability and Value Assessment (SIVVA) is a novel tool [23] that is useful for evaluating vulnerability, adaptive capacity, and conservation value of species and communities. It provides several improvements over previous vulnerability assessments, such as the explicit incorporation of sea level rise and the ability to account for uncertainty in the assessment process. It allows for inclusion of the ecological, evolutionary, and economic value of taxa, and offers a flexible and modular approach applicable to a broad range of taxonomic groups, including terrestrial, freshwater aquatic, and marine species. SIVVA allows for the evaluation of multiple interacting threats, both independently and collectively, to measure vulnerability and adaptive capacity, allowing for a more comprehensive assessment that is particularly useful for setting conservation priorities in regions that face numerous threats to biodiversity.

In this study we incorporated a multi-step approach to evaluate vulnerability and adaptive capacity of federally endangered subspecies in Florida. We focused on subspecies pairs that differed in endangered status, because their phylogenetic relatedness provided the opportunity to ask questions about ecological traits related to adaptation and vulnerability. Because each subspecies in the taxon pair represented a closely related but distinct population, potential variation in ecological traits may help explain the relative ability to respond to environmental change. This framework allowed for a robust test to examine potential differences in traits for endangered taxa, especially given that controlled experiments on these subspecies are not feasible. Although comparing traits among related taxa and accounting for phylogenetic similarity is not uncommon in ecological studies, we are unaware of other regional-scale studies designed to investigate trait differences among many pairs of closely related taxa (subspecies) that vary in conservation status. The target group for this study included all of the federally endangered subspecies located in Florida.

We examined reproductive output, home range size, dispersal ability, and survival because they can affect population persistence and extinction risk [24,25,26]. Our approach coupled SIVVA with in-depth literature surveys of ecological traits to evaluate vulnerability and adaptive capacity of endangered subspecies under future scenarios of sea level rise, habitat fragmentation, and climate. Because endangered taxa often have restricted ranges, small population sizes, and reduced genetic variation, we expected endangered subspecies to demonstrate higher vulnerability and lower adaptive capacity compared to closely related, non-endangered subspecies. We also developed four specific *a priori* hypotheses related to vulnerability criteria important in Florida, and predicted that endangered subspecies would have greater vulnerability to sea level rise, habitat fragmentation, and altered temperature and precipitation regimes. Lastly, we expected that endangered subspecies would have higher vulnerability and decreased adaptive capacity because of lower reproductive output, greater home range size, shorter dispersal, and lower survival rates.

Materials and Methods

1. Study Taxa

Study taxa comprised 12 taxon-pairs ($n = 8$ mammals, $n = 4$ birds). Each pair consisted of a federally endangered subspecies and a closely related non-endangered subspecies in the same parent species group (in a few cases, two non-endangered subspecies were used for comparison; Table 1). Although the target group included taxon-pairs encompassing all of the federally threatened and endangered subspecies in Florida ($n = 23$ animals, $n = 2$ plants), analyses were conducted on 12 taxon-pairs due to limited data for 13 taxon-pairs. Scientific nomenclature for endangered subspecies followed the U.S. Fish and Wildlife Service (USFWS) Endangered Species Program [11]; nomenclature for non-endangered subspecies followed the Integrated Taxonomic Information System (ITIS) [27] and supplemental literature when indicated. Although there were a few cases of disputed taxonomy (Table 1), all federally endangered subspecies represent distinct populations that do not interact with the non-federally endangered subspecies. Non-endangered subspecies were chosen based on information availability, with preference given to those with the greatest geographic similarity.

2. Vulnerability Assessments (SIVVA)

Vulnerability assessments using the SIVVA framework were conducted for 12 endangered subspecies and their closely related non-endangered subspecies ($n = 23$ total assessments; *Peromyscus polionotus phasma* and *Peromyscus polionotus niveiventris* have the same parent species, *Peromyscus polionotus*; Table 1). The SIVVA framework consists of four modules: vulnerability (sensitivity+exposure to threats), adaptive capacity (ability to adjust to threats), conservation value, and information availability. Each module contains a set of criteria ($n = 30$ total SIVVA criteria) that describe key threats and factors relevant to conservation planning. For example, the vulnerability module includes 12 different criteria describing potential threats to species persistence, including sea level rise, habitat fragmentation, and altered temperature and precipitation (Table S1).

Each taxon was assessed independently by two experts with knowledge regarding the taxon of interest. Experts were provided with detailed taxon range maps, projections (e.g., sea level rise, human population growth), and a summary of published literature for the taxon. The accuracy of expert opinion greatly improves when it is coupled with consultation of relevant literature [28,29]. For each assessment, the taxon was scored on a continuous scale from zero to six for each of 30 criteria. In general, higher scores corresponded to higher vulnerability, lower adaptive capacity,

Table 1. Study species.

Federally listed subspecies	Common name	Federal status	Non-listed subspecies
Mammals			
Neotoma floridana smalli	Key Largo woodrat	Endangered	*floridana*
Odocoileus virginianus clavium	Key deer	Endangered	*virginianus*
Oryzomys palustris natator[a]	Silver rice rat	Endangered	*natator, palustris*[b]
Puma concolor coryi	Florida panther	Endangered	*couguar*[c]
Peromyscus gossypinus allapaticola	Florida salt marsh vole	Endangered	*gossypinus*
Peromyscus polionotus niveiventris	Southeastern beach mouse	Threatened	*rhoadsi, subgriseus*
Peromyscus polionotus phasma	Anastasia Island beach mouse	Endangered	*rhoadsi, subgriseus*
Sylvilagus palustris hefneri	Lower Keys marsh rabbit	Endangered	*paludicola*
Birds			
Ammodramus maritimus mirabilis	Cape Sable seaside sparrow	Endangered	*maritimus, peninsulae*
Ammodramus savannarum floridanus	Florida grasshopper sparrow	Endangered	*pratensis, perpallidus*
Polyborus plancus audubonii	Audubon crested caracara	Threatened	*cheriway*[d]
Rostrhamus sociabilis plumbeus	Everglade snail kite	Endangered	*sociabilis*

Study species comprised 12 taxon-pairs, each included a federally threatened or endangered subspecies and a closely related non-listed subspecies in the same parent species group (in some cases, two non-listed subspecies were used for comparison). Scientific nomenclature for listed subspecies followed the U.S. Fish and Wildlife Service Endangered Species Program [11]; nomenclature for non-listed subspecies followed the Integrated Taxonomic Information System [27] and supplemental literature when indicated.

[a]The endangered taxon *Oryzomys palustris natator* [11] is also referred to as *Oryzomys argentatus* [66].

[b]Subspecies taxonomy for *O. palustris* has undergone several revisions. For this study, we compared the federally endangered population (*O. p. natator* aka *O. argentatus*) to two mainland subspecies of *O. palustris* located in Florida and the southeastern United States. According to [66], *O. p. natator* is a subspecies that occurs in central Florida; it does not interact with the endangered population.

[c]Subspecies taxonomy for *Puma concolor* has undergone several revisions. According to [46], *P. c. couguar* refers to cougars throughout North America.

[d]*Polyborus plancus audubonii* [11] is included as a sub-population of the species *Caracara cheriway* (northern crested caracara) according to ITIS [27] nomenclature. We compared the *P. p. audubonii* population to other non-interacting individuals with the most geographic similarity in the *C. cheriway* complex.

greater conservation value, and greater information availability (e.g., we assumed greater information availability enhances the probability of conservation success). For example, vulnerability criterion scores of one and two corresponded to positive responses to projected environmental change (one was most positive), a score of three corresponded to no effect, and scores of four, five, and six indicated increasingly negative responses to the environmental threat. Scores reflect expert opinion, but have quantitative guidelines, for example, a score of six for vulnerability to sea level rise corresponded to "50% or more of known range being inundated by 1 m of sea level rise by 2100." For all criteria, a score of zero indicated insufficient information to evaluate the criterion. Detailed descriptions of score metrics were provided with each criterion, and although the specific metrics related to each score varied depending on the criterion, the relative scale was constant.

Summary scores were calculated for each module as the total number of points divided by the total number of possible points (Table S1). Note that a higher adaptive capacity score equates to reduced adaptive potential. We also applied a pre-determined weighting scheme that reflected the relative importance of the criteria for terrestrial vertebrate taxa [23]. Weights were randomly permuted 1000 times to assess the influence of the weighting scheme on each module score for each taxon. The level of expert certainty was also incorporated via a checked box next to criterion scores with low confidence. To account for the uncertainty, values of zero or one were added or subtracted from each checked criterion (while maintaining a score range of 1–6), and each module score was recalculated using 1000 Monte Carlo simulations. From these simulations, we created 95% confidence intervals around each module score that summarized potential deviations in the overall score resulting from potential variability in criterion scores due to assessor uncertainty. For a more detailed description of SIVVA, refer to Reece and Noss [23].

3. Ecological Traits

We gathered information on ecological traits from published literature for 12 taxon-pairs and 4 trait categories: litter or clutch size, home range size, dispersal distance, and annual adult survival (Dataset S1). Observations from published literature were classified as independent if they contained data without overlapping individuals, locations, and/or time periods. Within each taxon-pair x trait combination we controlled for factors such as gender, age class (e.g., juvenile vs. adult), sample duration (e.g., monthly vs. annual survival), and measurement unit. When applicable, data were converted to maintain constant sample units for each taxon-pair x trait combination (e.g., home range data reported as hectares were converted to km^2). We obtained information for at least one trait for each of the 12 taxon-pairs. Territory size was used as the home range metric in *Ammodramus* taxa. We also compiled data that quantified geographic distance to the coast for each taxon, determined by calculating the Euclidean distance between the geographic range centroid (geometric center) and the closest coastline. Geographic range maps were obtained from NatureServe [30]. Geographic distance to the coast was transformed using the natural log to meet normality assumptions.

4. Statistical Analyses

4.1. Vulnerability assessments (SIVVA). We tested whether each SIVVA module score (vulnerability, adaptive capacity, conservation value, information availability) differed between endangered and non-endangered subspecies using linear mixed effects models. Each model contained a fixed effect of taxon status (endangered vs. non-endangered) and random effects of assessor identity and taxon-pair. We tested for the effect of taxon status on SIVVA scores by comparing models with and without the fixed effect using a chi-square statistic [31]. Because each taxon was assessed by two independent experts, pairwise differences within taxon-pairs were evaluated using 95% confidence intervals around each SIVVA module score, calculated for each taxon using 1000 Monte Carlo simulations based on scoring uncertainty.

We developed four *a priori* hypotheses regarding vulnerability to sea level rise, habitat fragmentation, altered temperature, and altered precipitation, therefore we also evaluated these vulnerability criteria independently (criteria 1,3,4, and 5 of the vulnerability module; Table S1) using a mixed effects model with fixed and random effects as described for the overall SIVVA analyses. Owing to recent criticisms of the Bonferroni correction including the inflation of type II error and loss of power [32,33] we accounted for the false discovery rate when conducting multiple tests by controlling for the proportion of false positives among rejected null hypotheses [34,35]. Scores of the individual vulnerability criteria were non-normal, so we specified a Poisson error distribution in all models except the model for sea level rise. We tested for differences in sea level rise vulnerability using the proportion of habitat lost to 1 m of sea level rise by 2100. The proportion of habitat inundated was calculated via taxon range maps obtained from NatureServe [30]. The sea level rise scenario map was generated using ArcGIS v10 software and a "bathtub inundation" approach based on a 10 m resolution digital elevation map. This scenario represents a modest estimate of sea level rise that will likely occur by 2100 [36,37,38]. Because vulnerability to sea level rise was scored as a proportion, we used a logit transformation for these data [39]. The test for vulnerability associated with habitat fragmentation was based on SIVVA criterion scores that assessed potential habitat limitations (including dispersal and migration paths) resulting from natural barriers and human land use practices. This was evaluated using taxon range maps (which included natural features) in conjunction with Florida 2060 population projections and associated forecasts estimating future development and land use changes [40]; this source represented the only accessible statewide projections at the time of this research. To assess variation in vulnerability to altered temperature and precipitation, we compared SIVVA vulnerability criteria scores evaluating taxon dependence on narrow temperature or precipitation regimes. As before, significance of all models was tested by comparing models with and without the fixed effect (endangered vs. non-endangered) using a chi-square statistic [31]. Because each taxon was evaluated by two independent assessors, differences within taxon pairs were evaluated using 95% confidence intervals around each score, as determined from Monte Carlo simulations based on scoring uncertainty.

4.2. Ecological traits. To test for differences in ecological traits between endangered and non-endangered subspecies across all taxon-pairs, trait data were converted to proportions of the maximum value for each taxon-pair x trait combination (which allowed for comparisons across taxon-pairs with data that varied in scale). Proportion data were logit transformed to account for non-normality, and the minimum non-zero proportion in the data set was added to the numerator and denominator of the logit function to correct for proportion values equal to one [39]. Data were analyzed using a linear mixed effects model with trait nested within taxon status (endangered vs. non-endangered); taxon-pair was included as a random effect. Tukey's HSD contrasts were calculated to determine pairwise differences between all endangered and non-endangered subspecies for each trait. In addition to the 'global' analysis across all taxon-pairs (which tested for overall trait differences between endangered and non-endangered subspecies), trait differences within individual taxon-pairs were calculated using Wilcoxon rank sum tests (non-transformed data) for taxon-pair x trait combinations with at least three independent observations per taxon.

Long distance dispersal can affect species range shifts [41,42], and dispersal capacity may affect the ability to utilize fragmented landscapes [43]. Therefore, we conducted Spearman rank correlations (non-transformed data) between mean and maximum dispersal estimates. Because there was a significant positive correlation between mean and maximum dispersal distance ($r = 0.898$, $p < 0.001$), and analyses conducted with mean and maximum dispersal metrics produced equivalent results, we chose to report analyses using only mean dispersal distance.

4.3. Relating ecological traits to SIVVA. To determine whether ecological characteristics were associated with SIVVA modules, we conducted Pearson product-moment correlation analyses between ecological traits (transformed data) and both the vulnerability and adaptive capacity SIVVA module scores. Because vulnerability to sea level rise, habitat fragmentation, and altered temperature and precipitation are important in this region, correlation analyses were also conducted between each ecological trait and these four vulnerability criteria. We accounted for the false discovery rate when conducting multiple tests by controlling for the proportion of false positives among rejected null hypotheses [34,35]. Because vulnerability to habitat fragmentation, altered temperature, and altered precipitation were evaluated on the SIVVA scale from 1–6, we conducted Spearman's rank correlations for these data, while maintaining Pearson correlations for the percent habitat lost to 1 m of sea level rise.

Statistical analyses were conducted in R [44]. Mixed model analyses were performed using the lme4 package [45], Tukey's contrasts were calculated using the multcomp package [46], and all other analyses were conducted using the base package in R ($\alpha = 0.05$ for all analyses).

Results

1. Vulnerability Assessments (SIVVA)

Endangered subspecies as a group had higher SIVVA vulnerability than non-endangered subspecies ($\chi^2 = 44.931$, df = 1, $p < 0.001$; Figure 1), and pairwise differences within each taxon-pair indicated that all endangered subspecies had significantly higher vulnerability compared to their closely related non-endangered subspecies ($p < 0.05$). For individual vulnerability criteria, endangered subspecies showed greater percent habitat inundation under 1 m of sea level rise ($\chi^2 = 13.654$, df = 1, $p < 0.001$) and higher vulnerability to habitat fragmentation ($\chi^2 = 13.560$, df = 1, $p < 0.001$) compared to non-endangered subspecies. Endangered subspecies did not show significantly greater vulnerability to altered temperature ($p = 0.559$) or precipitation ($p = 0.278$). Vulnerability to sea level rise, habitat fragmentation, and altered temperature and precipitation for each endangered subspecies are shown in Table 2.

Endangered subspecies exhibited lower overall adaptive capacity as a group (higher SIVVA adaptive capacity scores) compared to non-endangered subspecies ($\chi^2 = 36.436$, df = 1, $p < 0.001$; Figure 1). Although each endangered subspecies had lower adaptive capacity than closely related non-endangered subspecies (higher adaptive capacity score), pairwise differences within each taxon-pair indicated this was significantly lower for only 6 out of 12 endangered subspecies ($p < 0.05$). Conservation value was higher for endangered subspecies as a group ($\chi^2 = 71.519$, df = 1, $p < 0.001$; Figure 1), and pairwise differences indicated that all

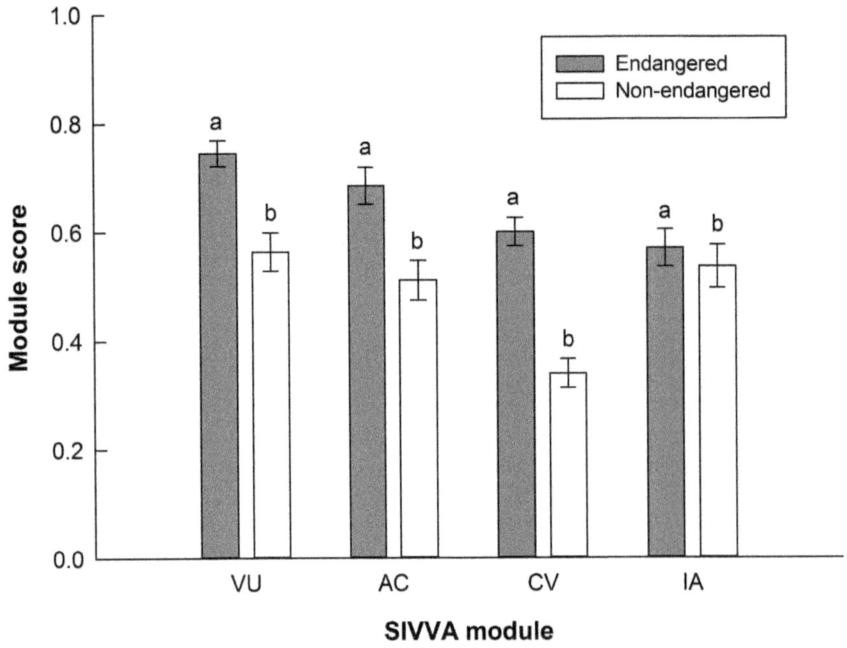

Figure 1. Standardized Index for Vulnerability and Value Assessment (SIVVA) module scores. Mean SIVVA module scores (± SE) for threatened and endangered subspecies (shaded bars) and non-listed subspecies (white bars) for vulnerability (VU), adaptive capacity (AC), conservation value (CV), and information availability (IA). Note that a higher AC module score corresponds to lower adaptive capacity. For each module, bars with different letters are significantly different ($p<0.05$).

endangered subspecies had significantly higher conservation value than their non-endangered subspecies pair ($p<0.05$). Endangered subspecies in our study group also had slightly greater information availability compared with non-endangered subspecies ($\chi^2=4.207$, df = 1, $p=0.040$; Figure 1), and pairwise differences showed that 4 out of 12 endangered subspecies had greater information availability.

Table 2. Threatened and endangered subspecies vulnerability.

	Vulnerability			
Scientific name	**Sea level rise**	**Fragmentation**	**Temperature**	**Precipitation**
Mammals				
Neotoma floridana smalli	High	High	Neutral	Neutral
Odocoileus virginianus clavium	Very high	Very high	Moderate	Very high
Oryzomys palustris natator	Very high	High	Neutral	Neutral
Puma concolor coryi	Moderate	High	Moderate	Low
Peromyscus gossypinus allapaticola	High	High	Neutral	Neutral
Peromyscus polionotus niveiventris	High	Moderate	Neutral	Neutral
Peromyscus polionotus phasma	High	High	Neutral	Neutral
Sylvilagus palustris hefneri	High	Very high	Moderate	Very high
Birds				
Ammodramus maritimus mirabilis	High	Very high	Neutral	Neutral
Ammodramus savannarum floridanus	Low	Very high	Moderate	Moderate
Polyborus plancus audubonii	Low	High	Positive[a]	Neutral
Rostrhamus sociabilis plumbeus	Low	High	Moderate	High

Vulnerability of 12 federally threatened and endangered subspecies to sea level rise, habitat fragmentation, altered temperature, and altered precipitation. Sea level rise vulnerability was derived from the percent of habitat inundated under 1 m of sea level rise (Low = 0–25%, Moderate = 26–50%, High = 51–75%, Very high = 76–100%). Vulnerability to habitat fragmentation, altered temperature, and altered precipitation were based on the Standardized Index for Vulnerability and Value Assessment (SIVVA) criteria scores (Neutral = 3, Low = 3–3.75, Moderate = 3.75–4.5, High = 4.5–5.25, Very high = 5.25–6).
[a]*P. p. audubonii* was evaluated to respond positively to altered temperature.

2. Ecological Traits

Across all taxon-pairs, we detected significant variation among traits ($\chi^2 = 80.671$, df = 6, $p<0.001$), and Tukey's HSD contrasts indicated that endangered subspecies had larger home range sizes ($p = 0.002$) and shorter dispersal distances ($p = 0.033$) than non-endangered subspecies (Figure 2). There was no difference in reproductive output per reproductive event or annual adult survival between endangered and non-endangered subspecies across all taxon-pairs.

Within taxon-pairs, *Puma concolor coryi* ($W = 6$, $n_1 = 3$, $n_2 = 12$, $p = 0.048$) and *Ammodramus savannarum floridanus* ($W = 0$, $n_1 = 3$, $n_2 = 4$, $p = 0.029$) had lower reproductive output than their closely related non-endangered subspecies (Figure 3). Reproductive output did not differ in the other four taxon-pairs with sufficient data to conduct within-taxon-pair comparisons (Figure 3). For the two taxon-pairs with sufficient home range information to conduct within-taxon-pair comparisons, *Neotoma floridana smalli* had a larger home range than *Neotoma floridana floridana* ($W = 9$, $n_1 = n_2 = 3$, $p = 0.050$), while home range size of *P. c. coryi* did not differ from other cougars in North America (*Puma concolor couguar*; Table 1; Figure 4) [47]. For dispersal distance there was only one taxon-pair with sufficient information, whereby *P. c. coryi* showed shorter dispersal than *P. c. couguar* ($W = 0$, $n_1 = 3$, $n_2 = 8$, $p = 0.006$; Figure 4). Adequate data was not available to conduct within-taxon-pair comparisons for survival.

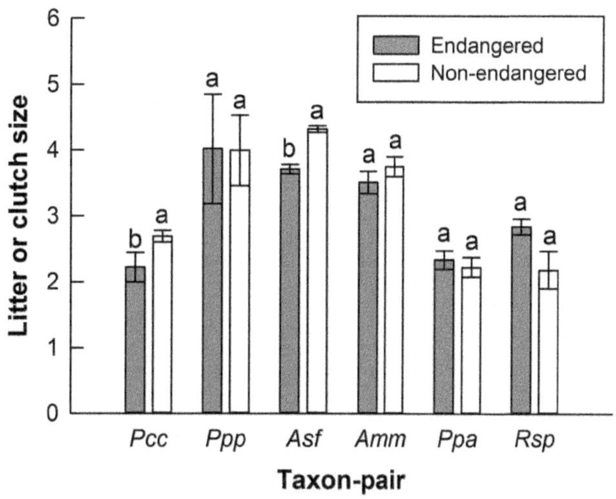

Figure 3. Reproduction within taxon-pairs. Mean litter or clutch size (± SE) for taxon-pairs with a minimum of three independent observations per taxon. For each taxon-pair, bars with different letters are significantly different (shaded bars = threatened and endangered subspecies; white bars = non-listed subspecies; $p<0.05$). Taxon-pairs are labeled according to the federally listed subspecies: *Pcc = Puma concolor coryi, Ppp = Peromyscus polionotus phasma, Asf = Ammodramus savannarum floridanus, Amm = Ammodramus maritimus mirabilis, Ppa = - Polyborus plancus audubonii, Rsp = Rostrhamus sociabilis plumbeus.*

3. Relating Ecological Traits to SIVVA

There was no significant correlation between most of the ecological traits and the vulnerability or adaptive capacity SIVVA modules. Not surprisingly, there was a negative correlation between geographic distance to the coast and vulnerability, such that taxa closer to the coast showed higher vulnerability ($r = -0.842$, $p<0.001$; Figure 5). We also observed associations between some of the ecological traits and the four *a priori* vulnerability criteria important in this region, where geographic distance to the coast was negatively correlated to both the percent habitat lost to sea level rise ($r = -0.675$, $p<0.001$) and habitat fragmentation vulnerability ($r = -0.672$, $p<0.001$; Figure 5), indicating that taxa located closer to coast are more susceptible to seal level rise and habitat fragmentation effects. There was a negative correlation between dispersal distance and vulnerability

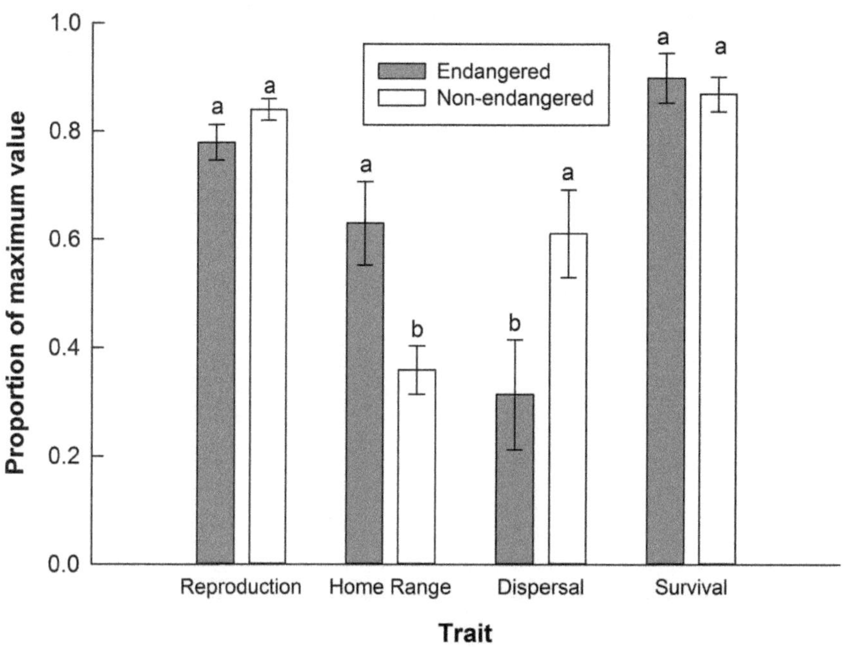

Figure 2. Ecological trait values. Mean ecological trait values (± SE) in threatened and endangered subspecies (shaded bars) and non-listed subspecies (white bars) for litter or clutch size, home range size, dispersal distance, and annual adult survival. Data were transformed to proportions of the maximum value for each taxon-pair x trait combination. For each trait, bars with different letters are significantly different ($p<0.05$).

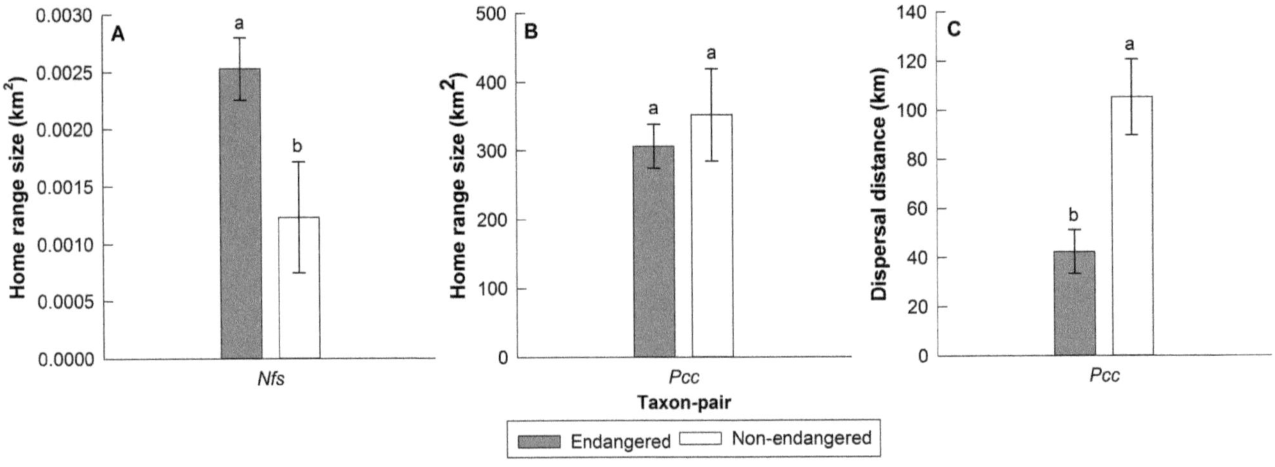

Figure 4. Home range and dispersal within taxon-pairs. Mean home range size and dispersal distance (± SE) for taxon-pairs with a minimum of three independent observations per taxon. Taxon-pairs are labeled according to the federally listed subspecies. (**A**) home range for the *Nfs* = *Neotoma floridana smalli* taxon-pair, (**B**) home range and (**C**) dispersal distance for the *Pcc* = *Puma concolor coryi* taxon-pair. For each taxon-pair x trait combination, bars with different letters are significantly different (shaded bars = threatened and endangered subspecies; white bars = non-listed subspecies; $p<0.05$).

to altered precipitation ($r = -0.880$, $p = 0.021$), where taxa with shorter dispersal also showed greater vulnerability to changes in precipitation.

Discussion

The high threat of habitat inundation from sea level rise for endangered subspecies is of primary concern because approximately 10% of Florida's land area lies less than 1 m above current sea level. The mean percentage of habitat loss under a 1 m sea level rise scenario was 52% for endangered subspecies, compared to 11% for non-endangered subspecies. Additionally, all but one endangered subspecies showed high vulnerability to habitat fragmentation (*P. p. niveiventris* showed moderate vulnerability). Florida also has high rates of human population growth [20] and land use conversion [48]. Sea level rise vulnerability coupled with landscape fragmentation may limit habitat availability and inhibit dispersal into new areas, and highlights the challenge of managing human land use activities in conjunction with biodiversity conservation [49,50,51].

Subspecies with the highest proportion of habitat inundation from sea level rise were located in the Florida Keys (*Oryzomys palustris natator*, *Odocoileus virginianus clavium*, *Sylvilagus palustris hefneri*, *Peromyscus gossypinus allapaticola*, *N. f. smalli*) or in coastal areas (*P. p. phasma*, *P. p. niveiventris*). Subspecies in the Florida Keys facing habitat inundation are particularly vulnerable to dispersal limitations imposed by both the island archipelago and human development. Similarly, the entire range of *P. p. phasma* is located on Anastasia Island, and rising sea levels may exacerbate dispersal limitations because of the limited size of the island and its separation from the mainland. More inland species located in peninsular Florida, such as *Ammodramus maritimus mirabilis* and *P. c. coryi*, also exhibited relatively high levels of percent habitat lost to 1 m of sea level rise (showing 57% and 50%, respectively), indicating that sea level rise vulnerability is not only an issue for taxa located in coastal areas.

In addition to high vulnerability, endangered subspecies exhibited lower adaptive capacity and higher conservation value compared to non-endangered subspecies. Subspecies with significantly lower adaptive capacity were located in the Florida Keys (*O. v. clavium*, *N. f. smalli*, *P. g. allapaticola*), coastal areas (*P. p. phasma*, *P. p. niveiventris*), and central Florida (*A. s. floridanus*). These taxa not only have greater vulnerability (exposure+sensitivity), but also a compromised ability to adapt to environmental change (adaptive capacity). The higher conservation value observed in the endangered subspecies was attributable to phylogenetic and geographic distinctiveness, endemism, and their endangered status. Even when the endangered status criterion was removed, all endangered subspecies retained their higher conservation value. Additionally, many of the endangered subspecies were rated by experts as having a low probability of recovery (e.g., *P. c. coryi*, *O. v. clavium*, *A. s. floridanus*, *A. m. mirabilis*, *Polyborus plancus audubonii*, *S. p. hefneri*, *P. p. niveiventris*; conservation value module, criterion 7). The assessment of conservation priority for these and other taxa in Florida is extremely valuable because there are limited resources available for conservation efforts.

Trait-based filtering operates when abiotic or biotic factors exclude species with certain ecological traits from a community, while allowing species with other traits to exist [52]. In the context of environmental change (e.g., climate or land use), species with life history traits that make them more sensitive or less adaptable to change are more vulnerable to extinction. The ecological trait data considered in our study helps explain differences in extinction risk, where endangered subspecies showed larger home range sizes and greater dispersal limitation compared to non-endangered subspecies. Larger home range size is often associated with greater extinction risk [53] because it increases sensitivity to fragmentation and exposure to human exploitation or persecution. Larger home range size may be indicative of higher resource requirements or lower habitat quality for endangered subspecies in Florida, which may increase vulnerability. Lower dispersal capacity in endangered subspecies could limit their ability to track changing climate and colonize new areas (e.g., in response to rising sea levels). The compromised life history traits in the endangered subspecies may represent inherent characteristics in these taxa or may be a consequence of the degraded landscape. In either case, it may affect their ability to adapt to rapid environmental change, especially given that many endangered taxa in Florida have

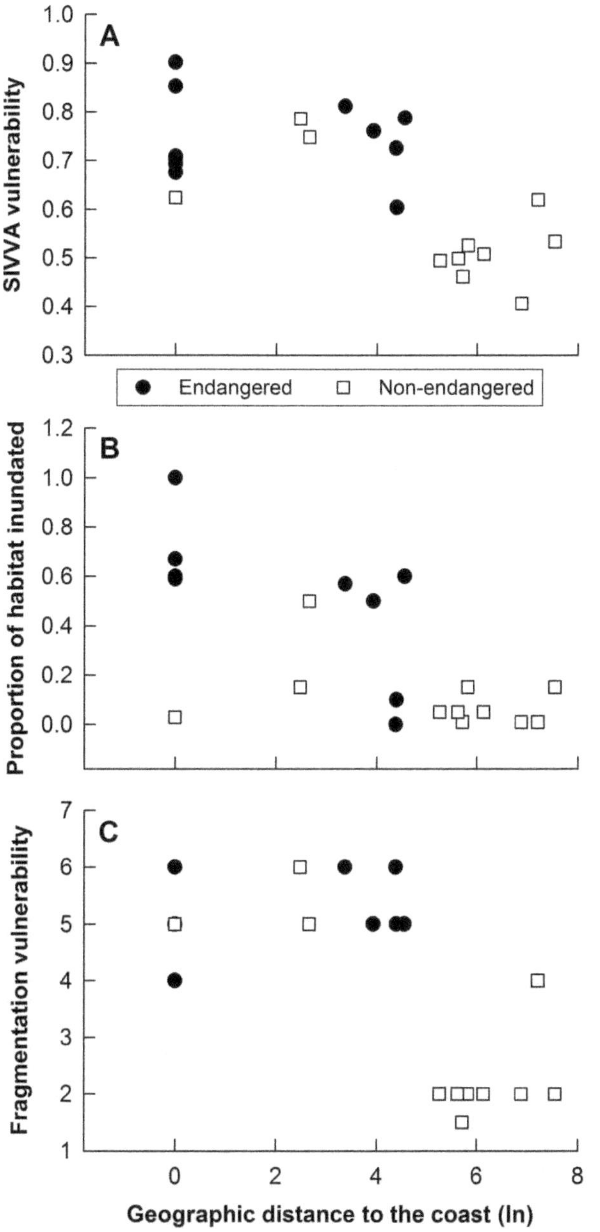

Figure 5. Associations between geographic distance to the coast and vulnerability. Correlation between geographic distance to the coast (natural log) and (**A**) the vulnerability module of the Standardized Index for Vulnerability and Value Assessment (SIVVA) ($r=-0.842$, $p<0.001$), (**B**) the proportion of habitat inundated under 1 m of sea level rise ($r=-0.675$, $p<0.001$), and (**C**) vulnerability to habitat fragmentation ($r=-0.672$, $p<0.001$). Black circles = threatened and endangered subspecies; white squares = non-listed subspecies.

already experienced environmental pressures such as habitat degradation and fragmentation.

The overall trend of larger home range size in endangered subspecies was driven by greater home range and territory sizes in small mammals (*N. f. smalli*, *O. p. natator*) and sparrows (*A. m. mirabilis*, *A. s. floridanus*), which showed a high magnitude of difference between endangered and non-endangered subspecies. The endangered large mammals (*P. c. coryi*, *O. v. clavium*) had smaller home ranges than non-endangered subspecies, likely due to habitat fragmentation effects, but the magnitude of difference was smaller. For the taxa with sufficient data to conduct within taxon-pair-comparisons, *N. f. smalli* showed a larger home range than its non-endangered subspecies, and the home range size *P. c. coryi* did not differ from its non-endangered subspecies.

All of the endangered subspecies included in the dispersal analyses were more dispersal limited than their closely related non-endangered subspecies, however there was enough data to conduct only one within-taxon-pair comparison, whereby *P. c. coryi* exhibited a shorter dispersal distance than *P. c. couguar*. This is of particular concern because the current geographic range of *P. c. coryi* is a small fragment of a once widespread distribution [54], which is characterized by several barriers to dispersal, including the Caloosahatchee River to the north, extensive mangrove swamps to the south, and urban development to the east and west [55]. Several males were recently documented north of the Caloosahatchee River, but no females have been recorded in central Florida since the early 1970s [56].

Because there were no overall differences across all taxon-pairs in reproduction or survival between endangered and non-endangered subspecies, these may not be crucial factors affecting vulnerability for most of our study taxa. However, the lower reproductive output in *P. c. coryi* and *A. s. floridanus* may indicate these subspecies are more vulnerable to environmental change, because slow life history (e.g., small litter size, long gestation time) is associated with greater extinction risk [57], and reproductive output affects population growth and recovery from disturbance. Due to limited data, we did not incorporate survival rates at earlier life stages (e.g., hatchling, juvenile), although these factors may be crucial for long-term population persistence.

The negative association between geographic distance to the coast and overall vulnerability has implications for conservation, because endangered taxa situated closer to the coast also showed greater habitat inundation under 1 m of sea level rise and greater vulnerability to habitat fragmentation. High population growth [20] and coastal development in this region [58] will affect conservation success for coastal and near-coastal taxa. Our results are consistent with recent studies highlighting the vulnerability of Florida's coastal species to sea level rise [59,60,61]. Additionally, the negative correlation between dispersal distance and vulnerability to changes in precipitation may limit the ability of some taxa to track changing climate (e.g., *P. c. coryi*, *O. v. clavium*, *R. s. plumbeus*) if they need to disperse to find more suitable environments.

This study highlights the importance of assessing ecological traits of endangered subspecies, because roughly 50% of the target group was not included in the study owing to limited availability of information. Although we found slightly greater information availability for endangered subspecies in 4 out of 12 taxon-pairs that were included in the study, we observed a general trend of limited data to conduct within-taxon-pair comparisons. Our framework allowed for high levels of phylogenetic control, especially given that each taxon-pair was comprised of non-interacting closely related taxa, with preference given to those with greatest geographic proximity and similarity. Therefore, despite data limitations, we were able to provide assessments of vulnerability and adaptive capacity of species that often have limited information, are not available for controlled experiments, and are in need of conservation. Our ecological trait assessment along with SIVVA provided a robust approach to evaluating species attributes important for conservation.

Many taxa face multiple interacting threats, which are important to evaluate jointly. The use of SIVVA in conjunction with detailed literature surveys of ecological traits provides an in-depth approach to understand the potential drivers of vulnerability and adaptive capacity, differentiate relative impacts of different threats, and prioritize conservation efforts. This type of approach is crucial to help understand the loss of biodiversity and create conservation management practices that attain the most successful outcomes. According to a recent global assessment, the current rate of biodiversity loss is estimated between 100 and 1000 times greater than the historical or "background" rate, and is attributable to human activities such as land use change [62]. The evaluation of species vulnerability and adaptive capacity is extremely valuable given the importance of biodiversity in ecosystem functioning [62,63] and sustaining human populations [62]. Studies with more narrow taxonomic or regional focus are more informative for practical conservation [64,65], and the flexible and modular nature of SIVVA is applicable to a broad range of taxonomic groups and ecological systems. The higher vulnerability and lower adaptive capacity of endangered subspecies in Florida is of particular concern in this region, especially given that many of these subspecies have larger home range sizes and shorter dispersal distances compared to their closely related non-endangered subspecies. The inevitable threat of sea level rise and continued habitat fragmentation in this region underscore the need to create and implement conservation plans that maximize conservation success.

Acknowledgments

We thank numerous species experts and staff at the Florida Natural Areas Inventory for valuable help with SIVVA assessments. We also thank C. Speroterra and D. N. Bucklin for providing comments on a previous version of this manuscript.

Supporting Information

Table S1 List of criteria in each module for the Standardized Index for Vulnerability and Value Assessment (SIVVA).

Dataset S1 Ecological trait data for threatened and endangered subspecies and their closely related non-listed subspecies. Subspecies common name, taxon status (endangered vs. non-endangered), subspecies.

Author Contributions

Conceived and designed the experiments: AMB JSR RFN LAB FJM SRM JIW. Performed the experiments: AMB JSR RFN JIW. Analyzed the data: AMB JSR JIW. Contributed reagents/materials/analysis tools: LAB FJM SSR JIW. Wrote the paper: AMB JSR JIW.

References

1. Brooks TM, Mittermeier RA, Mittermeier CG, Da Fonseca GAB, Rylands AB, et al. (2002) Habitat loss and extinction in the hotspots of biodiversity. Conserv Biol 16: 909–923.
2. McGeoch MA, Butchart SHM, Spear D, Marais E, Kleynhans EJ, et al. (2010) Global indicators of biological invasion: species numbers, biodiversity impact and policy responses. Divers Distrib 16: 95–108.
3. Smith KF, Acevedo-Whitehouse K, Pedersen AB (2009) The role of infectious disease in biological conservation. Anim Conserv 12: 1–12.
4. Loehle C, Eschenbach W (2012) Historical bird and terrestrial mammal extinction rates and causes. Divers Distrib 18: 84–91.
5. Intergovernmental Panel on Climate Change (2007) Climate Change 2007: Synthesis Report. Contribution of Working Groups I, II and III to the Fourth Assessment Report of the Intergovernmental Panel on Climate Change. Core Writing Team: Pachauri RK, Reisinger A, editors. Geneva: Intergovernmental Panel on Climate Change.
6. Parmesan C (2006) Ecological and evolutionary responses to recent climate change. Annu Rev Ecol Evol Syst 37: 637–669.
7. Parmesan C, Yohe G (2003) A globally coherent fingerprint of climate change impacts across natural systems. Nature 421: 37–42.
8. Root TL, Price JT, Hall KR, Schneider SH, Rosenzweig C, et al. (2003) Fingerprints of global warming on wild animals and plants. Nature 421: 57–60.
9. Knight GR, Oetting JB, Cross L (2011) Atlas of Florida's Natural Heritage – Biodiversity, Landscapes, Stewardship, and Opportunities. Tallahassee: Florida State University.
10. Stein BA, Kutner LS, Adams JS (2000) Precious Heritage: The Status of Biodiversity in the United States. Oxford: Oxford University Press.
11. U.S. Fish and Wildlife Service (2012) U.S. Fish & Wildlife Service endangered species program. http://www.fws.gov/endangered/species/index.html. Accessed 2012 July 12.
12. Weiss J, Overpeck J (2003) Maps of areas susceptible to sea level rise. http://www.geo.arizona.edu/dgesl/research/other/climate_change_and_sea_level/sea_level_rise/sea_level_rise.htm. Accessed 2012 April 15.
13. Ross MS, O'Brien JJ, Ford RG, Zhang K, Morkill A (2009) Disturbance and the rising tide: the challenge of biodiversity management on low-island ecosystems. Front Ecol Environ 7: 471–478.
14. Zhang K, Dittmar J, Ross M, Bergh C (2011) Assessment of sea level rise impacts on human population and real property in the Florida Keys. Clim Change 107: 129–146.
15. Noss RF (2011) Between the devil and the deep blue sea: Florida's unenviable position with respect to sea level rise. Clim Change 107: 1–16.
16. Christensen JH, Hewitson B, Busuioc A, Chen A, Gao X, et al. (2007) 2007: regional climate projections. In: Solomon S, Qin D, Manning M, Chen Z, Marquis M, et al., editors. Climate Change 2007: The Physical Science Basis. Contribution of Working Group I to the Fourth Assessment Report of the Intergovernmental Panel on Climate Change. Cambridge and New York: Cambridge University Press. 847–940.
17. Von Holle B, Wei Y, Nickerson D (2010) Climatic variability leads to later seasonal flowering of Floridian plants. PLoS ONE 5: 1–9.
18. Misra V, Carlson E, Craig RK, Enfield D, Kirtman B, et al. (2011) Climate scenarios: a Florida-centric view. Florida Climate Change Task Force. Center for Ocean-Atmospheric Prediction Studies 14: 1–61.
19. Parmesan C, Root TL, Willig MR (2000) Impacts of extreme weather and climate on terrestrial biota. B Am Meteorol Soc 81: 443–450.
20. Mackun P, Wilson S (2011) Population Distribution and Change: 2000 to 2010. U.S. Department of Commerce, Economics and Statistics Administration, U.S. Census Bureau C2010BR-01.
21. Pimentel D, Zuniga R, Morrison D (2005) Update on the environmental and economic costs associated with alien-invasive species in the United States. Ecol Econ 52: 273–288.
22. South Florida Water Management District (2010) Executive summary 2010 south Florida environmental report. West Palm Beach: South Florida Water Management District.
23. Reece JS, Noss RF (2014) Prioritizing species by conservation value and vulnerability: a new index applied to species threatened by sea-level rise and other risks in Florida. Nat Areas J (In press).
24. Cardillo M (2003) Biological determinants of extinction risk: why are smaller species less vulnerable? Anim Conserv 6: 63–69.
25. Fritz SA, Bininda-Emonds ORP, Purvis A (2009) Geographical variation in predictors of mammalian extinction risk: big is bad, but only in the tropics. Ecol Lett 12: 538–549.
26. Massot M, Clobert J, Ferriere R (2008) Climate warming, dispersal inhibition and extinction risk. Glob Change Biol 14: 461–469.
27. Integrated Taxonomic Information System (2012) Integrated Taxonomic Information System. http://www.itis.gov/index.html. Accessed 2012 September 14.
28. Clevenger AP, Wierzchowski J, Chruszcz B, Gunson K (2002) GIS-generated, expert-based models for identifying wildlife habitat linkages and planning mitigation passages. Conserv Biol 16: 503–514.
29. Martin TG, Burgman MA, Fidler F, Kuhnert PM, Low-Choy S, et al. (2012) Eliciting expert knowledge in conservation science. Conserv Biol 26: 29–38.
30. NatureServe (2010) NatureServe: A network connecting science with conservation. www.natureserve.org. Accessed 2012 April 10.
31. Bolker BM, Brooks ME, Clark CJ, Geange SW, Poulson JR, et al. (2009) Generalized linear mixed models: a practical guide for ecology and evolution. Trends Ecol Evol 24: 127–135.
32. Moran MD (2003) Arguments for rejecting the sequential Bonferroni in ecological studies. Oikos 100: 403–405.
33. Nakagawa S (2004) A farewell to Bonferroni: the problems of low statistical power and publication bias. Behav Ecol 15: 1044–1045.
34. Benjamini Y, Yekutieli D (2001) The control of the false discovery rate in multiple testing under dependency. Ann Stat 29: 1165–1188.
35. García LV (2004) Escaping the Bonferroni iron claw in ecological studies. Oikos 105: 657–663.
36. Pfeffer WT, Harper JT, O'Neel S (2008) Kinematic constraints on glacier contributions to 21st-century sea-level rise. Science 321: 1340–1343.

37. Strauss BH, Ziemlinski R, Weiss JL, Overpeck JT (2012) Tidally adjusted estimates of topographic vulnerability to sea level rise and flooding for the contiguous United States. Environ Res Lett 7: 014033.
38. Vermeer M, Rahmstorf S (2009) Global sea level linked to global temperature. Proc Natl Acad Sci USA 106: 21527–21532.
39. Warton DI, Hui FKC (2011) The arcsine is asinine: the analysis of proportions in ecology. Ecology 92: 3–10.
40. Zwick PD, Carr MH (2006) Florida 2060: a population distribution scenario for the state of Florida. Gainesville: Geoplan Center at the University of Florida.
41. Pearson RG, Dawson TP (2005) Long-distance plant dispersal and habitat fragmentation: identifying conservation targets for spatial landscape planning under climate change. Biol Conserv 123: 389–401.
42. Phillips BL, Brown GP, Travis JM, Shine R (2008) Reid's paradox revisited: the evolution of dispersal kernels during range expansion. Am Nat 172: S34–S48.
43. Van Teeffelen AJA, Vos CC, Opdam P (2012) Species in a dynamic world: consequences of habitat network dynamics on conservation planning. Biol Conserv 153: 239–253.
44. R Development Core Team (2012) R: A language and environment for statistical computing. http://www.R-project.org. Accessed 2011 December 1. Vienna: R Foundation for Statistical Computing.
45. Bates D, Maechler M, Bolker B (2012) lme4: Liner mixed-effects models using S4 classes. R package version 0.999999-0. http://CRAN.R-project.org/package = lme4. Accessed 2012 January 4.
46. Hothorn T, Bretz F, Westfall P (2008) Simultaneous inference in general parametric models. Biom J 50: 346–363.
47. Culver M, Johnson WE, Pecon-Slattery J, O'Brien SJ (2000) Genomic ancestry of the American puma (*Puma concolor*). J Hered 91: 186–197.
48. Walker RT, Solecki WD (1997) Land use dynamics and ecological transition: the case of south Florida. Urban Ecosyst 1: 37–47.
49. Brussaard L, Caron P, Campbell B, Lipper L, Mainka S, et al. (2010) Reconciling biodiversity conservation and food security: scientific challenges for a new agriculture. Environ Sustain 2: 34–42.
50. Noss RF, Murphy DD (1995) Endangered species left homeless in Sweet Home. Conserv Biol 9: 229–231.
51. Tscharntke T, Clough Y, Wanger TC, Jackson L, Motzke I, et al. (2012) Global food security, biodiversity conservation and the future of agricultural intensification. Biol Conserv 151: 53–59.
52. Weiher E, Keddy PA (1995) Assembly rules, null models, and trait dispersion – new questions from old patterns. Oikos 74: 159–164.
53. Woodroofe R, Ginsberg JR (1998) Edge effects and the extinction of populations inside protected areas. Science 280: 2126–2128.
54. Kautz R, Kawula R, Hoctor T, Comiskey J, Jansen D, et al. (2006) How much is enough? Landscape-scale conservation for the Florida panther. Biol Conserv 130: 118–133.
55. Thatcher CA, van Manen FT, Clark JD (2006) Identifying suitable sites for Florida panther reintroduction. J Wildl Manage 70: 752–763.
56. Thatcher CA, van Manen FT, Clark JD (2009) A habitat assessment for Florida panther population expansion into central Florida. J Mammal 90: 918–925.
57. Purvis A, Gittleman JL, Cowlishaw G, Mace GM (2000) Predicting extinction risk in declining species. Proc R Soc Lond B Biol Sci 267: 1947–1952.
58. Finkl CW, Charlier RH (2003) Sustainability of subtropical coastal zones in southeastern Florida: challenges for urbanized coastal environments threatened by development, pollution, water supply, and storm hazards. J Coast Res 19: 934–943.
59. Maschinski J, Ross MS, Liu H, O'Brien JO, von Wettberg EJ, et al. (2011) Sinking ships: conservation options for endemic taxa threatened by sea level rise. Clim Change 107: 147–167.
60. Saha AK, Saha S, Sadle J, Jiang J, Ross MS, et al. (2011) Sea level rise and South Florida coastal forests. Clim Change 107: 81–108.
61. Schmidt JA, McCleery R, Seavey JR, Cameron Devitt SE, Schmidt PM (2012) Impacts of a half century of sea-level rise and development on an endangered mammal. Glob Change Biol 18: 3536–3542.
62. Rockström J, Steffen W, Noone K, Persson Å, Chapin III FS, et al. (2009) A safe operating space for humanity. Nat 461: 472–475.
63. Naeem S, Duffy JE, Zavaleta E (2012) The functions of biological diversity in an age of extinction. Science 336: 1401–1406.
64. Cardillo M, Georgina MM, Gittleman JL, Jones KE, Bielby J, et al. (2008) The predictability of extinction: biological and external correlates of decline in mammals. Proc R Soc Lond B Biol Sci 275: 1441–1448.
65. Fisher DO, Owens IPF (2004) The comparative method in conservation biology. Trends Ecol Evol 19: 391–398.
66. Indorf JL (2010) Phylogeography of the marsh rice rat (*Oryzomys palustris*) in wetlands of the southeastern United States. Doctor of philosophy dissertation. Coral Gables: University of Miami.

4

Conus: First Comprehensive Conservation Red List Assessment of a Marine Gastropod Mollusc Genus

Howard Peters[1]*, Bethan C. O'Leary[1], Julie P. Hawkins[1], Kent E. Carpenter[2], Callum M. Roberts[1]

1 Environment Department, University of York, York, United Kingdom, **2** International Union for Conservation of Nature, Global Marine Species Assessment, Biological Sciences, Old Dominion University, Norfolk, Virginia, United States Of America

Abstract

Marine molluscs represent an estimated 23% of all extant marine taxa, but research into their conservation status has so far failed to reflect this importance, with minimal inclusion on the authoritative Red List of the International Union for the Conservation of Nature (IUCN). We assessed the status of all 632 valid species of the tropical marine gastropod mollusc, *Conus* (cone snails), using Red List standards and procedures to lay the groundwork for future decadal monitoring, one of the first fully comprehensive global assessments of a marine taxon. Three-quarters (75.6%) of species were not currently considered at risk of extinction owing to their wide distribution and perceived abundance. However, 6.5% were considered threatened with extinction with a further 4.1% near threatened. Data deficiency prevented 13.8% of species from being categorised although they also possess characteristics that signal concern. Where hotspots of endemism occur, most notably in the Eastern Atlantic, 42.9% of the 98 species from that biogeographical region were classified as threatened or near threatened with extinction. All 14 species included in the highest categories of Critically Endangered and Endangered are endemic to either Cape Verde or Senegal, with each of the three Critically Endangered species restricted to single islands in Cape Verde. Threats to all these species are driven by habitat loss and anthropogenic disturbance, in particular from urban pollution, tourism and coastal development. Our findings show that levels of extinction risk to which cone snails are exposed are of a similar magnitude to those seen in many fully assessed terrestrial taxa. The widely held view that marine species are less at risk is not upheld.

Editor: David L. Roberts, University of Kent, United Kingdom

Funding: The Natural Environment Research Council (NERC) and Economic and Social Research Council (ESRC) generously supported the Red List research under a studentship grant to Howard Peters http://www.nerc.ac.uk/funding/. The authors gratefully acknowledge funding from the Biodiversity Synthesis Center of the Encyclopedia of Life (EOL) at the Field Museum of Natural History, Chicago for the Conus Synthesis Workshop http://synthesis.eol.org/center/funding; also Tom Haas and the New Hampshire Charitable Foundation and the Thomas W. Haas Foundation for their support of the Global Marine Species Assessment under the IUCN Global Species Programme. The funders had no role in study design, data collection and analysis, decision to publish, or preparation of the manuscript.

Competing Interests: The authors have declared that no competing interests exist.

* E-mail: hp510@york.ac.uk

Introduction

Extinction risk of marine organisms has attracted little attention compared to that of terrestrial taxa, with a widely held view that such risk is inconsequential due to high dispersal ability and large geographic ranges [1,2] especially when taking reference from the fossil record [1,3]. These beliefs are particularly prevalent when considering marine invertebrates, where a decline in abundance of the important phylum *Mollusca* has been overshadowed by the collapse in many exploited vertebrates, especially finfish [4]. This is primarily due to their relatively minor contribution to human protein requirements and the generally held belief that molluscs possess greater resilience to extinction through their perceived wide distribution and a likelihood of hidden pockets of survivors [5]. Marine invertebrates in general are seriously under-represented within the IUCN Red List [6]. Only cuttlefish, lobsters and scleractinian corals have been fully assessed and published [6,7]. Although limited research on the impact of habitat loss and fishing pressure on marine gastropod molluscs has been undertaken on a regional scale including for shell fisheries [8,9], there have been no comprehensive assessments of trends in species abundance, commercial and environmental impacts and extinction risk to any genera with a global biogeographical distribution.

Cone snails of the genus *Conus* offer an excellent opportunity to explore global threats to marine molluscs owing to their exceptional diversity [10], wide distribution, high degree of endemism, varied depth distribution [11], and an established global market in their trade from amateur shell collectors to commercial traders [12]. In addition, cone snails are used in some communities in the Pacific as an occasional foodstuff [13] but, more importantly, they are actively targeted by international drug companies and researchers as a potential pharmacological resource [14].

Cone snails constitute the family *Conidae*, which together with the *Turridae* (turrid snails) and *Terebridae* (auger snails) comprise the superfamily *Conoidea* otherwise known as *Toxoglossa* ('poison tongue') owing to the venom apparatus they deploy for immobilising prey [15]. The *Conoidea* form part of the order *Neogastropoda* in the sub-class *Prosobranchia* of the class *Gastropoda* of the phylum *Mollusca* [11].

Cone snails live throughout the world's tropical coastal waters with a steep latitudinal diversity gradient away from the tropics, extending into cooler regions that include southern California, northern Gulf of Mexico, Florida and the Carolinas, North Africa, the Mediterranean, South Africa, Australia, southern Japan and

China [16]. Distribution varies widely with some species occurring across the entire tropical Indo-Pacific but others restricted to a single bay or seamount [11,17].

The genus *Conus* is taxonomically challenging. Although morphological characteristics of the shell remain the initial means of species identification [11], more recently, other traits have also been employed to differentiate among species, in particular the radular teeth used in the capture of prey, whose shape and structure not only reflects the dietary preferences of the species [18] but may also be specific to a single species [19]. Separation of species through DNA sequence variations provides even greater reliability, but more recently, character-based DNA barcoding has been highly effective in distinguishing among closely-related species [20]. For this assessment we relied upon expertise from taxonomists in *Conus* to create a dataset of valid species.

The fossil record indicates that the first *Conus* appeared in a sea that covered what is now England and France during the Lower Eocene around 55 mya [21]. During subsequent radiations the genus expanded around the globe and by the Holocene had formed into four biogeographical regions: Indo-Pacific (IP), Eastern Pacific (EP), Western Atlantic (WA), and Eastern Atlantic (EA). Although widening of the Atlantic during the Cretaceous and Cenezoic has today created an impermeable barrier to *Conus* crossing the ocean, there have been some migrations in the past, as witnessed from the fossil record and more recently by *C. ermineus* extant in both the EA and WA and *C. chaldaeus, C. ebraeus* and *C. tessulatus* found in both the IP and EP [22].

The majority of the 632 species of *Conus* assessed (53.6%) occurs in the infralittoral zone of 5 m deep or less, with most of the remaining species (27.7%) at 50 m deep or less. However, there are some species such as *C. teramachii* that live in deeper parts of the continental shelf extending to 1,000 m where they may be brought to the surface as bycatch of demersal fisheries. The bathymetric ranges of individual species vary considerably with some shallow water species living within a one or two metre depth range and some deep-water species being found within a 500 m range or more [11].

Microhabitats vary by species and most often consist of sand or mud into which the cone snail may burrow, but may also include inter-tidal limestone benches (the smooth remains of reef structures from earlier geological periods when sea levels were higher [23]) with sand or algal turf, sub-tidal reef platforms with living and dead corals, or boulders with sandy layers [24]. They may also be found among coral rubble and occasionally among mangroves and seagrasses.

Cone snails are generally nocturnal in their feeding habit [25] and group-specific in their preference for worms, molluscs or fish (Fig. 1) although some species have a mixed diet [26]. The smallest groupings by diet are the obligate piscivores with around 50 species [11,27], and obligate molluscivores with approximately 80 [28]. The majority of *Conus* are vermivorous with polychaetes representing the largest dietary component, that can be the exclusive source of food for some species [11]. All cone snails use venom to immobilize their prey. The diversity of venoms employed by a particular species in the capture of prey is a reflection of the degree of specialisation in its diet [29].

From the earliest civilizations, people have prized cone shells for their exceptional beauty, with examples discovered among prehistoric artefacts used for personal adornment extending back 5,000 years [30]. Their striking patterns and wide range of colours and shades continue to attract collectors today with rare examples in perfect condition changing hands for thousands of dollars with common and abundant species traded for cents to a dollar or two each [12].

Over millions of years *Conus* has evolved a battery of peptide toxins (conopeptides/conotoxins) for immobilizing prey [31]. The venom of each species is a cocktail mixed from between 50 and 200 different peptides each of only 10 to 35 amino acids in length and is generally targeted at voltage-gated or ligand-gated ion channels [30]. These conopeptides have become a focus for biomedical research worldwide [14]. Indeed, with the probability that there are on average over 100 distinct toxins for each species [30], as a whole, the *Conidae* can probably synthesize in excess of 50,000 toxins with little, if any, replication [32].

Cone snails are therefore important to: a) biodiversity; they have evolved into one of the largest of all marine genera, b) biopharmaceutics; they offer unparalleled opportunities in the development of novel drugs, and c) economics; their shells provide income to poor fishing communities through sales to tourists, traders and a global business in the specimen shell trade.

Habitat loss is considered by many malacologists to be the primary risk factor facing tropical marine mollusc species (Bouchet pers. comm. 2011) and there is plenty of hard evidence to support this view. In Queensland, Australia, for example, abundance and species richness of mollusc assemblages have been shown to be adversely affected by removal of subtropical mangrove forests, with population declines of 83% recorded [33]. In San Diego, Southern California the endemic horn snail, *Cerithidia fuscata*, that lived along intertidal mudflats was last seen in 1935 after pollution and dredging had driven it to extinction [34]. Where coral cover has been extensively damaged or degraded through pollution, sedimentation, coastal development and destructive fishing, as witnessed throughout much of the tropics, coral-associated molluscs such as the *Conidae* are being usurped by bivalve crevice dwellers [35].

In this paper we report one of the first comprehensive extinction risk assessments of a taxonomically well-resolved marine taxon. Our research assesses the extinction risk to the global populations of *Conus*, examining each species' distribution, current and projected threats from disturbance to habitats, pollution, coastal development, and shell gathering. We have examined where possible the effects of fragmentation of populations and the likely impact of demersal fisheries on deeper water species. The assessment enables us to reappraise whether marine taxa are less extinction prone than terrestrial. In addition, our aim is to provide data in support of conservation measures for those species at the greatest risk of extinction over the short to medium term.

Methods

Red List Assessment

We used the assessment standards and procedures of the International Union for the Conservation of Nature (IUCN) Red List of Threatened Species to assess extinction risk to 632 species of *Conus*. This is the world's leading resource for describing the global conservation status of plants and animals and uses a standard methodology to classify species into one of nine categories, together with a codified set of criteria [36]. The assessment includes examination of the effects of both ecological change and commercial exploitation on the subject taxa. Data derived during the research and discovery process for each species is compiled to a standard format together with maps, images and other supporting documentation.

Following taxonomic review, we divided valid species into 12 biogeographical working sets for detailed assessment. A comprehensive assessment was not possible for those species where data was substantially absent. For example, species endemic to areas of protracted civil unrest such as the Horn of Africa may not have

Figure 1. Diet and toxicity. Left: *C. geographus* Linnaeus, 1758; piscivorous, 65–165 mm; intertidal to 20 m; significant fatality risk to humans. Centre: *C. textile* Linnaeus, 1758; molluscivorous, 40–150 mm; intertidal to 50 m; handle with extreme caution. Right: *C. betulinus* Linnaeus, 1758; vermivorous, 55–177 mm; intertidal to 20 m; minimal risk to humans; note operculum. All species Indo-Pacific.

been researched in the field for many years. Coincidentally, these regions are not generally subject to intensive coastal development, harbour works and refineries and so may offer a degree of protection to marine taxa. Similarly, species occurring in deep water, where recovery is most commonly through fisheries by-catch, often suffer a paucity of data including extent of distribution and habitat types. Furthermore, bathymetry data will often rely on the questionable estimation of fishers. Wherever possible for deep-water *Conus*, we have focussed our attention on the level of demersal fishing in the area, including destructive methods such as dredging that may seriously affect mollusc assemblages.

Most *Conus* species, however, occur in shallow water where impacts such as coastal development, pollution and habitat destruction can be more easily recorded. Such threats can give rise to population fragmentation leading to a serious decline in abundance which may be difficult to quantify until it has become extreme. However, indicators including market prices for specimen shells provide a useful guide to increasing scarcity. Knowledge voids are common for *Conus* but where they occur we have, where possible, used estimation or inference using suboptimal data permitted under Red List standards [36]. Despite this, 13.8% of *Conus* species were found to be so deficient in data we were unable to make an assessment with any degree of reliability.

Key indicators of risk

Distribution. A key indicator of potential risk to a species is the size of its geographical distribution. The Red List standard assessment uses two measures: Extent of Occurrence (EOO) and Area of Occupancy (AOO). EOO for marine species is the area within a polygon drawn around the boundary of the species' range, excluding land areas. This will include areas which may not be physically occupied by the taxon, e.g. deep water, but which could contribute to larval dispersal. AOO is the physical area within the EOO in which the taxon is known to occur. For shallow water species, this may be calculated from the perimeter of an island or length of coastline, extended by the width of habitat calculated from the known or inferred bathymetric range of the species over the area under review. However, for 'linear' habitats such as rivers and coastlines, IUCN suggests that their standard habitat width of 2 km should be used in computing AOO [36] and we have adopted this approach in the assessment for *Conus*. It should be noted that both the AOO and the EOO are only of major significance in assessing the level of threat if the species has a restricted range, and that for most wide-ranging *Conus* species that fall outside the parameters of the assessment criteria these range sizes are not calculated. However, spatial data, derived from range maps created for every species assessment, were projected using ArcGIS version 10.1 (Environmental Systems Research Institute) to generate species richness maps for a) all species and b) only those species with a range less than the median mapped range size. All data were standardised onto 1° grid cells and projected to world cylindrical equal area.

Number of Locations. It is possible that a catastrophic event could have a profound effect on the population size of some species. Although marine molluscs are resilient in being able to endure physical forces such as extreme weather events, small populations may be extirpated as a result of sudden habitat loss caused by catastrophic events such as major oil spills. The 'location' count indicates the number of areas in which a single catastrophic event could affect all individuals of the taxon present, events that may cumulatively drive a species into extinction. The value of this measure is another key factor in determining the level of risk a global population faces.

Literature search

We conducted a comprehensive search through published papers and other literature for data relating to *Conus* species' populations, depth, distribution, habitats, trade in animals and shells, use for foodstuff, pharmaceuticals, etc., together with any conservation measures in place, including indirect conservation as may be offered by marine protected areas. We sought information on current and possible future threats, including coastal development for tourism, industry or port construction, nutrient loading from agricultural run-off, pollution from domestic and industrial effluent, intensive trawling, siltation from land-based sources, dredging for shipping channels and mineral extraction. Data on activities such as these can often only be found in trade publications, contract award notifications etc.

We also examined the market in shells to determine 'collectability', pricing fluctuations, scarcity and demand. Some shells with exceptional colour and form will achieve iconic status, and if they are also rare like *C. gloriamaris* or *C. milneedwardsi*, it adds to their cachet. Species that live within a highly restricted range, within a single bay for example, are often at heightened risk from human activity. This particularly applies to shallow water species which may be gathered as curios in areas where new beach tourism projects are being developed or planned. We synthesised distribution data including observed fragmentation, location counts, marketability, population declines and threats for each species to apply one of the nine categories listed below.

Assessment categories

There are three categories of extinction risk: critically Endangered (CR), Endangered (EN) and Vulnerable (VU) that broadly define 'extremely high', 'very high, or 'high' risk of extinction respectively. In addition, there are two extinct categories, Extinct (EX) and Extinct in the Wild (EW), and three other categories: Near Threatened (NT) for species that will be elevated to a threatened category in the short term unless the potential risk is removed; Data Deficient (DD) where there is insufficient data to determine a category, and Least Concern (LC) where current and projected population levels indicate the species is not at risk. As this was a comprehensive assessment, we did not use the category Not Evaluated (NE), where the species has been recorded but no assessment has been carried out.

For *Conus*, the criteria in support of the selected category are primarily derived from a range of variables based on estimated population size and/or level of decline together with species range size and location count.

Synthesis and pre-publication checks

Following our research and assessment, the results were reviewed by a panel of fourteen international experts, each with specialist knowledge of the *Conus* species within their allotted biogeographical working sets. The review took the form of a five day synthesis workshop with teams comprising leading academics together with renowned specialists from the commercial sector with comprehensive field knowledge of species' distribution, scarcity and threats, and a facilitator experienced in Red List standards and procedures. This peer-review process confirmed or modified findings of the original assessment authors, and allowed inclusion of supplementary field-based knowledge from the participating experts. All reports were checked for consistency by the Mollusc Specialist Group of the IUCN Species Survival Commission before final approval and submission for publication through the IUCN Red List Unit.

Results

The greatest species richness for *Conus* is in the Philippines and countries to the south and east towards Papua New Guinea, the Solomon Islands, New Caledonia and Fiji (Fig. 2A). Species occurring in the Atlantic including West Africa and the Gulf of Mexico and Caribbean are fewer in number. However, when very wide-ranging species are omitted, i.e. those equal to or greater than the median mapped range size, centres of endemism are revealed in scattered locations including the Caribbean, the Marquesas, New Caledonia and the Eastern Atlantic, in particular Cape Verde, Senegal and Angola (Fig. 2B).

Global threats

Three of 632 *Conus* species assessed were considered to be Critically Endangered (CR), 11 Endangered (EN) and 27 Vulnerable (VU), which together represent 6.5% of all global species (Fig. 3), with a further 26 species (4.1%) categorised as Near Threatened (NT). Over one in ten of all *Conus* species is therefore considered at risk or may become so in the near future. 87 species (13.8%) were categorised as Data Deficient (DD) of which 75 (86.2%) occur in the Indo-Pacific.

All 14 CR and EN species occur in the waters off Cape Verde and Senegal, West Africa (Table 1). Of the 27 assessed as VU, eight are from Cape Verde and Senegal with three from Angola (Table 1), seven from the Western Atlantic (Table 2), and nine from the Indian Ocean, including two from Western Australia (Table 3). Only three threatened species occur east of longitude 60 (Oman to Mascarenes): *C. rawaiensis* from Western Thailand and *C. compressus* and *C. thevenardensis* from Western Australia – all VU. According to this assessment procedure, there are no threatened species in the Pacific (Fig. 4). Of the 26 Near-Threatened species (NT), Cape Verde and Senegal are again over-represented with 14 of the 17 species from the Eastern Atlantic. Of the remainder in this category, five are from the Western Atlantic, one from the Western Indian Ocean, and three from the Pacific (Fig. 4).

Analysis by Region

Marine molluscs that are wide-ranging are likely to be more resilient against threats than those that are range-restricted, with dispersed populations providing a reservoir for re-colonization in the event of local extirpations [2]. The Eastern Atlantic species occupy a limited length of coast with few islands when compared to the Western Atlantic and, more particularly, the Indo-Pacific. It is also intersected by large rivers draining the tropical land mass of Africa which render substantial areas of coastal water unsuitable for many marine molluscs. Conversely, islands of the tropical Indo-Pacific and Caribbean contribute substantial areas of shallow water habitat suitable for taxa such as *Conus* and do not generally suffer any significant flux of freshwater. Figure 5 shows the percentage distribution of species' range sizes within each of the four oceanic regions. This graphically illustrates that wide-ranging *Conus* species, i.e. AOO >2,000 km^2, are uncommon within the Eastern Atlantic compared to the other regions.

Eastern Atlantic. Ninety-eight species of *Conus* occur along the Eastern Atlantic seaboard from the Mediterranean and Morocco south to Namibia, with associated island archipelagos including the Canaries, Azores, and Cape Verde (plus one: *C. ermineus*, that also occurs in the Western Atlantic and was included in that region). There is one species from the island of St Helena, *C jourdani*, within this grouping although no live specimens have been observed and it is categorised as DD. With three CR, 11 EN and 11 VU species, representing 25.5% of the Eastern Atlantic species, and a further 17 species NT (Table 1), 42.9% of Eastern Atlantic

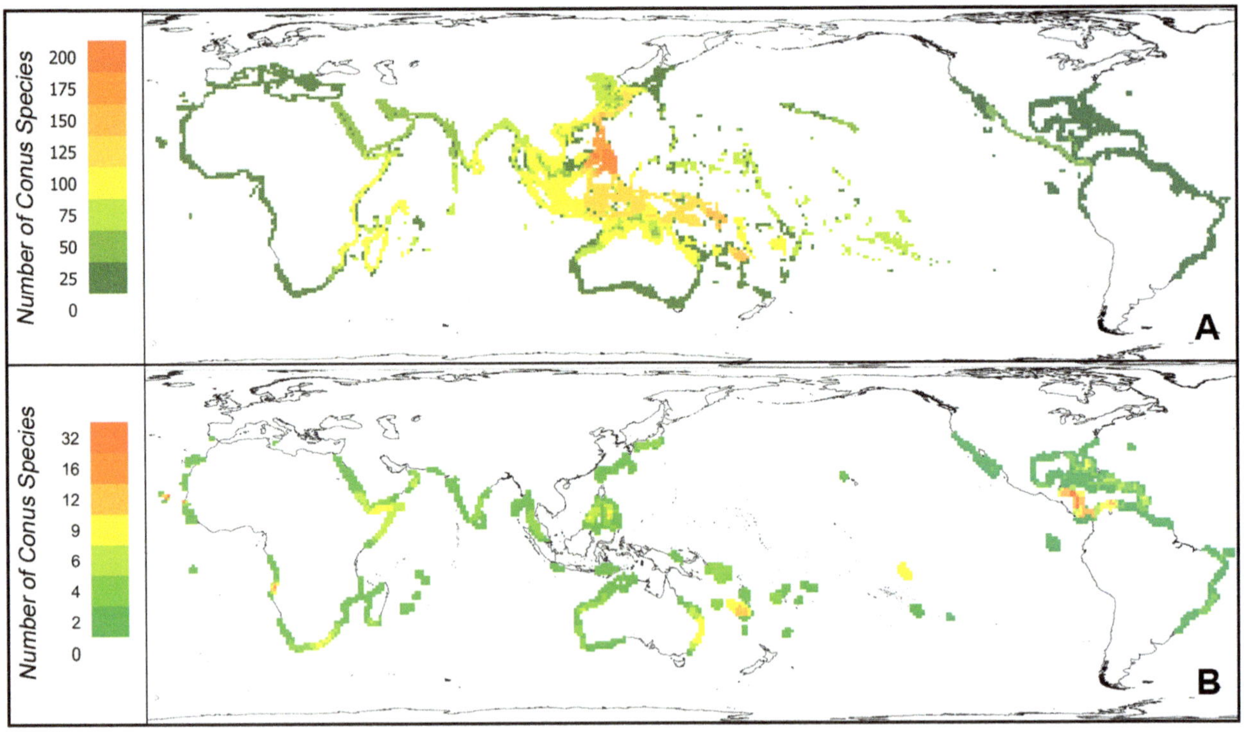

Figure 2. ***Conus* species distribution.** Species richness from a composite of individual species maps (Fig. 2A), and only species with mapped area less than the median indicating regions of potential endemism (Fig. 2B).

Conus are considered at risk of extinction or liable to become so. This exceptional concentration of threatened species is found nowhere else across the genus' wide distribution and the disproportionate contribution of species from Cape Verde and Senegal demands further explanation.

Cape Verde is home to 8.9% of all *Conus* species. With 53 species endemic from a total of 56 present in the archipelago, endemism is exceptionally high at 94.6%. Forty-three species are

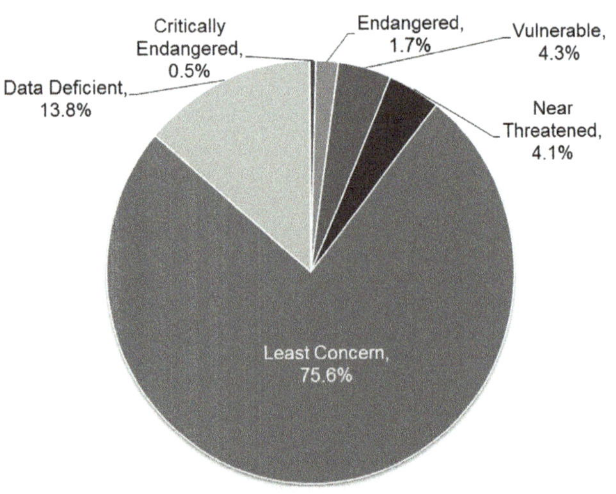

Figure 3. Global extinction risk to *Conus*. The percentage contribution for each assessed category to the global diversity of 632 *spp* of *Conus*. These are represented by 3 Critically Endangered species; 11 Endangered; 27 Vulnerable; 26 Near Threatened; 87 Data Deficient, and 478 of Least Concern.

each restricted to a single island. All three CR species are found in Cape Verde, *C. lugubris*, *C. mordeirae*, and *C. salreiensis*, together with four EN and five VU (Table 1). There are also 12 NT species. With 24 species in either a threatened or near-threatened category, Cape Verde has 45.3% of its *Conus* diversity at risk compared to 7.4% for the remainder of the world. Angola and Senegal contribute the next largest numbers of endemic *Conus* species with 22 and 13 respectively, which together with 53 species endemic to Cape Verde account for 89.8% of all 98 species within the Eastern Atlantic. Senegal contributes seven EN and three VU species with Angola contributing three VU (Table 1).

Western Atlantic. We assessed 113 species of *Conus* from the Western Atlantic where they occur from the Carolinas and Bermuda south to Brazil and throughout the Gulf of Mexico and Caribbean. Species are widely variable in their distribution across the region. There are six threatened species, all categorised as VU (see Table 2), representing 5.3% of the total and a further four NT, together resulting in 8.8% of *Conus* species within this region considered at immediate or potential risk.

Indo-Pacific. We assessed 390 species of *Conus* from the Indo-Pacific where they occur across the tropics and subtropics, from East Africa south to South Africa and north to the Red Sea and the Persian Gulf and across the whole of the Indian Ocean and the Western and Central Pacific, south to Australia and New Zealand, north to Japan, east to French Polynesia and Easter Island and northeast to Hawaii.

Only nine species were found to be VU. All occur within the Indian Ocean with six species from the western flank: two from Oman, one from the southern Red Sea, two from the Mascarenes and one from South Africa. From the eastern flank there is one species from Thailand and two from Western Australia. There are also four NT species including one from Oman with the other

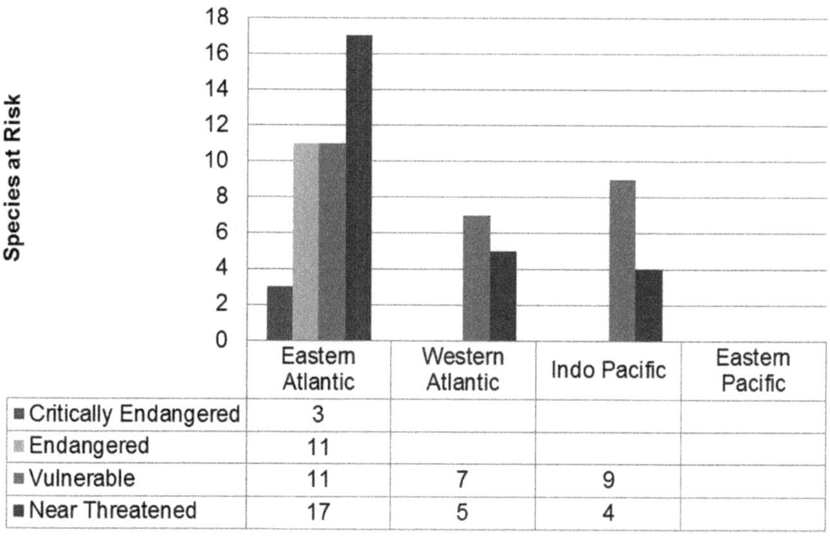

Figure 4. Number of *Conus* species at risk by ocean basin. The number of species at risk by ocean basin for each threatened category. There are no species at risk in the Eastern Pacific.

three being the only *Conus* species potentially at risk in the Pacific – one each from Queensland, the Philippines and the Marquesas.

Eastern Pacific. We assessed 31 species of *Conus* from the Eastern Pacific where they occur from Southern California south along the Pacific coast of Meso-America to Southern Ecuador including the Galapagos and other island groups of the region.

No species were assessed as threatened or near threatened in the Eastern Pacific.

Threats

The nature of threats to those species of *Conus* at risk of extinction are varied and depend primarily, but not exclusively, on the proximity and nature of human habitation and development adjacent to coastlines where the molluscs occur. This alone, however, will not normally create a scenario for species extinction. Wide-ranging species are capable of maintaining their viability through resilience from multiple sub-populations. Although most threatened *Conus* species are range-restricted, this is not always the case: two species from the USA, *C. anabathrum* and *C. stearnsii* occur along the west coast of Florida where their ranges are substantially fragmented by shoreline development. However, restricted range, coupled with shallow water habitat, magnifies the impact of stressors such as coastal development or pollution. Of the 41 *Conus* species globally assessed as threatened with extinction, 32 (78.0%) occur within an AOO of 250 km^2 and a minimum depth of 5 m or less. In the Eastern Atlantic, of the 25 threatened species, this rises to 100%.

Threats to those *Conus* species assessed within one of the three threatened categories can be classified into four causal groups: 1. pollution, either from proximity to actual or potential petrochemical spills, or urban and industrial effluent; 2. disturbance to habitat from coastal development either resulting from human population increases, e.g. sea defences, residential and commercial structures, including aquaculture facilities, and port construction, or tourism infrastructure. Also included in this group is damage to habitat caused by damaging and extensive demersal fishing; 3. shell gathering, and 4. environmental change e.g. elevated sea-surface temperatures (Fig. 6). There will frequently be a

Table 1. Threatened *Conus* of the Eastern Atlantic (EA).

Critically Endangered (CR)		**Endangered (EN)**		**Vulnerable (VU)**	
Cape Verde	*C. lugubris*	Cape Verde	*C. ateralbus*	Angola	*C. allaryi*
Cape Verde	*C. mordeirae*	Cape Verde	*C. crotchii*	Angola	*C. cepasi*
Cape Verde	*C. salreiensis*	Cape Verde	*C. cuneolus*	Angola	*C. xicoi*
		Cape Verde	*C. fernandesi*	Cape Verde	*C. decoratus*
		Senegal	*C. belairensis*	Cape Verde	*C. felitae*
		Senegal	*C. bruguieresi*	Cape Verde	*C. fontonae*
		Senegal	*C. cloveri*	Cape Verde	*C. regonae*
		Senegal	*C. echinophilus*	Cape Verde	*C. teodorae*
		Senegal	*C. hybridus*	Senegal	*C. cacao*
		Senegal	*C. mercator*	Senegal	*C. guinaicus*
		Senegal	*C. unifasciatus*	Senegal	*C. tacomae*

Table 2. Threatened *Conus* of the Western Atlantic (WA).

Vulnerable (VU)	
Aruba	*C. hieroglyphus*
Florida	*C. anabathrum*
Florida	*C. stearnsii*
Bahamas	*C. richardbinghami*
Brazil	*C. henckesi*
Martinique	*C. hennequini*
Venezuela	*C. duffyi*

Table 3. Threatened *Conus* of the Indo-Pacific (IP).

Vulnerable (VU)	
Oman	*C. ardisiaceus*
Oman	*C. melvilli*
S Red Sea	*C. cuvieri*
Mascarenes	*C. julii*
Réunion	*C. jeanmartini*
SE South Africa	*C. immelmani*
W Thailand	*C. rawaiensis*
Australia	*C. compressus*
Australia	*C. thevenardensis*

combination of causes, for example tourism infrastructure may also increase shell gathering. Similarly, the proximity of shanty towns devoid of planning regulations poses an elevated risk of effluent discharge into the marine environment. Finally habitat destruction from sand removal, beach nourishment works and recreational use of the sea may all result in disturbance to local mollusc populations.

Cape Verde is experiencing a major structural change from a largely services and fisheries based economy supported by development aid and remittances from its diaspora to one of beach tourism [37]. This is accompanied by a myriad of threats from road and resort construction, unlawful removal of beach sand for cement [38] and casual shell gathering by tourists. All three CR species occur in Cape Verde where their populations are already reduced. *C. lugubris* and *C. mordeirae* live in areas where habitat has already been lost to development and *C. salreiensis* which is restricted to a single bay has had observable declines in population since a harbour was constructed. Each is found in an area along a shallow coastal strip of less than 11 km in length. Harbour expansion and the accidental discharge of engine fuel increase the pressures on small, range-restricted *Conus* populations such as *C. fernandesi*, *C. fontonae* and *C. regonae*. With so many *Conus* species occupying highly restricted ranges within the archipelago, modest threats such as these could have a profound impact.

Around the Dakar peninsula, Senegal, it has been observed that species restricted to its highly polluted coastal waters are showing a marked decline in abundance coupled with an overall diminution of shell size including *C. echinophilus*, *C. hybridus*, *C. mercator* and *C. unifasciatus*. In common with many maritime cities in developing countries, Dakar suffers from a burgeoning population with largely inadequate waste-processing infrastructure. South in Angola, *Conus* species categorised as at risk face similar threats to those in Senegal.

In the Western Atlantic some disturbance to *Conus* can be traced to human migration to the Florida coast. Tourism and retirement have driven large-scale construction projects for condominiums and other coastal infrastructure leading to significant loss of habitat for *C. anabathrum* and *C. stearnsii*. Tourism also represents the underlying risk to the vulnerable species *C. hennequini* in Martinique and *C. hieroglyphus* in Aruba. Shell collecting in the Bahamas threaten *C. richardbinghami*. General coastal development in Bahia, Brazil threatens *C. henckesi* where it occurs only off two small islands. The Venezuelan government has voiced plans for substantial development on the islands of Los Roques which will place the shallow water species, *C. duffyi* at risk.

The *Conus* species of the Indo-Pacific are at less risk. In the north-western Indian Ocean, the Persian Gulf, the southern Red Sea including the Gulf of Aden and the Horn of Africa, civil wars, poverty, piracy and the security situation offer some degree of protection from coastal development. However, there are still concerns from oil spillage in the region and two scarce species from Oman, *C. ardisiaceus* and *C. melvilli* together with *C. cuvieri* from Djibouti are categorised as VU. In Southern Natal and the Mascarene islands of Mauritius and Réunion respectively, *C. immelmani* and *C. julii*, have both declined in numbers almost certainly from over-collecting, with *C. jeanmartini* also from Réunion being subject to intensive trawling in its deep-water habitat.

In the Eastern Indian Ocean, *C. rawaiensis* occurs only in an area estimated at less than 35 km^2 in a single location off the western shores of Thailand in a region zoned for tourism. In Western Australia, an extreme localized warm-water event in 2011 from La Niña, in the region around Geraldton to Shark Bay including the Abrolhos Islands, resulted in a catastrophic decline of marine molluscs including *Conidae*. *C. compressus*, a restricted range species, possibly suffered a 50% decline in abundance. Also in Western Australia, *C. thevenardensis*, already rare, is subject to a range of threats including a large oil installation, tourism and dredging.

Other Red List categories

The results for the three threatened categories paint an incomplete picture. There are also 87 species assessed to be Data Deficient and 26 as Near Threatened, together representing 17.9% of the global diversity. Many of the Data Deficient species are considered to be scarce in the wild even though the causes and extent of the threats they face cannot yet be determined with sufficient accuracy. Twice as many of the 87 DD species, (39.1%) occur only at depths of 50 m or more, i.e. below gleaning, scuba, and (many) artisanal fishing gears, compared to 19.4% for the remaining 545 species globally. Specimens may be brought to the surface from these depths as by-catch from fisheries. However, demersal gear such as dredges may also contribute substantially to the endangerment of the species recovered through destruction of their habitat, especially for those that are also of restricted range. Occurrence in deep water does not automatically result in a DD categorisation. Despite paucity of data, taxa with a known distribution greater than 2,000 km^2 but with no known threat would normally be assessed as Least Concern. At the other extreme, there are also a number of DD species where there is an almost total absence of recent sightings but extinction cannot be proven, i.e. there is reasonable doubt that the last individual has died [39]. This is exemplified by species such as *C. jourdani* from St Helena which is only known from 'dead' shells washed onto the beach in one small bay; *C. bellulus* and *C. luteus* which have not

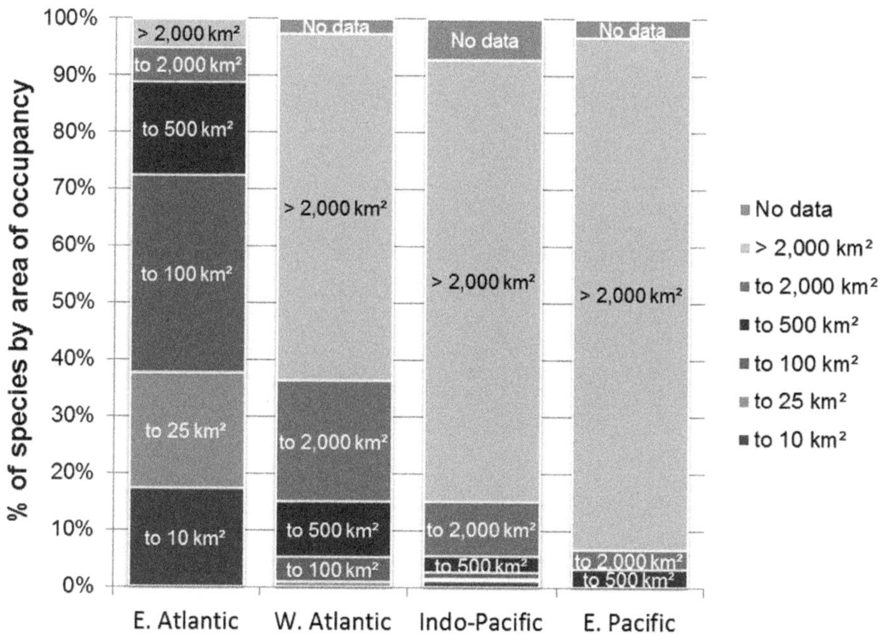

Figure 5. Contribution of range-restricted species to *Conus* biodiversity within each ocean basin. This illustrates by region the percentage of total species by area of occupancy, with wide-ranging species, i.e. >2,000 km^2 being minimal in the Eastern Atlantic but the major contributor to the Indo-Pacific and Eastern Pacific *Conus*. The abbreviated key describes the band sizes, e.g. to 10 km^2 = 0–10 km^2, to 25 km^2 = 11–25 km^2, to 100 km^2 = 26–100 km^2, etc.

been reported since the 1970s; *C. splendidulus* which has not been seen in 20 years and *C. sauros* which is possibly extinct.

Threatened species frequently have small ranges. Of 478 *Conus* species categorised as LC, 103 have an AOO of 2,000 km^2 of less (21.5%). However, of the 67 species with threatened and near-threatened categories, all except four occur within an AOO of 2,000 km^2 or less (94.0%), with 40 (59.7%) restricted to a range of 100 km^2 or less (Fig. 7). By comparison, 27 of the 59 DD species (45.8%) where it had been possible to approximate their AOO, were recorded at 2,000 km^2 or less (Fig. 7). Furthermore, many of

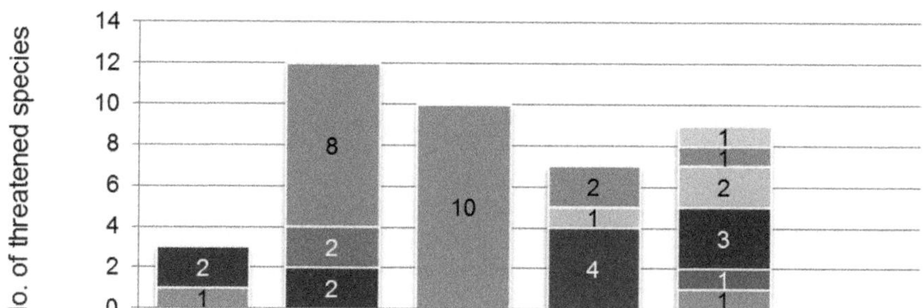

	Eastern Atlantic					
	Angola	Cape Verde	Senegal	Western Atlantic	Indo-Pacific	Eastern Pacific
Dem Fishing					1	
Tourism		8		2	1	
Shell Coll				1	2	
Ports etc		2				
Petro-chem	2	2			3	
Elev SST					1	
Effluent	1		10		1	
Coast Dev				4		

Figure 6. Main threats to *Conus* by ocean basin. The number of *Conus* species at risk (consolidation of CR, EN and VU) indicating primary causes of endangerment, being demersal fishing, tourism, shell collecting, ports and harbours, petro-chemical spills, elevated sea-surface temperatures, effluent discharge and runoff, and coastal development.

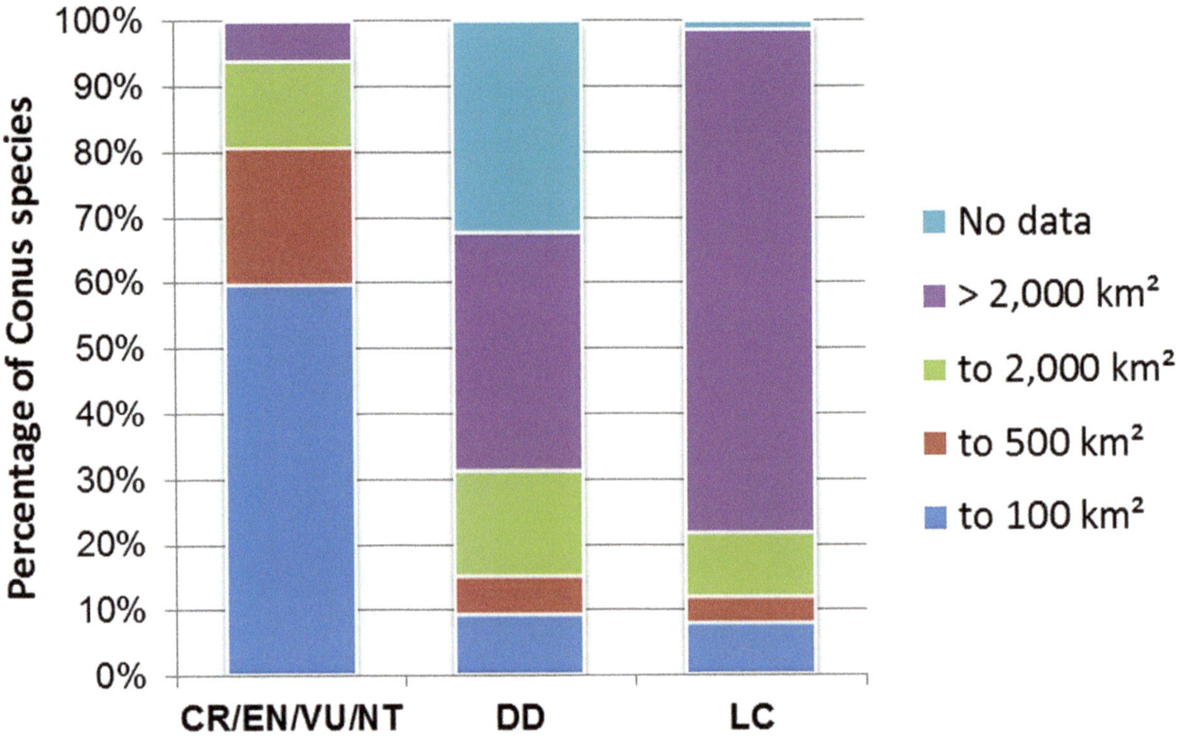

Figure 7. Contribution of range-restricted *Conus* species within assessment categories. This shows the percentage of species by assessment category, with DD species having a higher proportion of taxa with small ranges compared to LC species and also with a large number of species with no distribution data. All threatened and near-threatened categories have been grouped. Key: CR Critically Endangered, EN Endangered, VU Vulnerable, NT Near Threatened, DD Data Deficient and LC Least Concern

the 36.8% of DD species that are wider-ranging (AOO >2,000 km^2) are based on infrequent sightings. It would therefore be reasonable to suppose that a substantial proportion of the DD species are potential candidates for listing as threatened.

DD species are overwhelmingly found in the Indo-Pacific. There are three in the Eastern Atlantic, eight in the Western Atlantic, 75 in the Indo-Pacific and one in the Eastern Pacific. Outside the Indo-Pacific there is just a single species with a known AOO of more than 2,000 km^2.

Discussion

It is widely believed that extinction risk in the sea is less likely than in the terrestrial environment and that this is supported by the fossil record [2,3]. This view is based largely on perceived high fecundity, greater dispersal ability and geographic range size [40]. With 6.5% of *Conus* species at risk globally this would appear to follow this perception, however, in regions offering reduced dispersal opportunity, such as the whole of the Eastern Atlantic, 25.5% of species are threatened. Cone snails here have a similar level of extinction risk to species in well-assessed terrestrial taxa, such as freshwater invertebrates (34% of 7,784 species assessed at risk), lepidoptera (from 8.5% of butterflies in Europe to 17% in the U.S. at risk), European terrestrial molluscs (20% at risk) [41,42] and bryophyte flora from the Canaries (21% at risk) [43]. Contributing to the pattern seen, many cone snails have limited dispersal ability, small geographic ranges and/or are rare. The level of extinction risk is similar in other well assessed marine taxa, including corals (27% of species at risk) [7,44] and scombrid and billfish (11% of 61 species at risk) [45]. Given the rapid escalation of threats to the marine environment [46], if the pattern seen in these groups is typical of marine species generally, then there is a high risk that extinctions will soon become common in the sea, just as they now are on land.

Our global assessment of the conservation status of all 632 cone snails shows that three-quarters (75.6%) of species are classified as Least Concern under IUCN Red List standards. However, beneath this relatively optimistic result lies a picture of substantial regional variations with indicators signalling wider concerns. In the Eastern Atlantic along the shores of Senegal, Cape Verde and Angola, species restricted in their range and subject to the effects of industrialisation and urbanisation face an elevated risk of extinction. Endemism for marine species occurs most commonly in isolated island groups where the original dispersal was assisted by a pelagic larval stage or by transport on rafting matter [47]. Endemics may also be found where there may be non-reversing currents transporting water away from the tropics towards higher latitudes [40]. All Cape Verde endemic *Conus* have a non-planktonic larval stage having lost the ability during speciation to feed during larval dispersal [48]. This conforms to the hypothesis that non-planktonic, i.e. lecithotrophic, species of *Conus* commonly originate from planktotrophic species [49]. All three species assessed as Critically Endangered occur in the waters off Cape Verde where they are exposed to habitat degraded through coastal development primarily driven by tourism. Similarly, of the 11 Endangered species, four are found in Cape Verde with the remaining seven occurring off the coast of Senegal, in particular the Dakar peninsula, where high levels of pollution from industrial and residential effluent is thought to be the driver of declining abundance and observable reductions in body size. A further 11 species (40.7%) of the 27 assessed as Vulnerable occur in Cape Verde, Senegal and Angola, making West Africa home to 61% of the 41 *Conus* species threatened with extinction. Of the remaining 16 species categorised as Vulnerable, seven are found in the Western Atlantic where they are primarily exposed to coastal development, tourism and shell collecting. The remaining nine occur in the Indian Ocean where petrochemicals, shell-collecting and elevated sea-surface temperatures represent the principal causes of decline.

Threats

Overfishing. The effect of overfishing on the abundance of fish stocks has been extensively reported in both the scientific and general press over many years [50,51]. However, threats to invertebrates from fishing are seldom equated with extinction, especially marine molluscs. Although extremely unusual, near-extinctions in this group have occurred in the recent past; for example in the white abalone *Haliotus sorenseni* from southern California and Baja California by the mid-1990s had been fished to the edge of extinction [52]. Once counted in the millions there are now probably less than 1,600 individuals remaining.

Amongst marine molluscs, most species are sought by shell collectors [12]. Although this does not threaten the survival of the vast majority of molluscs, shell collecting has undoubtedly caused the decline and endangerment of some species, particularly 'trophy' shells. Throughout the Indo-Pacific, the spectacular giant triton (*Charonia tritonis*), has been extensively fished and in many areas has been extirpated [53]. Similarly, although primarily removed for its adductor muscles, the giant clam (*Tridacna gigas*) the largest of all bivalve molluscs, has met the same fate [54]. In Zanzibar, East Africa, the cowries *Cypraea tigris, C. histrio* and *C. lynx* were found to be up to 18 times less abundant in exploited tourist areas [8]. We identified three rare species of *Conus* threatened by shell collecting: *C. richardbinghami* from the Bahamas and *C. immelmani* and *C. julii* from the Mascarenes. Taxa already facing pressures from factors such as pollution may be pushed further towards extinction by gathering for shells, yet warning indicators such as sudden price inflation on the shell market may not alone warrant inclusion to a threatened category.

Bioprospecting. *Conus* is exceptionally important to biomedical science, although there is dispute about the number of animals taken for their bioactive compounds. To protect their intellectual property, pharmaceutical companies are silent on the issue, but researchers are adamant that volumes are negligible. In their dialogue in *Science* Chivian et al. (2003; 2004) raised important concerns about the quantity of cone snails taken from the wild, indicating that thousands were then collected to satisfy research demands [55,56]. This was forcefully rebutted by Duda et al. (2004) who reviewed recent conotoxin research from which they determined that a maximum of 20 research groups were working on *Conus* toxins at that time, and that any single characterisation required fewer than 21 animals to be sacrificed [57]. Regardless of where the true determinant lies, balancing the legitimate needs of medical research without further compromising natural resources is essential. Fortunately, alternative, more sustainable options are now available including milking venom without killing the animal [58], polymerase chain reaction (PCR) sequencing of DNA fragments that requires just one specimen [59] and more recently digital marine bioprospecting using massive parallel deep sequencing of transcriptomes that requires only minute samples of bioactive material [60].

Habitat loss. It has been shown that habit loss leads to declines in species richness, reduced biomass and loss of complexity [61,62], often accompanied by colonisation by species that inhibit recovery [63]. Virtually all of the world's 'trawlable' area of continental shelf has already been altered, and about half the area of all the continental shelves is hit by trawls every year [64], changing the structure and function of habitats, destroying assemblages and resulting in homogenisation of the seabed [65]. Of 133 marine species that have been recorded as having gone extinct either regionally or globally, 37% were attributed entirely or in part to habitat loss [66]. Extinctions of marine gastropod molluscs from loss of habitat are set to continue and include the horn snail *Cerithidea fuscata* from southern California last seen in 1935, the eelgrass limpet *Lottia alveus alveus* from the northwest USA last collected in 1929, and from the 19th century the rocky shore limpet '*Colisella*' *edmitchelli* also from southern California and the periwinkle *Littoraria flam*mea from China; all driven to extinction through loss of habitat from anthropogenic causes, with the possible exception of the eelgrass limpet that lost its habitat from a slime mould that may have been introduced from ships' ballast [2,34].

Our assessment found that with the exception of three species made vulnerable by shell collecting (see above), all 38 other *Conus* species threatened with extinction are impacted to some degree by habitat loss, either directly from coastal and port development or indirectly from pollution or from human exacerbated natural occurrences such as El Niño/La Niña–Southern Oscillation (ENSO) warm-water events (Fig. 6).

Red List comparatives

Our *Conus* assessment is the first global study for the IUCN Red List for any marine gastropod mollusc genus and one of the few for marine invertebrates. Other marine invertebrates that have been the subject of a global assessment include 845 reef-building corals, 247 lobsters and 195 cuttlefishes [44]. Data Deficiency is a common thread throughout each of these studies with 17%, 35% and 76% of species for each respective grouping [44] compared to 14% for *Conus*. Preliminary results available for oceanic squid show that 57% of this group are of Least Concern with the remaining 43% Data Deficient. As with the data deficient cone snails, many of these cephalopods are deep-water species that have only been captured on a few occasions [44].

Of the 845 corals that have been globally assessed 27.3% fall into a threatened category with a further 20.8% near threatened, although prior to the massive bleaching event of 1998 it has been estimated that 95.3% of non-DD species would have been categorised as Least Concern [7]. The exceptional ENSO event which resulted in this bleaching largely devalues any post-event comparison, although it has been shown that La Niña can impact some mollusc assemblages through stress, changes in productivity and availability of dietary preferences [67]. In Australia, the La Niña event of 2010-11 gave the highest monthly Southern Oscillation Index values on record accompanied by elevated sea-surface temperatures in Western Australia [68]. In the region around Geraldton to Shark Bay including the Abrolhos Islands, this coincided with an estimated 50% mortality in molluscs that included *Conus compressus* (H. Morrison pers. comm. 2011). Scleractinian corals and molluscs, including *Conus*, also share the threat of ocean acidification with the prospect of arrested development in their aragonite-forming structures [69,70]. Of all the threats faced by these fauna this is the most intractable and one that could even determine their continued existence.

For freshwater molluscs, Red List assessments have been completed for 1,500 of the 5,000 described species [6,42]. Results show that out of 7,784 freshwater invertebrates assessed to date, gastropods are the most threatened group with a threat range of 33% (if no DD species are threatened) to 68% (if all DD species are threatened) [42]. In common with *Conus*, the threatened species include range-restricted habitat specialists that are particularly at risk from loss of habitat and pollution.

Further Research and Conservation Priorities

As a global assessment for conservation has not been undertaken on any other marine gastropod it is not possible to explore relationships between different gastropod genera to identify commonality of risks. Further research is urgently needed to address this issue.

Data Deficient *Conus* species can normally be characterised by minimal sightings and a lack of data on distribution. This may result from their bathymetric profile (39% of DD species occur only at 50 m or more) and/or genuine rarity. However, DD species share a common trait with those categorised as threatened in having a higher percentage with restricted range than those from the general population, with the implication that there could be a significant proportion of DD species at risk. Bearing the title 'Data Deficient' or 'Near Threatened' places these 113 species jointly away from the spotlight afforded the 41 species in the three threatened categories. However, with the potential to double the number of species at risk it is essential that the taxa that make up these categories should not be ignored but instead benefit from further research into their true status.

One of the primary sources of information on species distribution, habitats, populations and threats for our Red List assessment has been the specimen shell trade. In many parts of the developing world, trade in shells provides valuable additional revenue to some of the poorest families living along tropical coastlines. Research is needed to assess the threat from rare shell collecting towards mollusc population decline to determine what measures should be taken to enable this activity to continue sustainably while at the same time allowing for protection of vulnerable species.

The need to identify conservation strategies for all species at risk is compelling, although for developing nations, snail conservation is unlikely to become a driver for environmental improvements. In the absence of in situ conservation measures, captive breeding programmes may ultimately be necessary, such as those undertaken for tree snails of the genus *Partula* from the Pacific Islands [71]. At present, except possibly for species such a *C. pennaceus* and *C. textile* that emerge as mature veligers, this is not a viable option, as the complexity of plankton essential for developing larvae cannot be easily replicated [28].

Over half (53.6%) of all *Conus* species occur at depths of 5 m or less where they are susceptible to gleaning, and nearly three-quarters (74.5%) occur at or above recreational SCUBA diving depths of 30 m or less. Marine Protected Areas (MPAs) offer one of the few sanctuaries, but regional authorities need encouragement to strengthen enforcement and to erect prominent signage against shell gathering within MPAs.

Cape Verde presents a special case in *Conus* conservation. With 45.3% of its species at risk there is a strong argument for legislating against export of both animals and shells, and with manageable borders the country is ideally suited to export controls. This small archipelago also signals a warning to other nations developing their coastal infrastructure: new roads bring visitors to areas previously protected by their isolation, and illegal sand removal for construction from beaches and shallow water of the littoral zone [72] pose a constant threat to habitat. Regional authorities should be required to undertake environmental impact assessments that take account of these issues when planning new developments.

The toxins that make *Conus* so successful are generally unique to each species [59] and any extinction in the genus could in turn deprive science of a potential pharmacological resource. The extraordinary number of species and the global distribution of these tropical snails make them an important contributor to marine biodiversity, and with the appeal of their shells they help support some of the world's poorest people.

Finally, there exists a well-defined community of cone snail aficionados who together are highly influential in the trade in cone shells. This includes leading academics as well as collectors and dealers. A positive first step from our Red Listing is that following a preliminary presentation of our findings at their international convention, a core of members has been motivated to explore a voluntary embargo in trade of critically endangered species and to consider this also for other *Conus* species at risk [73].

Acknowledgments

We would like to thank: Mark Westneat, Audrey Aronowsky, Sarah Kim and Beth Sanzenbacher at the Biodiversity Synthesis Center, Chicago for their organisation of the *Conus* synthesis workshop at the Field Museum in Chicago; Philippe Bouchet, José Coltro, Tom Duda, Alan Kohn, Eric Monnier, Hugh Morrison, Ed Petuch, Guido Poppe, Gabriella Raybaudi-Massilia, Sheila Tagaro, Manuel Jiménez Tenorio, Stephan Veldsman and Fred Wells for volunteering their time and expertise during the assessment and at the synthesis workshop; Monika Böhm, Heather Harwell, Andrew Hines, Suzanne Livingstone, Jonnell Sanciangco and Mary Seddon for facilitating at the synthesis workshop; Mike Filmer for helping to resolve the many taxonomic issues; Hannah Cubaynes, Zarozinia Sheriff and all the interns who assisted with species research; Klaus & Christina Groh of ConchBooks for use of images and maps for the Red List; Mia Theresa Comeros and Angela Goodpaster for their work on the Red List maps; and Caroline Pollock and Janet Scott of the IUCN Red List Unit for bringing the assessment to publication.

Author Contributions

Conceived and designed the experiments: HP BCO JPH CMR. Analyzed the data: HP BCO CMR. Contributed reagents/materials/analysis tools: KEC. Wrote the paper: HP JPH CMR. Revised the manuscript: JPH KEC CMR. Supported the synthesis workshop: HP KEC CMR.

References

1. Harnik PG, Simpson C, Payne JL (2012) Long-term differences in extinction risk among the seven forms of rarity. Proc Biol Sci 279: 4969–4976. doi:10.1098/rspb.2012.1902.
2. Roberts CM, Hawkins JP (1999) Extinction risk in the sea. Trends Ecol Evol 14: 241–246.
3. McKinney ML (1998) Is marine biodiversity at less risk? Evidence and implications. Divers Distrib 4: 3–8.
4. McManus JW (1997) Tropical marine fisheries and the future of coral reefs: a brief review with emphasis on Southeast Asia. Coral Reefs 16: S121–S127. doi:10.1007/s003380050248.
5. Jamieson GS (1993) Marine Invertebrate Conservation: Evaluation of Fisheries. Am Zool 33: 551–567.
6. IUCN (2013) IUCN Red List of Threatened Species. IUCN Red List Threat Species Version 20122. Available: http://www.iucnredlist.org. Accessed 8 February 2013.
7. Carpenter KE, Abrar M, Aeby G, Aronson RB, Banks S, et al. (2008) One-third of reef-building corals face elevated extinction risk from climate change and local impacts. Science 321: 560–563. doi:10.1126/science.1159196.
8. Newton LC, Parkes EVH, Thompson RC (1993) The Effects of Shell Collecting on the Abundance of Gastropods on Tanzanian Shores. Biol Conserv 63: 241–245.
9. Queensland Government Dept of Primary Industries and Fisheries (2007) Annual status report 2007 Queensland Marine Specimen Shell Collection Fishery. Available: http://www.daff.qld.gov.au/documents/Fisheries_SustainableFishing/AnnualStatusReport-QLDMarineSpecimen-ShellCollectionFishery-2007.pdf.
10. Bouchet P (1990) Turrid genera and mode of development: the use and abuse of protoconch morphology. Malacologia: 69–77.
11. Röckel D, Korn W, Kohn AJ (1995) Manual of the Living Conidae, Vol 1. Verlag Christa Hemmen.

12. Rice T (2007) A Catalog of Dealers' Prices for Shells: Marine, Land and Freshwater, 23rd edition. Of Sea and Shore Publications.
13. Chadwick A, Olivera BM (2009) Cone Shells and Human Culture. Cone Collect 12: 16–21. Available: http://www.theconecollector.com/.
14. Garber K (2005) Peptide leads new class of chronic pain drugs. Nat Biotechnol 23: 399. doi:10.1038/nbt0405-399.
15. Taylor JD, Kantor YI, Sysoev AV (1993) Foregut anatomy, feeding mechanisms, relationships and classification of the Conoidea (=Toxoglossa) (Gastropoda). Bull Nat Hist Museum London (Zoology) 59: 125–170.
16. Kohn AJ, Perron FE (1994) Life History and Biogeography – Patterns in Conus. Oxford Science Publications.
17. Monteiro A, Tenorio MJ, Poppe GT (2004) A Conchological Iconography. The Family Conidae. The West African and Mediterranean Species of Conus. Germany: ConchBooks, Hackenheim.
18. Tenorio MJ, Tucker JK, Chaney HW (2012) The Families Conilithidae and Conidae – The Cones of the Eastern Pacific. In: Poppe GT, Groh K, editors. A Conchological Iconography. ConchBooks, Hackenheim. p. 200.
19. Franklin JB, Fernando SA, Chalke BA, Krishnan KS (2007) Radular morphology of Conus (Gastropoda: Caenogastropoda: Conidae) from India. Molluscan Res 27: 111–122.
20. Zou S, Li Q, Kong L, Yu H, Zheng X (2011) Comparing the Usefulness of Distance, Monophyly and Character-Based DNA Barcoding Methods in Species Identification: A Case Study of Neogastropoda. PLoS One 6: e26619. doi:10.1371/journal.pone.0026619.
21. Kohn AJ (1990) Tempo and Mode of Evolution in Conidae. Malacologia 32: 55–67.
22. Duda TF, Kohn AJ (2005) Species-level phylogeography and evolutionary history of the hyperdiverse marine gastropod genus Conus. Mol Phylogenet Evol 34: 257–272. doi:10.1016/j.ympev.2004.09.012.
23. Kohn AJ (1983) Marine Biogeography and Evolution in the Tropical Pacific: Zoological Perspectives. Bull Mar Sci 33: 528–535.
24. Kohn AJ (1968) Microhabitats, Abundance and Food of Conus on Atoll Reefs in the Maldive and Chagos Islands. Ecology 49: 1046–1062.
25. Kohn AJ (1959) The Ecology of Conus in Hawaii. Ecol Monogr 29: 47–90.
26. Duda TF, Kohn AJ, Palumbi SR (2001) Origins of diverse feeding ecologies within Conus, a genus of venomous marine gastropods. Biol J Linn Soc 73: 391–409. doi:10.1006/bijl.2001.0544.
27. Olivera BM, Walker C, Cartier GE, Hooper D, Santos AD, et al. (1999) Speciation of cone snails and interspecific hyperdivergence of their venom peptides. Potential evolutionary significance of introns. Ann N Y Acad Sci 870: 223–237.
28. Livett BG, Gayler KR, Khalil Z (2004) Drugs from the sea: conopeptides as potential therapeutics. Curr Med Chem 11: 1715–1723.
29. Remigio EA, Duda TF (2008) Evolution of ecological specialization and venom of a predatory marine gastropod. Mol Ecol 17: 1156–1162. doi:10.1111/j.1365-294X.2007.03627.x.
30. Terlau H, Olivera BM (2004) Conus venoms: a rich source of novel ion channel-targeted peptides. Physiol Rev 84: 41–68. doi:10.1152/physrev.00020.2003.
31. Olivera BM (1997) E.E.Just Lecture, 1996. Conus venom peptides, receptor and ion channel targets, and drug design: 50 million years of neuropharmacology. Mol Biol Cell 8: 2101–2109.
32. Craig AG, Bandyopadhyay P, Olivera BM (1999) Post-translationally modified neuropeptides from Conus venoms. Eur J Biochem 264: 271–275.
33. Skilleter GA, Warren S (2000) Effects of habitat modification in mangroves on the structure of mollusc and crab assemblages. J Exp Mar Bio Ecol 244: 107–129. doi:10.1016/S0022-0981(99)00133-1.
34. Carlton JT (1993) Neoextinctions of Marine Invertebrates. Am Zool 33: 499–509.
35. Zuschin M, Hohenegger J, Steininger FF (2001) Molluscan assemblages on coral reefs and associated hard substrata in the northern Red Sea. Coral Reefs 20: 107–116. doi:10.1007/s003380100140.
36. IUCN Standards and Petitions Subcommittee (2010) Guidelines for Using the IUCN Red List Categories and Criteria. Version 8.0.: 85. Available: http://www.iucnssg.org/tl_files/Assets/pdf/RL Docs/RedListGuidelines.pdf.
37. AfDB OECD, UNDP UNECA (2012) African Economic Outlook 2012, Western African Countries. African Development Bank. Dakar, Senegal. Available: http://www.africaneconomicoutlook.org/en/countries/west-africa/cape-verde/.
38. Irwin A, Wilson C (2011) Cape Verde. Bradt Travel Guides Ltd.
39. IUCN (2012) IUCN Red List Categories and Criteria: Version 3.1. Second edition. Gland, Switzerland and Cambridge, UK: IUCN. iv.
40. Roberts CM, McClean CJ, Veron JEN, Hawkins JP, Allen GR, et al. (2002) Marine biodiversity hotspots and conservation priorities for tropical reefs. Science 295: 1280–1284. doi:10.1126/science.1067728.
41. Gerlach J, Hoffman Black S, Hochkirch A, Jepsen S, Seddon M, et al. (2012) Terrestrial invertebrate life. In: Collen B, Böhm M, Kemp R, Baillie JEM, editors. Spineless: Status and trends of the world's invertebrates. Zoological Society of London, United Kingdom. pp. 46–57.
42. Darwall W, Seddon M, Clausnitzer V, Cumberlidge N (2012) Freshwater invertebrate life. In: Collen B, Böhm M, Kemp R, Baillie JEM, editors. Spineless: Status and trends of the world's invertebrates. Zoological Society of London, United Kingdom. pp. 26–33.
43. González-Mancebo JM, Dirkse GM, Patiño J, Romaguera F, Werner O, et al. (2012) Applying the IUCN Red List criteria to small-sized plants on oceanic islands: conservation implications for threatened bryophytes in the Canary Islands. Biodivers Conserv 21: 3613–3636. doi:10.1007/s10531-012-0385-0.
44. Kemp R, Peters H, Allcock L, Carpenter KE, Obura D, et al. (2012) Marine invertebrate life. In: Collen B, Böhm M, Kemp R, Baillie JEM, editors. Spineless: Status and trends of the world's invertebrates. Zoological Society of London, United Kingdom. pp. 34–45.
45. Collette BB, Carpenter KE, Polidoro BA, Juan-Jorda MJ, Boustany A, et al. (2011) High Value and Long Life — Double Jeopardy for Tunas and Billfishes. Science (80-) 333: 291–292. doi:10.1126/science.1208730.
46. Roberts CM (2012) Ocean of Life. Allen Lane/Penguin Books.
47. Devantier LM (1992) Rafting of tropical marine organisms on buoyant coralla. Mar Ecol Prog Ser 86: 301–302.
48. Cunha RL, Castilho R, Rüber L, Zardoya R (2005) Patterns of cladogenesis in the venomous marine gastropod genus Conus from the Cape Verde islands. Syst Biol 54: 634–650. doi:10.1080/10635159100747l.
49. Duda TF, Palumbi SR (1999) Developmental shifts and species selection in gastropods. Proc Natl Acad Sci U S A 96: 10272–10277.
50. Myers RA, Hutchings JA, Barrowman NJ (1997) Why do fish stocks collapse? The example of cod in the Atlantic Canada. Ecol Appl 7: 91–106.
51. Thurstan RH, Brockington S, Roberts CM (2010) The effects of 118 years of industrial fishing on UK bottom trawl fisheries. Nat Commun 1: 15. doi:10.1038/ncomms1013.
52. National Marine Fisheries Service (2008) White Abalone Recovery Plan (Haliotis sorenseni). Available: http://www.nmfs.noaa.gov/pr/pdfs/recovery/whiteabalone.pdf.
53. Moore A, Ndobe S (2008) Reefs at risk in Central Sulawesi, Indonesia – status and Outlook. Proceedings of the 11th International Coral Reef Symposium, Ft. Lauderdale, Florida, 7-11 July 2008 Session number 18. pp. 840–844. Available: http://nova.edu/ncri/11icrs/proceedings/files/m18-35.pdf.
54. Wells S (1997) Giant clams: Status, trade and mariculture, and the role of CITES in management. IUCN—the World Conservation Union (Gland, Switzerland).
55. Chivian E, Roberts CM, Bernstein AS (2003) The threat to cone snails. Science 302: 391. doi:10.1126/science.302.5644.391b.
56. Chivian E, Roberts CM, Bernstein AS (2004) Response to: How much at risk are cone snails? Science 303: 955–957.
57. Duda TF, Bingham J-P, Livett BG, Kohn AJ, Raybaudi Massilia G, et al. (2004) How much at risk are cone snails? Science 303: 955–957.
58. Hopkins C, Grilley M, Miller C, Shon KJ, Cruz LJ, et al. (1995) A new family of Conus peptides targeted to the nicotinic acetylcholine receptor. J Biol Chem 270: 22361–22367.
59. Livett BG, Sandall DW, Keays D, Down J, Gayler KR, et al. (2006) Therapeutic applications of conotoxins that target the neuronal nicotinic acetylcholine receptor. Toxicon 48: 810–829. doi:10.1016/j.toxicon.2006.07.023.
60. Urbarova I, Karlsen BO, Okkenhaug S, Seternes OM, Johansen SD, et al. (2012) Digital marine bioprospecting: mining new neurotoxin drug candidates from the transcriptomes of cold-water sea anemones. Mar Drugs 10: 2265–2279. doi:10.3390/md10102265.
61. Airoldi L, Balata D, Beck MW (2008) The Gray Zone: Relationships between habitat loss and marine diversity and their applications in conservation. J Exp Mar Bio Ecol 366: 8–15. doi:10.1016/j.jembe.2008.07.034.
62. Munday PL (2004) Habitat loss, resource specialization, and extinction on coral reefs. Glob Chang Biol 10: 1642–1647. doi:10.1111/j.1365-2486.2004.00839.x.
63. Thrush SF, Dayton PK (2002) Disturbance to Marine Habitats by Trawling and Dredging: Implications for Marine Biodiversity. Annu Rev Ecol Syst 33: 449–473. doi:10.1146/annurev.ecolsys.33.010802.150515.
64. Watling L, Norse EA (1998) Effects of Mobile Fishing Gear on Marine Benthos. Conserv Biol 12: 1178–1179. doi:10.1046/j.1523-1739.1998.0120061178.x.
65. Gray JS, Dayton P, Thrush S, Kaiser MJ (2006) On effects of trawling, benthos and sampling design. Mar Pollut Bull 52: 840–843. doi:10.1016/j.marpolbul.2006.07.003.
66. Dulvy NK, Sadovy Y, Reynolds JD (2003) Extinction vulnerability in marine populations. Fish Fish 4: 25–64. doi:10.1046/j.1467-2979.2003.00105.x.
67. Riascos JM, Heilmayer O, Laudien J (2007) Population dynamics of the tropical bivalve Cardita affinis from Málaga Bay, Colombian Pacific related to La Niña 1999–2000. Helgol Mar Res 62: 63–71. doi:10.1007/s10152-007-0083-6.
68. Australian Government (2012) Record breaking La Niña events; An analysis of the La Niña life cycle and the impacts and significance of the 2010–11 and 2011–12 La Niña events in Australia. Melbourne, Australia: Bureau of Meteorology.
69. Doney SC, Fabry VJ, Feely RA, Kleypas JA (2009) Ocean Acidification: The Other CO2 Problem. Ann Rev Mar Sci 1: 169–192. doi:10.1146/annurev.-marine.010908.163834.
70. Rodolfo-Metalpa R, Houlbrèque F, Tambutté É, Boisson F, Baggini C, et al. (2011) Coral and mollusc resistance to ocean acidification adversely affected by warming. Nat Clim Chang 1: 308–312. doi:10.1038/nclimate1200.
71. Tonge S, Bloxam Q (1991) A review of the captive-breeding programme for Polynesian tree snails. Int Zoo Yearb 30: 51–59.
72. Lopes EP (2010) Recent data on marine bivalves (Mollusca, Bivalvia) of the Cape Verde Islands, with records of six species new to the archipelago. Zool Caboverdiana 1: 59–70.
73. Monteiro A, Malcolm G, Herndl G (2012) Endangered cone species. Cone Collect (Ed A Monteiro): 36. Available: http://www.theconecollector.com/. Accessed 1 January 2013.

5

Tourism Revenue as a Conservation Tool for Threatened Birds in Protected Areas

Rochelle Steven*, J. Guy Castley, Ralf Buckley

International Centre for Ecotourism Research, Griffith University, Gold Coast, Queensland, Australia

Abstract

Many bird populations worldwide are at risk of extinction, and rely heavily on protected area networks for their continued conservation. Tourism to these areas contributes to conservation by generating revenue for management. Here we quantify the contribution of tourism revenue for bird species in the IUCN Red List, using a simple accounting method. Relevant data are available for 90 (16%) of the 562 critically endangered and endangered species. Contributions of tourism to bird conservation are highest, 10–64%, in South America, Africa, and their neighbouring islands. Critically endangered bird species rely on tourism more heavily than endangered species ($p<0.02$). Many protected areas could also enhance their management budgets by promoting birdwatching tourism specifically.

Editor: Brock Fenton, University of Western Ontario, Canada

Funding: This project was supported by funding from the Griffith Scool of Environment, Griffith University (www.griffith.edu.au). The funders had no role in study design, data collection and analysis, decision to publish, or preparation of the manuscript.

Competing Interests: The authors have declared that no competing interests exist.

* E-mail: r.steven@griffith.edu.au

Introduction

Approximately 13% of extant bird species are threatened with extinction [1,2] and protected area (PA) networks are the mainstay of global efforts to conserve them [2,3,4]. These networks include: formal gazetted public national parks and reserves (e.g. IUCN Category II); internationally designated areas such as Biosphere Reserves; communal conservancies; and private reserves [5,6]. There are >120,000 public protected areas worldwide, covering ~13% of global terrestrial habitats [4]. Some bird species remain extant only as a result of reserves designated and managed specifically to conserve them [7,8].

Despite previous efforts, current protected area networks provide incomplete protection for global biodiversity, and species continue to decline in diversity and abundance [3,9]. Reversing these trends of decline requires a thorough understanding of species ecology, threatening processes, as well as the expansion and active management of global protected area networks [3,9,10,11]. These approaches require funds for conservation measures such as habitat restoration, control of invasive species, reintroductions of threatened species, and anti-poaching efforts [3,9,10,12,13]. Generally, funds for conserving biodiversity in protected areas worldwide remain inadequate to meet these needs [9,14].

Many protected area managers seeking to expand and diversify their funding portfolios consider tourism revenue to be an increasingly significant fiscal source for protected area management and conservation. Revenue is raised from entrance and activity charges, accommodation, concession and lease fees, and sales of tourist commodities [6,13,15,16–22]. Additional indirect contributions to conservation may occur through various social mechanisms. These include educating tourists and changing their behaviours; providing benefits to local communities to reduce dependence on natural resources [23–26]; and improving local awareness and attitudes towards conservation [12,24,25].

However, tourism can also create negative impacts in protected areas, affecting both the environment and species within them [16,27,28]. The scale, extent and severity of these impacts depend on the nature of the tourism activity. Previous studies have highlighted the variable impacts to birds from both motorised and non-motorised tourism activities [27,28]. So long as these impacts are managed effectively, tourism can deliver net benefits to species and ecosystems [6,15,20,29,30] but these are rarely quantified.

The relationship between tourism revenue to protected areas, and conservation of the threatened species which occur in them, has been demonstrated recently for mammals and frogs at the global scale [29,30], but not for any other taxa. Here, therefore, we examine firstly, the contributions of protected areas networks to conservation of threatened bird species; and secondly, the degree to which tourism revenue to protected areas contributes to the conservation of critically endangered and endangered bird species. We hypothesise that for threatened birds in developing countries, where government funding for conservation is scarce, tourism revenue plays a key role in conservation funding.

Methods

The degree to which individual threatened species depend on tourism revenues to protected areas can be quantified using simple accounting approaches, either for species populations [29] or species ranges [30]. Here we apply the population accounting approach [29], to all 562 bird species classified as critically endangered (CR) or endangered (EN) in the IUCN Red List of Threatened Species [31]. We use the widely adopted IUCN codes, CR and EN respectively, to refer to these threat classifications when reporting results [1,4,29,30]. For each species, this approach calculates T, the proportion of global population protected by tourism, as $T = (\Sigma_1^n S_n R_n)/G$ where S_n are the sizes of subpopulations in parks, R_n are proportions of parks revenue from tourism, and G is the global population for the species concerned [29].

We compiled data on the population status and distributions for each of these 562 threatened bird species from the IUCN Red List and BirdLife International databases [32]. As in previous studies [1,4,10,29,30] we used these data sources and classifications as they provide a consistent and reliable measure of threat status at the global scale, despite their inherent limitations. In calculating S_n, we included only resident subpopulations [31,32] within each protected area. For migratory species, we included only those protected areas which are significant breeding sites. We included all protected areas which support known populations of critically endangered or endangered bird species, including those where tourism revenue, $R = 0$. We obtained data on R from parks agency financial statements, annual reports and published compendia [21,22,29,30] for ~35 of the world's 196 sovereign nations (Table S1). Such financial data for individual protected areas across global scales are rarely available or consistent. Therefore, we followed approaches used for mammals and frogs [29,30] and calculated R at national scale, since protected area agencies routinely transfer funds between individual parks.

Results

There are 190 CR and 372 EN bird species in the IUCN Red List of Threatened Species, 562 species in total. For 109 CR and 132 EN species, 57% and 35% respectively, there are no records of occurrence in any protected area worldwide. The distributions of the remaining 81 CR and 240 EN species intersect 520 protected areas in 77 countries. Of these, 17 CR and 11 EN species survive in only a single protected area. Four species survive only in Galapagos National Park in Ecuador; three on Norfolk Island in Australia; two in Junin National Reserve in Peru; and two in Alakai Wilderness Preserve in Hawaii, USA (Table S2). All remaining CR and EN bird species, 34% and 62% respectively, currently survive in at least two individual protected areas.

Worldwide, 413 protected areas conserve only a single CR or EN bird species, and 107 protected areas in 33 countries support two or more species (Table S3). Galapagos National Park supports eight CR and EN species; and Pedra Talhada Biological Reserve in Brazil, and Gough Island Nature Reserve in the southern Atlantic Ocean each support six (Table 1). This highlights the relatively high numbers of threatened bird species supported in island protected areas worldwide.

Tourism revenue contributions to protected area budgets, R, range from zero in some countries to 100% for a number of private reserves. North Island in the Seychelles, and Cerro Batipa Private Reserve in Panama, provides examples of the latter (Table S1). For most of the countries relevant to this analysis, R at national scale ranges from 5–80%. For Australia, New Zealand and the USA, $R<10\%$, and for Canada, $R = 14\%$. For the 19 South and Central American countries where data are available, $R>20\%$ for five, $10\%<R<20\%$ for five, and $R<10\%$ for nine. Of the six continental African countries where population data S_n are known for CR and EN bird species, data on R are available for five, where R ranges from between 36–81%. Globally, R is higher on average for developing than developed nations.

Subpopulation data for threatened bird species in protected areas, S_n, are available for 55 CR and 85 EN species, comprising 216 subpopulations in all. Data for both S and R are available for 91 species in 131 subpopulations. For four species, data on R are available for only some of the relevant protected areas. These species are the Seychelles Magpie-robin *Copsychus sechellarum*, Seychelles White-eye *Zosterops modestus*, Black-footed Albatross *Phoebastria nigripes*, and Sooty Albatross *Phoebetria fusca*. Calculations of T for these species thus reflect only some of the known subpopulations.

For these 91 species, T ranges from 0–64% (Table 2, Fig. 1). For nine species, $T = 0$, as the protected areas in which these species occur do not receive income from tourism. For 41 species: $0\%<T<5\%$; for 21 species, $5\%\leq T<10\%$; for 8 species, $10\%\leq T<20\%$; and for 12 species, $T>20\%$ (Fig. 1). Mean T is significantly greater for CR than EN bird species (arcsine square-root transformation, $t = 0.082$, d.f. = 89, $p<0.02$). For two species, $T\geq 50\%$: the Seychelles Magpie-robin ($T = 64\%$) and the African Penguin ($T = 50\%$). It is not only small populations that rely on tourism revenue since there is no correlation between T and global populations G ($R^2 = 0.008$, $d.f. = 89$, $p = 0.397$) (Fig. 2). For example, the Seychelles Magpie-robin and African Penguin with $T\geq 50\%$ have global population sizes G of 100–500 and 50,000–100,000 respectively.

Discussion

Our findings as presented above fall into two major categories: firstly, on the degree of protection or otherwise for threatened bird species; and secondly, on the relative contributions of revenue from tourism. In the first category, we show that 57% of critically endangered and 35% of endangered bird species do not occur at all inside the current global protected area network, highlighting the global extinction risk facing many species. We also show that 14 individual protected areas each protect the last remaining populations for more than one critically endangered or endangered bird species (Table 1), and are hence of particular significance for bird conservation. Half of these protected areas are in South America, reflecting the high diversity and restricted ranges of neotropical avifauna. Three are on islands, particularly vulnerable to some threats but relatively protected from others [33], and sometimes amenable to direct conservation interventions, with a number of successful examples [34–36]. These findings confirm longstanding concerns over inadequate protection of bird species worldwide [9,10].

These findings are subject to limitations of available data on populations, distributions and threats to bird species. These limitations influence both the IUCN Red List classifications of species as critically endangered or endangered, and the reliability and completeness of numerical estimates for protected subpopulations, S_n. However, these limitations are unlikely to compromise our findings for several reasons. The first relates to the global scale of our assessment. A number of countries, and in some case subsidiary jurisdictions such as states and provinces, maintain their own legislation for protection of threatened species, which do not necessarily match the IUCN Red List either in the species included, the categories of threat status, or the threat classifications allocated. These differences, however, are not consistent. In some countries, national or subsidiary legislation may be more comprehensive, stringent or up to date than the IUCN Red Lists, but in other countries and jurisdictions the reverse applies, especially in developing nations. In this analysis, therefore, we relied on the IUCN Red List as the only standardised international database available.

Furthermore, our assessment made no attempt to evaluate the accuracy of the threatened species classification used but focused simply on where threatened species occurred in protected areas and the degree to which these areas were funded through tourism. Similar global assessments using these data sets have recently been

Table 1. Protected areas where four or more critically endangered (CR) or endangered (EN) bird species are recorded.

Reserve Name	Country	Total number of CR or EN bird species recorded in this reserve	Number of CR or EN species recorded only in this reserve
Galapagos National Park	Ecuador	8	4
Pedra Talhada Biological Reserve	Brazil	6	0
Gough Island Nature Reserve	Saint Helena	6	1
Munchique National Natural Park	Colombia	5	0
South East Island (Rangatira) Nature Reserve	New Zealand	5	1
Emas National Park	Brazil	4	0
Frei Caneca Private Reserve	Brazil	4	0
Sooretama Biological Reserve	Brazil	4	0
Sierra Nevada de Santa Marta Biosphere Reserve	Colombia	4	1
French Southern and Antarctic Lands	France	4	0
Black River Gorges National Park	Mauritius	4	0
Fiordland National Park	New Zealand	4	0
Mangere Island Nature Reserve	New Zealand	4	0
Mt Canlaon Natural Park	Philippines	4	0

completed for mammals and frogs [29,30] and these data are widely used for other global threatened species assessments [1,4,10]. This is less of a limitation for birds than for mammal and frog species [29,30], because in the case of birds, data from IUCN are supplemented by those from BirdLife International. For frogs [30], the shortcomings of subpopulation data forced the authors to rely on an alternative and less precise accounting metric, proportion of known range. For mammals [29], data on S and R were available jointly for only 90 of 1131 candidate species, which included vulnerable as well as critically endangered and endangered species. The calculations of T presented here for threatened bird species are thus the most comprehensive yet available for any major taxonomic group.

For the future conservation of threatened bird species, two key issues emerge from this first section of our analyses. (a) The limitations in available data demonstrate the importance of

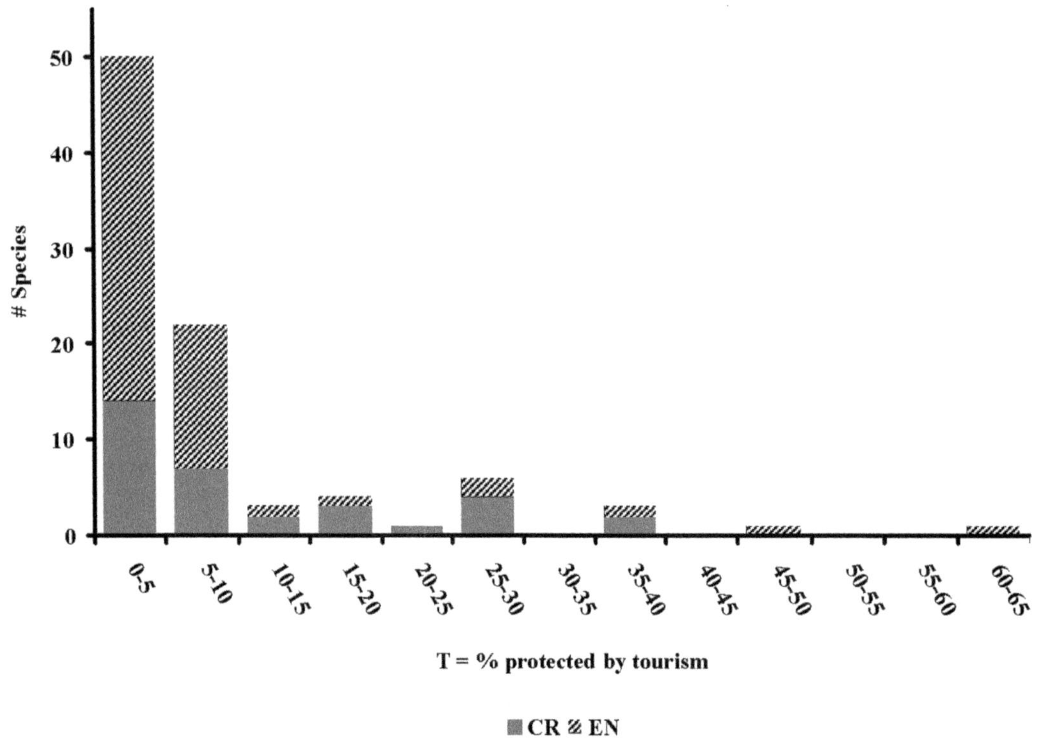

Figure 1. Proportions T of CR and EN bird species with conservation funding from tourism.

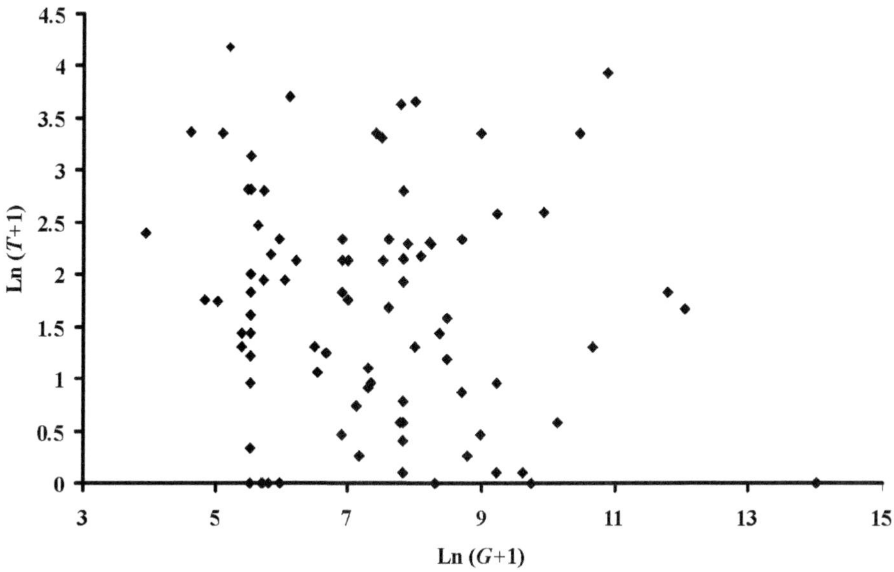

Figure 2. Relation between *T* and *G*, natural log scales, for CR and EN bird species (n = 91).

continuing research on bird species populations, distributions, ecology and threats. (b) The high proportions of threatened bird species (43%) which do not currently occur in any protected area worldwide demonstrates the importance of promoting biodiversity conservation across both public and private lands, even where expansion of the formal protected area network is not feasible. (c) Data on the joint distribution of critically endangered and endangered bird species, as highlighted here, could be used to direct conservation funding efficiently, thereby maximising overall threat reduction per unit expenditure.

Our second set of findings relates to the role of tourism revenue contributions to conservation of threatened bird species occurring at least in part within protected areas. Our results identify 41 species where at least 10% of each global population relies on tourism revenue. Naturally, these calculations are also subject to limitations in the scope, accuracy and reliability of financial data in the budgets of protected area agencies. Only a few countries produce and publish transparent, timely and externally audited annual financial statements for their protected area agencies that distinguish different revenue sources. Therefore, if anything, our findings underestimate the importance of tourism revenue for threatened birds in protected areas. When financial data become available for more protected area networks in more countries, a revised analysis would potentially identify a greater number of threatened species that are dependent on tourism revenues.

Despite these shortcomings in detailed data, our findings are broad and robust enough to demonstrate the important role of tourism in funding the conservation of threatened bird species. This reflects changing patterns in protected area funding more generally, over recent decades. Globally, government budget allocations are still the principal source of protected area management funds [13,21]. However, many protected area agencies increasingly rely on tourism revenues to supplement or replace government allocations (Table 2) [19]. Tourism thus contributes to the conservation of threatened bird species, especially in developing nations with high biodiversity.

The other side of this coin, however, is that threatened bird species protected in parks are now subject to risks associated with the volatility of global tourism markets in addition to those already imposed by direct disturbance impacts [12,28]. This is more of an issue for parks agencies in developing nations operating with less reliable government funding [13,21,37–40]. In addition, raising revenue from tourism to public protected areas may rely on charging user fees for visitors, and this is subject to equally volatile political and economic constraints [12,13,21]. In addition, for some countries, tourism revenues raised in protected areas not retained by the parks agencies at all. Such challenges are less problematic when tourism and conservation action occurs on private land, as has been demonstrated in southern Africa [16,41–44]. For threatened bird species, the contributions of private enterprise have been quantified for only a few cases, notably the Seychelles White-eye on North Island and the Resplendent Quetzal (*Pharomachrus mocinno*) in the Monteverde Cloud Forest Reserve [6,34].

The broad accounting approach taken here simply quantifies the degree to which tourism to protected areas contributes to funding conservation of threatened bird species. A number of more detailed or sophisticated approaches may also be feasible. 1. Differentiation by land tenure, distinguishing public, communal and private lands. 2. An accounting approach which calculates net contributions [45], allowing for negative environmental impacts [27,28] as well as positive conservation measures, and distinguishing high-impact park-based mass tourism from low-impact park-based nature tourism. 3. A more detailed approach which relies on ecological modelling rather than purely on accounting. 4. A narrower focus, on the bird-watching or avitourism subsector specifically, as below.

Birdwatching is a significant and expanding subsector of the tourism industry. Avitourists travel either to see particular bird species, especially those that are rare or threatened; or to visit areas with high endemism and high diversity of bird species [46–52]. As shown here, many threatened bird species currently occur only outside protected areas, and this includes species attractive to birdwatchers. Specialist avitourism may thus create further incentives to expand the protected area network, or improve bird conservation on lands outside that network. In particular, this includes Important Bird Areas, designated in part due to the presence of threatened bird species. The role of avitourism in

Table 2. Proportions of critically endangered (CR) and endangered (EN) bird populations protected through tourism-related revenue.

Scientific Name	Common Name	IUCN	Global population (*G*)	Number* of protected subpopulations with known *S, R*	Number of individuals protected through tourism (*SR*)	Percentage of global population protected through tourism (*T*)
Copsychus sechellarum	Seychelles Magpie-robin	EN	178	2	114	64.0
Spheniscus demersus	African Penguin	EN	52000	6	26040	50.1
Zosterops modestus	Seychelles White-eye	EN	450	3	178	39.6
Sephanoides fernandensis	Juan Fernandez Firecrown	CR	3000	1	1137	37.9
Malaconotus alius	Uluguru Bush-shrike	CR	2400	1	881	36.7
Mimus trifasciatus	Floreana Mockingbird	CR	100	1	28	28.0
Camarhynchus pauper	Medium Tree-finch	CR	1660	1	460	27.7
Phoebastria irrorata	Waved Albatross	CR	34700	1	9612	27.7
Mimus melanotis	San Cristobal Mockingbird	EN	8000	1	2216	27.7
Camarhynchus heliobates	Mangrove Finch	CR	160	2	44	27.5
Spheniscus mendiculus	Galapagos Penguin	EN	1800	1	474	26.3
Artisornis moreaui	Long-billed Tailorbird	CR	250	1	55	22.0
Buteo ridgwayi	Ridgway's Hawk	CR	240	1	38	15.8
Polioptila clementsi	Iquitos Gnatcatcher	CR	250	1	39	15.6
Laterallus tuerosi	Junin Rail	EN	2500	1	388	15.5
Podiceps taczanowskii	Junin Grebe	CR	304	1	47	15.5
Hylonympha macrocerca	Scissor-tailed Hummingbird	EN	20000	1	2480	12.4
Pterodroma phaeopygia	Galapagos Petrel	CR	10000	4	1219	12.2
Terpsiphone corvina	Seychelles Paradise-flycatcher	CR	278	2	30	10.8
Zosterops albogularis	White-chested White-eye	CR	50	1	5	10.0
Grus americana	Whooping Crane	EN	382	1	36	9.4
Manorina melanotis	Black-eared Miner	EN	1000	1	94	9.4
Papasula abbotti	Abbott's Booby	EN	6000	1	564	9.4
Zosterops tenuirostris	Slender-billed White-eye	EN	2000	1	188	9.4
Ara ambiguus	Great Green Macaw	EN	3700	3	334	9.0
Pardalotus quadragintus	Forty-spotted Pardalote	EN	3800	2	339	8.9
Cacatua haematuropygia	Philippine Cockatoo	CR	2700	2	239	8.9
Aceros narcondami	Narcondam Hornbill	EN	340	1	27	7.9
Threskiornis bernieri	Madagascar Sacred Ibis	EN	3250	1	254	7.8
Myiotheretes pernix	Santa Marta Bush-tyrant	EN	2500	1	190	7.6
Anas laysanensis	Laysan Duck	CR	1100	2	82	7.5
Acrocephalus familiaris	Millerbird	CR	1000	1	74	7.4
Myadestes palmeri	Puaiohi	CR	500	1	37	7.4
Oreomystis bairdi	Akikiki	CR	1840	1	136	7.4
Crax blumenbachii	Red-billed Curassow	EN	250	3	16	6.4
Aratinga brevipes	Socorro Parakeet	EN	300	1	18	6.0
Mimus graysoni	Socorro Mockingbird	CR	420	1	25	6.0
Anas wyvilliana	Hawaiian Duck	EN	2525	1	148	5.9
Leptoptilos dubius	Greater Adjutant	EN	1000	2	52	5.2
Petroica traversi	Black Robin	EN	250	2	13	5.2
Phoebastria nigripes	Black-footed Albatross	EN	130000	3	6700	5.2
Strigops habroptila	Kakapo	CR	124	2	6	4.8
Pterodroma axillaris	Chatham Petrel	EN	1100	1	53	4.8
Pterodroma magentae	Magenta Petrel	CR	150	2	7	4.7
Psephotus chrysopterygius	Golden-shouldered Parrot	EN	2000	2	87	4.4
Eudyptes sclateri	Erect-crested Penguin	EN	170000	2	7392	4.3

Table 2. Cont.

Scientific Name	Common Name	IUCN	Global population (*G*)	Number* of protected subpopulations with known *S*, *R*	Number of individuals protected through tourism (*SR*)	Percentage of global population protected through tourism (*T*)
Anas nesiotis	Campbell Islands Teal	CR	250	3	10	4.0
Fregata andrewsi	Christmas Frigatebird	CR	4800	1	188	3.9
Heteroglaux blewitti	Forest Owlet	CR	250	1	8	3.2
Pyrrhura griseipectus	Grey-breasted Parakeet	CR	250	1	8	3.2
Odontophorus strophium	Gorgeted Wood-quail	EN	4300	1	137	3.2
Porphyrio hochstetteri	Takahe	EN	220	2	7	3.2
Thinornis novaeseelandiae	Shore Plover	EN	220	1	6	2.7
Cyanoramphus malherbi	Malherbe's Parakeet	CR	663	3	18	2.7
Rollandia microptera	Titicaca Grebe	EN	3000	2	81	2.7
Phoebetria fusca	Sooty Albatross	EN	42000	1	1133	2.7
Carpodectes antoniae	Yellow-billed Cotinga	EN	794	1	20	2.5
Mergus octosetaceus	Brazilian Merganser	CR	250	1	6	2.4
Penelope albipennis	White-winged Guan	CR	250	3	6	2.4
Megadyptes antipodes	Yellow-eyed Penguin	EN	4800	3	110	2.3
Geothlypis beldingi	Belding's Yellowthroat	CR	1500	1	30	2.0
Torreornis inexpectata	Cuban Sparrow	EN	700	1	13	1.9
Conothraupis mesoleuca	Cone-billed Tanager	CR	250	1	4	1.6
Callaeas cinereus	Kokako	EN	1538	1	24	1.6
Pipile jacutinga	Black-fronted Piping-guan	EN	10000	1	156	1.6
Ardea humbloti	Madagascar Heron	EN	1500	1	22	1.5
Ardeola idae	Madagascar Pond-heron	EN	6000	1	85	1.4
Amazona vinacea	Vinaceous Amazon	EN	2500	3	30	1.2
Penelope perspicax	Cauca Guan	EN	1235	2	14	1.1
Apteryx mantelli	Northern Brown Kiwi	EN	25000	3	202	0.8
Amazona rhodocorytha	Red-browed Amazon	EN	2500	1	19	0.8
Eleothreptus candicans	White-winged Nightjar	EN	2400	1	18	0.8
Pedionomus torquatus	Plains-wanderer	EN	8000	2	50	0.6
Phytotoma raimondii	Peruvian Plantcutter	EN	1000	1	6	0.6
Curaeus forbesi	Forbes's Blackbird	EN	2500	1	12	0.5
Haliaeetus vociferoides	Madagascar Fish-eagle	CR	250	1	1	0.4
Eutriorchis astur	Madagascar Serpent-eagle	EN	250	1	1	0.4
Cistothorus apolinari	Apolinar's Wren	EN	1300	1	4	0.3
Anodorhynchus hyacinthinus	Hyacinth Macaw	EN	6500	1	18	0.3
Brotogeris pyrrhoptera	Grey-cheeked Parakeet	EN	15000	2	22	0.1
Anas bernieri	Madagascar Teal	EN	2500	1	2	0.1
Terenura sharpei	Yellow-rumped Antwren	EN	10000	1	8	0.1
Ara rubrogenys	Red-fronted Macaw	EN	4000	1	1	0.0
Diomedea sanfordi	Northern Royal Albatross	EN	17000	1	2	0.0
Thalassarche melanophrys	Black-browed Albatross	EN	1220000	2	77	0.0
Eleoscytalopus psychopompus	Bahia Tapaculo	CR	250	1	0	0.0
Zosterops chloronothus	Mauritius Olive White-eye	CR	296	1	0	0.0
Foudia rubra	Mauritius Fody	EN	328	1	0	0.0
Nesoenas mayeri	Pink Pigeon	EN	395	1	0	0.0
Psittacula eques	Mauritius Parakeet	EN	300	1	0	0.0
Trichocichla rufa	Long-legged Thicketbird	EN	250	3	0	0.0

*Other subpopulations lacking data on *S* and *R* may also exist.

contributing to bird conservation through mechanisms such as Important Bird Areas merits further attention.

In conclusion, despite the dependence of tourism revenues on market conditions, and the risk associated with such dependence, we demonstrate that tourism revenue to protected areas makes a significant contribution to the conservation of threatened bird species, comparable to that for threatened mammals and frogs [29,30]. While government funding remains a critical source of funding for many protected areas and conservation globally, we propose that this could be increased by promoting tourism on both public and private lands. Specialist niche tourism markets such as avitourism may provide further incentives to promote sustainable conservation tourism, particularly in those habitats identified as Important Bird Areas.

Acknowledgments

We thank staff from protected area agencies worldwide for providing data on *R*; our colleagues for assistance in translating non-Anglophone reports and currency measures; and D. Jones and C. Morrison for comments on the manuscript prior to submission.

Author Contributions

Conceived and designed the experiments: JGC RB. Performed the experiments: RS JGC. Analyzed the data: RS JGC. Contributed reagents/materials/analysis tools: JGC RB. Wrote the paper: RS JGC RB.

References

1. Hoffmann M, Hilton-Taylor C, Angulo A, Böhm M, Brooks TM, et al. (2010) The impact of conservation on the status of the world's vertebrates. Science 330: 1503–1509.
2. Hilton-Taylor C, Pollock C, Chanson J, Butchart SHM, Oldfield T, et al. (2009) Status of the world's species. In: Vié J-C, Hilton-Taylor C, Stuart SN, editors. Wildlife in a changing world. An analysis of the 2008 IUCN Red List of Threatened Species. IUCN: Gland. pp 15–42.
3. Rodrigues ASL, Andelman SJ, Bakarr MI, Boitani L, Brooks TM, et al. (2004) Effectiveness of the global protected area network in representing species diversity. Nature 428: 640–643.
4. Butchart SHM, Scharlemann JPW, Evans MI, Quader S, Aricò S, et al. (2012) Protecting important sites for biodiversity contributes to meeting global conservation targets. PLoS ONE 7: e32529.
5. Langholz JA, Krug W (2004) New forms of biodiversity governance: Non-state actors and the private protected area plan. Journal of International Wildlife Law and Policy 7: 9–29.
6. Buckley RC (2010) Conservation Tourism. Wallingford: CAB International. 214 p.
7. Department of Conservation New Zealand (DOC) (2011) Kākāpō. Available: http://www.doc.govt.nz/conservation/native-animals/birds/land-birds/kakapo/docs-work/. Accessed 2011 Aug.
8. Department of Conservation New Zealand (DOC) (2011) Black Robin. Available: http://www.doc.govt.nz/conservation/native-animals/birds/land-birds/black-robin/docs-work/. Accessed 2011 Aug.
9. Rodrigues ASL, Akçakaya HR, Andelman SJ, Bakarr MI, Boitani L, et al. (2004) Global gap analysis: Priority regions for expanding the global protected area network. BioScience 54: 1092–1100.
10. Beresford AE, Buchanan GM, Donald PF, Butchart SHM, Fishpool LDC, et al. (2011) Poor overlap between the distribution of protected areas and globally threatened birds in Africa. Animal Conservation 14: 99–107.
11. Lockwood M (2010) Good governance for terrestrial protected areas: A framework, principles and performance outcomes. J Environ Manage 91: 754–766.
12. Eagles PFJ, McCool SF, Haynes CD (2002) Sustainable tourism in protected areas. Guidelines for planning and management. Gland: IUCN. 191 p.
13. Emerton L, Bishop J, Thomas L (2006) Sustainable financing of protected areas: A global review of challenges and options. Gland: IUCN. 109 p.
14. McCarthy DP, Donald PF, Scharlemann JPW, Buchanan GM, Balmford A, et al. (2012) Financial costs of meeting global biodiversity conservation targets: Current spending and unmet needs. Science DOI: 10.1126/science.1229803.
15. Buckley RC (2009) Parks and tourism. PLoS Biology 7(6), e1000143.
16. Castley JG (2010) Southern and Eastern Africa. In: Buckley RC, Conservation tourism. Wallingford: CAB International. pp. 145–175.
17. Buckley RC (2011) Tourism and environment. Ann Rev Enviro Resources 36: 397–416.
18. Bookbinder MP, Dinerstein E, Rijal A, Cauley H, Rajouria A (1998) Ecotourism's support of biodiversity conservation. Conserv Biol 12: 1399–1404.
19. Buckley RC (2003) Pay to play in parks. J Sust Tour 11: 56–73.
20. Tisdell C, Wilson C (2012) Nature-Based Tourism and Conservation. Cheltenham: Edward Elgar. 511p.
21. Mansourian S, Dudley N (2008) Public funds to protected areas. Gland: WWF International. 43 p.
22. Bovarnick A, Fernandez-Baca J, Galindo J, Negret H (2010) Financial sustainability of protected areas in Latin America and the Caribbean: Investment policy guideline. United Nations Development Programme (UNDP) and The Nature Conservancy (TNC). 162 p.
23. Mmopelwa G, Blignaut JN (2006) The Okavango Delta: The value of tourism. South African Journal of Economic and Management Sciences 9: 113–127.
24. Sekhar NU (2003) Local people's attitudes towards conservation and wildlife tourism around Sariska Tiger Reserve, India. J Environ Manage 69: 339–347.
25. Walpole MJ, Goodwin HJ (2001) Local attitudes towards conservation and tourism around Komodo National Park, Indonesia. Environ Conserv 28: 160–166.
26. Mbaiwa JE (2003) The socio-economic and environmental impacts of tourism development on the Okavango Delta, north-western Botswana. J Arid Environ 54: 447–467.
27. Buckley RC (2004) Environmental Impacts of Ecotourism. Wallingford: CAB International. 389 p.
28. Steven R, Pickering C, Castley JG (2011) A review of the impacts of nature based recreation on birds. J Environ Manage 92: 2287–2294.
29. Buckley RC, Castley JG, Pegas FdeV, Mossaz AC, Steven R (2012) A population accounting approach to assess tourism contributions to conservation of IUCN-Redlisted mammal species. PLoS ONE 7(9): e44134. doi: 10.1371/journal.pone.0044134.
30. Morrison C, Simpkins CA, Castley JG, Buckley RC (2012) Tourism and the conservation of critically endangered frogs. PLoS ONE 7(9): e43757.
31. IUCN (2011) The IUCN Red List of Threatened Species. Available: http://www.iucnredlist.org/. Accessed 2011 March.
32. BirdLife International (2011) IUCN Red List for birds. Available: http://www.birdlife.org Accessed September 2011.
33. Whittaker RJ (1995) Disturbed island ecology. TREE 10: 421–425.
34. Rocamora G, Skerrett A (2001) Seychelles. In: Fishpool L, Evand I editors. Important Bird Area in Africa and associated islands. Cambridge: Birdlife International. pp 751–768.
35. Shah NJ (2001) Eradication of alien predators in the Seychelles: An example of conservation action on tropical islands. Biodivers Conserv 10: 1219–1220.
36. Department of Conservation New Zealand (DOC) (2011) Offshore islands and conservation. Mangere Island. Available: http://www.doc.govt.nz/conservation/land-and-freshwater/offshore-islands/mangere-and-rangatira-islands/mangere-island/. Accessed 2011 Aug.
37. Benitez SP (2001) Visitor use fees and concession systems in protected areas: Galápagos National Park Case Study. Ecotourism Program Technical Report Series Number 3. Arlington: The Nature Conservancy. 19 p.
38. Shareef R, McAleer M (2008) Modelling international tourism demand and uncertainty in Maldives and Seychelles: A portfolio approach. Mathematics and Computers in Simulation 78: 459–468.
39. Smeral E (2010) Impacts of the world recession and economic crisis on tourism: Forecasts and potential risks. Journal of Travel Research 49: 31–38.
40. Balmford A, Beresford J, Green J, Naidoo R, Walpole M, et al. (2009) A global perspective on trends in nature-based tourism. PLoS Biol 7: 1–6.
41. Cousins JA, Sadler JP, Evans J (2008) Exploring the role of private wildlife ranching as a conservation tool in South Africa: Stakeholder Perspectives. Ecology and Society 13: 43.
42. Castley G, Knight M, Gordon J (2009) Making Conservation Work: Innovative Approaches to Meeting Conservation and Socio-economic Objectives (an

Example from the Addo Elephant National Park, South Africa) In: Suich H, Child B, Spencelely A (eds) Evolution and Innovation in Wildlife Conservation: Parks and Game Ranches to Transfrontier Conservation Areas. SASUSG, IUCN, London: Earthscan. pp. 307–324.
43. Buckley R (2010) Safaris can help conservation. Nature 467: 1047.
44. Buckley R, Pabla HS (2012) Tourism ban won't help Indian tigers. Nature 489: 33.
45. Buckley R (2012) Sustainable tourism: research and reality. Annals of Tourism Research 39: 528–546.
46. Biggs D, Turpie J, Fabricius JC, Spenceley A (2011) The value of avitourism for conservation and job creation – an analysis from South Africa. Conservation and Society 9: 80–90.
47. Puhakka L, Salo M, Sääksjärvi IE (2011) Bird diversity, birdwatching tourism and conservation in Peru: A geographic analysis. PLoS ONE 6, e26786.
48. Veríssimo D, Fraser I, Groombridge J, Bristol R, MacMillan DC (2009) Birds as tourism flagship species: a case study of tropical islands. Animal Conservation 12: 549–558.
49. Booth JE, Gaston KJ, Evans KL, Armsworth PR (2011) The value of species rarity in biodiversity recreation: A birdwatching example. Biol Conserv 144: 2728–2732.
50. Hvenegaard GT (2002) Birder specialization differences in conservation involvement, demographics, and motivations. Human Dimensions of Wildlife 7: 21–36.
51. Şekercioğlu ÇH (2002) Impacts of birdwatching on human and avian communities. Environ Conserv 29: 282–289.
52. Connell J (2009) Birdwatching, twitching and tourism: Towards an Australian perspective. Aust Geogr 40: 203–217.

6

Optimized Spatial Priorities for Biodiversity Conservation in China: A Systematic Conservation Planning Perspective

Ruidong Wu[1,2]*, Yongcheng Long[3], George P. Malanson[4], Paul A. Garber[5], Shuang Zhang[3], Diqiang Li[6], Peng Zhao[3], Longzhu Wang[3], Hairui Duo[7]

1 Institute of International Rivers and Eco-security, Yunnan University, Kunming, Yunnan, China, **2** Yunnan Key Laboratory of International Rivers and Transboundary Eco-security, Yunnan University, Kunming, Yunnan, China, **3** The Nature Conservancy China Program, Kunming, Yunnan, China, **4** Department of Geography, The University of Iowa, Iowa City, Iowa, United States of America, **5** Department of Anthropology, University of Illinois, Urbana, Illinois, United States of America, **6** Institute of Forest Ecology, Environment, and Protection, Chinese Academy of Forestry, Beijing, China, **7** School of Nature Reserve, Beijing Forestry University, Beijing, China

Abstract

By addressing several key features overlooked in previous studies, i.e. human disturbance, integration of ecosystem- and species-level conservation features, and principles of complementarity and representativeness, we present the first national-scale systematic conservation planning for China to determine the optimized spatial priorities for biodiversity conservation. We compiled a spatial database on the distributions of ecosystem- and species-level conservation features, and modeled a human disturbance index (HDI) by aggregating information using several socioeconomic proxies. We ran Marxan with two scenarios (HDI-ignored and HDI-considered) to investigate the effects of human disturbance, and explored the geographic patterns of the optimized spatial conservation priorities. Compared to when HDI was ignored, the HDI-considered scenario resulted in (1) a marked reduction (~9%) in the total HDI score and a slight increase (~7%) in the total area of the portfolio of priority units, (2) a significant increase (~43%) in the total irreplaceable area and (3) more irreplaceable units being identified in almost all environmental zones and highly-disturbed provinces. Thus the inclusion of human disturbance is essential for cost-effective priority-setting. Attention should be targeted to the areas that are characterized as moderately-disturbed, <2,000 m in altitude, and/or intermediately- to extremely-rugged in terrain to identify potentially important regions for implementing cost-effective conservation. We delineated 23 primary large-scale priority areas that are significant for conserving China's biodiversity, but those isolated priority units in disturbed regions are in more urgent need of conservation actions so as to prevent immediate and severe biodiversity loss. This study presents a spatially optimized national-scale portfolio of conservation priorities – effectively representing the overall biodiversity of China while minimizing conflicts with economic development. Our results offer critical insights for current conservation and strategic land-use planning in China. The approach is transferable and easy to implement by end-users, and applicable for national- and local-scale systematic conservation prioritization practices.

Editor: Duccio Rocchini, Fondazione Edmund Mach, Research and Innovation Centre, Italy

Funding: This study was funded by the National Natural Science Foundation of China (No. 31260148) and the National Key Technologies R&D Program of China (No. 2011BAC09B07). The funders had no role in study design, data collection and analysis, decision to publish, or preparation of the manuscript.

Competing Interests: The authors have declared that no competing interests exist.

* Email: rdwu@ynu.edu.cn

Introduction

Anthropogenic effects have resulted in the loss of biodiversity at an unprecedented rate, while resources for biodiversity conservation remain constrained in terms of both human and financial capacity [1]. That is why the systematic planning of priority areas is crucial to achieve the most cost-effective conservation, such as identifying large-scale biodiversity hotspots or assembling fine-resolution portfolios of conservation priorities [2–4]. In the last two decades, systematic conservation planning (SCP) has emerged as an effective approach for identifying conservation priorities [3–6]. SCP aims to identify a network of priority areas so as to effectively achieve explicit conservation goals in terms of representing the full range of biodiversity and sustaining their long-term survival [5]. Efficient conservation priorities can be identified through an optimized planning algorithm for meeting conservation goals at the minimum land area or other costs (e.g., land prices, management and opportunity costs [7]). SCP provides an operational framework for minimizing land-use conflicts between conserving natural environments and economic development, and thus increase the likelihood of implementing the proposed conservation actions [3,6]. Here, we present the first national-scale SCP study for China to determine the optimized spatial priorities for biodiversity conservation.

China – one of the world's "megadiversity countries" – is home to many globally valued conservation priorities [2]. However, China's biodiversity is under severe threat due to the increasing pressure resulting from the country's historically unprecedented economic growth [8]. Meanwhile, China's conservation investment is considerably lower compared to developed and other developing countries [9]. Thus, the systematic conservation priority-setting has been emphasized in China during last two decades [8,10–11].

During this period, China has developed several templates of national-scale conservation priorities, which were based principally on the species (e.g., endemic, threatened, and/or other indicator species) richness patterns as well as expert judgments (e.g., [10,12–15]). These templates are crucial in guiding China's national-level conservation decisions; however, we think there are several critical limitations in previous priority-setting studies.

First, the effects of human disturbance are not incorporated in previous studies, whereas we believe explicit inclusion of human disturbance in priority-setting can minimize land-use conflicts and lower costs for meeting conservation goals [3,6]. Second, the scoring procedure in these studies is inefficient for achieving the goal of full representation of all biodiversity targets [16], i.e. the goal for representativeness – one of the core principles for designing an efficient reserve system [5]. The current scoring procedure requires a greater amount of land (and increases other costs) to achieve the same conservation goals and these greater demands are unlikely to get support from local authorities. Third, a study designed to systematically integrate conservation features at both ecosystem- and species-level is still lacking, as the conservation features used in previous studies are either species or ecosystem based. By incorporating biodiversity features from multiple organization levels, the resulting portfolio of conservation priorities is more efficient in representing the full range of biodiversity concerns and in maintaining the ecological integrity of ecosystems [5,16]. SCP can overcome this inefficiency in scoring procedure by employing the principle of between-site complementarity that serves to boost the efficient representation of all biodiversity targets, and provide mechanisms for integrating human disturbance and conservation features at multiple organization levels [17].

This study aims to determine the optimized national-scale spatial priorities in China and to ensure effectively fulfilling biodiversity conservation goals given the constraints of human disturbance by implementing a SCP approach. Taiwan, Hong Kong and Macao are not included in our analysis due to lack of required information. Specifically, we are trying to address two questions: (1) How will the inclusion of human disturbance affect the result of conservation priority-setting? (2) What are the geographical patterns of the optimized conservation priorities in China? In this analysis, we integrated human disturbance, conservation features at both the ecosystem- and species-level, and the principles of complementarity and representativeness. We used the software Marxan [17] to determine each unit's conservation value and to identify priorities with regular hexagons (100 km^2 per cell) as the planning units. We investigated the effects of human disturbance using two Marxan scenarios – a disturbance ignored scenario and a disturbance considered scenario. For the second scenario, human disturbance was included as a penalty function by aggregating information on several socioeconomic proxies as an index layer. We then explored the spatial patterns of the priority units, irreplaceable areas, and primary large-scale priority areas (i.e., the large clustered regions of high-conservation-value units). The analysis is limited to the data available at national-scale and applicable resolution, and misses some variability within the range of human disturbances. We believe this study is applicable to national- and local-scale conservation and other sustainable land-use planning for systematically evaluating each site's conservation value and identifying spatially optimized priority areas.

Methods

Conservation Features Mapping

Considering the complexity of biodiversity and severe lack of detailed spatial distribution data, surrogates (e.g., endangered/endemic species, key habitat types and environmental features) are often used in conservation planning [3,7,16,18]. Integrating conservation features from multiple levels can ensure the efficient representation of biodiversity [5] and compensate for limitations in the data [16]. In this analysis, we used both ecosystem- and species-level features as the surrogates.

The ecosystem-level features included were: (1) priority natural ecosystems as defined by Li, Song & Ouyang [13], including 129 natural ecosystems of forests, grasslands, meadows, deserts and wetlands, and (2) natural vegetation types derived from the national 1: 1,000,000 vegetation map, including 559 natural vegetation formations [19]. This study considered wetlands and lakes (in the priority natural ecosystems and natural vegetation types), but data on aquatic systems and species was lacking. We expected that China's key ecological elements, processes and services were covered with priority natural ecosystems and that basic habitat types were represented by finer-scale classifications of natural vegetation types. The species-level features were endangered species of plants, mammals, and birds. Endangered mammals and birds were identified according to China's "National List of Key Protected Wildlife" and the IUCN Red List Categories of critically endangered, endangered and vulnerable species [20]. Endangered plants were defined in the "China Plant Red Data Book: Endangered and Rare Plants" [21].

Previous studies often use county-level species distribution data derived from the published literatures [14–15], while our analysis was performed using a finer-scale resolution. For plants and mammals, we mapped each species' geographic range by combining its distribution data for counties, preferred habitat types and elevation range. For a bird species, the range was derived by intersecting only counties and habitat types, because knowledge of the altitude distribution of most avian species is lacking. This mapping process included: (1) collecting each species' attribute information, i.e. species name, taxonomy, endangered category, distribution across counties, preferred habitat types, and elevation range, (2) mapping each species' distributions across counties, habitat types and elevation range, respectively, and (3) identifying the overlap region among these distribution layers as each species' current range.

We collected the attribute information using the following resources. For plants, we used "National Key Protected Wild Plant Resources Survey" [22] as the primary source and other supplementary sources including "Subject Database of China Plants" [23], "China Species Information Services" [24] and "China Plant Red Data Book: Endangered and Rare Plants" [21]. For mammals and birds, we used "National Key Terrestrial Wildlife Resources Survey" [25] as the primary source and other supplementary sources including "Database of Fauna Sinica" [26], "Distributions of China Mammal Species" [27] and "China Red Data Book of Endangered Animals: Mammals" [28].

The datasets on county boundaries and habitat types were derived from the national 1: 1,000,000 geographic databases and the national 1: 1,000,000 vegetation map [19], respectively. The elevation range for each species was extracted from the Shuttle Radar Topography Mission (SRTM) 90 m Digital Elevation Model (DEM) [29]. We mapped the species' geographic ranges for 373 plant, 115 mammal and 81 bird species.

Human Disturbance Index Mapping

We used several socioeconomic proxies, including proportion of land converted by human use, human population density, gross domestic product (GDP) and road density to calculate the human disturbance index (HDI) or human footprint [4,30]. The basic planning units were regular hexagons, each sized 100 km^2. The analysis included three steps. First, we calculated an individual HDI (IHDI) for each of these proxies. For proportion of converted land, we calculated the IHDI as the percent area of human-developed-land use – including croplands, plantations, rural settlements and urban/industrial areas – within each hexagon. For human population density and GDP, we calculated the IHDIs as their mean values per square kilometer within each unit. For road density, we considered four transportation levels (i.e., railway, expressway, national-provincial road and other-level roads), and calculated an IHDI for each level as the total road length within each unit. Second, we normalized the data ranges of all IHDIs on a scale of from 0.00 to 1.00, and then summed them to get the total HDI. Finally, we empirically transformed the data range of the total HDI on a scale of from 10.00 to 300.00 (Figure 1) so as to clearly demonstrate the overall human disturbance pattern.

We obtained datasets on land uses, human population density and GDP from the Data Center for Resources and Environmental Sciences of the Chinese Academy of Sciences [31], and all are 1 km×1 km resolution grid files. The road networks were derived from the national 1: 1,000,000 geographic database.

Conservation Priority-setting

We used the software Marxan (v2.0.2) to implement the conservation priority-setting process. Marxan was developed to cost-effectively solve an optimization problem of representing a suite of biodiversity targets [17]. To ensure that all conservation features were captured across their ranges of environmental and genetic variations [32], we first stratified their ranges with China's 53 terrestrial ecoregions [33], and then defined a quantitative conservation target for each feature per ecoregion. Due to limited data available for setting up appropriate conservation targets [34], we defined the target for each conservation feature as a uniform percentage area of its distribution range as suggested in previous studies (e.g., [3,32]). Specifically, the quantitative targets were selected based on expert opinions as follows: 30% for endangered species, 20% for priority natural ecosystems and 10% for natural vegetation types. An internationally recognized lowest target of 10% was set for natural vegetation types because they were assumed to represent the variety of basic habitat types.

We ran Marxan with two scenarios – a HDI-considered scenario and a HDI-ignored scenario. For the HDI-considered scenario, we integrated HDI values as a penalty function in Marxan analysis, i.e. a unit having a higher degree of disturbance would receive a greater penalty. For the HDI-ignored scenario, we used a uniform penalty of 1.0 per unit. The units with greater HDI values exhibit a more highly degraded ecological condition and should offer less potential from a conservation perspective [4]. Therefore, Marxan's algorithm sought to identify the optimized priority areas by minimizing the total HDI score in the HDI-considered scenario or the total land area in the HDI-ignored scenario. For the Marxan configurations, we: (1) generated 1000 solutions; (2) included a boundary length file and a modifier factor to control the compactness of priority areas; (3) implemented Simulated Annealing followed by Iterative Improvement; and (4)

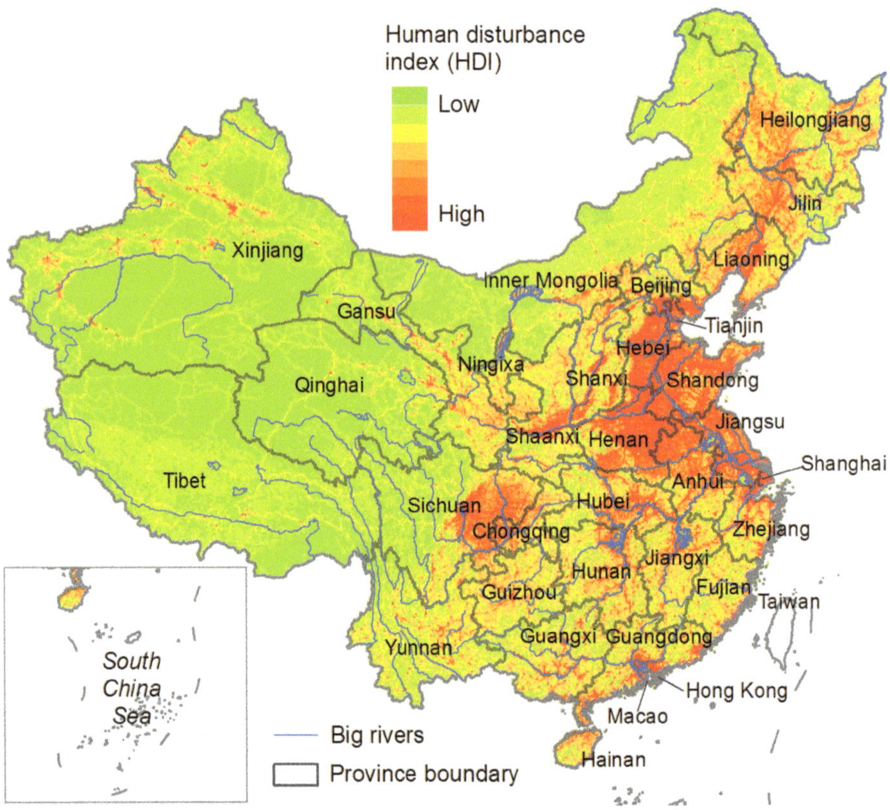

Figure 1. Human disturbance index (HDI). HDI was modeled by aggregating information on several socioeconomic proxies, including proportion of land converted by human use, human population density, gross domestic product, and road density.

used the default values for Number of Iterations (1,000,000) and Temperature Decreases (10,000).

We derived the conservation value that reflects the relative priority or irreplaceability of each planning unit [3] from the frequency of solutions selected, and used the best of the 1,000 solutions as the most cost-effective portfolio of priority units. We then identified the irreplaceable units as those selected in more than 800 solutions; 80% is often used for accuracy assessment for spatial data (e.g., [35]).

Effects of Human Disturbance

We compared the total HDI score and total area of the two portfolios of priority units generated by the HDI-ignored and HDI-considered scenarios, respectively. The changes in irreplaceable areas between the two scenarios were assessed in terms of the total area and the proportional area changes by province.

To assess the effects of human disturbance at a finer-scale, we further investigated the distributions of the priority units and irreplaceable areas on different environmental zones of HDI, elevation, and terrain ruggedness. We derived seven zones for each variable as follows: (1) We classified HDI zones by applying the Quantile Classification Scheme on HDI values; (2) We derived elevation zones from the SRTM 90 m DEM according to studies on geomorphology [36] (the elevation classification schemes were <200, 200–500, 500–1,000, 1,000–1,500, 1,500–2,000, 2,000–4,000, and >4,000 m); (3) We calculated a terrain ruggedness index (TRI) as the average difference in elevation between a center cell and its eight neighboring cells using the SRTM DEM, and the Quantile Classification Scheme was then used to break the TRI values into seven terrain categories, i.e. level, near-level, slightly-rugged, intermediately-rugged, moderately-rugged, highly-rugged and extremely-rugged [37–38].

Spatial Patterns of Conservation Priorities

Using the outputs from the HDI-considered scenario, we analyzed the spatial distributions of priority units and irreplaceable areas on HDI, elevation, TRI zones and provinces. We then delineated the primary large-scale priority areas as the large clusters of high-conservation-value planning units through an expert-based visual interpretation process.

Results

Effects of Human Disturbance

We presented the conservation value (based on a scale of from 0 to 1,000) of individual 100 km^2 hexagon units distributed throughout China (Figure 2). The portfolio of priority units in the HDI-ignored scenario (Figure 3) covered 24.6% of China's land area. By explicitly including the HDI as an additional penalty, we achieved the same conservation targets with a small increase (~7%) in the total area of priority units compared to when HDI was ignored, meanwhile a clear reduction of ~9% in the total HDI score was observed. The overlapping region (Figure 3) covered 46.3% and 43.2% of the priority units in the HDI-ignored and HDI-considered scenarios, respectively. A strong and positive spatial correlation exists (Spearman's rank correlation, $r = 0.871$, $p << 0.001$) between the two conservation value layers.

The irreplaceable units in the HDI-ignored scenario (Figure 2A) covered 2.8% of China's landmass, while an increase of ~43% in the total irreplaceable area was observed in the HDI-considered scenario (Figure 2B). The overlapping region occupied 82.7% and 57.7% of the irreplaceable areas in the HDI-ignored and HDI-considered scenarios, respectively. High proportional increases in irreplaceable area occurred principally in provinces located in the eastern coastal region, middle-lower Yangtze River Basin and northeastern China, whereas provinces in western and southwestern China had the fewest changes (Figure 4). Several provinces in the eastern highly-disturbed regions (Figure 1), including Guangdong, Jiangxi, Henan and Hebei, also were found to have small changes in their irreplaceable areas (Figure 4).

Compared to the results in the HDI-ignored scenario, the portfolio of priority units in the HDI-considered scenario contained: (1) fewer units in the three highest HDI zones and more units in the four lower HDI zones, (2) fewer units only in the lowest (<200 m) elevation zone and more units in the other six zones, and (3) fewer units in the level TRI zone and more units in each of the other TRI zones (Figure 5). The HDI-considered scenario identified a greater number of irreplaceable units in almost all environmental zones than did the HDI-ignored scenario, with the sole exception of the highest HDI zone (Figure 6).

Spatial Patterns of Conservation Priorities

We analyzed the spatial patterns of conservation priorities using the outcomes from the HDI-considered scenario. The priority units consistently decreased with increasing HDI value (Figure 5A), with the majority (~76%) located in the four lower HDI zones. The <200 m elevation zone included only 5.8% of all priority units, and the zones of 200–1,000, 1,000–2,000 and > 2,000 m contained 29.6%, 25.8% and 38.8% of the priority units, respectively. The priority units generally had an increasing distribution trend on TRI zones from level to extremely-rugged terrain (Figure 5C), with the vast majority located in slightly- to extremely-rugged zones, and only 3.2% identified in level zone and 11.5% in near-level zone. All provinces included some units that were required for meeting the conservation targets (Figure 3), with the greatest proportion occurring in Xinjiang followed by Tibet, Inner Mongolia, Qinghai, Sichuan and Yunnan. These six western provinces contained 72.5% of the total priority units.

The irreplaceable units had a normal-like distribution on the HDI zones that peaked in the fourth zone (Figure 6A). Compared to the distribution of priority units, greater proportions of irreplaceable units were selected in lower elevation zones, with 10.7%, 39.3%, 26.7% and 23.3% of the total irreplaceable area located at <200, 200–1,000, 1,000–2,000 and >2,000 m zones, respectively. In particular, the highest zone (>4,000 m) contained the smallest proportion of irreplaceable areas (Figure 6B) although the greatest number of priority units occurred there (Figure 5B). In addition, over 75% of the irreplaceable areas were located in intermediately- to extremely-rugged TRI zones (Figure 6C). Provinces with the greatest number of irreplaceable areas were Yunnan followed by Guangxi, Tibet, Xinjiang, Inner Mongolia and Sichuan, and they contained 51.5% of the total irreplaceable area.

Overall, many more units in western China were assigned higher conservation values compared to eastern and southern regions, where the distributions of high-value units were severely fragmented (Figure 7). Based on the conservation value data and expert knowledge, we visually delineated the boundaries of 23 primary large-scale priority areas and excluded many small isolated areas (Figure 7). These large-scale priority areas covered ~28% of China's landmass and were mainly distributed in remote regions at high elevation and/or rugged terrain. Regions that have experienced high-intensity disturbances, e.g. Northeast China Plain, North China Plain, South Huaihe and Middle-lower Yangtze River Plain, Sichuan Basin and Pearl River Delta Area, did not contain any large-scale priority areas (Figure 7).

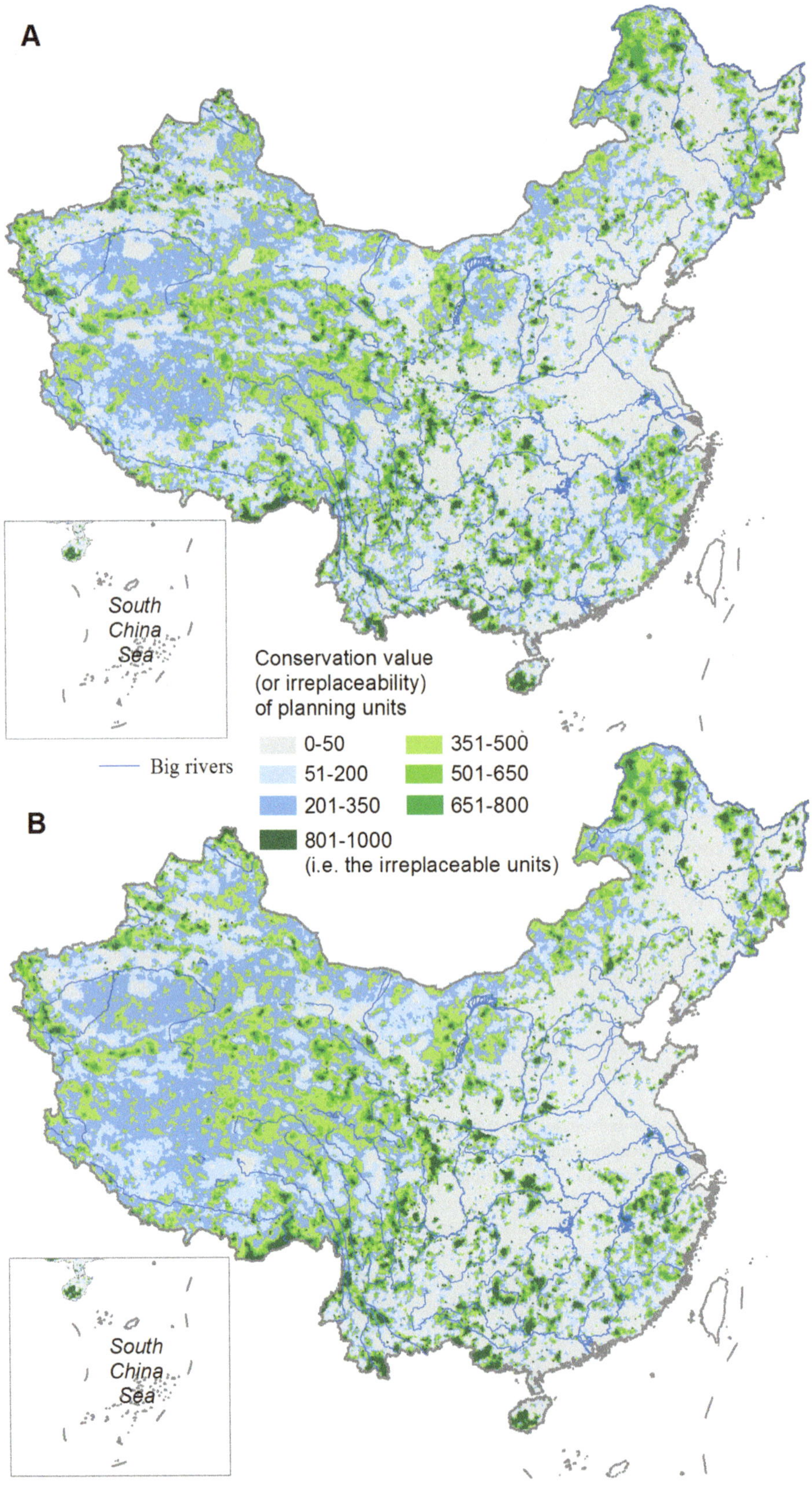

Figure 2. The conservation value of 100 km^2 hexagon units for achieving the defined conservation targets. (A) HDI-ignored scenario and (B) HDI-considered scenario.

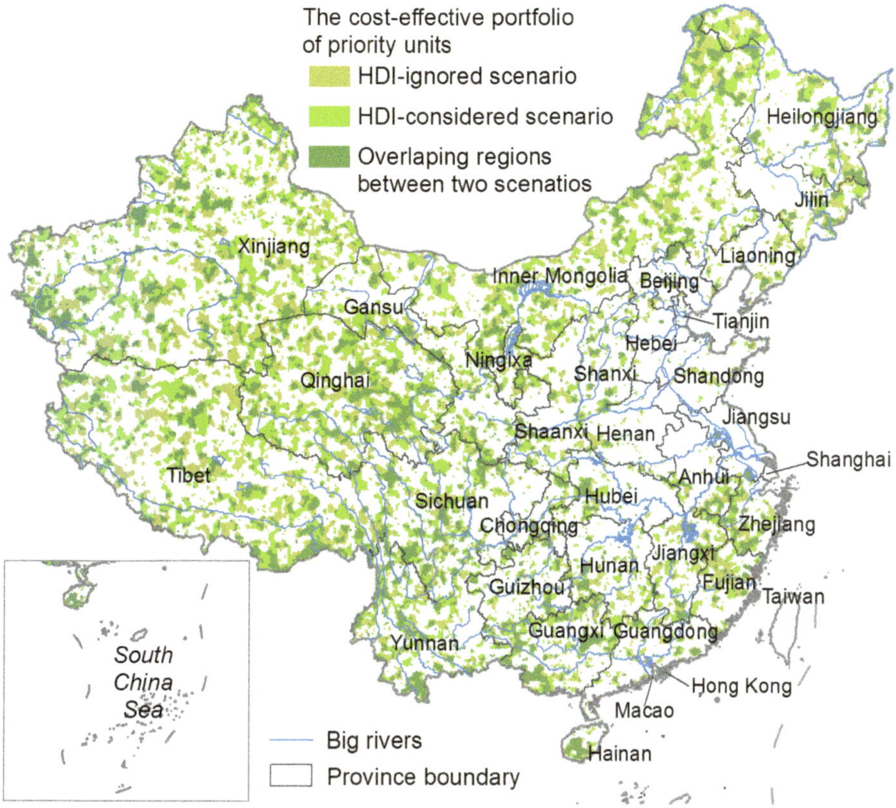

Figure 3. The cost-effective portfolios of priority units identified by the HDI-ignored and HDI-considered scenarios, respectively.

Discussion

In this study we implement a rigorous planning framework to identify the optimized national-scale conservation priorities in China. Our framework addresses several key features overlooked in previous studies, i.e. human disturbance, integration of ecosystem- and species-level conservation features, and principles of complementarity and representativeness.

Effects of Human Disturbance

Due to a lack of site-specific data on the ecological integrity of most biodiversity features [16], a HDI (or suitability index) is often modeled by aggregating human disturbance data to provide an indirect measure of ecological condition [4,30]. By explicitly considering HDI, our goal is to direct conservation towards the least-disturbed regions while still fully meeting conservation goals. We feel that this approach will promote conservation success and more efficiently achieve conservation goals [3,6]. Moreover, areas with higher disturbances offer less conservation potential as they have lower habitat suitability for sustaining conservation features [4].

Our result indicates that the portfolio of priority units in the HDI-considered scenario is characterized by a marked reduction in the total HDI score and a slight increase in the total area, and in addition, more priority units are identified at less-disturbed, higher and/or rugged regions (Figure 5). Such effects are derived from implementing Marxan's algorithm for identifying an optimized portfolio of priority areas that has the minimum total penalty score [17]. Therefore, many priority units identified in the HDI-ignored scenario, especially those distributed as fragments on highly-disturbed lands, were excluded or devalued in the HDI-considered scenario so as to minimize the total HDI score of the portfolio. This requires the HDI-considered scenario to select a greater number of priority units with lower HDI values to achieve the same conservation goals, because each of these units contains relatively fewer conservation features and/or covers smaller areas within their distribution ranges.

The total irreplaceable area in the HDI-considered scenario increased significantly (~43%) and more irreplaceable units were selected in almost all HDI, elevation, and TRI zones except the highest HDI zone (Figure 6). We think the increase results from the fact that Marxan solutions favor those units with relatively lower penalty scores, which also was reported by Carwardine et al. [3]. This indicates that human disturbance can partly degrade the potential options available for implementing cost-effective conservation. Our result, that the most highly-developed provinces had the greatest proportional increases in irreplaceable area while western less-disturbed provinces had smaller changes (Figure 4), also supports this perspective. However, we also found that several highly-developed provinces had only small changes in irreplaceable area. We think this is because those provinces contain relatively fewer conservation features and limited overlap exists between the distributions of conservation features and areas of human disturbances.

A fundamental concern in including human disturbance is that priority areas may be biased to remote, higher and more rugged places. Such a biased distribution has been a severe problem resulting in the existing reserve networks failing to adequately represent the overall biodiversity [34,38]. Does our analysis further increase the existing biases in the location of established reserves? We feel it does not, because our framework implements 'representativeness' as a core principle in identifying priority areas

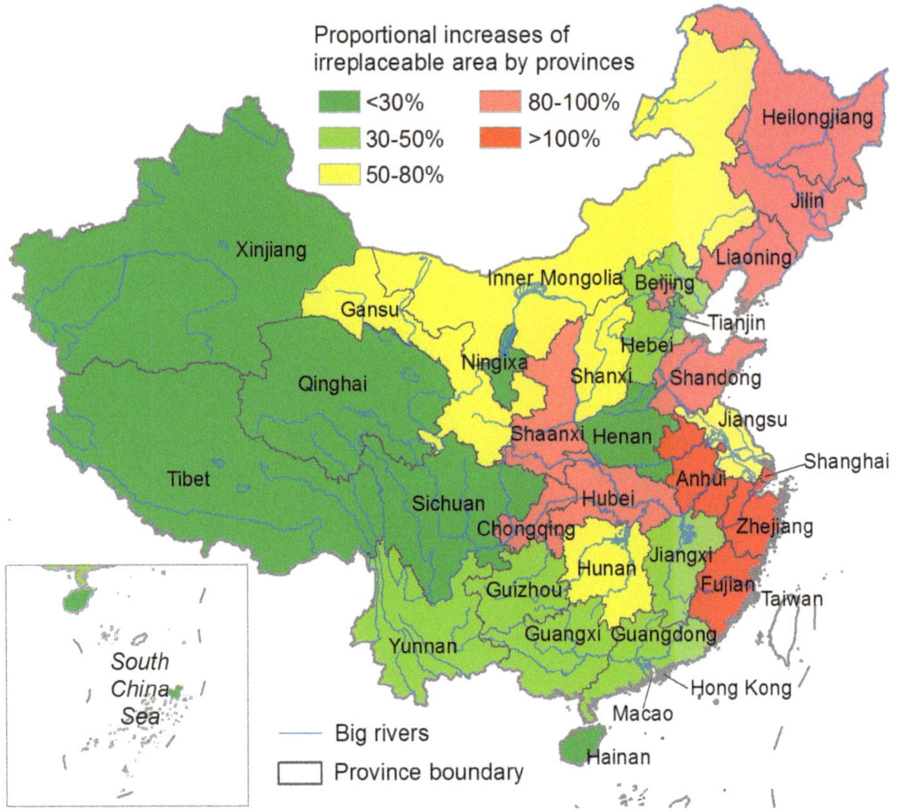

Figure 4. Proportional changes in irreplaceable area between the HDI-ignored and HDI-considered scenarios by province.

and defines explicit conservation targets for all selected conservation features. The goal for representing the full range of biodiversity requires that the priority-setting process also focuses on disturbed landscapes of high biodiversity conservation significance [5]. Similar to Linke et al. [4], we integrated human disturbance as a discounting factor for ecological condition so as to ensure that the resulting portfolio was optimized for maximizing conservation achievements.

Although apparent shifts of priority units towards less-disturbed zones were observed (Figure 5A), the HDI-considered scenario only selected fewer priority units in the <200 m elevation zone (Figure 5B) and level zone (Figure 5C), and identified more irreplaceable units in almost all HDI, elevation, and TRI zones except the highest HDI zone (Figure 6). The lowest/level zone may provide less conservation potential because of limited current biodiversity in response to long-term human disturbance [33]. We also found considerable overlap, and strong and positive pairwise associations between the portfolios of priority units and the portfolios of irreplaceable areas identified by the HDI-ignored and HDI-considered scenarios, respectively. These results demonstrate that our analysis is conservation target based, and the inclusion of human disturbance did not result in the biased distribution of conservation priorities.

Spatial Patterns of Conservation Priorities

Recognizing the advantages of including human disturbance in priority-setting, we analyzed the spatial patterns of conservation priorities using the results from the HDI-considered scenario. Human disturbance has caused severe degradation of natural ecosystems and many species extinctions, which can greatly diminish the conservation value of a region that was historically rich in biodiversity [14,33]. Therefore, the higher the disturbance intensity, the lower the proportion of priority units was allocated in a region (Figure 5A). Rugged terrain often serves as a natural barrier for human development, and these mountainous areas have become refuges for many endangered species; These areas also are preferred as conservation priorities because they maintain more diverse habitats and higher animal and plant biodiversity [14].

We found higher percentages of irreplaceable area occurred in lower elevation zones (Figure 6B) compared to the distribution of priority units (Figure 5B). This implies that there are relatively fewer cost-effective options for fulfilling conservation targets in lowland regions, whereas the highland areas have greater flexibility in priority-setting. As moderately-disturbed and/or intermediately- to extremely-rugged zones contain the majority of irreplaceable areas (Figure 6), these habitats should be targeted to identify potentially important areas for implementing cost-effective conservation. These habitats are mainly found in western provinces, which include the vast majority of both priority units and irreplaceable areas, and therefore we consider those provinces to be of great significance in conserving China's biodiversity.

Previous researches have revealed that the remaining natural landscapes in eastern and southern China are highly fragmented, and western China supports more intact natural ecosystems and endangered species [14,33]. This study similarly found that western China contains more high-value units clustered in relatively larger patches, while the high-value units in eastern and southern regions are severely fragmented and principally located in mountainous areas (Figure 7). Our result shows that the

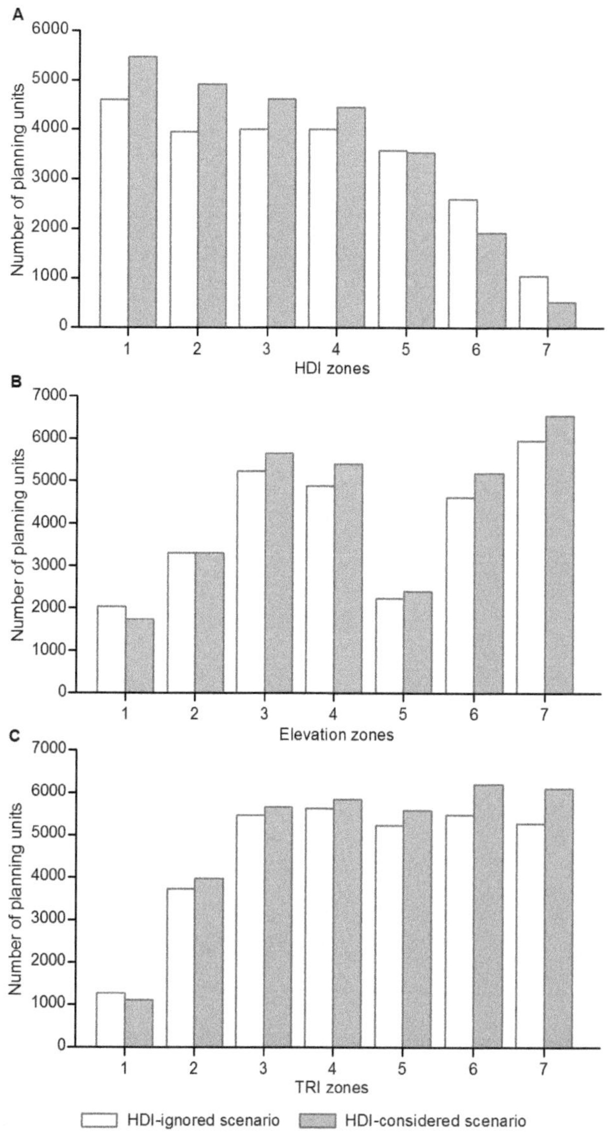

Figure 5. The distribution of priority units on (A) HDI, (B) elevation, and (C) TRI zones. The numbers 1 to 7 on the horizontal axes represent (A) low to high HDI value classifications, (B) elevation zones of <200, 200–500, 500–1,000, 1,000–1,500, 1,500–2,000, 2,000–4,000, and >4,000 m, and (C) terrain categories of level, near-level, slightly-rugged, intermediately-rugged, moderately-rugged, highly-rugged, and extremely-rugged.

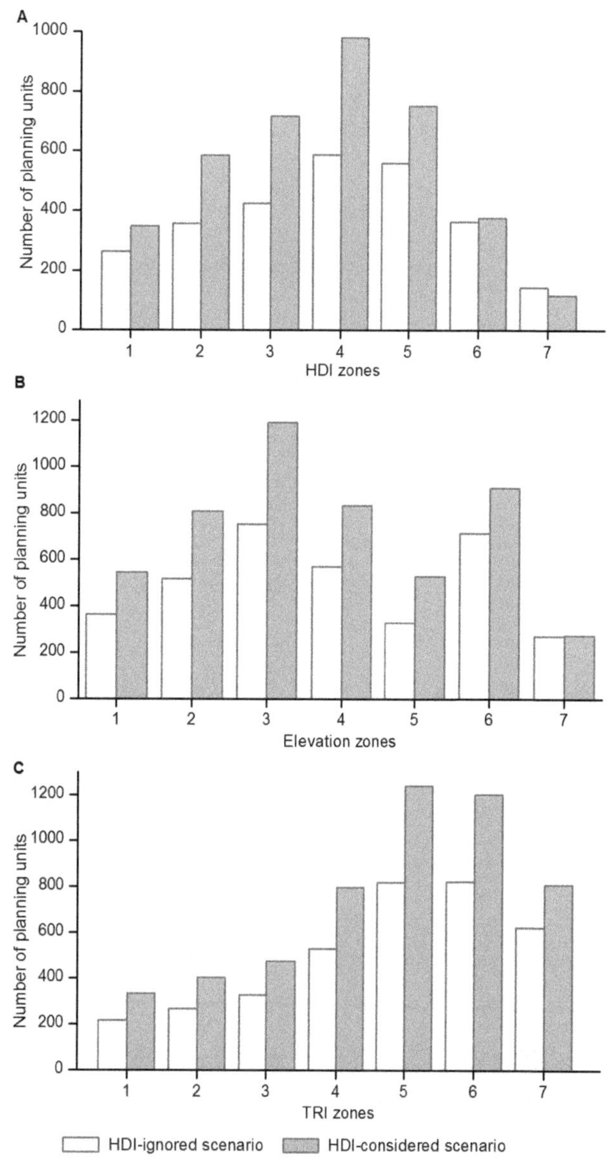

Figure 6. The distribution of irreplaceable units on (A) HDI, (B) elevation, and (C) TRI zones. See Figure 5 for the explanation of numbers 1 to 7 on the horizontal axes.

primary large-scale priority areas are mainly distributed in remote places with high elevation and rugged terrain (Figure 7). This finding is generally consistent with the results in previous studies (e.g., [10,12–15]).

However, we also identified several priority areas that were rarely considered before, including the Hulunbuir Grassland, Xilingol Grassland, Alashan-Ordos Region, Altai Mountain, and Pamirs Plateau (Figure 7). All are located in Inner Mongolia and Xinjiang, covering grassland, semi-desert, alpine and tundra biomes. These areas are not rich in species diversity, but they are valued for maintaining several important ecosystems that sustain many endemic species and critical ecosystem services [39]. Our result agrees with the limited number of studies that have considered goals for ecosystem conservation. For example, the Alashan-Ordos Region and Altai Mountain are recognized as the key areas for protecting priority terrestrial ecosystems [39], and each of these five areas exhibits some overlap with the global 200 priority ecoregions, including Daurian/Mongolian Steppe, Altai-Sayan Montane Forests, and Middle Asian Montane Woodlands and Steppe [40].

The primary large-scale priority areas are the centers of biodiversity and evolution as they provide refuges for many species and sustain important ecosystem services [8,12], and in these areas it is usually simpler and less expensive to implement conservation actions. In addition to establishing reserves, these areas should be subject to a variety of sustainable management approaches that seek to balance extractive uses with the retention of natural resources and ecosystem functions, such as the various ecosystem service policies currently implemented in China [9]. However, these large-scale priority areas are not sufficient to fulfill China's overall conservation goals [15], because many species,

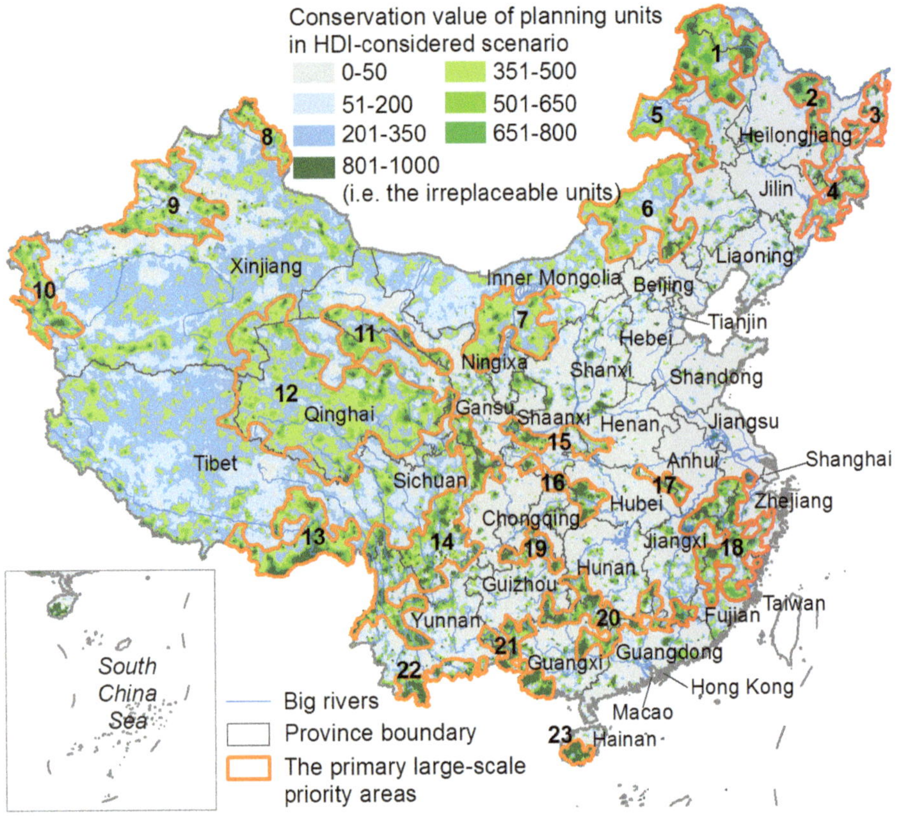

Figure 7. The distribution of the 23 primary large-scale priority areas. 1 – Daxing'anling Mountain, 2 – Xiaoxing'anling Mountain, 3 – Sanjiang Plain, 4 – Changbai Mountain, 5 – Hulunbuir Grassland, 6 – Xilingol Grassland, 7 – Alashan-Ordos Region, 8 – Altai Mountain, 9 – Tianshan Mountain, 10 – Pamirs Plateau, 11 – Qilian Mountain, 12 – Sanjiangyuan-Qiangtang Region, 13 – Southeast Himalaya Mountain, 14 – Hengduan Mountain, 15 – Qinling Mountain, 16 – Daba Mountain, 17 – Dabieshan Mountain, 18 – Mountain Region connecting Fujian-Zhejiang-Jiangxi-Anhui, 19 – Wuling Mountain, 20 – Nanling Mountain, 21 – Mountain Region in western Guangxi, 22 – Xishuangbanna, and 23 – Southern Hainan Island.

particularly those that exploit special microhabitats, may only occur in places close to developed landscapes and are already highly threatened [5]. Therefore, the priority units distributed in highly-disturbed regions that are not included in the large-scale priority areas should be included in local conservation actions. This may be even more urgent in order to prevent the immediate loss of biodiversity [41].

The Priority-setting Framework

Using this priority-setting framework, we are trying to ensure the identification of a comprehensive and cost-effective portfolio of conservation priorities for China. The process is driven by explicitly delineating spatial distributions and quantitatively defined targets for representative conservation features. We believe the analysis is rigorous, objective, transparent, and replicable.

We acknowledge that the availability and accuracy of spatial data on biodiversity and disturbances are a primary constraint for national-scale priority-setting. Therefore, our results can be further refined as more comprehensive data become available. This study has used the most up-to-date national survey data on key protected plant and animal species [22,25], as well as highly recognized information sources that have been used in previous studies (e.g., [14]). The ecosystem-level features represent meaningful biodiversity surrogates because they are the emergent entities of unique species assemblages and easily mapped; moreover, they are useful indicators of ecological processes and ecosystem services [16]. The integration of conservation features from different levels and multiple taxa can improve the effectiveness of priority areas in representing the overall biodiversity [5]. The HDI, a coarse simplification of current ecological condition, could be improved when better disturbance data and modeling methods are developed.

Hence we suggest that China increase its budget for improving the GIS-based conservation decision-making platform and enhance data sharing mechanisms. Moreover, the integration of ecological processes, ecosystem services, socioeconomic objectives and climate projections represents future research priorities in SCP [4,6,7].

Application

Systematic conservation priority-setting has significant implications in assisting China in achieving its cost-effective conservation goals as a megadiversity country. For instance, this approach has been applied in conservation priority-setting for China nationwide, and the work is a key component for developing the National Biodiversity Strategies and Action Plans (NBSAPs) [8]. China's Ministry of Environmental Protection requires all provinces, major river watersheds, and counties develop their Local Biodiversity Strategies and Action Plans (LBSAPs) [8]. Thus, we applied this priority-setting approach to come up with the first provincial LBSAPs for Sichuan [42], and now this approach is in high demand in China.

Not only is this approach useful in its direct application to conservation planning, but it also has important applicability for strategic land-use planning and sustainable development practices; e.g. the Ecological Function Regionalization and Major Function Oriented Zoning [43–44] could be further refined using our approach. Such planning seeks to optimize the spatial patterns of economic development and environment protection by investigating the synergies and trade-offs between their distributions [45]. Areas recognized as conservation priorities should be primarily preserved for sustaining biodiversity and ecosystem services. As China is now adopting a new paradigm of sustainable development by undertaking a transition from conventional industrialization to ecological civilization [46] numerous redlines on natural resources and environment management (e.g., the Key Ecological Function Regions and Development Prohibited/Restricted Zones) have been established to ensure the country's ecological security [47]. This priority-setting approach is of great significance for determining the spatially optimized conservation network or redlines for strategic land-use planning.

Conclusions

This study presents optimized national-scale spatial priorities for biodiversity conservation in China by implementing a systematic priority-setting approach with the integration of human disturbances, ecosystem- and species-level conservation features, and principles of complementarity and representativeness. Inclusion of human disturbance is essential for a cost-effective priority-setting – maximizing conservation achievement while minimizing conflicts with economic development. Such an approach will ensure the optimal spatial distribution of priority areas and reduce biases in conservation investment and/or land-use planning. The majority of priority units we identified are located in relatively remote, high and/or rugged places, however, areas that are moderately-disturbed, <2,000 m in altitude, and/or intermediately- to extremely-rugged in terrain should be targeted to identify potentially important regions for implementing cost-effective conservation. To achieve the overall biodiversity conservation goal in China, we delineate 23 primary large-scale priority areas, as well as recognize many isolated priority units in disturbed regions that need even more urgent conservation so as to prevent the immediate loss of biodiversity.

While requiring further refinement, our results provide valuable insights for current conservation and strategic land-use planning in China. This approach uses publicly available information, and is transferable and easy to implement by end-users, and applicable for national- and local-scale systematic conservation prioritization practices. Improved data, especially in the details of human disturbance and for aquatic systems at national-scale, will further enhance its applicability.

Acknowledgments

We thank Siobhan Kenney and Bastian Bertzky at UNEP-WCMC for valuable comments and English editing. We thank Bei Huang, Rui Zhang and Shan Sun, master students in Southwest Forestry University for their valuable help in building the spatial database, and the Data center for Resources and Environmental Sciences of the Chinese Academy of Sciences for providing data on land uses, human population density and GDP. PAG acknowledges the support of Chrissie, Sara, and Jenni.

Author Contributions

Conceived and designed the experiments: RDW YCL GPM PAG SZ. Performed the experiments: RDW SZ DQL PZ LZW HRD. Analyzed the data: RDW YCL LZW HRD. Contributed reagents/materials/analysis tools: RDW DQL LZW HRD. Wrote the paper: RDW YCL GPM PAG SZ.

References

1. McCarthy DP, Donald PF, Scharlemann JPW, Buchanan GM, Balmford A, et al. (2012) Financial costs of meeting global biodiversity conservation targets: current spending and unmet needs. Science 338: 946–949.
2. Brooks TM, Mittermeier RA, da Fonseca GAB, Gerlach J, Hoffmann M, et al. (2006) Global biodiversity conservation priorities. Science 313: 58–61.
3. Carwardine J, Wilson KA, Ceballos G, Ehrlich PR, Naidoo R, et al. (2008) Cost-effective priorities for global mammal conservation. Proceedings of the National Academy of Sciences of USA 105: 11446–11450.
4. Linke S, Kennard MJ, Hermoso V, Olden JD, Stein J, et al. (2012) Merging connectivity rules and large-scale condition assessment improves conservation adequacy in river systems. Journal of Applied Ecology 49: 1036–1045.
5. Margules CR, Pressey RL (2000) Systematic conservation planning. Nature 405: 243–253.
6. Venter O, Possingham HP, Hovani L, Dewi S, Griscom B, et al. (2013) Using systematic conservation planning to minimize REDD+ conflict with agriculture and logging in the tropics. Conservation Letters 6: 116–124.
7. Withey JC, Lawler JJ, Polasky S, Plantinga AJ, Nelson EJ, et al. (2012) Maximising return on conservation investment in the conterminous USA. Ecology Letters 15: 1249–1256.
8. The Ministry of Environmental Protection of China (2011) China national biodiversity conservation strategy and action plans 2011–2030. Beijing: Chinese Environmental Science Press.
9. Liu JG, Ouyang ZY, Yang W, Xu WH, Li SX (2013) Evaluation of ecosystem service policies from biophysical and social perspectives: the case of China. In: Levin SA, editor. Encyclopedia of biodiversity, second edition, volume 3. pp. 372–384. Waltham: Academic Press.
10. Chen LZ (1993) China's biodiversity and conservation strategy. Beijing: Science Press.
11. Ma KP (2001) Hotspots assessment and conservation priorities identification of biodiversity in China should be emphasized. Acta Phytoecologica Sinica 25: 125.
12. Chen CD (1998) Biodiversity of China: a country study. Beijing: Chinese Environmental Science Press.
13. Li DQ, Song YL, Ouyang ZY (2003) Research on the national forestry nature reserve system plan. Beijing: China Land Press.
14. Tang ZY, Wang ZH, Zheng CY, Fang JY (2006) Biodiversity in China's mountains. Frontiers in Ecology and the Environment 4: 347–352.
15. Zhang YB, Ma KP (2008) Geographic distribution patterns and status assessment of threatened plants in China. Biodiversity and Conservation 17: 1783–1798.
16. Groves CR (2003) Drafting a conservation blueprint: a practitioner's guide to planning for biodiversity. Washington DC: Island Press.
17. Ardon JA, Possingham HP, Klein CJ (2010) Marxan good practices handbook, Version 2. Victoria: Pacific Marine Analysis and Research Association.
18. Crous CJ, Samways MJ, Pryke JS (2013) Exploring the mesofilter as a novel operational scale in conservation planning. Journal of Applied Ecology 50: 205–214.
19. Zhang XS (2007) Vegetation map of People's Republic of China (1: 1,000,000). Beijing: Geology Press.
20. IUCN (2012) The IUCN red list of threatened species. Available: http://www.iucnredlist.org. Accessed 10 November 2012.
21. Fu LG (1992) China plant red data book: endangered and rare plants. Beijing: Science Press.
22. State Forestry Administration of China (2009) National key protected wild plant resources survey. Beijing: China Forestry Publishing House.
23. Institute of Botany (2009) Subject database of China plants. Available: http://www.plant.csdb.cn. Accessed 25 October 2012.
24. WCS (2005) China species information services. Available: http://www.baohu.org. Accessed 20 May 2012.
25. State Forestry Administration of China (2009) National key terrestrial wildlife resources survey. Beijing: China Forestry Publishing House.
26. Institute of Zoology (2009) Database of Fauna Sinica. Available: http://www.zoology.nsdc.cn. Accessed 15 September 2012.
27. Zhang RZ (1997) Distributions of China mammal species. Beijing: China Forestry Publishing House.
28. Wang S (1998) China red data book of endangered animals: mammals. Beijing: Science Press.
29. USGS (2004) Shuttle Radar Topography Mission DEM. Available: http://glcf.umiacs.umd.edu/data/srtm. Accessed 16 July 2012.
30. Sanderson EW, Jaiteh M, Levy MA, Redford KH, Wannebo AV, et al. (2002) The human footprint and the last of the wild. BioScience 52: 891–904.
31. Yang XH, Ma HQ (2009) Natural environment suitability of China and its relationship with population distributions. International Journal of Environmental Research and Public Health 6: 3025–3039.

32. Chan KMA, Shaw MR, Cameron DR, Underwood EC, Daily GC (2006) Conservation planning for ecosystem services. PLoS Biology 4: e379.
33. Wu RD, Zhang S, Yu DW, Zhao P, Li XH, et al. (2011) Effectiveness of China's nature reserves in representing ecological diversity. Frontiers in Ecology and the Environment 9: 383–389.
34. Rodrigues ASL, Andelman SJ, Bakarr MI, Boitani L, Brooks TM, et al. (2004) Effectiveness of the global protected area network in representing species diversity. Nature 428: 640–643.
35. Lea C, Curtis AC (2010) Thematic accuracy assessment procedures: National Park Service vegetation inventory, version 2.0. Natural resource report NPS/2010/NRR-2010/204. Fort Collins: National Park Service, US Department of the Interior.
36. Li BY, Pan BT, Han JF (2008) Basic terrestrial geomorphological types in China and their circumscriptions. Quaternary Sciences 28: 535–543.
37. Riley S, DeGloria SD, Elliot R (1999) A terrain ruggedness index that quantifies topographic heterogeneity. Intermountain Journal of Sciences 5: 23–27.
38. Wu RD, Ma GZ, Long YC, Yu JH, Li SN, et al. (2011) The performance of nature reserves in capturing the biological diversity on Hainan Island, China. Environmental Science and Pollution Research 18: 800–810.
39. Xu WH, Ouyang ZY, Huang H, Wang XK, Miao H, et al. (2006) Priority analysis on conserving China's terrestrial ecosystems. Acta Ecologica Sinica 26: 271–280.
40. Olson DM, Dinerstein E (2002) The global 200: priority ecoregions for global conservation. Annals of the Missouri Botanical garden 89: 199–224.
41. Joppa LN, Pfaff A (2009) High and far: biases in the location of protected areas. PLoS One 4: e8273.
42. The Ministry of Environmental Protection of China (2011) Issue of Sichuan biodiversity conservation strategy and action plans. Available: http://www.zhb.gov.cn/zhxx/gzdt/201112/t20111215_221379.htm. Accessed 15 July 2013.
43. The Ministry of Environmental Protection of China, Chinese Academy of Sciences (2008) Announcement on the issue of "National Ecological Function Regionalization". Available: http://www.mep.gov.cn/info/bgw/bgg/200808/t20080801_126867.htm. Accessed 15 May 2013.
44. The State Council of China (2011) Circular of the State Council on the issue of "Major Function Oriented Zoning". http://www.gov.cn/zwgk/2011-06/08/content_1879180.htm. Accessed 20 May 2013.
45. Fan J, Li P (2009) The scientific foundation of Major Function Oriented Zoning in China. Journal of Geographical Sciences 19: 515–531.
46. Pan J (2012) From industrial toward ecological in China. Science 336: 1397.
47. Lü Y, Ma Z, Zhang L, Fu B, Gao G (2013) Redlines for the greening of China. Environmental Science & Policy 33: 346–353.

7

Conservation Investment for Rare Plants in Urban Environments

Mark W. Schwartz[1,2]*, **Lacy M. Smith**[2], **Zachary L. Steel**[3]

1 John Muir Institute of the Environment, University of California Davis, Davis, California, United States of America, 2 Department of Environmental Science & Policy, University of California Davis, Davis, California, United States of America, 3 Graduate Group in Ecology, University of California Davis, Davis, California, United States of America

Abstract

Budgets for species conservation limit actions. Expending resources in areas of high human density is costly and generally considered less likely to succeed. Yet, coastal California contains both a large fraction of narrowly endemic at-risk plant species as well as the state's three largest metropolitan regions. Hence understanding the capacity to protect species along the highly urbanized coast is a conservation priority. We examine at-risk plant populations along California's coastline from San Diego to north of San Francisco to better understand whether there is a relationship between human population density and: i) performance of at-risk plant populations; and ii) conservation spending. Answering these questions can help focus appropriate strategic conservation investment. Rare plant performance was measured using the annualized growth rate estimate between census periods using the California Natural Diversity Database. Human density was estimated using Census Bureau statistics from the year 2000. We found strong evidence for a lack of a relationship between human population density and plant population performance in California's coastal counties. Analyzing US Endangered Species expenditure reports, we found large differences in expenditures among counties, with plants in San Diego County receiving much higher expenditures than other locations. We found a slight positive relationship between expenditures on behalf of endangered species and human density. Together these data support the argument that conservation efforts by protecting habitats within urban environments are not less likely to be successful than in rural areas. Expenditures on behalf of federally listed endangered and threatened plants do not appear to be related to proximity to human populations. Given the evidence of sufficient performance in urban environments, along with a high potential to leverage public support for nature in urban environments, expenditures in these areas appear to be an appropriate use of conservation funds.

Editor: David L. Roberts, University of Kent, United Kingdom

Funding: These authors have no support or funding to report.

Competing Interests: The authors have declared that no competing interests exist.

* E-mail: mwschwartz@ucdavis.edu

Introduction

Determining the degree to which limited resources should be applied to conservation in urban environments remains a critical challenge [1–3]. This challenge is acute in California where a large number of at-risk plant species are restricted to regions of high human density [4–6]. Limited conservation resources require strategic investment in conservation (e.g., [7]). The land cost associated with urban environments is typically far higher than rural areas [8]. Further, conservation opportunities within an urbanized landscape often consist of small, isolated fragments [9–11]. Small, isolated fragments of natural habitats may have lower species richness than larger natural areas [12], and populations within these sites may be at an elevated risk of extinction [13–16]. Rare species, particularly plants and insects, are often spatially associated with urban environments [8,15,17]. Further, the social value of conserving biodiversity that is accessible to urban populations may be high [18–22]. Thus, Lawson et al. [23] raised considerable interest in demonstrating that extant populations of plants in urban areas do not seem to have different population growth rates than those in rural environments.

A constraint of the Lawson et al. [23] study is that it did not distinguish performance of species among habitat types. We might predict that particular types of habitats (e.g., serpentine outcrops) may be relatively resilient to the kinds of changes in urban environments that put rare plant species at risk. For example, we might expect that the threat of invasive plant species to native populations may be higher in wetland habitats than in edaphically stressful environments (e.g., serpentine). Alternatively, other habitat types (e.g., agricultural landscapes with low human density) may also be vulnerable to similar degradation as in urbanized environments [15]. Protected wetlands, for example, may be threatened by invasive species, nutrient additions, environmental toxins, and other impacts across gradients of human density.

We have two distinct objectives in this paper. First, we further test the hypothesis of Lawson et al. [23] —that urbanization has no detectable effect on performance across a rural to urban gradient. We expand the evaluation of this hypothesis by taking a much closer look at potential confounding factors that may mask a relationship, including asking whether habitat types express differential performance relationships across the rural to urban gradient. We predict that species in some urban habitat types may be more resilient to urbanization than others. If so, then focusing on habitat types that are resilient to urbanization can help increase effectiveness of urban conservation efforts. To assess this hypothesis, we replicate the methods of Lawson et al. [23] to examine population performance of plant species across a human population density gradient. Specifically, we used the California Natural Heritage plant observation data (California Natural Diversity Database, CNDDB [24]) to determine population trends where repeated population size data are available. We then classified species into different habitat types and life forms to examine performance as a function of human density (people/km^2) in more detail.

Our second objective is to assess the degree to which conservation investment in California may be biased by human density. One argument is that the high cost of urban conservation places too high a demand on limited funds relative to the benefits obtained [16]. The results of Lawson et al. [23] challenge the notion that investment in urban conservation is wasted investment. We cannot, unfortunately, relate government spending directly to plant population outcomes because our plant performance data are neither temporally nor spatially linked to expenditures. We can, however, assess patterns in spending on behalf of federally listed endangered plants to determine if it appears inappropriate, given plant performance. We hope that comparing and contrasting these two independent data sets help us to develop better strategies for how expenditures can be effectively applied to plant conservation in California, where there is a high fraction of urban-associated rare species and many of the resources for conservation are generated [25]. Together, these two pieces of information can help guide appropriate investment in conserving plant diversity within the urbanized and urbanizing California landscape.

Methods

We defined the study area to encompass the richest region of rare species in the state of California [6], focusing on at-risk plant species along coastal California from the Mexican border through the San Francisco Bay Area. Our study area was composed of the 17 counties that border the coastline, including San Francisco Bay, from San Diego to Sonoma County (Fig. 1a). Within this region, we aggregated all plant population census data available through the California Natural Diversity Database (CNDDB) [24]. The CNDDB is a spatially explicit database of rare plant and animal occurrences within California. Tracked plants are those defined by the California Native Plant Society [26] as at-risk. Plant occurrence records are based on sightings by both professional and amateur botanists. All records are then vetted by professional biologists before being entered into the database [24]. We use data on at-risk plant populations that contained species, location, and dates of at least two quantitative estimates of population size across a time interval of one year or more. Most CNDDB population locations do not contain repeated estimates of population size, a requirement in order to estimate performance.

We aggregated data on 1,682 population change estimates among 253 species and subspecies across 795 locations. Data included observations from 1897 through 2007. Although data span the 20th century, over 99% of population change estimates are since 1980 and 73% since 1990. Species are also sampled in a skewed distribution with 60% of species appearing in just one or two locations (range = 1 to 21; average 2.86, s.d. = 2.95; median = 2, Appendix S1).

We classified species into three classes of growth form: annual herbs (902 observations, 97 species and subspecies); perennial herbs (662 observations; 121 species or subspecies); and woody shrubs and trees (118 observations; 35 species or sub-species). We also assigned species into one or more of 19 habitat types based on *Jepson Manual* species' habitat descriptions and predominant site type descriptions from the CNDDB records. Habitat types are general (e.g., deciduous woodland, grassland, sage scrub) for habitats that are dominant in the landscape, but more specific for edaphically extreme environments in which numerous rare species occur (e.g., vernal pools, alkali sinks, serpentine). Some species were lumped into a non-differentiated "other" category when habitat descriptions were vague (e.g., roadsides) or difficult to place into a particular habitat type (e.g., swales) These difficult to classify habitats are often indicative of stressful environments.

For each population, assessed plant population performance was estimated using an annualized lambda (λ_{ann}) value following Akçakaya et al.[27];

$$\lambda_{ann} = \left[(N_{t+y} + 1) \ / \ (N_t + 1)\right]^{1/y} \quad (1)$$

where N_t is the first population estimate, N_{t+y} is the second population estimate, and y is the number of years between the estimates. Because individual locations may have population estimates from more than two years, each location may have multiple annualized λ estimates. A growing population is indicated by $\lambda_{ann}>1.0$, while values less than one indicate a declining population. Owing to a strong skew (a small fraction of observations experience very large population change) in the resulting population growth measures, we transformed these measures by taking a natural log of the annual growth measure. We assessed λ_{ann} in three ways; using: i) all observed population changes (n = 1,682), ii) all individual populations (n = 795), and iii) the average of repeated estimates of population changes just for locations with multiple estimates (n = 341).

We categorized plant survey data by the specificity of the population estimate as follows: i) exact numbers less than 200; ii) exact numbers greater than 200 that likely represents a count or a close estimate (e.g., 1,347); or iii) round numbers that likely reflect an order of magnitude estimate (e.g., 1,000's). Exactness of a population estimate is likely strongly correlated with population size (e.g., small populations might be counted, large ones are estimated). Inexact estimates, however, can lead to spurious estimates of population change. For example, the difference between 5 and 10 individuals is likely to be exact and accurate while estimates of a large population may vary substantially among estimators. Hence, there is a trade-off between precision and bias in population data, so we tested for response differences in λ_{ann} depending on estimate specificity and initial population size.

Human density was assessed using census tracts as defined by the US Census Bureau statistics from the 2000 United States population census [28]. Census tracts are long-term county subdivisions containing between 2,500 and 8,800 people. We divided the census tract population by the area of each tract to get a human density (people/km^2) for each plant location. Human density varied from zero (uninhabited islands of the coast) to 12,217 people/km^2. Median density was 40.2 people/km^2. Urban centers are often defined as having population densities of upwards of 400 people/km^2 [29]. Our distribution of human densities was right-skewed, with 24% of observations from urban locations (>400 people/km^2). We used the natural log of human density to partially normalize this distribution from which to predict performance and also used a non-parametric goodness-of-fit test, dividing the data into groups based on human density and λ_{ann}.

We examined the relationship between λ and human density across the entire data set, by growth form (annual, perennial herbaceous plant, and woody plants), by habitat types (n = 19), and individually for the 25 species represented by 20 or more λ_{ann}

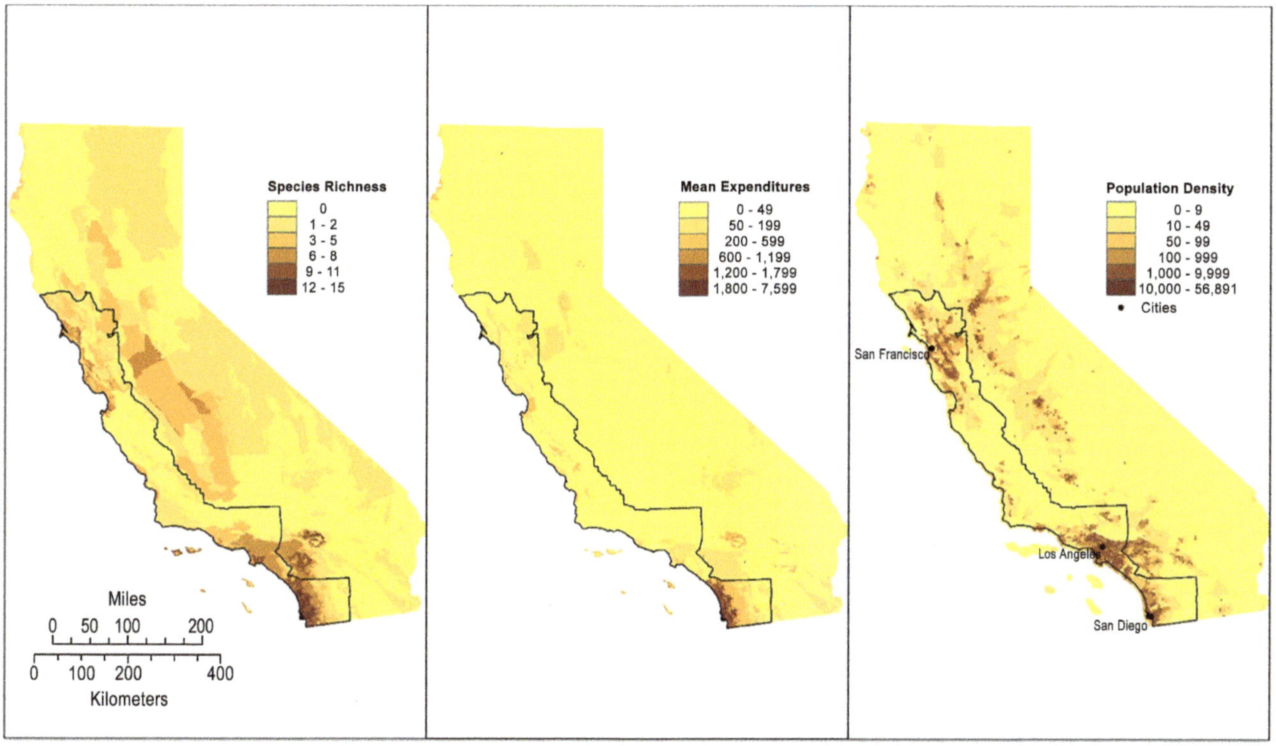

Figure 1. Maps of California highlighting population and expenditure attributes. The outline represents the study area for examining λ_{ann} relative to human density. Color shades represent: (A) species richness of federally listed plant species in California (number of species per square kilometer); and (B) mean annual expenditures for federally listed plant species (dollars per square kilometer per year). Locations of major California cities are included for reference on map (C).

estimates. We used the Human Threat Hypothesis (HTH) of Lawson et al. [23] as a basis for assessing whether λ_{ann} values decline with increasing human density. We expand beyond the work of Lawson et al [23] to test the the relationship between λ_{ann} and human density for specific growth forms (e.g., annual plants) and habitat types (e.g., wetlands). Our alternative hypothesis is that the variability in growth across growth form and habitat type to determine if some growth forms (e.g., annuals) exhibit more variability in population size. In general, greater variability in population size is expected in annuals than in perennial herbs or woody plants. In addition, we assessed whether mean growth rate is lower for particular growth forms or habitat types that are particularly prone to environmental degradation, (e.g., wetlands). This test serves to assess the sensitivity of the data to detect change. Finally, we combined the HTH with a null hypothesis of no difference among habitat types to test the hypothesis that relationship of λ_{ann} to human density does not differ among habitat types.

Urban populations change through time and densities estimated by the 2000 census may reflect impacts at the time of the CNDDB plant surveys with variable accuracy. Therefore, we tested for a relationship between λ_{ann} and human density using the data for the time period across which plant population change was estimated as a covariate. Our analysis consists of linear models of log transformed λ_{ann} as a function of growth form and habitat type, and log transformed human density and interactions among these variables. All analyses were done using JMP 9.0 (SAS institute).

Separately, we sought to understand conservation support across the gradient of high to low population density. We aggregated all recovery expenditures on behalf of 175 listed plant species in California using three years (2006–2008) of reported expenditures [30–32]. Eight species were not included as a consequence of recent listing action or taxonomic change. These 175 species represent over 95% of the total three-year endangered species expenditures. These expenditures are spent largely on managing populations and not on habitat acquisition, which is tracked separately. A separate treatment by Underwood et al. [33] reported on spending on behalf of acquisition, and is discussed in the context of our analysis. With a far smaller number of listed (n = 183) [34] rather than tracked rare species (n>2,000) [26], the geographic distribution of federally listed plant species (Fig. 1a) is less coastal than the distribution of California endemic plants [6].

We used CALJEP [35], a geospatial database of plant species distributions in California, to identify the distribution of these 175 federally listed species. We placed occurrences within a 1 km^2 grid over the entire state. We counted polygons identified as "present" and "probably present" as occurrences, and "possible" and "not recorded" as absences [35]. We averaged the per 1 km^2 expenditures across the three years based on an assumption that expenditures were evenly distributed across the range of a species. Most listed species are narrow endemics found in a very small portion of the state [4,6], hence we find this an acceptable abstraction of the geographic distribution of spending on behalf of listed species. We summed the total expenditures within grid cells based on all listed species occurring in each grid cell when more than one species were found in a cell. We then mapped human density, based on census tract numbers, at the same spatial scale across the entire state, and estimated the relationship between expenditures and density at this 1 km^2 grid cell scale for our coastal study region. In addition, we summed expenditures within census tracts and compared expenditures by human density at the census tract scale for the coastal counties for which we assessed λ_{ann} .

Finally, we also analyzed these data treating species as replicates. Here, we compared the average human density in all tracts in which a species occurs to the total three years of expenditures on behalf of the species. Results were similar using either species or census tract as the unit of observation. We focused on the data that depict spending by census tract as it relates more directly to our geographic depiction of the distribution of spending. All variables were log transformed to better fit the expectation of normality.

Results

Plant population growth by human density

Our results agree with those of Lawson et al. [17] in finding no relationship between human population density and λ_{ann} ($n = 1,682$, $F = 0.038$, $p = 0.85$, coefficient $= -0.003$) (Table 1, Fig. 2). This strongly non-significant result persists irrespective of how we assessed the data including across parametric versus nonparametric tests and all subsets of data based on sampling, sampling date, or sample specificity (Table 1). We found no bias in the examination of residuals for any of these tests. The effect size of human density on λ_{ann}, as estimated by 1,000 replicates of a randomized bootstrap sample of 1,000 observations, was nearly zero (mean correlation coefficient $= -0.006$; s.d. $= 0.0235$).

We observed no effect of year on λ_{ann} ($n = 1,682$; $F = 0.42$; $p = 0.84$). There was a slight positive relationship between initial plant population size and human density ($n = 1,682$; $F = 10.3$; $p = 0.001$; coefficient $= 0.095$) and a negative correlation between λ_{ann} and initial population size ($n = 1,682$; $F = 141.7$; $p < 0.001$; coefficient $= -0.154$). This combination of results should make it more likely to find an adverse impact of human population on λ_{ann}, yet we do not find this to be the case.

We also conducted a goodness-of-fit test on data classified into categories based on human density and λ_{ann}. Human density

Table 1. Plant population growth response as predicted by human density using 13 different tests of the relationship performed to assure consistency of results depending on how we treated (A) repeat sampling within a plant population; (B) violations of normal distributions driving non-parametric considerations; (C) initial population size and the ease of gaining an accurate population assessment; (D) when the plant population was assessed; and (E) growth form.

Criteria	N	P	Coefficient
A. Unit of observations			
All population change estimates	1682	0.845	−0.0032
Average of observations from each population	795	0.519	−0.0083
Average of observations with multiple estimates	341	0.596	−0.008
B. Nonparametric correlation*			
All population change estimates, Kendall's tau	1682	0.4	−0.0014
All population change estimates, Spearmans p	1682	0.406	−0.0203
C. Population estimate specificity			
Population size less than 200	654	0.613	−0.011
Exact estimate, population larger than 200	504	0.51	0.232
Rounded number estimate, population >200	521	0/296	−0.032
D. Population count end date			
Population change end date prior to 1990	446	0.221	−0.0435
Population change end date 1990 or later	1234	0.527	0.0116
E. Life Form			
Annual herbaceous plants	902	0.814	0.0067
Perennial herbaceous plants	660	0.857	0.0029
Woody trees and shrubs	118	**0.046**	−0.0498

*-nonparametric versions of most of correlations by population for varying units of observations (A) and estimate specificity (B) were similarly, not significant and had correlations close to 0.

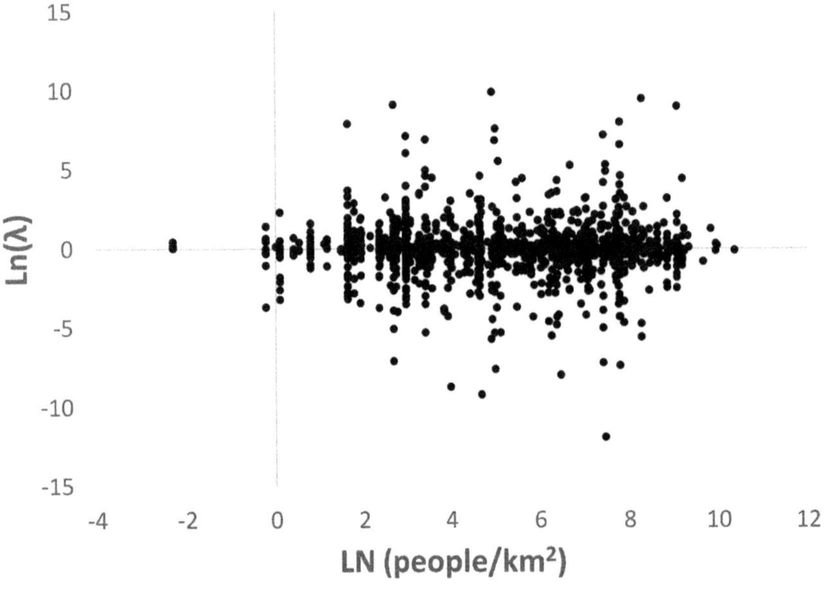

Figure 2. A scatterplot of the relationship between the natural log of annualized plant population growth (λ_{ann}) and human density (people/km^2). The plot shows no relationship between human density and λ_{ann}.

classes consisted of rural (<40 people/km^2, n = 785), peri-urban (40–400 people/km^2, n = 488), and urban (>400 people/km^2, n = 409). We classified λ_{ann} into shrinking ($\ln(\lambda_{ann}) < -0.2$; n = 560), stable ($-0.2 < \ln(\lambda_{ann}) < 0.2$, n = 595) and growing ($\ln(\lambda_{ann}) > 0.2$, n = 520) populations approximately by equal frequency categories. Goodness-of-fit test results were significant (chi square = 10.79; df = 4; p = 0.029; Table 2). The pattern of this significance suggests an increase in variability of plant growth responses in urban environments. We found more observations of both λ_{ann} growth and retraction in urban environments and fewer than expected stable transitions ($-0.2 > \lambda_{ann} > 0.2$). Conversely, the lowest human density sites exhibited fewer population declines and more stable transitions than expected (Table 2). We further examined this relation bycreating a more stringent criteria for growing and shrinking populations, where values between −1 and +1 were considered stable, and larger values were characterized as collapse ($\ln(\lambda_{ann}) < -1.0$; n = 218) or growth ($\ln(\lambda_{ann}) > 1$; n = 226). Although the general pattern observed above remained, the test result was not significant (chi-square = 4.46; p = 0.35)

Given that the chi-square tests suggest that urban populations may be more variable than rural ones, and that variability can lead to decreased persistence likelihoods, we further assessed variability in λ_{ann}. We restricted our assessment to observations with at least five repeat population change estimates (six survey dates). We found no relationship between the coefficient of variation (standard deviation/mean) as a function of human density (n = 80; F = 0.43; p = 0.51). An alternative explanation for increased variability of λ_{ann} in urban environments could be a differential distribution of life forms, with annual herbaceous plants found, on average, in higher human densities. This, in fact, is the case (one-way ANOVA, n = 1,682, df = 2, F = 25.7; p<0.001). Half (50.4%) of all observations were annual plants, yet 58.7% of urban λ_{ann} were from annual plants. These effects, however, are small. A linear model predicting λ_{ann} by growth form and human density yields no significant effects or interactions.

Assessing λ_{ann} by growth form indicated no differences among life forms in their population responses to human density for annual or perennial herbaceous plants (Table 1). Woody trees and shrubs, however, exhibited a significant decrease in λ_{ann} with increasing human density (Table1). This is the single significant result we observed relating λ_{ann} to human density. Examining the distribution of life form across human density showed a significant difference (ANOVA; n = 1,682, F = 25.70; p<0.001) with woody plants being found at significantly lower human densities.

Finally, we assessed λ_{ann} by species and by habitat type. We found no significant (p<0.1) patterns with human density, and observed an even balance of those with positive (n = 11) and negative (n = 11) coefficients relative to human density for the 22 species with 20 or more λ_{ann} estimates (Table 3). We constructed independent tests of λ_{ann} versus human density for 19 habitat types. Again, we found no significant (p = 0.1) correlations in either direction (Table 4). Further, we achieved the same results when we reduced the number of data points by removing data of lower overall quality and restricting estimates to those data with specific estimates of the number of plants (Table 4). We observed no effect of human density, nor an interaction of habitat type by population. We did not correct for family error rates in either case as not a single test was significant even at a 0.10 level. There were considerable differences in the distribution of habitat types with respect to human density, but no relationship between the distribution of human densities sampled within a habitat type and the overall mean performance for that habitat type (Table 4).

In summary, we found no support for the hypothesis that increasing human density results in decreasing plant performance other than for the case of woody plants. Among the 19 habitat types assessed, mean λ_{ann} varied with 12 habitat types exhibiting a positive net growth rate and seven habitat types with an average negative growth rate. No individual habitat type showed significant variation in λ_{ann} across a gradient of human density within habitats (Table 4).

Table 2. A contingency table of λ_{ann} relative to human density showing observed numbers of populations within each class and the expected numbers parenthetically; goodness-of-fit likelihood ratio = 10.8 (p = 0.03).

	Plant population change		
Human Population	**Shrink**	**Stable**	**Grow**
0–40 people/km^2	245 (261.4)	298 (277.7)	242 246.0)
40–400 people/km^2	110 (109.2)	125 (116.0)	93 (102.8)
>400 people/km^2	205 (189.4)	172 (201.3)	192 (178.3)

Plant Endangered Species Expenditures by Human Population Density

Our assessment of endangered species expenditures shows that there is a weak, but positive, relationship between the natural log of human density and the natural log of endangered species expenditures using both census tract (coefficient = 0.186;

Table 3. Correlations of λ_{ann} with human density for the 22 species with 20 or more individual observations.

Species	n	F	P	Coefficient
Acanthomintha ilicifolia	38	0.37	0.55	0.074
Amsinckia grandiflora	21	0.848	0.37	−0.167
Astragalus brauntonii	23	1.003	0.33	0.142
Blennosperma bakeri	27	0.027	0.87	−0.039
Camissonia benitensis	35	0.001	0.97	0.009
Clarkia franciscana	66	0.89	0.35	0.347
Cordylanthus maritimus ssp. *maritimus*	20	0.045	0.84	−0.063
Cordylanthus maritimus ssp. *palustris*	57	2.71	0.11	−0.101
Cordylanthus mollis ssp. *mollis*	26	0.032	0.86	0.045
Dichanthelium lanuginosum var. *thermal*	30	0.182	0.67	−0.049
Dudleya multicaulis	25	0.03	0.86	0.01
Dudleya setchellii	11	1.82	0.4	0.063
Fritillaria liliacea	54	0.126	0.72	−0.02
Gilia tenuiflora ssp. *arenaria*	22	0.297	0.59	−0.103
Helianthella castanea	20	1.77	0.2	0.219
Hesperolinon congestum	26	0.013	0.91	0.012
Holocarpha macradenia	101	0.552	0.46	0.083
Lasthenia conjugens	24	0.1	0.76	−0.184
Lupinus tidestromii	21	0.047	0.83	−0.055
Pentachaeta lyonii	34	0.566	0.46	0.083
Phacelia insularis var. *continentis*	26	0.001	0.97	−0.009
Triphysaria floribunda	35	0.301	0.59	−0.049

No significant (p<0.1) correlations and equal numbers of positive and negative correlation coefficients were observed.

Table 4. Summary statistics for correlation of population performance (natural log of annualized mean λ) with human density.

Habitat	Population Growth Mean ln(ë)	All Observations n	coefficient	p	Highly specific estimates n	coefficient	p	Distribution of occurrences by human population density Median People/km2	F	p
Conifer Forest	0.272	60	−0.014	0.81	49	−0.087	0.25	14.1	6.57	**0.011**
Alkali Sink	0.191	35	−0.08	0.68	18	−0.116	0.84	13.8	3.32	0.069
Desert	0.179	29	−0.008	0.9	22	−0.009	0.92	5.7	13.3	**0.003**
Deciduous Woodland	0.111	10	−0.001	0.97	8	0.026	0.63	35.3	0.902	0.342
Vernal Pool	0.103	162	0.04	0.64	54	−0.013	0.21	61	1.51	0.219
Sand Dunes	0.075	194	0.003	0.94	162	0.012	0.78	40	0.316	0.574
Freshwater Wetland	0.074	208	−0.024	0.56	147	−0.009	0.86	7.8	26.23	**0.0001**
Ocean Bluffs	0.069	64	−0.021	0.66	54	0	0.99	6.2	19.51	**0.0001**
Serpentine	0.04	300	−0.01	0.76	249	0.004	0.83	59	9.09	**0.003**
Sagescrub	0.012	249	0.032	0.48	215	0.034	0.47	38.6	7.04	**0.008**
Chaparral	0.011	288	0.016	0.68	227	0.046	0.92	59.3	2.04	0.154
Woodland	−0.015	161	−0.037	0.3	114	−0.026	0.5	20.4	34.63	**0.0001**
Riparian	−0.02	48	−0.03	0.57	16	0.052	0.56	7.8	2.69	0.101
Grassland	−0.03	752	0.005	0.83	511	0.033	0.26	59.3	6.71	**0.01**
Rocky Slopes	−0.035	104	0.007	0.78	77	0.017	0.57	40	0.351	0.553
Other (stress)	−0.104	60	0.047	0.31	51	0.047	0.4	7.4	0.455	0.5
SaltMarsh	−0.184	154	−0.094	0.16	91	−0.138	0.14	7.2	3.31	0.07

Each regression was conducted on two subsets of data: all observations; and data that meet our data quality criteria of having high precision. The key point to note is that there are no significant relationships. Also of note are the differences in mean λ_{ann} by habitat. Habitats are ranked from the highest mean λ_{ann} to the lowest (left-most data column). Many individual habitats were found in significantly more urban or rural environments (right columns), but the overall habitat performance was not predicted by either more urban or rural distributions. Values <0.05 are in bold face.

$r^2 = 0.029$; $p<0.001$ $n = 5{,}064$) and species (coefficient $= 0.112$; $r^2 = 0.025$; $p<0.039$; $n = 171$) as the replicate. We tried several additional transformations but the results did not substantively vary by transformation. Investigating the effect of range size on expenditures suggest that species with larger distributions have slightly more expenditures and that controlling for this effect reverses the effect of human density on expenditures (coefficient $= -0.17$; $r^2 = 0.037$; $p<0.01$ $n = 171$).

This coarse evaluation masks important detail, as San Diego County receives far more endangered species recovery expenditures than any other county in our study region (or likely, in the US) [26–28] (Table 5, Fig. 1). This county contains both regions of high and low human density. Excluding San Diego County actually increases both the coefficient and the fit of the positive relationship between human density and spending among census tracts (coefficient $= 0.234$; $r^2 = 0.081$; $p<0.001$ $n = 4{,}459$).

Among the 17 counties we included in this study, the relationship between human density and spending was significant in 10 counties, of which seven (Contra Costa, Los Angeles, Monterey, San Diego, San Francisco, Santa Cruz, Sonoma) were positive and three (Alameda, Orange, Santa Clara) were negative (Table 5). At a coarser scale, counties with very high average human densities tended to be on the highest end of expenditures (San Diego, Los Angeles Counties) or toward the low end of expenditures (Orange, Santa Clara Counties) with no apparent relationship between spending and human density.

Discussion

Our results both support and strengthen the conclusions of Lawson et al. [23] that plant populations perform equally well across the gradient from rural to urban locations. Using a larger dataset over a more extensive coastal region, and numerous additional analyses, we find no evidence that at-risk plant populations perform poorly in areas with high human density. Out of 65 tests of λ versus human population density, only a single result was significant. That result suggests that woody plants have lower average growth rates in areas of high human density than they do in more rural areas. However, this result is brought into question by the observation that only 16 of our 118 observations were from urban (>400 people/km) locations, and these represented just 6 of 35 woody species. Among the four woody species found in both urban (>400 people/km^2) and non-urban (<400 people/km^2) areas, three actually had higher average growth rates in urban populations than in their rural ones. In addition, this negative correlation is not driven by low λ_{ann} values in urban populations, but high λ_{ann} values in rural populations (Fig. 3).

The results of our study strongly suggest a potential for successful plant conservation efforts within the urban environments of California. We do not mean to imply, however, that populations of at-risk plants are not threatened by urban environments. Habitat loss, and the extirpation of populations, remains a critical issue. Our data suggest that protecting these populations from habitat loss may provide opportunities for protection of these at-risk biological resources. By focusing on those populations that remained extant across survey periods, we

Table 5. County level statistics of human population for coastal California from the San Francisco Bay metropolitan region southward and spending on behalf of federally listed endangered and threatened plant species from 2006–2008.

COUNTY	POPULATION	Mean $/km²	Coefficient	P
Alameda	1,443,741	57.47	−0.003	0.04
Contra Costa	948,816	58.87	0.008	0.0003
Los Angeles	9,453,140	85.69	0.002	<0.0001
Marin	246,104	90	−0.001	0.518
Monterey	407,907	372.35	0.035	0.0001
Napa	119,901	8.36	−0.003	0.133
Orange	2,852,710	64.61	−0.003	0.0019
San Benito	40,838	4.8	−0.003	0.157
San Diego	3,056,509	1591.99	0.012	0.0003
San Francisco	790,240	128.24	0.001	0.0004
San Luis Obispo	207,490	108.03	0.022	0.18
San Mateo	708,709	131.97	0.037	0.013
Santa Barbara	495,933	16.49	−0.001	0.435
Santa Clara	1,675,965	19.66	−0.003	<0.001
Santa Cruz	250,245	80.95	0.004	0.008
Solano	394,542	67.99	0.003	0.565
Sonoma	458,614	51.02	0.012	<0.0001
Ventura	806,420	50.99	0.002	0.412

Human density (people/km²) estimates are from the 2000 US census, summarized in the California Department of Finance (www.dof.ca.gov). Expenditure estimates are from the US Fish and Wildlife endangered and threatened species expenditure reports for 2006 through 2008. Expenditures are reported by species and we used species distribution maps to identify species within counties and averaged species expenditures within census tracts across distributions. Regression coefficients and p-values are for human density versus the expenditures by census tract.

found that λ_{ann} was no different across the spectrum from rural to urban environments.

A goal of plant conservation should be to protect viable populations over long time periods. The CNDDB data do not provide the capacity to conduct viability assessments using these data. Our observation of slightly greater rates of inter-annual variability for plants in areas of high human density, and high inter-annual variability generally, may result in higher long-term vulnerability of urban populations. Analyzing magnitude of this effect is beyond the scope of this project and the capacity of these data.

Our results expand on prior results. We consider variability in response in a variety of attributes, including habitat, growth form and time period, analyzing the CNDDB data in far more detail than previous studies. Second, we distinguish among growth forms and habitat types to assess performance across the gradient from rural to urban habitats. Our results suggest that specialized habitat types, such as alkali sinks, vernal pools and desert, may be more resilient to threats associated with urbanization as demonstrated by larger average λ values than other habitat types (Table 4). If homogenization of urban floras is driven by invasive species [15,16], then it stands to reason that habitat types that are less prone to invasion may be more resilient in urban systems. This does not help explain the relatively large net positive growth rates among species found in coniferous or deciduous forests, but may help explain the generally negative mean growth rates for riparian, grassland and salt marsh species (Table 4).

An alternative possible interpretation of the Lawson et al. [23] and these results is that the CNDDB data are not sufficiently specific and detailed to assess differences in λ_{ann}. The CNDDB data are haphazardly collected by a broad and diverse suite of individuals across a long timeframe using different methods. Our finding—that habitat types differ in performance in ways that we would predict based on habitat resilience—is encouraging because it supports the assertion that these data provide a signal of performance that can be assessed. If there were no habitat type

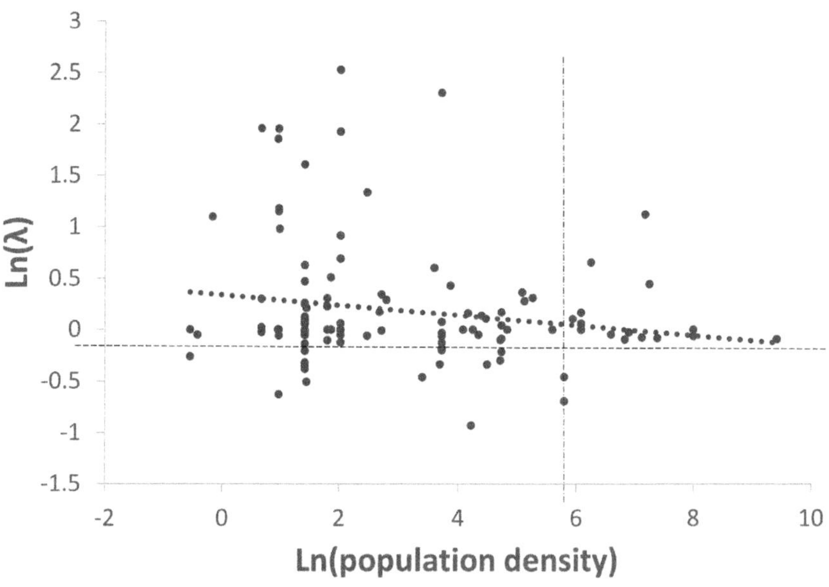

Figure 3. The correlation between human population density and plant population growth (λ) for woody plants. The only significant correlation observed between human density and plant performance (λ_{ann}) was for woody plants ($n = 118$, $F = 4.06$; $p = 0.046$; coefficient $= -0.050$). However this relationship is driven by high λ_{ann} in rural areas as opposed to strong negative λ_{ann} in urban settings. The horizontal dashed line represents no population change, the vertical dashed line separates urban (right) from non-urban (left) populations, with the dotted line representing the best fit correlation.

differences, then we might conclude that the data are simply too coarse to distinguish performance. However, since λ_{ann} values differ across habitat types, and those habitat types that are less prone to degradation through invasive species appear more resilient, this provides evidence that our data are, in fact, informative.

Finding no relationship between λ_{ann} and human density, we reinforce the assertion by Lawson et al. [23] that plant performance within protected habitats is not diminished simply by virtue of having a high local human density. This is encouraging given the significant conservation investment in and around major California urban areas [33]. Our study was not designed to determine why spending is allocated differentially across urban areas, with San Diego County receiving far more recovery funding than other California urban centers of high rare plant density. Nevertheless, we find it likely that some of the differential spending in California is driven by systematic conservation planning through tools such as the state's Natural Community Conservation Planning (NCCP) [36] and the Federal Habitat Conservation Planning (HCP) [37] .

Our geographic treatment of endangered species recovery expenditures is, necessarily, an abstraction of real expenditures. We do not know exactly where money was applied or to which species. However, our intent is to present an overview of the geography of federal expenditures in support of plant conservation, and not to assess the effectiveness of those reported expenditures.

Examining this result geographically, there is a complex relationship between human density (Fig. 1c), listed species density (Fig. 1a), and expenditures (Fig. 1b). Rural and medium density regions appear to uniformly receive little support for their endangered species (Fig. 1b, 1c). Similarly, the San Francisco and Los Angeles metropolitan regions have high human densities, high rare species richness, and modest expenditures (Fig. 1). In contrast, the San Diego region, with high human density, is rich in listed species and these species garner high levels of financial support (Fig. 1). This region appears unique in that respect. We speculate that the plants of San Diego County garner more funding because of the extensive multi-species conservation plans in place within the region [38].

The US Endangered Species expenditure data suggest that there are significant resources being applied to urban conservation in San Diego County, but not generally across the region. Underwood et al. [33], analyzing land protection expenditures in California, also found a large focus on spending along the urbanized coast of California. Schwartz et al. [4] made the case that providing for protection in urban environments can have collateral positive impacts on conservation through social engagement in the process of conservation. Whether by chance or design, evidence that λ_{ann} does not diminish with human density provides an endorsement of conservation expenditures in California. This is simply because such a large percentage of the unique flora of California is isolated to coastal, and often urban, regions of the state.

If plants are generally in need of protection, and human dominated landscapes are both rich in populations of rare species and under the most severe immediate threat, then it is sensible to skew investment toward these urban areas (San Francisco, Los Angeles, San Diego). However, this would be a poor investment of conservation resources if there were evidence that protected populations in urban environments were less likely to persist. Our study helps to assure conservation managers that these populations, once protected, do not appear to be at undue risk simply by virtue of urban proximity.

The case for investing in urban plant conservation [4] is further strengthened by examining extinction processes in urban environments [11,14,39]. The California Native Plant Society (CNPS) maintains a database on rare and endangered plants. This list includes 27 species considered extirpated in California [26]. Among these 27 extirpated species, 12 were formerly found within our study region. Among these 12 regional extirpated species, only two (*Ribes divericatum* var *parishii*, 1980, Los Angeles and San Bernardino counties; *Castilleja uliginosa*, 1987, Sonoma county) have been observed since 1954. In other words, just two of 12 documented extirpations have occurred since the onset of modern conservation measures such as the Endangered Species Act. These data do not indicate how many populations of plants may have been lost through time, nor do they report on individual county extirpations, but they do give an overall indication that California, with a high number of narrow endemics (defined as occurring in 1–2 counties [23]), has lost very few species through 30 years of conservation management. Given these arguments, we maintain that conservation investment in California urban centers can successfully protect rare plants over the medium term and provide strong incentives for local conservation value. Given the parallel conservation need to engage the urban population in protecting nature [21], investment of limited resources in urban plant conservation appears as likely as any to succeed.

Acknowledgments

We thank D. Lawson for her assistance and advice. We thank K. White and R. Bittman from the CA Department of Fish and Wildlife for their assistance with the CNDDB.

Author Contributions

Conceived and designed the experiments: MWS LMS ZLS. Performed the experiments: LMS ZLS. Analyzed the data: LMS ZLS MWS. Wrote the paper: MWS LMS ZLS.

References

1. Moilanen A, Anderson BJ, Eigenbrod F, Heinemeyer A, Roy DB, et al. (2011) Balancing alternative land uses in conservation prioritization. Ecological Applications 21: 1419–1426.
2. von der Dunk A, Gret-Regamey A, Dalang T, Hersperger AM (2011) Defining a typology of peri-urban land-use conflicts - A case study from Switzerland. Landscape and Urban Planning 101: 149–156.
3. Seabloom EW, Dobson AP, Stoms DM (2002) Extinction rates under nonrandom patterns of habitat loss. Proceedings of the National Academy of Sciences of the United States of America 99: 11229–11234.
4. Schwartz MW, Jurjavcic NL, O'Brien JM (2002) Conservation's disenfranchised urban poor. Bioscience 52: 601–606.
5. CDFG (2003) Atlas of the Biodiversity of California. Sacramento: California Department of Fish and Game.
6. Loarie SR, Carter BE, Hayhoe K, McMahon S, Moe R, et al. (2008) Climate Change and the Future of California's Endemic Flora. Plos One 3.
7. Ando A, Camm J, Polasky S, Solow A (1998) Species distributions, land values, and efficient conservation. Science 279: 2126–2128.
8. Schwartz MW (1999) Choosing the appropriate scale of reserves for conservation. Annual Review of Ecology and Systematics 30: 83–108.

9. Gibb H, Hochuli DF (2002) Habitat fragmentation in an urban environment: large and small fragments support different arthropod assemblages. Biological Conservation 106: 91–100.
10. Radeloff VC, Hammer RB, Stewart SI, Fried JS, Holcomb SS, et al. (2005) The wildland-urban interface in the United States. Ecological Applications 15: 799–805.
11. Williams NSG, Schwartz MW, Vesk PA, McCarthy MA, Hahs AK, et al. (2009) A conceptual framework for predicting the effects of urban environments on floras. Journal of Ecology 97: 4–9.
12. McKinney ML (2002) Urbanization, biodiversity, and conservation. Bioscience 52: 883–890.
13. Bastin L, Thomas CD (1999) The distribution of plant species in urban vegetation fragments. Landscape Ecology 14: 493–507.
14. Hahs AK, McDonnell MJ, McCarthy MA, Vesk PA, Corlett RT, et al. (2009) A global synthesis of plant extinction rates in urban areas. Ecology Letters 12: 1165–1173.
15. McKinney ML (2006) Urbanization as a major cause of biotic homogenization. Biological Conservation 127: 247–260.
16. Schwartz MW, Thorne JH, Viers JH (2006) Biotic homogenization of the California flora in urban and urbanizing regions. Biological Conservation 127: 282–291.
17. Dobson AP, Rodriguez JP, Roberts WM, Wilcove DS (1997) Geographic distribution of endangered species in the United States. Science 275: 550–553.
18. Dallimer M, Irvine KN, Skinner AMJ, Davies ZG, Rouquette JR, et al. (2012) Biodiversity and the Feel-Good Factor: Understanding Associations between Self-Reported Human Well-being and Species Richness. Bioscience 62: 47–55.
19. Fuller RA, Irvine KN, Devine-Wright P, Warren PH, Gaston KJ (2007) Psychological benefits of greenspace increase with biodiversity. Biology Letters 3: 390–394.
20. Miller JR, Hobbs RJ (2002) Conservation where people live and work. Conservation Biology 16: 330–337.
21. Schwartz MW (2006) How conservation scientists can help develop social capital for biodiversity. Conservation Biology 20: 1550–1552.
22. West P, Igoe J, Brockington D (2006) Parks and peoples: The social impact of protected areas. Annual Review of Anthropology. pp. 251–277.
23. Lawson DM, Lamar CK, Schwartz MW (2008) Quantifying plant population persistence in human-dominated landscapes. Conservation Biology 22: 922–928.
24. CNDDB (2009) Californa Natural Diversity Database. In: Game BDBDoFa, editor.
25. Press DM (2002) Saving Open Space: The Politics of Local Land Preservation In California. Berkeley: University of California Press.
26. CNPS (2011) Onlne Inventory of Rare and Endangered Plants. In: Society CNP, editor. 8 ed. Sacramento: California Native Plant Society.
27. Akcakaya HR, Burgman MA, Ginzberg LR (1999) Applied population ecology: principles and computer exercises. Sunderland Massachusetts: Sinauer Associates.
28. Bureau" USC Population Statistics. Washington DC: US Census Bureau.
29. Nations" U (2005) Demographic Yearbook 2005. In: Division P, editor. The Hague: United Nations.
30. USFWS (2006) Federal ad State Endangered and Threatened Species Expenditures; Fiscal years 2005 and 2006. In: Interior UDf, editor. Washigton DC: US Fish and Wildlife Service.
31. USFWS (2007) Federal and State Endangered and Threatened Species Expenditures; Fiscal year 2007. In: Interior Do, editor. Washington DC: Department of Interior.
32. USFWS (2008) Federal and State Endangered and Threatened Species Expenditures. In: Interior Do, editor. Washington DC: Department of Interior.
33. Underwood EC, Klausmeyer KR, Morrison SA, Bode M, Shaw MR (2009) Evaluating conservation spending for species return: A retrospective analysis in California. Conservation Letters 2: 130–137.
34. USFWS (2009) Environmental Conservation Online System. Washington DC: United States Fish and WIldlife Service.
35. Viers JH, Thorne JH, Quinn JF (2006) CalJep: a spatial distribution database of CalFlora and Jepson plant species. San Francisco Estuary and Watershed Science. Davis, CA: eSchlolarship, University of California.
36. Feldman TD, Jonas AEG (2000) Sage scrub revolution? Property rights, political fragmentation, and conservation planning in Southern California under the federal Endangered Species Act. Annals of the Association of American Geographers 90: 256–292.
37. Thomas CW (2001) Habitat conservation planning: Certainly empowered, somewhat deliberative, questionably democratic. Politics & Society 29: 105–130.
38. Franklin J, Regan HM, Hierl LA, Deutschman DH, Johnson BS, et al. (2011) Planning, implementing, and montoring multi-species habitat conservation plans. American Journal of Botany 98: 559–571.
39. Duncan RP, Clemants SE, Corlett RT, Hahs AK, McCarthy MA, et al. (2011) Plant traits and extinction in urban areas: a meta-analysis of 11 cities. Global Ecology and Biogeography 20: 509–519.

8

Short-Term Dispersal Response of an Endangered Australian Lizard Varies with Time of Year

Mehregan Ebrahimi[1,2]*, C. Michael Bull[1]

1 School of Biological Sciences, Flinders University, Adelaide, South Australia, Australia, **2** Department of Biology, Shiraz University, Shiraz, Fars Province, Iran

Abstract

Dispersal is an important component in the demography of animal populations. Many animals show seasonal changes in their tendency to disperse, reflecting changes in resource availability, mating opportunities, or in population age structure at the time when new offspring enter the population. Understanding when and why dispersal occurs can be important for the management of endangered species. The pygmy bluetongue lizard is an endangered Australian species that occupies and defends single burrow refuges for extended periods of time, rarely moving far from the burrow entrance. However, previous pitfall trapping data have suggested movement of adult males in spring and of juveniles in autumn of each year. In the current study we compared behaviours of adult lizards each month, over the spring-summer activity period over two consecutive field seasons, to provide deeper understanding of the seasonal dispersal pattern. We released adult pygmy bluetongue lizards into a central area, provided with artificial burrows, within large enclosures, and monitored the behaviour and movements of the released lizards over a four day period. There was a consistent decline in time spent basking, amount of movement around burrow entrances, and rates of dispersal from the central release area from early spring to late summer. Results could be relevant to understanding and managing natural populations and for any translocation attempts of this endangered lizard species.

Editor: Adam Stow, Macquarie University, Australia

Funding: There are no current external funding sources for this study. This research was supported by funding from the Australian Research Council: Linkage grant LP0883495 "Translocation as a conservation strategy for the Pygmy Bluetongue Lizard." Partner organizations who also provided funding to the same grant were Zoos SA (http://www.zoossa.com.au), the SA Department of the Environment, Water and Natural Resources (http://www.environment.sa.gov.au), the Field Naturalists Society of SA (http://www.fnssa.org.au), the SA Museum (http://www.samuseum.sa.gov.au), the Northern and Yorke NRM Board and the SA Murray-Darling NRM Board (http://www.naturalresources.sa.gov.au). Other research student grants were awarded to Mehregan Ebrahimi by the Holsworth Wildlife Research Endowment (http://www.anz.com.au/personal/private-banktrustees/trustees/granting/granting-programs/small-grants/) and the Sir Mark Mitchell Research Foundation (http://www.aetlimited.com.au). The Ministry of Sciences, Research and Technology of Iran (http://www.msrt.ir/) sponsored the PhD studies of Mehregan Ebrahimi. No individuals employed or contracted by any of the funders played any role in study design, data collection and analysis, decision to publish, or preparation of the manuscript.

Competing Interests: The authors have declared that no competing interests exist.

* Email: mehregan.ebrahimi@flinders.edu.au

Introduction

Animal dispersal is triggered by a range of seasonally variable ecological factors [1,2]. Animals may disperse to reduce kin competition (for example during natal dispersal) [3], to reduce the risk of predation [4,5], or to escape from a seasonally harsh environment [6]. They can also disperse to increase the chance of finding an appropriate mate [7–9]. The advantages for an individual disperser are balanced by energetic costs, costs of time spent during dispersal, costs of increased risk from exposure, and costs of lost opportunities that might have been available if the individual had not dispersed [10]. All of these costs might also vary with the time of the year.

In conservation management of endangered species, dispersal can be an important process. Dispersal between small populations allows gene flow and reduces the risks of inbreeding within populations [11,12]. However, for populations that persist in small isolated fragments of previously contiguous habitat, dispersing individuals may risk never locating another suitable habitat patch, or at least may risk prolonged exposure before locating suitable habitat. An imbalance between the number of individuals dispersing from a population that occupies a habitat fragment, and the number migrating into the population could lead to local population decline. Assisted migrations or translocations between isolated populations are a management strategy sometimes suggested to enhance gene flow without the risks of natural dispersal [13]. But the dispersal tendencies of the managed species may then diminish the effectiveness of this strategy if animals released into a population site disperse away from the release area, without contributing to the local population [14]. Thus an understanding of the dispersal behaviour, and how that varies with season, will be integral to management strategies for endangered species.

We investigated dispersal related behaviour at different times of year in an endangered Australian skink, the pygmy bluetongue lizard (*Tiliqua adelaidensis*). Populations of this lizard are now restricted to a few remnant patches of native grassland in a small geographic region in the mid-north of South Australia. This region has a Mediterranean climate, with cool wet winters and hot dry summers, and pygmy bluetongue lizard activity is restricted to the austral spring and summer (September–March) [15]. The species is normally solitary and individual lizards spend most of their time associated with their single entrance, vertical burrows, which have been constructed by mygalomorph and lycosid spiders [15,16]. They use the burrows as refuges from predators and from climatic extremes, and they partially emerge to sit at the burrow entrance to bask and to ambush passing invertebrate prey [15,16]. Although they aggressively defend their burrows from conspecifics, that defence is restricted to a region of less than 15 cm from the burrow entrance [17]. In natural populations they rarely emerge fully from their burrows, and their movements away from the burrow entrance are normally limited to defecation and catching prey. Occasionally adults move to find new burrows, males move to locate female partners in spring, and neonates and juveniles disperse to establish their own burrows in late summer [18,19]. When lizards move from their burrows they become vulnerable to predation from birds and snakes [20], and they risk moving out of the small patches of suitable habitat.

In this study we compared short-term dispersal related behaviours of lizards when they were released into novel burrows at different times within their activity season. We had three aims. The first was to document how dispersal behaviour of adult lizards varied across the lizard activity season under a standard set of conditions. We considered these results would help to interpret data from field populations and to understand the overall dynamics of burrow occupancy. Our second aim was to provide advice to conservation managers of this lizard to allow them to interpret population density surveys, and alert them about times in the season of highest risk from exposure to predators. Our third aim was to gain insights into the potential success of management programs of assisted migration or translocation, where an initial target would be to have released animals remain close to the area where they were released. This target might be more easily achieved during a time of the year when they are more sedentary.

Methods

The data that we use in this paper have already been reported as the control treatments of a series of experimental studies over the austral spring and summer of 2009–2010 and 2010–2011. Those studies investigated how variable conditions such as food availability, vegetation density and burrow density influenced the behaviour of the lizards and their tendency to disperse soon after being moved into a novel burrow. The methods have been previously reported in papers describing this series of studies [21–24]. In summary, we captured four male and four female pygmy bluetongue lizards from each of two populations near Burra, South Australia (33° 42′ S, 138° 56′ E) in September 2009, and used these same 16 lizards in all trials over the next 16 months. We were prevented from using different lizards in different trials by the capture permit conditions for this endangered species. The time over which we ran the trials included two successive spring and summer periods, when these lizards are normally active in the field. Lizards were held in individual plastic boxes (52.5 cm×38 cm×31 cm) before and between trials in a room with ambient conditions, and they were fed crickets and meal worms every third day. After the last trials in the 2009–2010 spring and summer field season, lizards were kept in the Animal Care Unit of Flinders University, Adelaide in a 25°C room with an initial 12:12 LD light regime. We reduced temperature and light gradually to 15°C and 10:14 LD over the austral winter, and then brought light and temperature back to 25°C and 12:12 LD by the following spring for the second field season of our trials.

For the trials, we established four circular cages (15 m diameter) in a line, about 5 m apart in the grounds of Monarto Zoo, South Australia (35° 06′ S, 139° 09′ E). Cages had 1 m high-galvanised iron walls and bird-proof wire roofs. We divided each cage into three areas; a central 4 m diameter circular area with artificial burrows provided where lizards were released, a 5 m wide matrix around the central area, considered to be unsuitable habitat as it did not have any refuge burrows or vegetation, and a ring, 0.5 m wide, around the inside cage perimeter with burrows that trapped any lizards that dispersed from the central area. We constructed artificial burrows from 30 cm lengths of 3 cm diameter wooden dowling with a 2 cm diameter hole drilled out of the centre. Artificial burrows were hammered into the ground until they were flush with the ground surface. Pygmy bluetongue lizards readily accepted these artificial burrows both in field populations [25] and in our cages [26]. We distributed 41 artificial burrows in the central area of each cage as previously described [21], one in the middle, and then 8, 16 and 16 burrows in three concentric rings. In this arrangement, each burrow was spaced an average 63 cm (SE = 0.01) from the next nearest burrow. We also distributed 30 artificial burrows evenly around the perimeter ring of each cage as refuges for lizards that dispersed from the central area. We ran a series of experimental trials during two spring and summer periods, from October 2009 to March 2010 and from October 2010 to January 2011. Each trial lasted four days, and each tested the impact of some experimental manipulation within the central cage area, for instance manipulation of food level [21], of vegetation density [22], or of the arrangement of the burrows [24]. In most trials two cages were used to apply an experimental treatment, and two were used as controls. The comparisons between experimental and control treatments have been documented in previous publications. In this paper we consider only the control treatment cages which were identical in all trials, with burrow conformation as described above, with no additional food supplementation, and with all vegetation in the central area cut to ground level. Normally there were two cages with these control treatments in each trial but sometimes only one. Thus the only differences among the different trials were the month (Oct–March) and field season (2009/10 or 2010/11) when the trials were run, and the increasing experience of the same group of lizards with the trial conditions over successive trials. In total we ran 17 trials across the two field seasons, with 28 control treatment cages considered in this analysis (Table 1).

In each trial we released four lizards in each cage (two males and two females) and confined them to the central area for one day using a 20 cm high black plastic wall [23]. We then removed the wall and observed lizard behaviour and movements for the next four consecutive days. To observe behaviour, we suspended four surveillance cameras (Longse: LICS23Hf, 3.5 mm lens) above each cage with a combined field of view that covered the central 4 m diameter area. Cameras recorded lizard behaviour during each day of trials from 0700 to 1800 h onto a 16 channel h.264 DVR (ESW26), powered by four 12 V batteries.

From our filmed records we derived seven behavioural parameters in each trial that allowed us to compare lizard behaviour among different months. In natural populations, pygmy bluetongue lizards spend most of their time associated with a single refuge burrow, retreating down the burrow to escape predators and climatic extremes, or sitting at the burrow entrance to bask and to ambush passing invertebrate prey [16]. Although lizards, in the field, occasionally emerge fully to bask, to capture prey, or to defecate, they rarely move further than a few centimetres from their burrow entrance [15,23,27]. The behaviours described below relate to this burrow centred focus of activity.

Table 1. The number of trials and the number of cages with the control treatment in each trial in each month of each field season.

	Field seasons			
	2009/2010		2010/2011	
Month	No. trials	Cages/trial	No. trials	Cages/trial
October	1	2	1	1
November	2	2	1	2
December	1	2	1	2
January	2	1	2	2
February	3	1	0	0
March	3	2	0	0

1) Activity time (h d^{-1}) was defined as the period from the first time the lizard head emerged from its burrow to the last time that lizard retreated completely into its burrow on that day. 2) Basking time (min h^{-1}) was defined as the amount of time when the lizard had at least partially emerged and remained at the entrance of its burrow. We called this basking because the lizard was exposed to solar radiation, but an additional function of this behaviour may have been to ambush passing invertebrate prey. We divided the total minutes spent basking each day by 11 (total hours of filming) to calculate the basking time as minutes per hour. 3) Number of movements around burrow. In some cases lizards fully emerged from their burrow, moved about, usually for a very short distance, and then retreated to the same burrow. These movements included lizards that just walked around the burrow entrance, lizards that basked while fully emerged, and lizards that moved away from the burrow entrance for defecation or foraging for prey. We recorded the number of these movements by each lizard on each day. 4) Number of burrow changes. In some other cases lizards fully emerged from their burrow and moved around to choose another burrow in the central area. We recorded the number of burrow changes for each lizard on each day. 5) Distance moved. If a lizard had moved to a different burrow within the central area during a day we measured the distance moved as the direct line distance between the burrow the lizard was in at the start of the day to the burrow it was in at the end of the day. 6) Number of dispersals. We recorded the mean number of cases per day when a lizard left the central area, moved across the habitat matrix, and was subsequently discovered occupying a burrow in the perimeter region. There could be no more than four dispersal events (from the four lizards) on any one day, but sometimes lizards subsequently dispersed themselves back to the central area and could be recorded dispersing again on another day. This behavioural parameter was not recorded in trials in February and March of the first season. 7) Number of fights. When two lizards approached each other on the ground surface, they always showed some agonistic interaction, either with the lizards scuffling together, or with one running away from the other. We defined all of these interactions as fights, and recorded the number of fights per lizard on each day.

In the trials, when a lizard moved into the perimeter area, it left the field of view of the cameras, and we had incomplete information about its behaviour on that day. For analyses, we derived one value of each behavioural parameter from the four lizards in each cage in each trial, using the average over all four days, from all lizards with complete data on each day.

Although the same 16 lizards were used in all 17 sets of experimental trials, in each trial we selected different combinations of four lizards for each of the control cages, and we treated each of the 28 sets of control cage results as independent replicates.

First we explored whether ambient thermal conditions during the trials influenced any of the behavioural parameters. There is an overall tendency for air temperatures to increase from spring to summer in South Australia. We derived average, minimum and maximum temperatures over each four day trial from temperature records at Pallamana Aerodrome weather station (35° 04′ S, 139° 13′ E), 10 km from Monarto Zoo. We then used one way ANOVAs to determine if the mean temperatures experienced differed among months, and Pearson correlations to examine the relationships between those temperatures and the average behavioural parameter values during each trial.

Next we took the results from October to January in each season, and conducted a two-way analysis of variance (ANOVA) for each behavioural parameter, with factors month (Oct–Jan) and field season (2009/10 or 2010/11). A significant month×season interaction would indicate that any behavioural change across months differed between the two seasons. One interpretation of that might be that they were becoming more familiar with the trial arenas in the second season, and adjusting their behaviours with experience. This might suggest that behavioural changes could be explained by increasing experience with the conditions rather than by any effect of season. Our results (see below) did not show any significant month×season interaction effects, so for the final analysis we pooled all trials across the two field seasons to derive mean values per month and used nested one-way ANOVAs to investigate the effect of month of release for each behavioural parameter. We used the month of the trial as a fixed factor with replicate control cages nested within months. We used the Bonferroni test for post hoc comparisons between pairs of months.

Ethics Statement

The study was conducted according to the guidelines of the Flinders University Animal Welfare Committee (approval no E206) in compliance with the Australian code of practice for the use of animals for scientific research and was conducted under Permit G25011 from the South Australian Department of Environment, Water and Natural Resources. This permit allowed us to captured 16 specimens (8 male and 8 female) from two populations near Burra (33° 42′ S, 138° 56′ E) and transfer them to Monarto Zoo (35° 06′ S, 139° 09′ E) and use them for the experiments described above. All 16 lizards were captured by hand after luring them from their burrows with mealworms, according to the method described by Fenner et al. [28]. This capture method was approved according to the guidelines in our permits. Monarto Zoo, the site of the experimental investigations, is privately owned by Zoos SA.

Results

Although there was a wide range of ambient temperatures experienced over the 17 trials (Fig. 1), there were no significant difference among months for the ambient thermal conditions we measured during the trials, either for the average ($F_{5,9} = 0.32$, $p = 0.87$), minimum ($F_{5,9} = 3.39$, $p = 0.13$), or maximum temperatures ($F_{5,9} = 0.44$, $p = 0.80$). Only one behavioural parameter was significantly correlated with temperature. Lizard basking time decreased when temperature increased (average temperature: $r = -0.434$, $p = 0.021$ (Fig. 1); minimum temperature: $r = -0.710$, $p < 0.001$; maximum temperature: $r = -0.518$, $p < 0.005$). There were no significant interactions between month and field season for any of the behavioural parameters (Table 2). The significant effect of month for six of those behaviours over the period October to January remained consistent between the two field seasons 2009/10 and 2010/11.

When the data were pooled across field seasons, and data from February and March 2010 were included in the analyses, five behavioural parameters retained significant differences among months (Table 3). For all five of those parameters there was a consistent trend for decreasing values as the field season progressed from spring to summer (Fig. 2). Lizards spent a shorter period of the day active (Fig. 2a), they basked for less time (Fig. 2b), moved around their burrows less often (Fig. 2c), dispersed from the central area less often (Fig. 2d), and were involved in fewer fights (Fig. 2e) as the season progressed from spring (October) to late summer (March).

Discussion

The data from our series of trials showed a consistent pattern across the two field seasons of monthly differences in the behaviours of pygmy bluetongue lizards. The lack of any significant month×season interaction effects suggests that those differences represent a real monthly change in behaviour, rather than an accumulating familiarity by our captive lizards for the experimental conditions. In that case we would have expected different levels of activity or behaviour in the spring of the second field season than in the first season. We cannot completely discount the possibility that the lizards had reset their experience level during the winter break, and that increasing familiarity with the experimental conditions accounted for the behavioural changes each season. However, we considered that a less likely

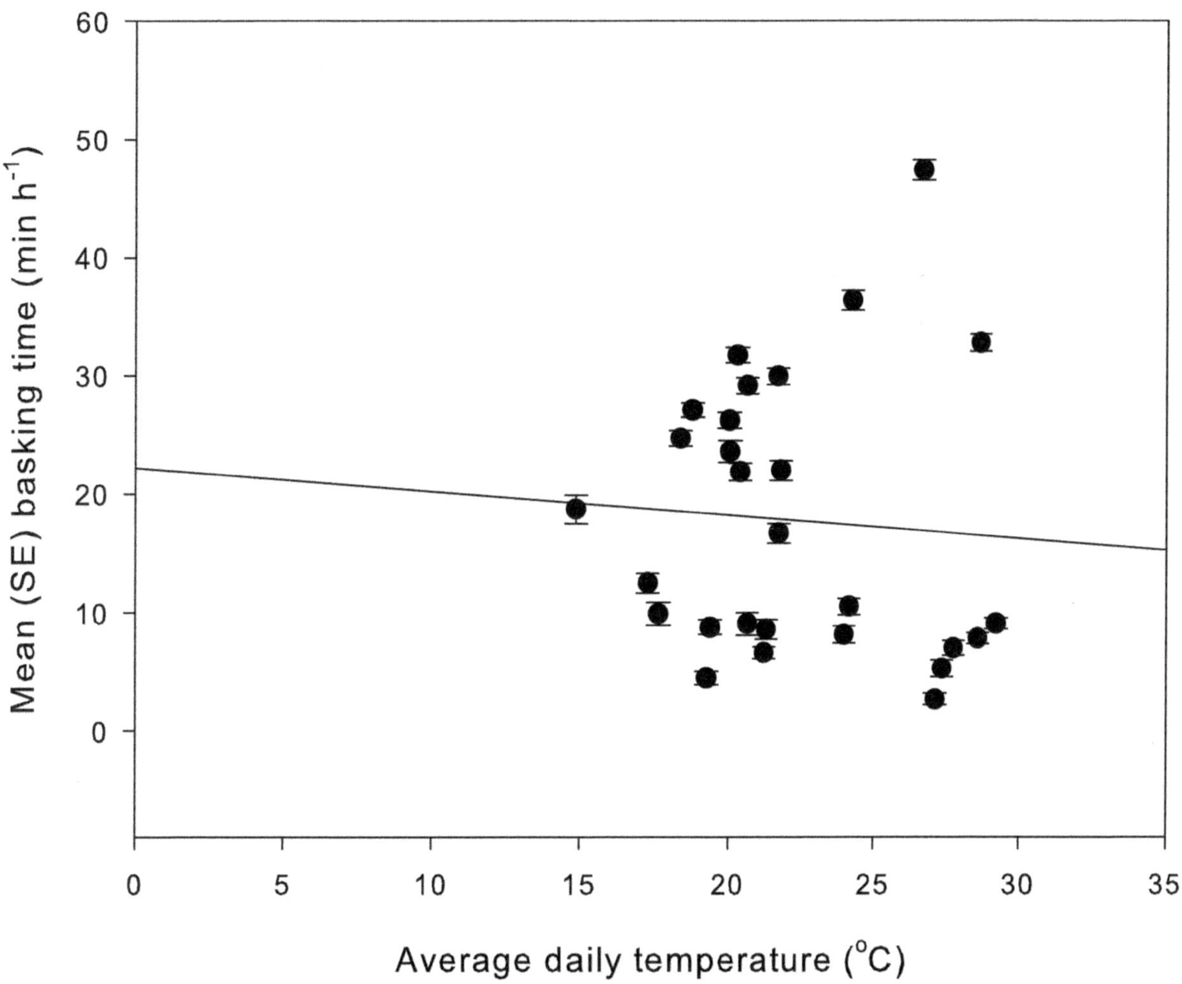

Figure 1. Relationship between the mean basking time of all lizards in a cage over a four day trial and the mean of the average daily temperature over the four day trial. The solid line indicates the significant negative correlation.

Table 2. Analyses of variance (ANOVA) considering the effect of month (Oct–Jan) and field season (2009/2010 and 2010/2011) on each of seven behavioural parameters, using months when trials were run in both field seasons.

Behavioural parameter	Field season			Month			Field season × month		
	df	*F*	p	df	*F*	p	df	*F*	p
Activity time	1, 12	1.68	0.21	3, 12	4.73	0.021*	2, 12	0.70	0.51
Basking time	1, 12	0.23	0.64	3, 12	4.33	0.027*	2, 12	3.16	0.08
Number of movements around burrows	1, 12	1.38	0.26	3, 12	16.99	0.001*	2, 12	0.36	0.70
Number of burrow changes	1, 12	1.02	0.33	3, 12	8.09	0.003*	2, 12	0.77	0.48
Distance moved	1, 12	0.03	0.85	3, 12	2.03	0.163	2, 12	0.91	0.42
Number of dispersals	1, 12	0.01	0.91	3, 12	7.49	0.005*	2, 12	0.44	0.52
Number of fights	1, 12	0.63	0.44	3, 12	4.28	0.028*	2, 12	0.45	0.64

Values with *indicate significant effects ($p<0.05$).

explanation of the trends we observed, given their consistency with field observations.

The field observations showed similar trends of higher levels of activity and movement in spring than later in summer. All of the incidental observations of predation of exposed lizards have occurred in October or November [20,29] and Schofield et al. [18] found that the number of adult pygmy bluetongue lizards captured in pitfall traps was highest in the spring, coinciding with the time when mating behaviour had previously been observed [30]. They suggested that lizards are most likely to leave their burrows and move around (to be trapped by pitfalls) at the time when they are seeking mating partners [18]. This may be reflected by the higher rates of movements and burrow changes in spring in our trials. Changes in some of our other behavioural parameters may be also correlated with these movements. For instance there may have been fewer fights in summer than spring because the lizards spent less time active, and moved less around their burrow entrances, so there were fewer opportunities for two individuals to encounter each other.

We considered whether the behavioural changes may have been temperature related. Temperatures in southern Australia generally increase from spring to summer, and one behavioural parameter, basking time, showed significant negative correlations with the mean ambient temperature, and with the minimum and maximum temperatures measured over the four day trials. This was consistent with the reduction in basking time with month within each field season. Perhaps lizards spent shorter periods emerged when it was warmer, to avoid overheating. However, other behaviours that also significantly declined over successive months, were not significantly related to temperature during the trials. Furthermore, there were no significant differences in temperature among months in the four day periods when individual trials were run, for any of the three temperature parameters that we considered. We concluded that, while some behaviours may be affected by the ambient temperature conditions, consistent behavioural changes in the lizards occurred from month to month independent of ambient temperature variation.

Management implications

Whatever the causal explanation of the trends we observed, the results have two implications for a conservation management program for this lizard species. First, within existing populations, behavioural insights derived from our results can alert managers about when lizards are at greatest risk from predators and from dispersal away from managed population sites. Dispersal has greater significance on populations of a species like the pygmy bluetongue lizard that persist in small fragments of previously more continuous native grassland habitat. Field populations of pygmy bluetongue lizards decline in size over the spring and summer period [31], and part of that attrition is likely to result from dispersal, that, in a fragmented landscape, is not balanced by migration in from other populations [32]. If artificial burrows are added to provide additional refuges for dispersing lizards, and to reduce both the time exposed on the surface, and the chance of moving away from suitable habitat, the seasonal population decline might be averted [33]. Our behavioural studies now indicate when this management intervention will be most effective.

The second implication of our results involves planning for possible translocations. The period immediately after release, while individuals become adjusted to and familiar with the novel release environment, is probably critical for the success of any translocation [34,35]. Animals may be stressed from the combined procedures of handling, holding and transportation [36], and they will be unfamiliar with refuge and foraging resources at the release

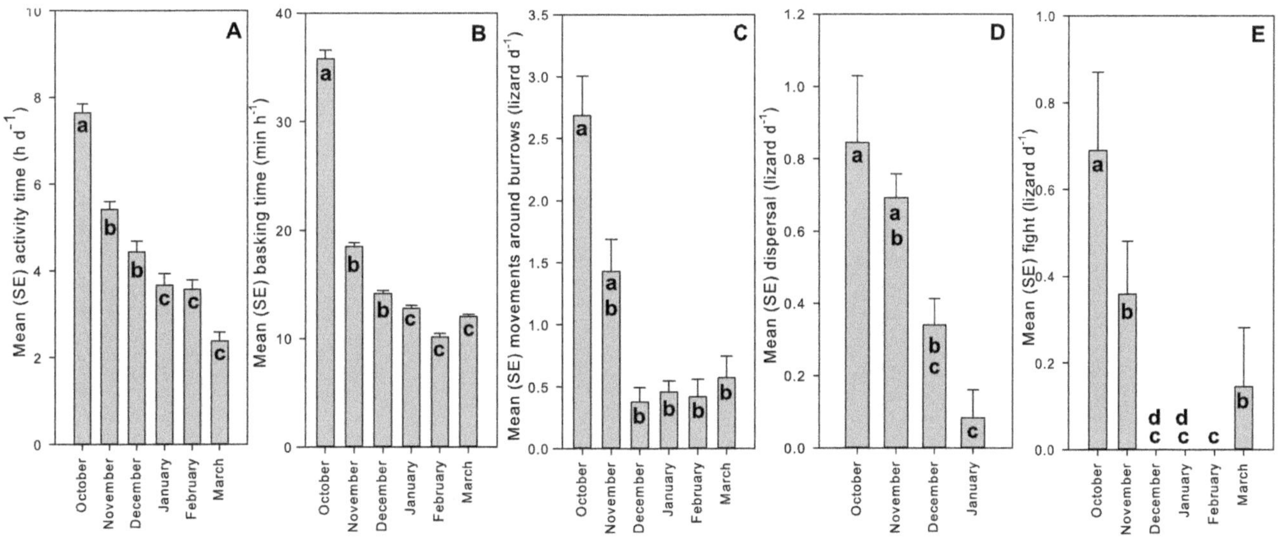

Figure 2. Mean and one standard error for five behavioural variables from trials in each month. Bars with different lower case letters were found to be significantly different in posthoc bonferroni pairwise comparisons. A) Mean activity time, B) Mean basking time, C) Mean number of movements around the burrow, D) Mean number of dispersals (no data available for February and March), E) Mean number of fights.

site [37–39], but if they remain close to where they are released until they have settled into the new conditions, then there is a chance they will stay [23,39,40].

Three aspects of our results suggest that lizards translocated later in the summer may have more chance of successfully establishing at the translocation site. First, lizards were less frequently active, moved about around their burrow entrance less, and spent less time basking at the burrow entrance later in the season. All of these behavioural changes would reduce the exposure of lizards to potential predators at the new site. Second, lizards interacted agonistically less often later in the season, avoiding increased levels of stress from intraspecific interactions. And third, the lizards dispersed less often from the central release area in our trials later in the season.

Ultimately the success of a translocation depends on the longer term responses of the translocated animals, although the first major first step is to prevent short-term dispersal away from the release area. This study has suggested that, for pygmy bluetongue lizards, the time of release could play a significant role in determining the possibility of dispersal from the translocation area, and that later in the activity season is a better time.

There are two broader conclusions from this study. One is that behavioural changes across different seasons might be critical for managing existing and translocated populations of endangered species. For our study system the remarkable seasonal differences in adult activity behaviour, from the current study and from results that have already been reported [18], have suggested that the time of mating might be important. Other reptiles and amphibians with narrow mating seasons, might have similar windows of time when they are more active, more exposed to predators, more likely to disperse away from secure habitat, and are subject to more stress. Conservation intervention, to reduce risks and stress related factors, may be most appropriate at these times. Counter to this, the monitoring of population density and individual body condition in these amphibians and reptiles, and the capture of individuals from source populations for translocation, might be easiest at the times when they are most active. Among mammals and birds, that have more stable social structures across longer periods of time, and where dispersal is often among subadults and juveniles, an equivalent period may be less obvious. The second conclusion is that understanding the basic behaviour of the target

Table 3. Analyses of variance (ANOVA) considering the effect of month on each behavioural parameter over all trials.

Behavioural parameter	Month			Nested factor (Cage(Month))		
	df	*F*	p value	df	*F*	p value
Activity time	5, 16	47.23	0.001*	6, 16	0.43	0.673
Basking time	5, 16	33.02	0.001*	6, 16	1.26	0.333
Number of movements around burrows	5, 16	5.03	0.036*	6, 16	2.55	0.062
Number of burrow changes	5, 16	1.42	0.336	6, 16	5.51	0.003*
Distance moved	5, 16	0.63	0.682	6, 16	2.67	0.054
Number of dispersals	3, 9	17.81	0.007*	6, 16	0.79	0.561
Number of fights	5, 16	591.3	0.001*	6, 16	0.08	0.966

Values with *indicate significance ($p<0.05$).

species is essential for optimizing management and translocation success [41,42].

Acknowledgments

Thanks go to the Mina Ansari for help during the field, and Monarto Zoo staff, Jeff Lugg, Ruth Reuter, Ian Smith and David McLelland for logistic support.

Author Contributions

Conceived and designed the experiments: ME CMB. Performed the experiments: ME. Analyzed the data: ME. Contributed reagents/materials/analysis tools: ME CMB. Contributed to the writing of the manuscript: ME CMB.

References

1. Macdonald DW, Johnson DDP (2001) Dispersal in theory and practice: Consquences for conservation biology. In: Colbert J, Danchin E, Dhondt AA, Nichols JD, editors. Dispersal. New York: Oxford University Press. 358–372.
2. Clobert J, Baguette M, Benton TG, Bullock JM, Ducatez S (2012) Dispersal ecology and evolution. United Kingdom: Oxford University Press.
3. Gandon S, Michalakis Y (2001) Multiple causes of the evolution of dispersal. In: Clobert J, Danchin E, Dhondt AA, Nichols JD, editors. Dispersal. USA, New York: Oxford University Press. 155–167.
4. Gilliam JF, Fraser DF (2001) Movement in corridors: Enhancement by predation threat, disturbance, and habitat structure. Ecology 82: 258–273.
5. Schaefer J (2001) Riffles as barriers to interpool movement by three cyprinids (*Notropis boops, Campostoma anomalum* and *Cyprinella venusta*). Freshwater Biology 46: 379–388.
6. Railsback SF, Lamberson RH, Harvey BC, Duffy WE (1999) Movement rules for individual-based models of stream fish. Ecological Modelling 123: 73–89.
7. Olsson M, Shine R (2003) Female-biased natal and breeding dispersal in an alpine lizard, *Niveoscincus microlepidotus*. Biological Journal of the Linnean Society 79: 277–283.
8. Tucker AD, McCallum HI, Limpus CJ, McDonald KR (1998) Sex-biased dispersal in a long-lived polygynous reptile (*Crocodylus johnstoni*). Behavioral Ecology and Sociobiology 44: 85–90.
9. Olsson M, Gullberg A, Tegelström H (1997) Determinants of breeding dispersal in the sand lizard, *Lacerta agilis*, (Reptilia, Squamata). Biological Journal of the Linnean Society 60: 243–256.
10. Bonte D, Van Dyck H, Bullock JM, Coulon A, Delgado M, et al. (2012) Costs of dispersal. Biological Reviews 87: 290–312.
11. Hanski I (1999) Metapopulation ecology: Oxford Uinversity Press.
12. Dale S (2001) Female-biased dispersal, low female recruitment, unpaired males, and the extinction of small and isolated bird populations. Oikos 92: 344–356.
13. Hedrick PW (1995) Gene flow and genetic restoration: The Florida Panther as a case study. Conservation Biology 9: 996–1007.
14. Hardman B, Moro D (2006) Optimising reintroduction success by delayed dispersal: Is the release protocol important for hare-wallabies? Biological Conservation 128: 403–411.
15. Milne T, Bull CM, Hutchinson MN (2003) Use of burrows by the endangered pygmy blue-tongue lizard, *Tiliqua adelaidensis* (Scincidae). Wildlife Research 30: 523–528.
16. Hutchinson MN, Milne T, Croft T (1994) Redescription and ecological notes on the pygmy bluetongue, *Tiliqua adelaidensis* (Squamata: Scincidae). Transactions of the Royal Society of South Australia 118: 217–226.
17. Fenner AL, Bull CM (2011) Central-place territorial defence in a burrow-dwelling skink: Aggressive responses to conspecific models in pygmy bluetongue lizards. Journal of Zoology 283: 45–51.
18. Schofield JA, Fenner AL, Pelgrim K, Bull CM (2012) Male-biased movement in pygmy bluetongue lizards: implications for conservation. Wildlife Research 39: 677–684.
19. Milne T, Bull CM, Hutchinson MN (2002) Characteristics of litters and juvenile dispersal in the endangered Australian skink *Tiliqua adelaidensis*. Journal of Herpetology 36: 110–112.
20. Fenner AL, Bull CM, Hutchinson MN (2008) Injuries to lizards: conservation implications for the endangered pygmy bluetongue lizard (*Tiliqua adelaidensis*). Wildlife Research 35: 158–161.
21. Ebrahimi M, Bull CM (2012) Food supplementation: a tool to increase the chance of successful translocation? A case study in the endangered pygmy bluetongue lizard (*Tiliqua adelaidensis*). Endangered Species Research 18: 169–178.
22. Ebrahimi M, Bull CM (2013) Behavioural changes in an endangered grassland lizard resulting from simulated agricultural activities. Journal of Arid Environments, Under review.
23. Ebrahimi M, Bull CM (2013) Determining the success of varying short-term confinement time during simulated translocations of the endangered pygmy bluetongue lizard (*Tiliqua adelaidensis*). Amphibia-Reptilia 34: 31–39.
24. Ebrahimi M, Bull CM (2014) Resources and their distribution can influence social behaviour at translocation sites: lessons from a lizard. Applied Animal Behaviour Science 156: 94–104.
25. Milne T, Bull CM, Hutchinson MN (2003) Fitness of the endangered pygmy blue tongue lizard *Tiliqua adelaidensis* in artificial burrows. Journal of Herpetology 37: 762–765.
26. Ebrahimi M, Fenner AL, Bull CM (2012) Lizard behaviour suggests a new design for artificial burrows. Wildlife Research 39: 295–300.
27. Fenner AL, Bull CM (2010) The use of scats as social signals in a solitary, endangered scincid lizard, *Tiliqua adelaidensis*. Wildlife Research 37: 582–587.
28. Fenner AL, Smales LR, Bull CM (2008) Pharyngodon wandillahensis n. sp (Nematoda: pharyngodonidae) from the endangered pygmy bluetongue lizard *Tiliqua adelaidensis* Peters, 1863 (Sauria: scincidae), South Australia, Australia. Comparative Parasitology 75: 69–75.
29. Fenner AL, Bull CM (2014) Accidental disturbance allows pygmy bluetongue lizards (*Tiliqua adelaidensis*) to survive predation attempts by eastern brown snakes (*Pseudonaja textilis*). Herpetofauna (in press).
30. Fenner AL, Bull CM (2009) *Tiliqua adelaidensis* (pygmy bluetongue lizard) mating behaviour. Herpetological Review 40: 91–92.
31. Fellows HL, Fenner AL, Bull CM (2009) Spiders provide important resources for an endangered lizard. Journal of Zoology 279: 156–163.
32. Smith AL, Gardner MG, Fenner AL, Bull CM (2009) Restricted gene flow in the endangered pygmy bluetongue lizard (*Tiliqua adelaidensis*) in a fragmented agricultural landscape. Wildlife research 36: 466–478.
33. Souter NJ, Bull CM, Hutchinson MN (2004) Adding burrows to enhance a population of the endangered pygmy blue tongue lizard, *Tiliqua adelaidensis*. Biological Conservation 116: 403–408.
34. Bodinof CM, Briggler JT, Junge RE, Beringer J, Wanner MD, et al. (2012) Postrelease movements of captive-reared Ozark Hellbenders (*Cryptobranchus alleganiensis bishopi*). Herpetologica 68: 160–173.
35. Scillitani L, Darmon G, Monaco A, Cocca G, Sturaro E, et al. (2013) Habitat selection in translocated gregarious ungulate species: An interplay between sociality and ecological requirements. The Journal of Wildlife Management 77: 761–769.
36. Dickens MJ, Delehanty DJ, M RL (2010) Stress: An inevitable component of animal translocation. Biological Conservation 143: 1329–1341.
37. Bright PW, Morris PA (1994) Animal translocation for conservation: performance of dormice in relation to release methods, origin and season. Journal of Applied Ecology 31: 699–709.
38. Rosatte R, Hamr J, Ranta B, Young J, Cool N (2002) Elk restoration in Ontario, Canada: Infectious disease management strategy, 1998–2001. Annals of the New York Academy of Sciences 969: 358–363.
39. Teixeira CP, de Azevedo CS, Mendl M, Cipreste CF, Young RJ (2007) Revisiting translocation and reintroduction programmes: the importance of considering stress. Animal Behaviour 73: 1–13.
40. Kleiman DG (1989) Reintroduction of captive mammals for conservation. Bioscience 39: 152–152.
41. Wallace MP (2000) Retaining natural behaviour in captivity for re-introduction programmes. In: Gosling LM, editor. Behaviour and conservation. Cambridge, United Kingdom: Cambridge University Press. 300–314.
42. Shier DM (2006) Effect of family support on the success of translocated black-tailed prairie dogs. Conservation Biology 20: 1780–1790.

9

The Implications of Habitat Management on the Population Viability of the Endangered Ohlone Tiger Beetle (*Cicindela ohlone*) Metapopulation

Tara M. Cornelisse*, Michelle K. Bennett, Deborah K. Letourneau

Environmental Studies Department, University of California Santa Cruz, Santa Cruz, California, United States of America

Abstract

Despite their role in providing ecosystem services, insects remain overlooked in conservation planning, and insect management approaches often lack a rigorous scientific basis. The endangered Ohlone tiger beetle (*Cicindela ohlone*) occurs in a 24-km^2 area in Santa Cruz County, California. The once larger metapopulation now consists of subpopulations inhabiting five patches of coastal prairie where it depends on bare ground for mating, foraging, and oviposition. Human activities have eliminated natural disturbances and spread invasive grasses, reducing *C. ohlone*'s bare-ground habitat. Management actions to restore critical beetle habitat consist of cattle and horse grazing, maintaining slow bicycle speeds on occupied public trails, and artificial creation of bare-ground plots. Recreational biking trails help maintain bare ground, but can cause beetle mortality if left unregulated. We tracked *C. ohlone* survivorship and estimated fecundity for three years. We then constructed a stage-structured population projection matrix model to estimate population viability among the five patches, and to evaluate the success of management interventions. We demonstrate that habitat creation, regulation of bicycle speed, and migration between patches increase *C. ohlone* survival and population viability. Our results can be directly applied to management actions for conservation outcomes that will reduce species extinction risk and promote recolonization of extirpated patches.

Editor: Don A. Driscoll, The Australian National University, Australia

Funding: Full funding for Open Access was provided by the University of California Santa Cruz Open Access Fund. TMC was supported by The Robert and Patricia Switzer Foundation (http://switzernetwork.org/) and NSFGK-12 DGE-0947923 (to G. Gilbert, I. Parker, D. Ash) (http://scwibles.ucsc.edu/). The funders had no role in study design, data collection and analysis, decision to publish, or preparation of the manuscript.

Competing Interests: The authors have declared that no competing interests exist.

* E-mail: tara.cornelisse@gmail.com

Introduction

Although the ultimate causes of species endangerment are most commonly habitat loss and invasive species introductions [1], proximate causes are reduced viability of small populations via genetic degeneration and demographic and environmental stochasticity [2–4]. Thus, while habitat protection is vital to species conservation, management is often needed to ensure viability of populations within protected habitat and across landscapes [5–8]. It is particularly important to understand the effect of management actions on tangible population viability goals of endangered species to ensure efficient and effective use of resources to prevent species extinctions [4,9,10].

Despite their significance in ecosystem functions, insects are frequently overlooked in conservation actions, and endangered insect recovery plans often lack quantitative population goals to ensure long-term viability [11,12]. More than 20 species of tiger beetles (Coleoptera: Cicindelinae) have been listed as threatened, endangered, or extinct worldwide and many more as US federal species of concern [13]. Pearson et al. [14] estimate that at least 33 (15%) of the 223 named species and subspecies of tiger beetles in the US and Canada may be declining at a rate that justifies their listing as threatened or endangered. Tiger beetles are associated with distinct disturbance-dependent bare-ground habitats needed to forage, find mates, and oviposit; thus, while they are sensitive to habitat degradation, they are increasingly dependent on anthropogenic disturbance [13,15].

The endangered Ohlone tiger beetle metapopulation (*Cicindela ohlone* Freitag and Kavanaugh) is endemic to the coastal prairies of Santa Cruz County, California. The remaining *C. ohlone* populations are present in habitat patches of a once more extensive metapopulation that consisted of 10–15 patches in the last 25 years [16]. Adult *C. ohlone* are generalist predators that stalk and chase down prey in open areas using visual cues, and larvae are sit-and-wait predators that construct a cylindrical burrow flush with the soil surface from which they lunge to capture passing arthropods; thus, both require bare ground for capturing prey [15]. The coastal prairie habitat evolved with disturbances such as large ungulate grazing and fires that created favorable conditions for the beetle's bare-ground habitat [17,18]; however, human activities have eliminated natural disturbances and spread invasive grasses, which form dense, extensive stands, reducing the incidence of bare ground [19,20].

Management of livestock grazing, recreation and artificial habitat creation currently maintain bare-ground in the remaining *C. ohlone* habitat patches. Creation of bare-ground plots by scraping the ground surface free of vegetation successfully augments egg-laying habitat for *C. ohlone* [21]. However, tiger beetle larval habitat augmentation within areas already limited by quality habitat could create a potential for negative density-dependence, with increased larval density leading to increased competition and reduce larval survivorship [22]. Recreational (i.e. hiking and cycling) trails also create bare ground, but high-speed cycling can disrupt the mating and foraging behaviors of *C. ohlone* adults (unpublished data). Thus, while systematic management of bare ground creation and recreation maintains *C. ohlone* habitat, it is unknown how these actions affect *C. ohlone* population viability.

The metapopulation dynamics of *C. ohlone* may be compromised because of habitat destruction and decline of habitat quality resulting in a few remnant populations; but because there have been two population turnover events in recent years, we have reason to believe *C. ohlone* has the capability of recolonization provided suitable habitat is available and managed in extirpated patches. However, recolonization of extirpated patches depends on asynchrony of *C. ohlone* population dynamics and their ability to migrate between populations [23], both of which are unknown.

We use population viability analysis (PVA) to model the effects of conservation management and metapopulation dynamics on *C. ohlone* viability. We predicted that artificial bare ground creation and managed recreation would augment the population growth rate of all *C. ohlone* populations when accounting for density-dependent effects, likely a significant factor in determining larval survival. We also predicted that *C. ohlone* vital rate dynamics would be asynchronous among populations and that any migration would reduce *C. ohlone* extinction risk. To test our predictions, we used PVA to model the growth rates of all *C. ohlone* populations and the associated vital rate sensitivities. We also determined how management strategies and metapopulation dynamics affected each rate effort to plan for the recovery of this endangered species.

Methods

Study sites

We conducted this study from January 2010 to August 2012 within the five remaining populations of *C. ohlone*, located in different coastal terrace prairie sites within a 24 km^2 area in Santa Cruz County, California: Lower Marshall (LM, 1.5 ha, 37.02°N 122.07°W) and Wilder Ranch (WR, 3.5 ha, 37.01°N 122.09°W), 1.3 km apart in the center of the range; Moore Creek (MC, 9.6 ha, 36.97°N 122.07°W) and University of California Campus (UC, 7.8 ha 36.98°N 122.07°W), 0.75 km apart in the south of the range; and Glenwood (GW, 2.9 ha, 37.07°N 121.99°W), 10 km north of the other sites. Santa Cruz County has a Mediterranean climate that receives an average of 77 (58–120) cm of rain, 95% of which falls from October to April. During this study, the annual precipitation was 99.8 cm, 72.6 cm, and 74.8 cm from 2010 to 2012, respectively. While the California coastal prairies can experience a rare fire or extreme drought, to our knowledge, the weather fluctuations experienced during this study represent the average conditions. This work was completed under USFWS permit #TE-39184A-0.

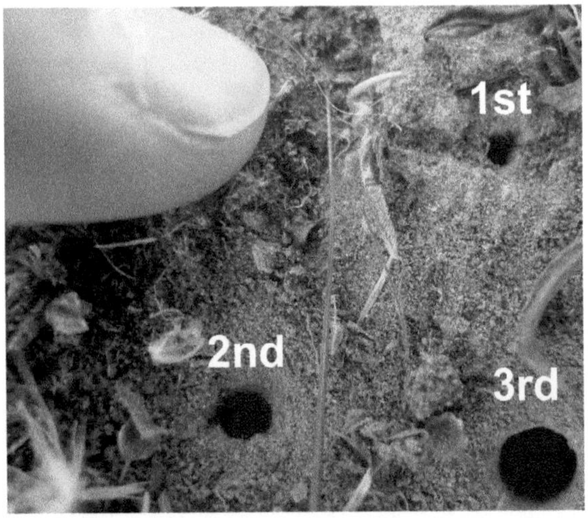

Figure 1. Burrows of the three Ohlone tiger beetle instars with an index finger for scale.

Cicindela ohlone lifecycle

Adult *C. ohlone* emerge from oviposition burrows, oviposit, and are active from late January to May. From February through early April, females deposit eggs singularly in the soil and the larvae develop at the site of oviposition. The first instars hatch in April through May, remaining in the first instar stage for four to six weeks [15,24], then progress to the second instar in May through June. Development to the third instar almost always occurs during the same summer, in July, after which the third instar plugs its burrow and pupates in late September through January, completing a one-year cycle. In addition, a few individuals have been observed to delay pupation until the following spring when the third instars unplug their burrows after winter inactivity [24].

Data collection

We surveyed adults once or twice per week late January to early June 2010 to 2012. We did not include UC in 2010 because we thought the beetles were extirpated from that site at the time. We estimated the number of *C. ohlone* adults and surveyed for larval burrows using a visual index count [25]. First instar larval burrows were surveyed in March to late April 2010 to 2012, identified by burrow diameter (Fig. 1) and assumed to be active if it was clearly delineated with a clean entrance, a sign of recent larval activity [15]. We overlaid the burrow(s) with a 0.25 m^2 gridded quadrat, marked the corners with a 3-cm wide metal washer and 5-cm long nail and the locations with a GPS to avoid disturbing larvae. We mapped all burrows in the quadrat on a gridded datasheet resembling the quadrat. In 2012, we identified and marked oviposition burrows in early March in the same manner as first instars. In 2010 and 2011 we marked 20 quadrats at each site and in 2012 we marked nine quadrats of oviposition burrows and 18 quadrats of first instar burrows, for a total of 27 quadrats at each site, which was dictated by the number of burrows found during surveys.

To determine if eggs survived to first instar, first to second, and second to third, or remained in the same stage, we revisited each site in late April and May, June, and July, respectively, sufficient time for all stage transitions to occur [15]. Quadrat markers were located via GPS and then either visually or using a metal detector. We lined up the gridded quadrat to the metal markers and considered the transitions to have occurred based on the increase in burrow entrance diameter, which the larva enlarges after molting to the next instar (Fig. 1). The following year, we revisited the locations of the previous year's third instar burrow once per week from late January though mid-March to check marked burrows for third instar larva survival and transition to adults by the presence of an irregular exit hole ≥6 mm.

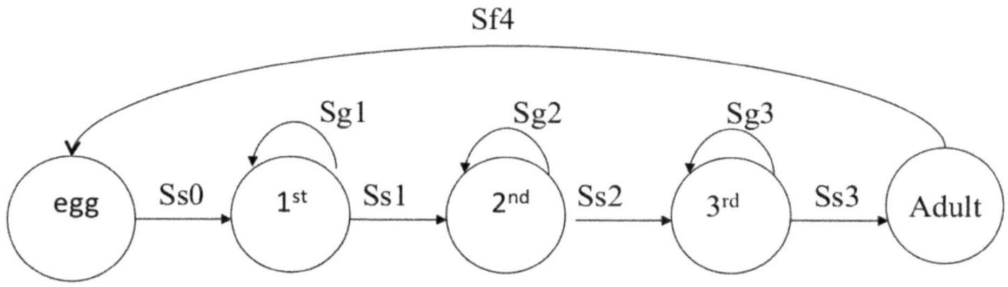

Figure 2. Ohlone tiger beetle stage class population model. The parameters presented are those we measured: survival of egg to 1st instar (Ss0), growth of 1st instar (Sg1), survival of 1st to 2nd instar (Ss1), growth of 2nd instar (Sg2), survival of 2nd to 3rd instar (Ss2), growth of 3rd instar (Sg3), survival of 3rd instar to adult (Ss3), and fertility of adults (Sf4).

Parameter estimates

We counted the total number of adults surveyed as females in the projection matrix model because visual index counts generally underestimate tiger beetle adults by ~50% [25] and sex ratios are not significantly different from 1:1 [16,25–27]. We assigned fertility rate (Sf4) as 40 because it is the best estimate of the mean total number of eggs a female *C. ohlone* oviposits in her lifetime [24]. We calculated larval growth rate within a stage (Sgi , i = instar stage 1, 2, or 3) as the proportion of individuals in the same stage at the next census. Transition rates (Ssi , i = the stage that transitioned to instar stage 1, 2, 3, or adult) were calculated as the proportion of individuals that transitioned from one stage to the subsequent stage in the next census [9]. Adult survival rates were included as zero because all adults senesce during the activity year. To estimate fecundity (F), we averaged three different estimates: (1) assuming a breeding pulse and mid-breeding census of eggs by multiplying the fertility rate by the square root of egg survivorship (Ss0); (2) assuming a constant breeding flow and mid-breeding census by multiplying the fertility rate by both the square root of (Ss0) and the square root of an estimate of adult survivorship; and (3) assuming a post-breeding census by multiplying the fertility rate by Ss0 (See [9], Ch. 6 for explanation of assumptions). We constructed a stage-class population model for *C. ohlone* (Fig. 2) and used the model parameters to construct a stage-class matrix model:

$$\begin{bmatrix} 0 & 0 & 0 & 0 & \mathrm{Sf4} * \mathrm{Ss0} \\ \mathrm{Ss0} & \mathrm{Sg1} & 0 & 0 & 0 \\ 0 & \mathrm{Ss1} & \mathrm{Sg2} & 0 & 0 \\ 0 & 0 & \mathrm{Ss2} & \mathrm{Sg3} & 0 \\ 0 & 0 & 0 & \mathrm{Ss3} & 0 \end{bmatrix}$$

Density dependence

To test for density dependent effects on larval survival, we used a random number generator to select an individual in each of the marked quadrats. We recorded survivorship from first to second and from second to third instar as survived (1) or dead (0) for each selected individual as well as the number of larval burrows in each quadrat surrounding the selected individual. Individuals still in the second instar stage during the last field visit were marked as survived (1). We tested the effect of larval density on the survival of selected individuals among sites using logistic regressions. We assumed prey was the most likely limiting factor involved in any inter-larval competition [28,29]. Larval density was log-transformed to fulfill assumptions of normality. Logistic regressions were carried out using SPSS v. 19.0.0 (SPSS, Inc., IBM).

Table 1. Number of Ohlone tiger beetle eggs and first instars marked (all quadrats combined), number of adults counted, and growth and survival parameters for each stage and site (see text for parameter explanation and calculation method).

	Glenwood (GW)			Lower Marshall (LM)			Wilder Ranch (WR)			Moore Creek (MC)			Campus (UC)	
Year	**2010**	**2011**	**2012**	**2010**	**2011**	**2012**	**2010**	**2011**	**2012**	**2010**	**2011**	**2012**	**2011**	**2012**
Eggs	na	na	58	na	na	31	na	na	36	na	na	27	na	28
1st instars	69	99	159	48	117	67	110	66	59	125	79	72	80	88
Adults	41	39	226	59	51	68	124	25	86	100	64	428	67	166
Ss0	na	na	0.59	na	na	0.32	na	na	0.06	na	na	0.41	na	0.54
Ss1	0.45	0.69	0.94	0.79	0.51	0.67	0.29	0.38	0.71	0.22	0.54	0.60	0.51	0.74
Sg2	0	0	0	0.05	0.03	0.27	0.03	0	0.05	0	0	0.02	0	0.2
Ss2	0.74	0.85	0.80	0.42	0.53	0.51	0.31	0.52	0.57	0.54	0.77	0.70	0.68	0.45
Sg3	0.74	0.21	0.20	0.75	0.69	0.69	0.3	0	0	0	0	0	0.14	0.14
Ss3	0.26	0.64	0.75	0.13	0.25	0.25	0.7	0.85	0.85	1.0	0.88	0.88	0.82	0.82

na indicates not marked that year.

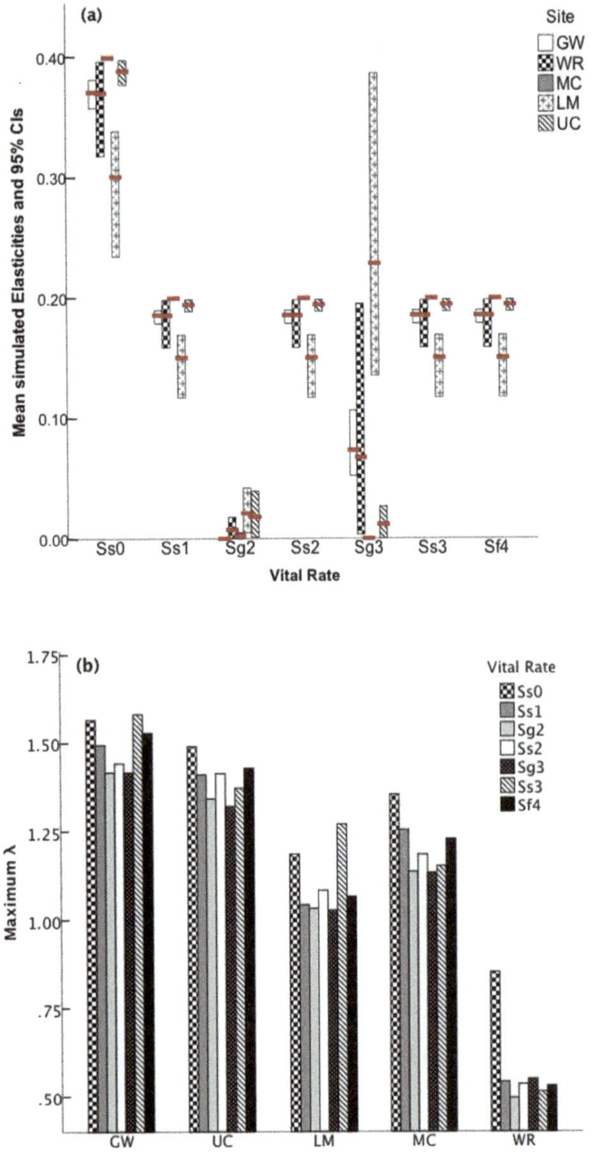

Figure 3. Vital rate elasticities as they relate to population growth rates (λ) (a) and the maximum possible population growth rate (λ) obtained when maximizing each vital rate (b), for each site. Red bars represent mean elasticities. See text for vital rate meanings.

Matrix modeling

All matrix analyses were done using Matlab Student Version 7.12 and methods described in Morris and Doak (2002) [9]. To account for demographic and environmental stochasticity, we constructed one matrix per study year and conducted the multiple matrices approach to estimate population growth rate [30]. Stochastic log growth rate, log λ_s, was determined both via simulation using the program stoc_log_lam and using Tuljapurkar's approximation (τ^2), which accounts for the covariance and variability of matrix elements among years. [9,31]. We assumed all matrices had equal probability of occurring and simulated 50,000 iterations. We used simext.m (Box 7.5 [9]), to evaluate the fraction of simulated populations that reach the quasi-extinction threshold after a designated time t_{max} set

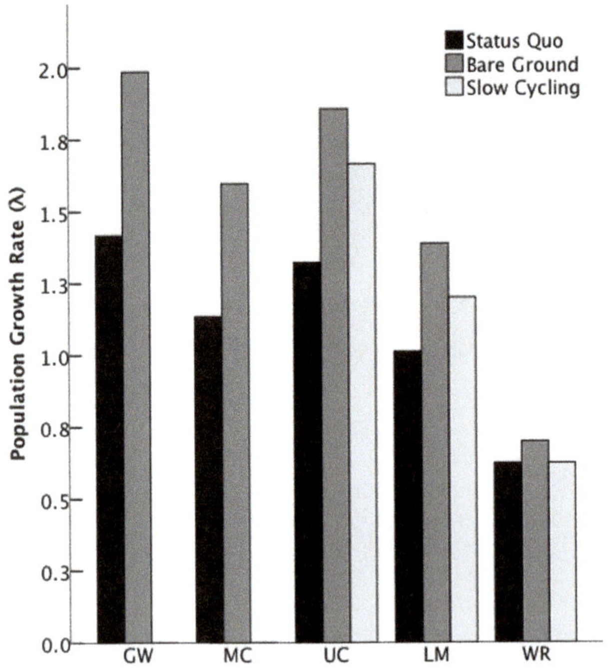

Figure 4. Population growth rates for all populations with current management (Status Quo (λ_1)), with increased bare ground, and with slower cycling where recreation is permitted.

to 25 and 50 years and the quasi-extinction threshold to 25, 10, and one individual(s) and weighted all matrices equally.

Sensitivity analysis

We conducted a stochastic sensitivity and elasticity analysis for each population by simulating multiple matrices using low, average, and high estimates of each vital rate using limitsens.m (Box 9.2 [9]). Maximum likelihood estimates of Ss1, Ss2, and Ss3 were calculated using Kendall.m [32] (Box 8.2 [9]) and we used the resulting confidence interval values as our high and low estimates of the survivorship vital rates in limitsens.m. Vital rates associated with the three measures of fecundity were used for fertility estimates; 40 for average, 60 for the high (highest estimated by [24]), and lowest fecundity value for the low estimate.

Sensitivity to management effects

To test the effect of creation of bare-ground and mandated slower cycling in *C. ohlone* habitats on population growth rates, we explicitly included a management scalar, *h*, in a deterministic matrix model using the program vitalsens.m (Box 9.1 [9]), in which we also calculate sensitivities and elasticities of *h* to each population growth rate. We averaged vital rates for all years and used 40 for the fertility value in the matrix.

Because female *C. ohlone* will lay up to 60 eggs in captivity [24], we assumed females would lay 60 eggs with increased bare ground. This is justified because creating bare ground in *C. ohlone* habitat will likely increase the number of eggs laid by females, as up to eight times more larval burrows were found in scraped ground compared with vegetation-covered controls [21]. Females will also obtain more food with more bare ground (e.g. hunting ground), increasing their fecundity and egg survivorship [15,29]. Furthermore, larvae that develop in bare ground experience increased prey availability and, in turn, reduced development time from 160 to 110 days, a 30% reduction [13,15,22,24,29]. Thus, we included bare ground

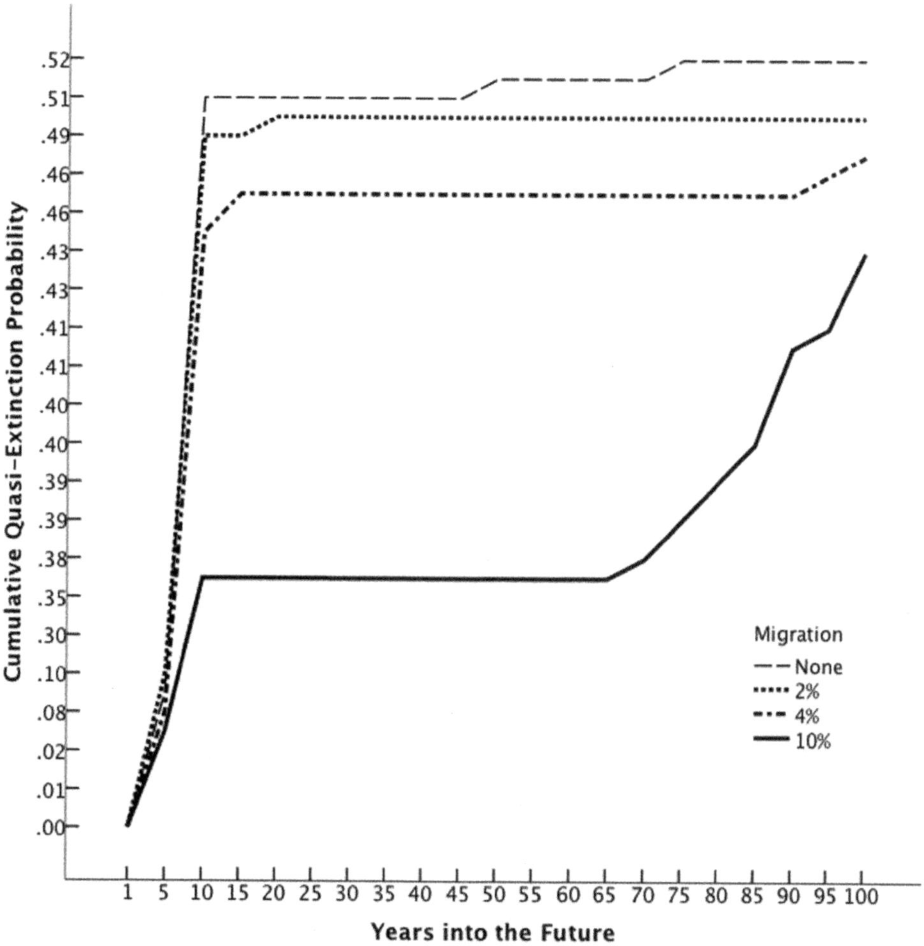

Figure 5. Cumulative quasi-extinction probability (set at ten individuals per stage class) for WR and LM populations combined with different migration scenarios.

creation as a management strategy that increases the survivorship and growth of eggs, 1st, 2nd, and 3rd instar by 30%, $h = 1.3$, and the number of eggs by 1.5 times (1.15*h). The resulting matrix is:

$$\begin{bmatrix} 0 & 0 & 0 & 0 & (Sf4 * 1.15h) * Ss0 * h \\ Ss0 * h & 0 & 0 & 0 & 0 \\ 0 & Ss1 * h & Sg2 * h & 0 & 0 \\ 0 & 0 & Ss2 * h & Sg3 * h & 0 \\ 0 & 0 & 0 & Ss3 * h & 0 \end{bmatrix}$$

We also modeled a hypothetical strategy that required all cyclists to slow down to speeds of 8–12 kph in *C. ohlone* habitats that allow bicycles: UC, LM, and WR. Reducing bicycle speed to 8–12 kph has the potential to reduce recreational disruptions to adult mating and foraging behavior (unpublished data), increasing adult survivorship and the number of eggs laid because of both increased fertility and increased reproductive period [15,29]. We incorporated the management strategy of reducing the bicycle speed allowed in *C. ohlone* habitat by including a scalar, h, to the matrix vital rates that increased the number of eggs laid to 80. This is justified in that tiger beetles are able to lay up to 200 eggs per female lifetime if a female lives for 30 days, an estimate for the average life span for adult tiger beetles [15]. We also increased the fecundity, or survivorship of eggs, by 30%, as justified above. The resulting matrix is:

$$\begin{bmatrix} 0 & 0 & 0 & 0 & (Sf4 * 1.5h) * Ss0 * h \\ Ss0 * h & 0 & 0 & 0 & 0 \\ 0 & Ss1 & Sg2 & 0 & 0 \\ 0 & 0 & Ss2 & Sg3 & 0 \\ 0 & 0 & 0 & Ss3 & 0 \end{bmatrix}$$

Metapopulation dynamics

We calculated vital rate correlation coefficients among sites from 2010–2012 using Pearson's correlation to check for asynchrony. To determine the overall metapopulation growth rate as well as the quasi-extinction risk, we created a metapopulation matrix composed of the individual population vital rates and used DemoMetaSim.m (Box 11.5 [9]). We capped the egg and larval stages at 100 individuals and adults at 300, high estimates of observed numbers, and the quasi-extinction thresholds to five, 10, and 20 individuals in each stage, and maximum time to 100 years for 500 runs. We ran the program first assuming no migration and including all sites and then by excluding sites one by one to test the extirpation or complete isolation of each site.

We also simulated quasi-extinction probability for LM and WR alone, as they are clumped in space. We simulated population growth and quasi-extinction risk for WR and LM assuming no migration, assuming one out of 50 adult females migrate between the two sites, or $m = 1/50 = 0.02$; one out of 25, or $m = 1/25 = 1.04$; and one out of 10, or 10% $m = 1.10$; thus, each adult vital rate (Sf4) was multiplied by 1.02, 1.04, and 1.10, respectively.

Results

The number of adults, marked burrows, and vital rates varied between sites and years (Table 1). Surviving first instar larvae never remained in the 1st instar stage between two successive censuses, thus we did not include a measure of Sg1 in the models. Egg survivorship estimates were similar between all populations except WR where we found the lowest estimate (Table 1), resulting in lower fecundity estimates. By averaging the three fecundity measurements described in the methods, we obtained the following fecundity (F) estimates for each population: GW 22.9 (±6.51); LM 14.8 (±5.81); WR 6.49 (±3.08); MC 19.7 (±4.17); UC 26.3 (±3.49).

Density dependence

The number of larvae in a quadrat ranged from 1–13, with an average of 3.3 (±2.4). There was no effect of larval density (number of larval burrows in a quadrat), site, or site × larval density interactions on survivorship of first instars to second ($R^2 = 0.029$, $p = 0.469$), nor second instars to third ($R^2 = 0.015$, $p = 0.692$). Thus, we found no evidence of density-dependent larval survival.

Matrix modeling

The simulated growth rates (λ_s) with 95% confidence intervals were: GW 1.41(1.405–1.415); LM 1.03 (1.028–1.032); WR 0.598 (0.596–0.601); MC 1.164 (1.153–1.174); and UC 1.163 (1.153–1.174). The growth rates calculated by Tuljapurkar's approximation were within the 95% confidence intervals of λ_s. For every population except WR, the stochastic quasi-extinction rate was zero in all time frames modeled. In all time frames and quasi-extinction thresholds, the extinction probability of the population at WR was 1.00, or definite extinction, by 21 years. The quasi-extinction probability or reaching 25 and 10 individuals was 1.00 by eight and 14 years, respectively.

Sensitivity analysis

The elasticities of Ss1, Ss2, Ss3, and Sf4 were equal with relatively small confidence intervals, whereas Sg2 elasticities were very small with narrow confidence intervals (Fig. 3a). Egg survivorship (Ss0) and 3rd instar growth (Sg3) elasticities varied greatly among populations and uncertainties of those parameters were high in both LM and WR, as indicated by the wide confidence intervals (Fig. 3a).

For UC, MC, and WR, maximizing the survivorship of eggs has the greatest potential to maximize population growth rate (Fig. 3b) with corresponding r^2 values (a measure of influence on the population growth, λ) of: 0.4437, 0.3941, and 0.4503, respectively. The population growth rate of WR remains below $\lambda = 1$ despite maximizing egg survivorship (Fig. 3b). For GW and LM, maximizing both egg survivorship ($r^2 = 0.5917$ and 0.6941) and survival of 3rd instars ($r^2 = 0.0224$ and 0.0918) maximized population growth rate (Fig. 3b).

Sensitivity to management effects

The simulated management strategies had positive effects on the growth rates of all populations, yet varied in their magnitude among the populations (Fig. 4). Sensitivities (and elasticities) for bare ground *h* were: GW 1.81 (1.19); MC 1.47 (1.20); UC 1.70 (1.19); LM 1.22 (1.14); WR 0.639 (1.19). Sensitivities (and elasticities) for slow cycling *h* were: UC 0.745 (0.582); LM 0.423 (0.458); WR 0.275 (0.572). Thus, while reducing bicycle speed resulted in a smaller increase in growth rates compared to increasing bare ground, both only marginally increased the WR growth rate (Fig. 4).

Metapopulation dynamics

The Pearson correlation analysis revealed that while vital rates were largely correlated among sites, Ss1, Ss2, and Ss3 were asynchronous between LM and all other sites, WR and UC, MC and GW, and MC and LM, respectively.

The simulated metapopulation growth rate ranged from 1.2935 to 1.6282 and the quasi-extinction probability was zero for all simulations containing the GW, MC, and UC populations. Since all combinations of sites that included GW, UC, and MC had a positive population growth rate and a zero chance of quasi-extinction at any threshold with no migration we did not simulate migration including these sites because migration only acts to decrease the risk of extinction, with our evidence of no negative density dependence.

The maximum population growth rate and probability of quasi-extinction for WR and LM in 100 years, assuming 10 individuals in each stage and no migration, was 0.9396 and 0.5160, respectively (Fig. 5). Assuming 2% migration, or one per 50 adult females migrate between the two sites, population growth was 1.0131 with quasi-extinction probability reduced to 0.4880 in 100 years; for 4% migration, the growth rate was 1.0115 and quasi-extinction probability 0.4640 in 100 years; for 10% migration, the growth rate was 1.0133 and quasi-extinction probability 0.4320 in 100 years (Fig. 5). Thus, migration increased the population growth rate for WR and LM to $\lambda > 1$and reduced the quasi-extinction probability by 5–10%.

Discussion

The endangered Ohlone tiger beetle represents the fragmented status of many threatened species for which coordinated, scientifically based and data driven management is desperately needed. By understanding population growth between and among populations as well as the effect of management strategies, we were able to evaluate the consequences of conservation actions on the recovery of *C. ohlone*.

Environmental and demographic stochasticity did not appear to be important factors of *C. ohlone* population growth, as matrix elements were not highly variable among years, Tuljakurpur's approximation fell within the narrow confidence intervals of the stochastic population growth, and we found no evidence of a density-dependent effect on larval survival. Despite these findings, environmental stochasticity should not be ignored as we only have three years of data, insufficient to predict insect population fluctuations [33]. In addition, the impact of environmental stochasticity on variable population growth will increase as global warming continues to alter climate patterns around the world [4].

The stochastic projection matrices revealed some large differences in growth rates among *C. ohlone* populations. GW had the highest growth rate, yet since it is the most physically isolated site, metapopulation theory would predict it to be at high risk of extirpation [23]. While several studies show that metapopulation theory is useful in explaining some patterns of extinction, many show that local scale, within patch, habitat characteristics are important for the conservation of insect species [34–40]. GW is grazed by horses year round and has a high percentage of bare ground and low standing vegetation (unpublished data). GW is also managed by the Land Trust of Santa Cruz County, which employs biological consultants to improve the land specifically for *C. ohlone* viability. Thus, GW is an example of how extinction risk of isolated patches can be reduced with increased habitat quality and management; yet, long-term persistence of GW as part of the metapopulation will require understanding of *C. ohlone* dispersal ability and rates between patches.

WR was the only population that had a negative population growth and high extinction risk across all projected models. While it seems that the reason for this was the low egg survivorship, all WR vital rates were generally lower than for other populations. Indeed, if we substitute the egg survivorship and mean fecundity values from LM (which is mostly likely a high estimate for true WR values) into the WR population projection matrix, the growth rate remains below 1.0 at 0.97. In contrast to GW, WR represented how a more connected habitat patch can become non-viable as habitat quality declines, suggesting that increased habitat management in WR could increase *C. ohlone* viability.

The sensitivity analyses revealed that, for all populations, the population growth rate was most sensitive to egg survivorship. The wide confidence intervals around the egg survivorship estimates for LM and WR indicated a greater uncertainty in those values and while we recognize the limited predictive power of a single estimate of egg survivorship, our estimates fall well within those published for tiger beetles [41]. The LM population growth rate was also highly sensitive to the growth of third instar larvae. This pattern is important for viability in that larvae will prolong their pupation without sufficient food, increasing their risk of mortality [15,25]. As the LM population growth rate was near 1.0, management that augments food availability, such as increased bare ground, may become important for population viability.

Increasing bare ground and requiring cyclists to slow down in *C. ohlone* habitat created large, positive changes in all populations. We based our assumed increase of 1.5 times the number of eggs laid on our previous study [21], which follows that as little as 9-m^2 of bare ground per site would be beneficial. Similarly but less so, management that requires a reduction in bicycle speed increased population growth rates in the three sites that allow bicycle recreation. We assumed that this management action only affected fecundity because recreation mainly affects the mobile adult stage. Our elasticity results showed that reducing bicycle speed to 8–12 kph increased population growth by 42–58%.

We found that any amount of migration among *all* sites ensured that the *C. ohlone* metapopulation would not go extinct within the next 100 years, assuming current management remains in effect. Despite the fact that vital rates were generally asynchronous among sites regardless of the distance separating them, *C. ohlone* dispersal dynamics are unknown and because remaining patches are fragmented, we are unsure if *C. ohlone* has sufficient colonization ability. Thus, future studies should focus on understanding *C. ohlone* dispersal ability and rates in order to quantify recolonization potential or the need for translocation.

When we removed the populations with high growth rates, GW, UC, and MC, equivalent to no migration between these populations and the other two, WR and LM had a 35% probability of quasi-extinction in the next 10 years, even with a high degree of migration. WR and LM are in a habitat cluster that has lost two populations in recent years [16,24]; this increased isolation could be contributing to the non-viable status of WR and low growth rate in LM via inbreeding depression. Fortunately, the coastal prairie habitat between these two sites is protected but management efforts focused on augmenting habitat quality in extirpated patches would be a necessary first step in promoting recolonization and maintaining *C. ohlone* viability in this area.

Cicindela ohlone declined across the landscape due to habitat destruction prior to its listing as an endangered species; however, as is the case with many threatened species, protection of occupied habitat alone is not enough to prevent its extinction [6,8]. At the site level, management actions that maintain bare ground and reduce incidental mortality of *C. ohlone* must be in place to maintain stable populations [13,21], whereas at a landscape level both recently extirpated sites and potential coast prairie habitat should be managed to maintain suitable *C. ohlone* habitat for future colonizations. We are currently analyzing the potential for unoccupied sites to contribute to *C. ohlone* habitat and range expansion. The results of this study clearly illustrate that *C. ohlone* has four viable populations that, with habitat management, could recolonize extirpated sites and avoid species extinction.

Acknowledgments

We thank the labs of Drs. K. Holl and S. Philpott as well as C.B. Knisley for comments in the planning phase of this study and manuscript. We thank M. Hash for assistance in the field. We also thank Don A. Driscoll and two anonymous reviewers for helpful comments that improved the manuscript. This work was completed under USFWS permit #TE-39184A-0.

Author Contributions

Conceived and designed the experiments: TMC. Performed the experiments: TMC MKB. Analyzed the data: TMC MKB. Contributed reagents/materials/analysis tools: TMC DKL. Wrote the paper: TMC MKB DKL.

References

1. Sodhi NS, Ehrlich PR, editors (2010) Conservation Biology for All. Oxford: Oxford University Press.
2. Simberloff D (1994) The ecology of extinction. Acta Palaeontol Pol 38: 159–174.
3. Lawton JH (1994) Population dynamic principles. Philos Trans R Soc Lond B Biol Sci 344: 61–68.
4. Brook BW, Sodhi NS, Bradshaw CJA (2008) Synergies among extinction drivers under global change. Trends Ecol Evol 23: 453–460.
5. Rodrigues ASL, Andelman SJ, Bakarr MI, Boitani L, Brooks TM, et al. (2004) Effectiveness of the global protected area network in representing species diversity. Nature 428: 640–643.
6. Fuller RA, McDonald-Madden E, Wilson KA, Carwardine J, Grantham HS, et al. (2010) Replacing underperforming protected areas achieves better conservation outcomes. Nature 466: 365–367.
7. Bonnot TW, Thompson FR III, Millspaugh JJ (2011) Extension of landscape-based population viability models to ecoregional scales for conservation planning. Biol Conserv 144: 2041–2053.
8. Laurance WF, Useche DC, Rendeiro J, Kalka M, Bradshaw CJA, et al. (2012) Averting biodiversity collapse in tropical forest protected areas. Nature 489: 290–294.
9. Morris WF, Doak DF (2002) Quantitative Conservation Biology. Sunderland: Sinauer Associates, Inc.
10. Traill LW, Brook BW, Frankham RR, Bradshaw CJA (2010) Pragmatic population viability targets in a rapidly changing world. Biol Conserv 143: 28–34.
11. Cardoso P, Erwin TL, Borges PAV, New TR (2011) The seven impediments in invertebrate conservation and how to overcome them. Biol Conserv 144: 2647–2655.
12. Losey JE, Vaughan M (2006) The economic value of ecological services provided by insects. Bioscience 56: 311–323.
13. Knisley CB (2011) Anthropogenic disturbances and rare tiger beetle habitats: benefits, risks, and implications for conservation. Terr Arthropod Rev 4: 41–61.

14. Pearson DL, Knisley CB, Kazilek CJ (2006) A field guide to the Tiger Beetles of the United States and Canada. New York, New York: Oxford University Press, Inc.
15. Pearson DL, Vogler AP (2001) Tiger Beetles. Ithaca: Cornell University Press.
16. Cooper D (2009) Ohlone tiger beetle (Cicindela ohlone) 5-Year Review: Summary and Evaluation. United States Fish and Wildlife Service.
17. Anderson MK (2007) Native American Uses and Management of California's Grasslands. In: Stromberg MR, Corbin JD, D'Antonio CM, editors. California Grasslands Ecology and Management. Berkeley: University of California Press. 57–67.
18. Wigand PE, Edwards SW, Schiffman PM (2007) Pleistocene and Pre-European Grassland Ecosystems. In: Stromberg MR, Corbin JD, D'Antonio CM, editors. California Grasslands Ecology and Management. Berkeley: University of California Press. 37–56.
19. Hayes GF, Holl KD (2003) Site-specific responses of native and exotic species to disturbances in a mesic grassland community. Appl Veg Sci 6: 235–244.
20. D'Antonio CM, Malmstrom C, Reynolds SA, Gerlach J (2007) Ecology of Invasive Non-native Species in California Grassland. In: Stromberg MR, Corbin JD, D'Antonio CM, editors. California Grasslands Ecology and Management. Berkeley: University of California Press. 67–86.
21. Cornelisse TM, Vasey MC, Holl KD, Letourneau DK (2013) Artificial bare patches increase habitat for the endangered Ohlone tiger beetle (Cicindela ohlone). J Insect Conserv 17: 17–22.
22. Takeuchi Y, Hori M (2007) Spatial density-dependent survival and development at different larval stages of the tiger beetle Cicindela japonica (Thunberg). Popul Ecol 49: 305–316.
23. Hanski I (1999) Metapopulation Ecology. Oxford: Oxford University Press.
24. Knisley CB, Arnold RA (2004) Biology and conseration of the Ohlone tiger beetle, *Cicindela ohlone*. Final Report to U.S. Fish and Wildlife Service. Ventura, California, USA. 34 p.
25. Knisley CB, Schultz TD (1997) The biology of tiger beetles and a guide to the species of South Atlantic States. Martinsville, Virginia: Virginia Museum of Natural History.
26. Tigreros N, Kattan GH (2008) Mating behavior in two sympatric species of Andean tiger beetles (Cicindelidae). Bol Mus Entomol Univ Valle 9: 22–28.
27. Shivashankar T, Pearson DL (1994) A comparison of mate guarding among five syntopic tiger beetle species from Peninsular India (Coleoptera: Cicindelidae). Biotropica 26: 436–442.
28. Knisley CB, Juliano SA (1988) Survival, development, and size of larval tiger beetles: effects of food and water. Ecology 69: 1983–1992.
29. Pearson DL, Knisley CB (1985) Evidence for food as a limiting resource in the life cycle of tiger beetles (Coleoptera: Cicindelidae). Oikos 45: 161–168.
30. Morris W, Doak D, Groom M, Kareiva P, Fieberg J, et al. (1999) A Practical Handbook for Population Viability Analysis. The Nature Conservancy.
31. Tuljapurkar S (1982) Population dynamics in variable environments. III. Evolutionary dynamics of r-selection. Theor Popul Biol 21: 141–165.
32. Kendall BE (1998) Estimating the magnitude of environmental stochasticity in survivorship data. Ecol Appl 8: 184–193.
33. Schultz CB, Hammond PC (2003) Using population viability analysis to develop recovery criteria for endangered insects: case study of the Fender's Blue Butterfly. Conserv Biol 17: 1372–1385.
34. Heisswolf A, Reichmann S, Poethke HJ, Schroder B, Obermaier E (2009) Habitat quality matters for the distribution of an endangered leaf beetle and its egg parasitoid in a fragmented landscape. J Insect Conserv 13: 165–175.
35. Fleishman E, Ray C, Sjogren-Gulve P, Boggs CL, Murphy DD (2002) Assessing the roles of patch quality, area, and isolation in predicting metapopulation dynamics. Conserv Biol 16: 706–716.
36. Beyer LJ, Schultz CB (2010) Oviposition selection by a rare grass skipper *Polites mardon* in montane habitats: Advancing ecological understanding to develop conservation strategies. Biol Conserv 143: 862–872.
37. Thomas JA, Bourn NAD, Clarke RT, Stewart KE, Simcox DJ, et al. (2001) The quality and isolation of habitat patches both determine where butterflies persist in fragmented landscapes. Proc R Soc Biol Sci 268: 1791–1796.
38. Poyry J, Paukkunen J, Heliola J, Kuussaari M (2009) Relative contributions of local and regional factors to species richness and total density of butterflies and moths in semi-natural grasslands. Oecologia 160: 577–587.
39. Hodgson JA, Moilanen A, Thomas CD (2009) Metapopulation responses to patch connectivity and quality are masked by successional habitat dynamics. Ecology 90: 1608–1619.
40. Collinge SK, Prudic KL, Oliver JC (2003) Effects of local habitat characteristics and landscape context on grassland butterfly diversity. Conserv Biol 17: 178–187.
41. Hoback WW, Golick DA, Svatos TM, Spomer SM, Higley LG (2000) Salinity and shade preferences result in ovipositional differences between sympatric tiger beetle species. Ecol Entomol 25: 180–187.

10

Predicting Environmental Suitability for a Rare and Threatened Species (Lao Newt, *Laotriton laoensis*) Using Validated Species Distribution Models

Amanda J. Chunco[1]*, Somphouthone Phimmachak[2,3], Niane Sivongxay[2], Bryan L. Stuart[4]

1 Department of Environmental Studies, Elon University, Elon, North Carolina, United States of America, **2** Department of Biology, National University of Laos, Vientiane, Lao PDR, **3** Department of Zoology, Kasetsart University, Bangkok, Thailand, **4** North Carolina Museum of Natural Sciences, Raleigh, North Carolina, United States of America

Abstract

The Lao newt (*Laotriton laoensis*) is a recently described species currently known only from northern Laos. Little is known about the species, but it is threatened as a result of overharvesting. We integrated field survey results with climate and altitude data to predict the geographic distribution of this species using the niche modeling program MAXENT, and we validated these predictions by using interviews with local residents to confirm model predictions of presence and absence. The results of the validated MAXENT models were then used to characterize the environmental conditions of areas predicted suitable for *L. laoensis*. Finally, we overlaid the resulting model with a map of current national protected areas in Laos to determine whether or not any land predicted to be suitable for this species is coincident with a national protected area. We found that both area under the curve (AUC) values and interview data provided strong support for the predictive power of these models, and we suggest that interview data could be used more widely in species distribution niche modeling. Our results further indicated that this species is mostly likely geographically restricted to high altitude regions (i.e., over 1,000 m elevation) in northern Laos and that only a minute fraction of suitable habitat is currently protected. This work thus emphasizes that increased protection efforts, including listing this species as endangered and the establishment of protected areas in the region predicted to be suitable for *L. laoensis*, are urgently needed.

Editor: Brian Gratwicke, Smithsonian's National Zoological Park, United States of America

Funding: The Seeding Postdoctoral Innovators in Research and Education (SPIRE) grant K12GM000678 from the Minority Opportunities in Research division of the National Institute of General Medical Sciences (NIGMS)/National Institutes of Health (NIH) and the National Science Foundation (DEB-1145922) provided funding for this research. The May 2012 locality of L. laoensis was obtained during a field survey supported by the Nam Ngum 3 Power Company Limited. The funders had no role in study design, data collection and analysis, decision to publish, or preparation of the manuscript.

Competing Interests: The Seeding Postdoctoral Innovators in Research and Education (SPIRE) grant K12GM000678 from the Minority Opportunities in Research division of the National Institute of General Medical Sciences (NIGMS)/National Institutes of Health (NIH) and the National Science Foundation (DEB-1145922) provided funding for this research. The May 2012 locality of *L. laoensis* was obtained during a field survey supported by the Nam Ngum 3 Power Company Limited.

* E-mail: achunco@elon.edu

Introduction

One-third of the world's amphibian species are threatened with extinction and nearly half (42%) are experiencing population declines [1]. A particular challenge in amphibian conservation is a lack of basic natural history information for many species. Nearly one-fourth of the world's amphibian species (23%) are so poorly known [1] that they are listed as Data Deficient in the IUCN Red List of Threatened Species [2]. Data are urgently needed for these taxa to prioritize the conservation of those that are threatened.

Southeast Asia has a highly diverse but highly threatened amphibian fauna [3]. New species of amphibians continue to be rapidly discovered in Southeast Asia [4], but this fauna is threatened from a high rate of deforestation and from over-harvesting for food, medicinal purposes, and the international pet trade [3]. The most critical conservation action needed for Southeast Asian amphibians is to identify, establish, and protect areas that are important for amphibian populations to mitigate this loss of biodiversity [3].

The Lao newt (*Laotriton laoensis*) was described in 2002 from a small geographic area in northern Laos [5]. After its description, commercial collectors used the publication to find and illegally harvest the species in Laos for sale into the international pet trade, where in Japan and Europe it commanded prices equivalent to >US$200 each [6]. Since then, commercial trade networks have become established within its range in northern Laos, threatening the species with extinction [7]. The species is also threatened to a lesser degree by habitat destruction and some consumption by local residents for food and medicinal purposes [7]. Although currently assessed as Data Deficient [8], the species has been recommended for upgrading to Endangered status in the IUCN Red List of Threatened Species on the basis of its restricted range and suspected population size reduction from overharvesting [7]. The species was also recommended [7] for listing in the Convention on International Trade in Endangered Species (CITES). In 2008, the government of Laos banned commercial trade in the species, although this has been inadequately enforced [7].

Table 1. Localities used to model habitat suitability for *Laotriton laoensis*.

Site Use	Province	District	Stream (Houay/Nam)	Coordinates*	Elev. (m)	Source
Model construction	Vientiane	Xaysomboun	H. Pa Tin	18°52′N 103°06′E	1,160	Stuart and Papenfuss 2002; Phimmachak et al. 2012
Model construction	Vientiane	Xaysomboun	H. Sang Kat	18°52′N 103°08′E	1,400	Stuart and Papenfuss 2002; Phimmachak et al. 2012
Model construction	Vientiane	Xaysomboun	H. Sang Kat	18°52′N 103°07′E	1,240	Phimmachak et al. 2012
Model construction	Vientiane	Xaysomboun	H. Thongkham	18°45′N 103°08′E	1,350	Phimmachak et al. 2012
Additional model validation	Vientiane	Xaysomboun	N. Pi	19°09′N 102°42′E	1,243	North Carolina State Museum of Natural Sciences (NCSM) 79785–86
Model construction	Xiengkhouang	Pek	H. Pieng	19°42′N 103°18′E	1,430	Phimmachak et al. 2012
Model construction	Xiengkhouang	Phoukout	H. Lueng	19°43′N 103°16′E	1,227	Stuart and Papenfuss 2002 (as "near Ban Nyot Phae")
Model construction	Xiengkhouang	Phoukout	H. Hao, H. Pee	19°42′N 103°13′E	1,144	Goldschmidt and Koehler 2007 (as "Ban Lae Village")
Model construction	Louangphabang	Phoukoun	H. Hin Rup	19°24′N 102°34′E	1,219	Phimmachak et al. 2012
Model construction	Louangphabang	Phoukoun	N. Madao	19°18′N 102°33′E	1,245	Phimmachak et al. 2012

*Coordinates were rounded to the nearest minute to prevent facilitating exploitation of the species, following Phimmachak et al. (2012).

A major remaining deficiency in knowledge of this species is the extent of its geographic range. *Laotriton laoensis* has been reported in the literature from only nine localities [5,7,9], with a verified extent of occurrence of approximately 4,800 km^2 [7]. The species is assumed to be endemic to Laos based on its known distribution, but the eastern extent of its range is uncertain and it could potentially extend beyond the border into adjacent northern Vietnam. The species is not known to occur within any of Laos' national protected areas [7], although several protected areas occur at the periphery of the species' known range in northern Laos. If the species does not occur in one of these national protected areas, then a new protected area needs to be established within its range to safeguard this unique species [7]. Verification of the extent of this species' geographic distribution is urgently needed.

Because field work is frequently resource intensive, one method for prioritizing areas for field surveys is species distribution modeling. Generally, species distribution models (SDMs) integrate observations of species occurrences with environmental variables to predict the range of the species of interest [10]. These models have been highly useful in correctly identifying suitable habitat for many species, even in understudied geographic regions [11,12]. SDMs are increasingly important in conservation biology and have been used to discover new populations of rare and endangered species [13], prioritize areas for conservation [14], and to predict levels of gene flow between populations [15]. As the distribution of *L. laoensis* remains poorly understood, SDMs are an ideal tool for predicting locations of additional, as of yet undiscovered, populations.

In this study, we use the species distribution modeling program MAXENT to predict the distribution of *L. laoensis*. We undertake this work with four goals: 1) to test the hypothesis that this species is restricted to northern Laos, 2) to determine whether or not any of the established protected areas in Laos provide suitable habitat for *L. laoensis*, 3) to characterize the environmental conditions of suitable areas to aid conservation efforts, and 4) to prioritize areas for future field surveys.

Methods

Ecological Niche Modeling

To identify suitable habitat for *Laotriton laoensis*, we used the ecological niche modeling program MAXENT (ver. 3.3.3e, [16]). We chose MAXENT over other modeling algorithms as it: 1) performs well compared to other niche modeling programs [17], 2) requires only presence data [16]), and 3) has previously been used successfully with very small sample sizes (i.e. as few as five records [12]. MAXENT integrates environmental data with species locality information to give a relative measure of suitability across a study area [16]. For environmental data, we used 19 bioclimatic variables and altitude data downloaded from World Clim Global Climate Data (www.worldclim.org ver. 1.4, [18]. Thus, suitable land was identified based on climate and altitude profiles. All environmental variables were at a resolution of 1 km×1 km.

For species localities, we used nine published field localities of *L. laoensis* [5,7,9] (Table 1). Because the variables included in any model can affect the model outcome [19], we ran several independent models using different suites of environmental variables to determine how sensitive the model results were to the specific parameters (Table 2). We employed four strategies to select variables for each model. First, we ran a complete model using all 19 Bioclimatic variables and altitude (the Full Model). Second, we used the results of the first model to select a subset of variables that contributed at least 5% to the Full Model to include in the second model (the Reduced Model). Third, because environmental variables are often highly correlated, we selected a subset of variables that were relatively uncorrelated for the next two models. To do so, we used Pearson's correlation coefficient (r)

Table 2. A description of the environmental data used in the five MAXENT models.

	Full Model	Reduced Model (Regional)	Reduced Model (Local)	Moderate Correlation (r <0.75)	Low correlation (r <0.5)	Expert Opinion
Annual Mean Temperature	x					x
Mean Diurnal Range in Temperature	x			x		
Isothermality	x		x	x	x	
Temperature Seasonality	x	x	x	x		
Max Temperature of Warmest Month	x					x
Min Temperature of Coldest Month	x					x
Temperature Annual Range	x					
Mean Temperature of Wettest Quarter	x					
Mean Temperature of Driest Quarter	x					
Mean Temperature of Warmest Quarter	x	x				x
Mean Temperature of Coldest Quarter	x					x
Annual Precipitation	x	x				x
Precipitation of Wettest Month	x					
Precipitation of Driest Month	x	x				
Precipitation Seasonality	x	x		x	x	
Precipitation of Wettest Quarter	x			x		
Precipitation of Driest Quarter	x					
Precipitation of Warmest Quarter	x	x		x	x	
Precipitation of Coldest Quarter	x			x	x	
Altitude	x		x	x	x	x

The same environmental parameters were used at the Full, Moderate Correlation, Low Correlation, and Expert Opinion models at both the Regional and the Local scale. For the Reduced Model, which was based on the results of the Full Model, a different suite of parameters was used at each scale as indicated below.

to measure the relative degree of correlation between each pairwise combination of variables using environmental data extracted from 100 randomly selected points each placed within 100 km of a known *L. laoensis* locality. We selected 100 km as the threshold to ensure points were reasonably close to known localities, while still providing enough distance to encompass the range of habitats in northern Laos. We randomly selected a set of relatively uncorrelated variables to include in the model using two different correlation thresholds, r>0.75 for the 'Moderate Correlation Model' and r>0.5 for the 'Low Correlation Model'. Finally, we used expert opinion; one of the authors (BLS) *a priori* selected a suite of variables likely to be most important based on his experience studying *L. laoensis* in the wild. These variables were used in the 'Expert Opinion Model'.

In selecting the geographic extent of any Maxent model, the background points chosen can strongly affect model outcome [20]. As we were interested in both the total predicted extent of the range and identifying other areas where *L. laoensis* is most likely to be found in the vicinity of where the species is known to occur, we ran each model at two different geographic scales. At the regional scale, the study area encompassed the entire Indochinese peninsula. Predicting across this entire area allowed us to determine the likelihood that the range of *L. laoensis* extended beyond the current observed range in northern Laos. At the local scale, the study area included only northern Laos and adjacent parts of Thailand and Vietnam. By drawing background points from this much smaller points, we were able to created more refined predictions without extrapolating predictions to areas well outside the likely range of the species (e.g. [20]).

Four of the models (the Full Model, Moderate Correlation, Low Correlation, and Expert Opinion) used the same set of environmental variables at both the Regional and Local scale. In the fifth model (i.e., the Reduced Model), we used a different set of parameters at each scale (Table 2) because the parameters used in the Reduced Model were specifically chosen based on the results of the Full Model, and these results differed at each scale (Table 2).

Under the logistic output setting, MAXENT returns a grid where each location in the study area is assigned a value between 0 and 1, where 0 represents a low probability of occurrence (i.e. low habitat suitability), and 1 represents a high probability of occurrence (i.e. high habitat suitability) [21]. To visualize these results, we projected the MAXENT output for each model in ArcGIS 9.3 using a UTM 48 N projection. All models were run nine times using the cross validation re-sampling method. Specifically, we ran a jackknife procedure with eight of the nine localities being used to train the model, with a different locality left out in successive iterations of the model [12]. The average of these nine runs was used in all analyses.

We used the default values of MAXENT with one exception: for the Full model, we further tested two alternative regularization multipliers (0.5 and 5) in addition to the default value of 1 to determine whether adjusting the regularization parameter would improve model performance ([21]).

To compare between the different models at each spatial scale, we calculated the pairwise correlations coefficient between each pair of models. To do so, we extracted the logistic value from 5000 random points throughout the area modeled at the regional spatial scale, and 1000 random points at the local spatial scale. We then

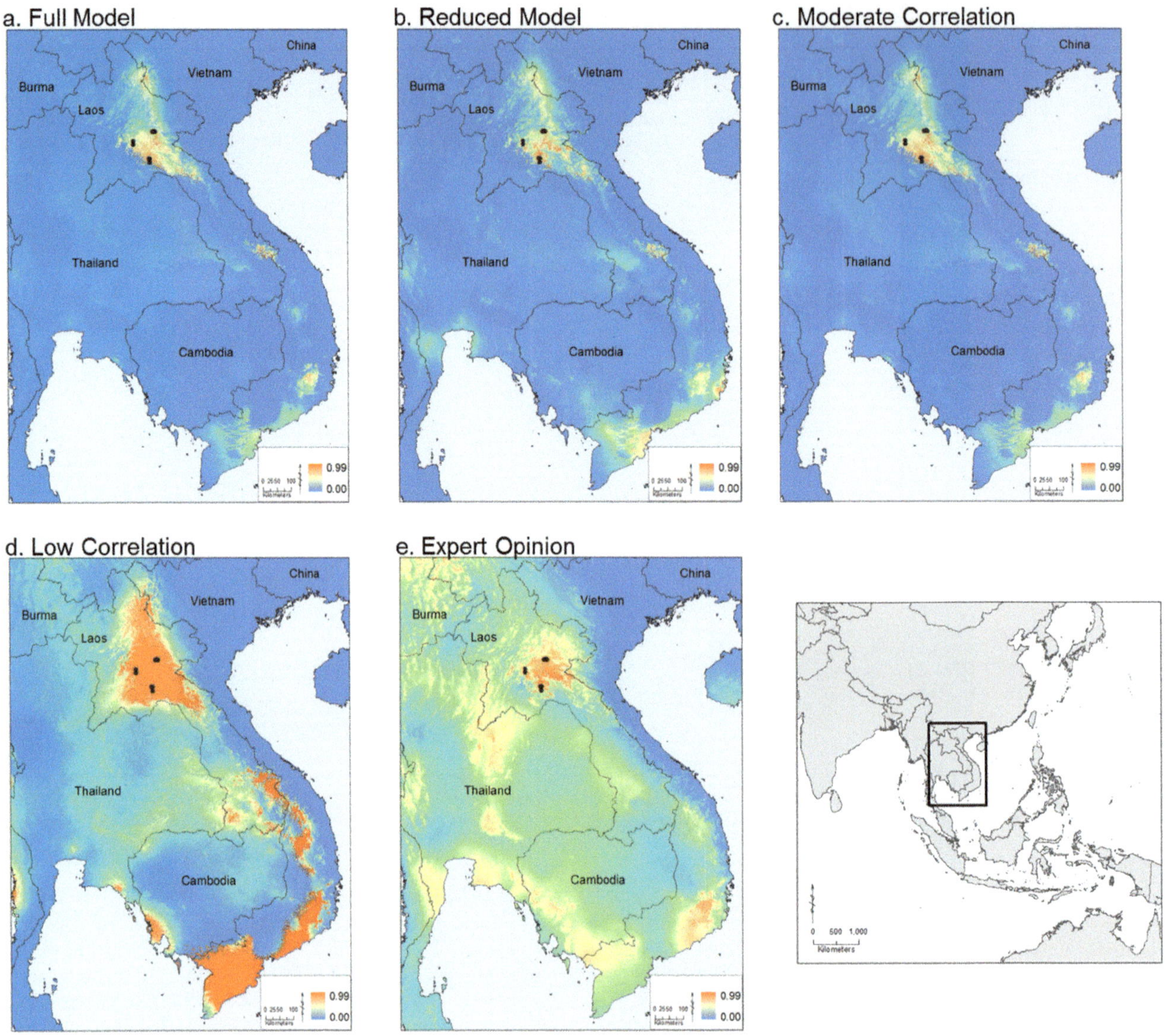

Figure 1. Regional MAXENT model results. These results show the habitat suitability for *L. laoensis*, under five unique combinations of environmental variables: a) Full Model, b) Reduced Model, c) Moderate Correlation (r <0.75), d) Low Correlation (r <0.5), e) Expert Opinion (see Table 2). The study area is shown as the boxed area in the inset map on the lower right.

calculated the Pearson's correlation coefficient (r) of the logistic values between each pair of models.

All necessary permits were obtained for the described field studies. The Nam Ngum 3 Power Company Limited provided permission for fieldwork. The CITES Management Authority, Department of Forestry, Ministry of Agriculture and Forestry, Vientiane, Laos, provided a specimen export permit (No. 035/12) to the North Carolina State Museum of Natural Sciences, and the

Table 3. The pairwise Pearson's correlation coefficient between all pairs of models at the regional scale.

	Full Model	Reduced Model	Moderate Correlation (r<0.75)	Low Correlation (r<0.75)	Expert Opinion
Full Model		0.857	0.899	0.619	0.364
Reduced Model			0.806	0.629	0.347
Moderate Correlation (r<0.75)				.651	0.276
Low Correlation (r<0.75)					0.384
Expert Opinion					

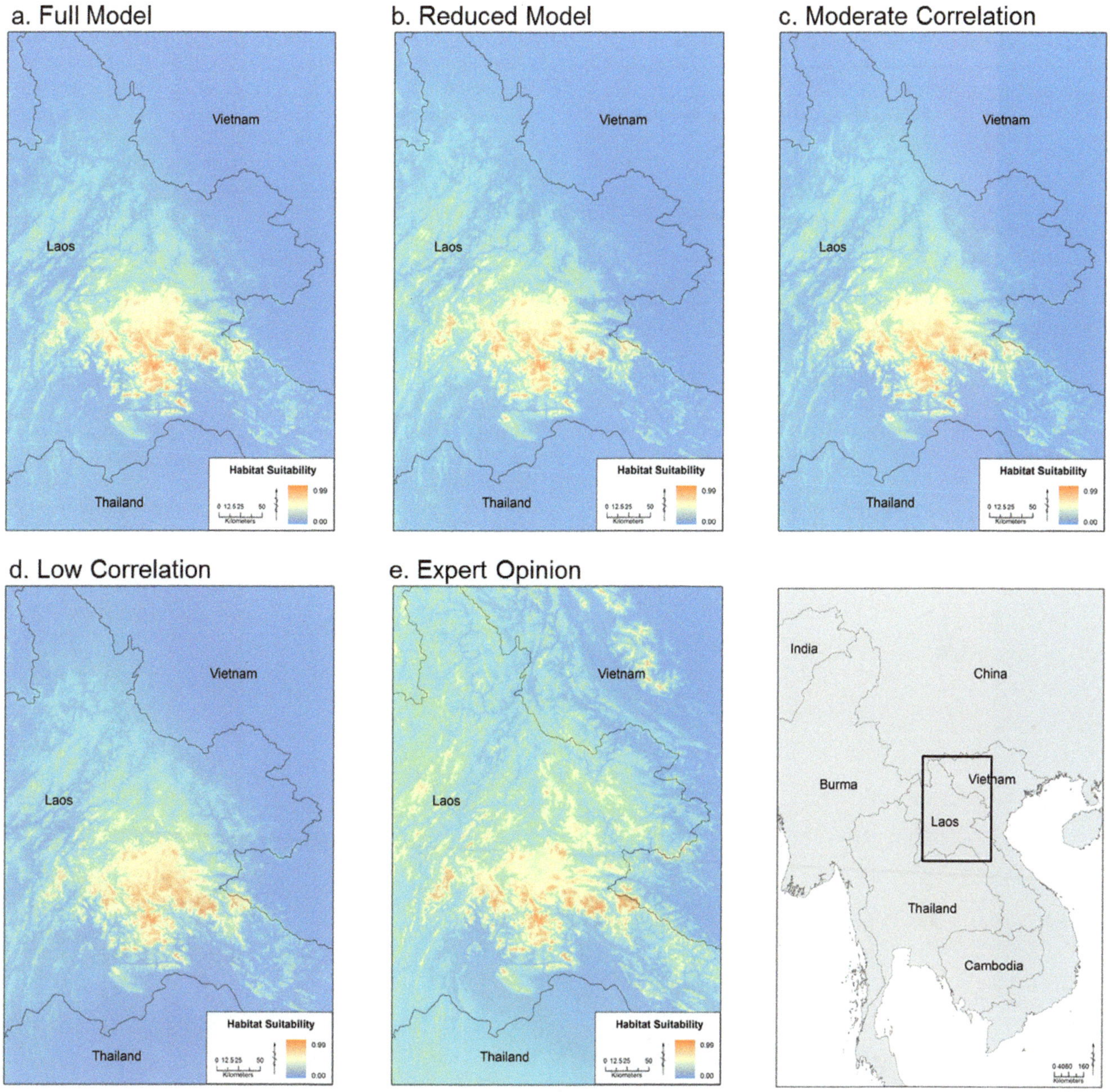

Figure 2. Local MAXENT model results. These results show the habitat suitability for *L. laoensis*, under five unique combinations of environmental variables: a) Full Model, b) Reduced Model, c) Moderate Correlation (r <0.75), d) Low Correlation (r <0.5), e) Expert Opinion (see Table 2). The study area is shown as the boxed area in the inset map on the lower right.

Table 4. The pairwise Pearson's correlation coefficient between all pairs of models at the local scale.

	Full Model	Reduced Model	Moderate Correlation (r<0.75)	Low Correlation (r<0.75)	Expert Opinion
Full Model		1.000	0.991	0.986	0.948
Reduced Model			0.991	0.986	0.948
Moderate Correlation (r<0.75)				0.995	0.957
Low Correlation (r<0.75)					0.962
Expert Opinion					

Table 5. The mean area under the curve (AUC) for each MAXENT model at both the regional and local scale.

	Regional Models AUC	Local Models AUC
Full Model	0.997	0.969
Reduced Model	0.996	0.970
Moderate Correlation (r<0.75)	0.998	0.970
Low Correlation (r<0.75)	0.989	0.969
Expert Opinion	0.993	0.970

U.S. Fish and Wildlife Service cleared importation into the United State (eDec 2012991011). Live vertebrates were handled in accordance with the Institutional Animal Care and Use Committee of the North Carolina State Museum of Natural Sciences Protocol No. NCSM 2011-01.

Model Validation

To validate the models, we compared the results of the MAXENT model run using only field localities to presence and absence data available from interviewing local residents. *Laotriton laoensis* is distinctive, diurnal, and frequently encountered by rural people in Laos who use the streams in which it occurs to fish and extract natural resources. Thus, the species is usually well known to people who live in the same geographic area. Local residents at each survey locality were shown life-sized color photographs exhibiting dorsal and ventral views of live *L. laoensis* and asked whether they knew this animal [7]. Those who recognized it were asked where it lives, how it moves, and to estimate its size in order to verify positive responses. At 13 sites, interviewees reported that the species was present, while at 23 sites, interviews were unfamiliar with the species [7].

To compare the model results with interview data, we converted the continuous logistic output to a binomial classification of either 'suitable' or 'unsuitable' land using a threshold based on the model results. We selected the 'lowest presence threshold' as the threshold for differentiating habitat types. The lowest presence threshold is the smallest logistic value associated with one of the observed species localities [12]. Any area with a logistic value above the threshold was considered suitable, while any area with

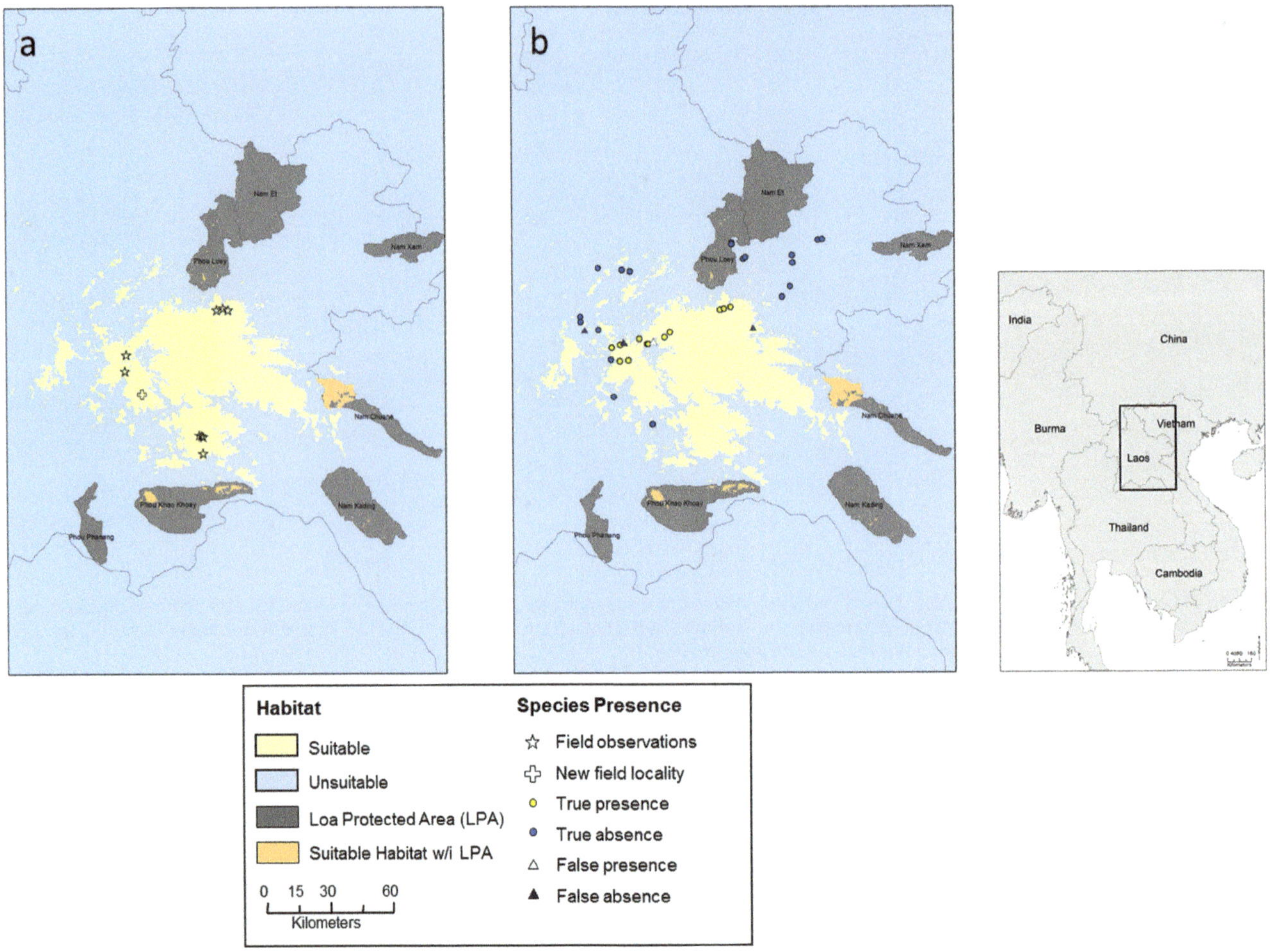

Figure 3. Binomial local MAXENT model results. These results show habitat suitability for *L. laoensis* under a binomial classification of habitat as either suitable (good) or unsuitable (poor). The map was based on the results of the Full Model (see Figure 2a) with a threshold of 0.35. Lao protected areas are also shown in gray, and areas of suitable habitat that fall within the borders of an LPA are in orange. Figure (a) shows localities where *L. laoensis* was found during field surveys while (b) shows locations where interview data was used to identify the presence or absence of *L. laoensis*.

Table 6. Contingency table of the 36 locations where *L. laoensis* presence or absence was determined by interview data.

		Species Known to Interviewees	
		Yes	**No**
Species predicted present by MAXENT model	Yes	12	4
	No	1	19

Each site fell in an area where MAXENT predicted either presence or absence.

a value below the threshold was considered unsuitable. This threshold (like all threshold approaches) is admittedly arbitrary. This threshold was, however, relatively conservative [12] and excluded a significant fraction of the study area while still categorizing each of the nine known localities as occurring in suitable habitat.

We then used this binomial classification to determine whether each of the 36 interview sites occurred in suitable or unsuitable areas. Each interview site thus fell into one of four categories: 1) MAXENT correctly predicted the presence of *L. laoensis* (i.e., true presence), 2) MAXENT correctly predicted absence of *L. laoensis* (i.e. true absence), 3) MAXENT predicted presence in areas where interview data showed the species is absent (i.e., false presence) and 4) MAXENT predicted absence where interview data showed presence (false absence). We used a chi-square test to determine the association between MAXENT predictions and species presence or absence.

We also used the interview sites where the species was recorded as present to further validate the model by rerunning the full model at the local scale using only interview presences as locality data. In addition to the interview data, we used an additional, tenth locality record of *L. laoensis* that was discovered in Laos by SP, NS, and BLS during field surveys in May 2012 immediately following the completion of the initial MAXENT modeling (Table 1). We both compared this test locality to the model predictions, and re-ran the field locality models incorporating this new site to determine whether or not it affected the model results.

Description of Suitable Environment

A good description of suitable habitat is essential for both targeting field surveys and for conservation strategy. To broadly characterize the climatic and altitudinal conditions of areas that were identified as suitable, we used the Full MAXENT model at the local scale. This model contained all the environmental data we used and was focused on the region where this species is known to

Table 7. The percent contribution of each environmental variable to each of the 5 local MAXENT models run.

	% Contribution to Model				
Environmental Variable	**Full Model**	**Reduced Model**	**Moderate Correlation (r<0.75)**	**Low Correlation (r<0.5)**	**Expert Opinion**
Annual Mean Temperature	0.2	N/A	N/A	N/A	0
Mean Diurnal Range in Temperature	0	N/A	0	N/A	N/A
Isothermality	**23.4**	**23.5**	**22.8**	**43.1**	N/A
Temperature Seasonality	**30.6**	**32.1**	**31.2**	N/A	N/A
Maximum Temperature of Warmest Month	1.3	N/A	N/A	N/A	0.4
Minimum Temperature of Coldest Month	0	N/A	N/A	N/A	0
Temperature Annual Range	2.8	N/A	N/A	N/A	N/A
Mean Temperature of Wettest Quarter	0.3	N/A	N/A	N/A	N/A
Mean Temperature of Driest Quarter	0	N/A	N/A	N/A	N/A
Mean Temperature of Warmest Quarter	1.9	N/A	N/A	N/A	0.1
Mean Temperature of Coldest Quarter	0	N/A	N/A	N/A	**16.8**
Annual Precipitation	0	N/A	N/A	N/A	1
Precipitation of Wettest Month	0	N/A	N/A	N/A	N/A
Precipitation of Driest Month	0	N/A	N/A	N/A	N/A
Precipitation Seasonality	0.4	N/A	1.2	0.2	N/A
Precipitation of Wettest Quarter	0.6	N/A	1	N/A	N/A
Precipitation of Driest Quarter	0.1	N/A	N/A	N/A	N/A
Precipitation of Warmest Quarter	0.1	N/A	0.1	0	N/A
Precipitation of Coldest Quarter	0	N/A	0.5	**10.3**	N/A
Altitude	**38.3**	**44.3**	**43.2**	**46.4**	**81.7**

Values of N/A indicate variable that were not included in a particular model. Values over 5% have been bolded.

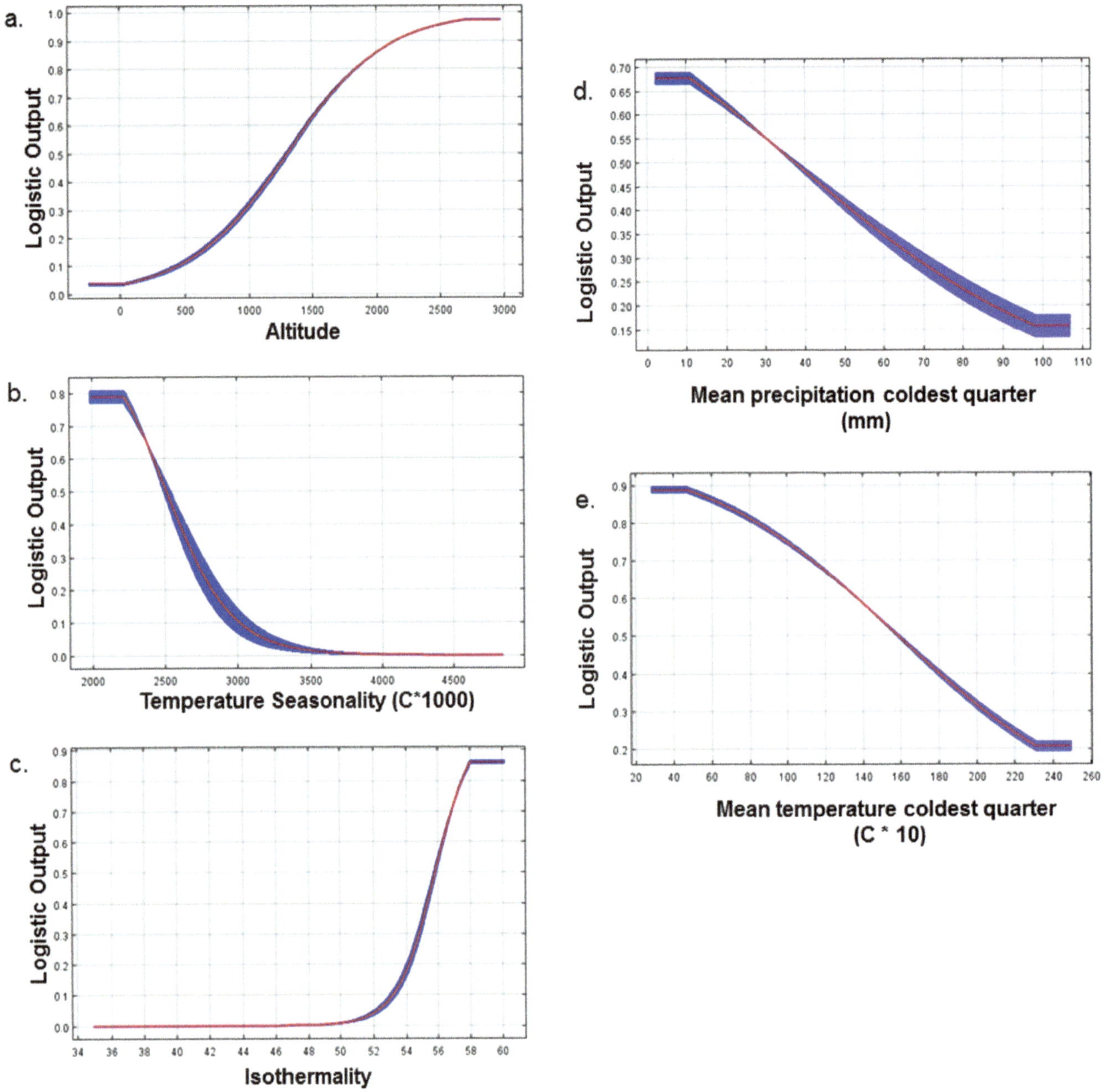

Figure 4. The variable response curves for the Full Maxent model at the local scale.

occur. We evaluated the response curves generated by Maxent for each environmental variable. The response curves show the relationship between the model prediction and each variable [22].

To identify the specific individual environmental factor that was limiting, we used a new tool developed by Elith *et al.* [20]. The limiting factor map identifies the specific environmental variable that most influences that model results at any given point, providing useful data for identifying the drivers of the species distribution (see Elith et al. [20] Appendix S3 for specific information and code). We performed a limiting factor analysis on the full model results at both the regional and the local scale.

To determine whether or not any area identified as suitable overlapped with a protected area, we overlaid the binomial map of suitable and unsuitable habitat with a map of the National Protected Areas in Laos. These were obtained from Protected Planet on March 23, 2012 (www.protectedplanet.net [23]). We then used the 'intersect' tool from Hawth's Tools [24] to determine whether there was any overlap between the predicted habitat and currently protected areas.

Results

Ecological Niche Modeling

We ran several Maxent models that highlighted the climatic and altitudinal suitability for *Laotriton laoensis* using nine published field localities as species presence data. At the regional scale, all five models identified areas of high suitability in northern Laos and additional areas in southern Laos and southern Vietnam (Figure 1). The first three models (Full, Reduced, and Moderate) showed

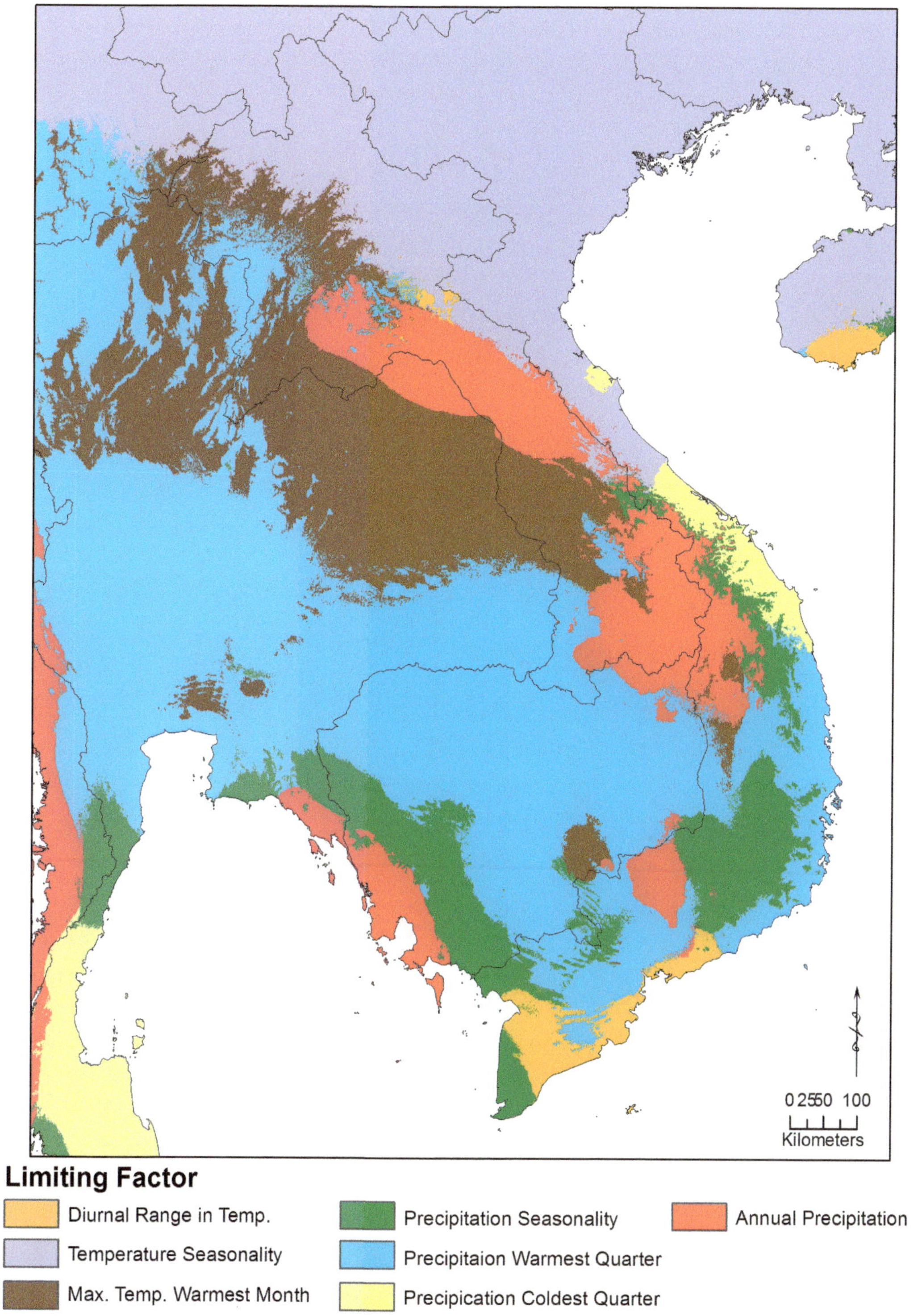

Figure 5. The limiting environmental factors identified at the regional scale.

strong similarity in the areas predicted as suitable (r >0.8, Table 3). The Low Correlation and the Expert Opinion models identified a much greater extent of habitat suitability and less similarity to any of the other models (r <0.7). In particular, the Low Correlation model highlighted several areas of very high suitability. However, because this model was based on the least amount of environmental data, (i.e., only 5 of the 19 environmental layers were included because of the stringent correlation requirements), this model was the least constrained and thus it is likely that this model over-predicted suitable habitat. While the relative amount of suitable habitat differed between the models, all models showed that suitable area in northern Laos is surrounded by unsuitable habitat.

Figure 6. The limiting environmental factors identified at the local scale.

For the local scale models, the areas of suitable habitat were highly similar between all five models, with all correlations above 0.945 (Figure 2, Table 4). Even the Reduced Model, with only three environmental variables, captured the same geographic extent of suitable habitat as the Full Model with 19 variables. Thus, within this more limited geographic extent, only a few key climatic variables are important in identifying suitable habitat. All models identified a limited area in northern Laos as suitable habitat (Figure 2). These models also suggested that there may be additional suitable habitat to the east of the area where this species has previously been documented.

Model Validation

For all models at both the regional and the local scale that were run using field locality data, the area under the curve (AUC) of the receiver operatic characteristic plot (ROC) was quite high (at least 0.969, Table 5). As AUC values above 0.75 are considered informative ([21]) our uniformly high values indicate that all models provided good discrimination between true positives and false positives [25,26]. The default parameters performed well; adjusting the regularization parameter resulted only slight differences in AUC for the full model at the local scale (AUC of 0.970 and 0.967 for regularization parameters of 0.5 and 5 respectively) and resulted in the same order of percent contribution of variables for all variables contributing at least 5% to the model. Thus, all results presented are from models using the default parameters.

To further validate the model results, we first converted the continuous logistic output to a binomial map of good versus poor habitat. We specifically chose the Full Model at the local scale for validation because the local scale covered the entire extent of the species' likely range and the Full Model used all available environmental data. Furthermore, as all models at the local scale were nearly identical in AUC (ranging from 0.969–0.970, Table 5), selecting the Full Model provides the greatest amount of environmental information without sacrificing model performance. The average lowest presence threshold for the Full Model was 0.35, so we used this value as the threshold for dividing suitable and unsuitable habitat (Figure 3).

Using this binomial map, we plotted the locations of both the presence data that went into the MAXENT model ($n = 9$), and the additional sites for which we had interview data ($n = 36$). Given that the field survey locations were used to set the threshold of suitable versus unsuitable habitat, all presence points fell within suitable habitat (Figure 3a). For interview data, points that fell within the suitable habitat (i.e. a logistic value of 0.35 or greater) were considered predicted presence, while those that occurred in unsuitable habitat were considered predicted absence (Figure 3b). When comparing the numbers of points that fell within each category, we found very strong support for the predictive ability of the MAXENT Model (Table 6). The model correctly predicted presence ($n = 12$) and absence ($n = 19$) at 31 of the 36 sites. The remaining five sites were divided between those where MAXENT predicted presence but interviewees were unfamiliar with the species (false presence, $n = 4$) and those where MAXENT predicted absence where interviewees were familiar with the species ($n = 1$). The MAXENT predictions are significantly better than random at predicting presence or absence ($\chi^2_{(1, 36)} = 15.97$, $p < 0.0001$).

When we compared the results of the full model at the local scale using either just interview presence data or field locality data, we found high similarity ($r = 0.854$). When comparing the newly obtained field locality of *L. laoensis* to the predictions of the local scale model using field locality data, we found that this site occurred in an area of high suitability, with a logistic value of 0.63– well above our threshold of 0.35 (Figure 3a). We also re-ran the MAXENT models including this new point, but found that the resultant maps and the percent contribution of the variables were nearly identical to the models that included the original nine localities (results not shown because the new models were indistinguishable from those shown in Figure 2, $r = 0.998$).

Description of Suitable Environment

When evaluating the percent contribution of each environmental variable to each model, altitude, isothermality, and temperature seasonality were uniformly important in all the local scale models in which they were included (Table 7). In addition to these three variables, both temperature and precipitation contributed at least 5% to one additional local scale model (Table 7). Thus, we used these five variables to characterize suitable habitat using the variable response curves.

Altitude always had the highest percent contribution in each of the local scale models (Table 7). Although altitude is often strongly correlated with other environmental variables, it is an easy metric for characterizing habitat while in the field and thus it was useful to include here. The response curve for altitude showed a roughly linear increase in logistic output (and thus habitat suitability) with increasing altitude (Figure 4a). Suitable habitat for *L. laoensis* is therefore almost certainly restricted to high altitude sites over 1,000 m elevation, as concluded from field surveys [7].

The next two most important environmental variables in terms of percent contribution to the models were temperature seasonality and isothermality (Table 7). Both temperature seasonality and isothermality are a measure of the variability in temperature over the course of the year. Logistic output decreases with temperature seasonality (Figure 4b) and increases with isothermality (Figure 4c). Finally, mean precipitation of the coldest quarter and mean temperature of the coldest quarter each had a relatively high percent contribution to one of the local models (the Low Correlation model and the Expert Opinion model respectively, Table 7). The response curves show decreasing logistic output with increasing precipitation (Figure 4d) and temperature (Figure 4e).

The limiting factors analysis shows that, at the regional scale, temperature seasonality, annual precipitation, maximum temperature of the warmest month, and precipitation of the warmest quarter are limiting in northern Laos (Figure 5). At the local scale, temperature seasonality and precipitation of the warmest quarter are still identified as limiting, along with precipitation seasonality, precipitation of the wettest quarter, temperature of the wettest quarter, and isothermality (Figure 6).

When we compared the distribution of suitable habitat for *L. laoensis* with the national protected area system of Laos, we found some suitable habitat patches in Phou Khao Khoay National Protected Area and Nam Chuane Proposed National Protected Areas (Figure 3).

When interpreting these data, it is important to consider that correlation between environmental variables is unavoidable in any study of this sort. Although we have minimized the issue of correlation between variables included in this model by selectively including variables that were only weakly correlated in the Moderate and Low Correlation models, it is possible that the variables reported as having a highly percent contribution to the model are not actually the drivers of the distribution of *L. laoensis*, but are instead important in the model only because those variables are correlated with environmental variables that were not included in the model. Regardless of whether these parameters directly shape the distribution of *L. laoenis* or are only correlated with the true drivers of the distribution, these results can still be used to identify likely areas for identifying other regions where *L.*

laoensis may be found and provide a starting point for experimental work to elucidate the environmental factors most important in driving the realized niche.

Discussion

MAXENT is increasingly being used to identify important areas of conservation for amphibian species [14,27]. Here, we used MAXENT to predict the distribution of a poorly known, threatened species. This work highlighted the use of MAXENT as a relatively inexpensive tool to study such species to establish priority sites for field sampling and implementing conservation measures. In particular, incorporating knowledge from local residents provided strong validation for our model results. While MAXENT has been increasingly used in conservation research, model results are often difficult to validate with field work because of the multiple constraints inherent in conservation. Using local knowledge could be one method for validating results when field surveys are difficult to undertake.

We had several goals in modeling the distribution of *L. laoensis*. First, we established that this species is likely endemic to northern Laos, although a small region of predicted habitat was identified in Vietnam near the border with Laos. Although it is possible that the species could have a larger range than our models suggest, the fact that all models show a restricted range (regardless of area chosen and scale used), the models were validated by interview data, field work has only identified this species in the regions highlighted by the model, and the uniqueness of the species all suggest that this is a highly geographically restricted species. This finding has important implications for establishing conservation policies to manage this potentially endangered species. Preventing the extinction of this species in the wild will depend entirely on conservation efforts within Laos (if it does occur in Vietnam, it likely does so only marginally). Also, the endemic status of the species can be emphasized in education and outreach efforts to promote awareness and national pride in this unique species.

Second, we established that only limited areas of suitable habitat are currently protected. Only two national protected areas, Phou Khao Khoay in Laos and Nam Chuane in Vietnam, had moderate sized patches of suitable land. In both cases, these habitat patches are entirely isolated by a swath of unsuitable habitat from the regions where *L. laoensis* has actually been observed. It is therefore possible that the range of *L. laoensis* may fall entirely outside of established national protected areas. Field surveys of Phou Khao Khoay and Nam Chuane are urgently needed to determine if the species is found within these protected areas (although note that Nam Chuane remains only a proposed protected area and is not yet officially designated as such). If either protected area does support *L. laoensis*, an important next step will be to determine the rate of gene flow between protected and unprotected populations. Given that only a small fraction of the potential range is protected, and potential habitat in these protected areas occurs as isolated patches at the periphery of the range, establishing a new protected area within the core of the range should be considered an essential part of mitigating extinction of this species from overexploitation, habitat degradation, and climate change. Isolated habitat patches were also identified in Nam Kading and Phou Loey National Protected Areas. However, because of the small size and relative isolation of these patches, it is less likely that *L. laoensis* occurs within these protected areas.

Third, in characterizing the habitat, we confirmed that this species is restricted to high elevations (above 1,000 meters) and that altitude, seasonality, and precipitation and temperature of the coldest quarter of the year are particularly important variables in predicting the distribution of this species. Although these models identified correlation rather than causation, it is interesting to note that the coldest quarter of the year is also the *L. laoensis* breeding season ([7]). Laboratory experiments to establish the effect of temperature on survival and development at early life stages might provide important insights into the factors that limit the distribution of this species. The restriction to high altitudes also suggests that this species will be vulnerable to increasing temperatures in SE Asia, as the potential for the species to move to higher elevations is limited.

Finally, we aimed to identify areas for targeting future field surveys for this rare species. All of the MAXENT models highlighted an area in northern Laos to the east of currently known populations as suitable for this species. Interview data strongly supported these model results. Only one site was found where MAXENT predicted absence but interview data revealed that the species was actually present. This point is very close to predicted suitable habitat (Figure 3b). Four sites were predicted present by MAXENT where this species was not recognized as present by interviewees. All four sites were found at the edge between suitable and unsuitable habitat suggesting that this might be marginal habitat for *L. laoensis*. Alternatively, a dispersal barrier would also explain the discrepancy between MAXENT predictions and the actual distribution. Either way, the high predictive power of this model suggests that these results will be highly useful for guiding future field surveys. Furthermore, our use of interview data to validate model results provided a relatively inexpensive way to collect enough data to validate our model in a short amount of time while simultaneously engaging with the local community. We suggest that other studies could take advantage of this methodology to quickly provide data that can be used to construct and validate species distribution models, particularly for species that are reliably recognized by local residents in interviews and that need immediate conservation action.

Laotriton laoensis is currently at risk of extinction from several different causes. Previous work has demonstrated that this species is occasionally consumed for food and used for traditional medicine, and is often sold to commercial collectors for the international pet trade [6,7]. In addition, our work here reveals that *L. laoensis* faces additional threats of a limited geographic range and a high altitude distribution rendering it vulnerable to climate change. It has previously been suggested that this species should be listed as Endangered by the IUCN and considered for listing by CITES [7]. This work reiterates the need for listing, as our results indicate that it is highly unlikely that this species will be found much beyond its currently known range. In addition, because suitable habitat is currently almost entirely unprotected, a new reserve should be established in core habitat to provide needed protection from harvesting and habitat degradation. Finally, our work serves as a case study of how modelers can work more closely with local people to create models that help to efficiently focus resources toward on the ground conservation.

Acknowledgments

We would like to thank Michelle Kovarik, Brooke Wheeler, Jen Costanza, and Jill Hurst for valuable feedback on the manuscript. We would also like to thank Brian Gratwicke, the PLoS One academic editor, and two reviewers for their helpful feedback and suggestions which greatly improved the manuscript.

Author Contributions

Conceived and designed the experiments: BLS AJC. Performed the experiments: BLS AJC SP NS. Analyzed the data: AJC BLS NS SP. Wrote the paper: BLS AJC.

References

1. Stuart SN, Hoffmann M, Chanson JS, Cox NA, Berridge RJ, et al. (2008) Threatened amphibians of the world. Barcelona, Spain; IUCN, Gland, Switzerland; and Conservation International, Arlington, Virginia, USA: Lynx Edicions.
2. IUCN (2008) The IUCN Red List of Threatened Species. Available: http://www.iucnredlist.org/initiatives/amphibians/analysis. Accessed 21 March 2012.
3. Rowley J, Brown R, Bain R, Kusrini M, Inger R, et al. (2010) Impending conservation crisis for southeast asian amphibians. Biology Letters 6: 336–338. 10.1098/rsbl.2009.0793.
4. Brown RM, Stuart BL (2012) Patterns of biodiversity discovery through time: An historical analysis of amphibian species discoveries in the Southeast Asian mainland and adjacent island archipelagos. In: Gower DJ, Johnson KG, Richardson JE, Rosen BR, Rüber L, Williams ST, editors. Biotic Evolution and Environmental Change in Southeast Asia. Cambridge University Press. 348–389.
5. Stuart BL, Papenfuss TJ (2002) A new salamander of the genus *Paramesotriton* (Caudata: Salamandridae) from Laos. J Herpetol 36: 145–148.
6. Stuart BL, Rhodin AGJ, Grismer LL, Hansel T (2006) Scientific description can imperil species. Science 312: 1137–1137.
7. Phimmachak S, Stuart BL, Sivongxay N (2012) Distribution, natural history, and conservation of the Lao newt (*Laotriton laoensis*) (Caudata: Salamandridae). J Herpetol 46: 120–128.
8. van Dijk PP, Stuart B (2004) *Paramesotriton laoensis*. In: IUCN 2011. IUCN Red List of Threatened Species. Version 2011.2. Available: http://www.iucnredlist.org/details/59461/0. Accessed 19 March 2012.
9. Goldschmidt T, Koehler G (2007) New species of the *Hygrobates salamandrarum*-group (Acari, Hydrachnidia, Hygrobatidae) from Southeast Asia. Zoologischer Anzeiger - A Journal of Comparative Zoology 246: 73–89.
10. Guisan A, Thuiller W (2005) Predicting species distribution: Offering more than simple habitat models. Ecol Lett 8: 993–1009.
11. Raxworthy CJ, Martinez-Meyer E, Horning N, Nussbaum RA, Schneider GE, Ortega-Huerta MA, Townsend PA (2003) Predicting distributions of known and unknown reptile species in Madagascar. Nature 426: 837–841.
12. Pearson RG, Raxworthy CJ, Nakamura M, Townsend Peterson A (2007) Predicting species distributions from small numbers of occurrence records: A test case using cryptic geckos in Madagascar. J Biogeogr 34: 102–117.
13. Bourg NA, McShea WJ, Gill DE (2005) Putting a cart before the search: Successful habitat prediction for a rare forest herb. Ecology 86: 2793–2804.
14. Urbina-Cardona J, Flores-Villela O (2010) Ecological-niche modeling and prioritization of conservation-area networks for Mexican herpetofauna. Conserv Biol 24: 1031–1041.
15. Tolley KA, De Villiers AL, Cherry MI, Measey GJ (2010) Isolation and high genetic diversity in dwarf mountain toads (*Capensibufo*) from South Africa. Biol J Linn Soc 100: 822–834.
16. Phillips SJ, Anderson RP, Schapire RE (2006) Maximum entropy modeling of species geographic distributions. Ecol Model 190: 231–259.
17. Elith J, Graham CH, Anderson RP, Dudík M, Ferrier S, et al. (2006) Novel methods improve prediction of species' distributions from occurrence data. Ecography 29: 129–151.
18. Hijmans RJ, Cameron SE, Parra JL, Jones PG, Jarvis A (2005) Very high resolution interpolated climate surfaces for global land areas. Int J Climatol 25: 1965–1978.
19. Roedder D, Loetters S (2009) Niche shift versus niche conservatism? Climatic characteristics of the native and invasive ranges of the Mediterranean house gecko (*Hemidactylus turcicus*). Global Ecol Biogeogr 18: 674–687.
20. Elith J, Kearney M, Phillips S (2010) The art of modelling range-shifting species. Methods in Ecology and Evolution 1: 330–342.
21. Phillips SJ, Dudik M (2008) Modeling of species distributions with Maxent: New extensions and a comprehensive evaluation. Ecography 31: 161–175.
22. Phillips S (2006) A brief tutorial on Maxent. AT & T Research Available: http://www.cs.princeton.edu/schapire/maxent/tutorial/tutorial.doc.
23. IUCN and UNEP (2010) The World Database on Protected Areas (WDPA). UNEP-WCMC. Cambridge, UK. Available: http://protectedplanet.net. Accessed 23 March 2012.
24. Beyer HL (2004) Hawth's analysis tools for ArcGIS. Available at http://www spatialecology com/htools.
25. Pearce J, Ferrier S (2000) Evaluating the predictive performance of habitat models developed using logistic regression. Ecol Model 133: 225–245.
26. Fielding AH, Bell JF (1997) A review of methods for the assessment of prediction errors in conservation presence/absence models. Environ Conserv 24: 38.
27. Embert D, Reichle S, Larrea-Alcázar DM, Cortez C, Muñoz A, et al. (2011) Priority areas for amphibian conservation in a neotropical megadiverse country: The need for alternative, non place based, conservation. Biodiversity & Conservation 20: 1557–1570.

11

The Genetic Integrity of the *Ex Situ* Population of the European Wildcat (*Felis silvestris silvestris*) Is Seriously Threatened by Introgression from Domestic Cats (*Felis silvestris catus*)

Kathrin A. Witzenberger[1,2], Axel Hochkirch[2*]

1 Zoo Hoyerswerda, Hoyerswerda, Germany, **2** Trier University, Department of Biogeography, Trier, Germany

Abstract

Studies on the genetic diversity and relatedness of zoo populations are crucial for implementing successful breeding programmes. The European wildcat, *Felis s. silvestris*, is subject to intensive conservation measures, including captive breeding and reintroduction. We here present the first systematic genetic analysis of the captive population of *Felis s. silvestris* in comparison with a natural wild population. We used microsatellites and mtDNA sequencing to assess genetic diversity, structure and integrity of the ex situ population. Our results show that the ex situ population of the European wildcat is highly structured and that it has a higher genetic diversity than the studied wild population. Some genetic clusters matched the breeding lines of certain zoos or groups of zoos that often exchanged individuals. Two mitochondrial haplotype groups were detected in the in situ populations, one of which was closely related to the most common haplotype found in domestic cats, suggesting past introgression in the wild. Although native haplotypes were also found in the captive population, the majority (68%) of captive individuals shared a common mtDNA haplotype with the domestic cat (*Felis s. catus*). Only six captive individuals (7.7%) were assigned as wildcats in the STRUCTURE analysis (at K = 2), two of which had domestic cat mtDNA haplotypes and only two captive individuals were assigned as purebred wildcats by NewHybrids. These results suggest that the high genetic diversity of the captive population has been caused by admixture with domestic cats. Therefore, the captive population cannot be recommended for further breeding and reintroduction.

Editor: William J. Murphy, Texas A&M University, United States of America

Funding: This work is part of a PhD project that is financially supported by the Scholarship Programme of the German Federal Environmental Foundation. The funders had no role in study design, data collection and analysis, decision to publish, or preparation of the manuscript.

Competing Interests: The authors have declared that no competing interests exist.

* Email: hochkirch@uni-trier.de

Introduction

Due to the ongoing threats to biodiversity, species conservation is still a challenging task [1]. Although in situ conservation represents by far the most effective way to protect endangered species, it is evident that not all species can be preserved in their natural habitats. Therefore, ex situ conservation and reintroductions have become common measures of species conservation [2,3]. Despite the great advances in organization and logistics (studbooks, online databases) and increasing expertise in veterinary medicine, inbreeding and outbreeding still pose serious threats for ex situ conservation [4–7]. Captive breeding may be compromised by unknown founder origin and relationships [8], undetected hybridisation [9], limited representation of natural genetic diversity [10] or studbook errors [11,12]. These problems can affect both the genetic integrity of the captive population as well as the success of reintroduction projects. Therefore, data on the genetic diversity and relatedness in zoo populations can provide valuable information for improving conservation programmes [13].

The influence of captive breeding on the genetic diversity of endangered species has already been subject to a number of theoretic and genetic studies [7,14–16,17]. However, the majority of the genetic studies focus on highly endangered species, for which studbooks or international breeding programmes already exist [18–20]. This circumstance has two drawbacks: (1) Species with small population size might have a reduced genetic diversity even in their wild populations. Therefore, the genetic diversity of captive populations of such highly endangered species might not reflect the effects of captive breeding. (2) Studbooks are usually kept in a way to avoid the loss of genetic diversity or inbreeding depression. Hence, it is likely that genetic analyses of ex situ populations will confirm the success of such coordinated breeding schemes [16,21]. However, studbooks exist only for ca. 850 species (WAZA 2010), while ca. 10.000 species are currently kept in zoos and bred without any coordination (ISIS 2010). Many of these species are still rather common in the wild, but nevertheless some of them are regionally endangered and sometimes even subject to reintroductions. In order to assess the influence of uncoordinated breeding on the genetic diversity of a captive population, it is useful to study one of these species rather than well managed captive populations. Comparing the genetic variability of ex situ populations and wild populations also requires that a target species has viable wild populations and a sufficiently large captive population. Furthermore, it should have been successfully bred for many generations in captivity as genetic erosion may accumulate over time [18].

The European wildcat (*Felis silvestris silvestris*) is an ideal study object for such an analysis. It is still rather common in Europe and classified as Least Concern on the IUCN Red List of Threatened Species [22]. However, the populations of this species are fragmented and in some countries declining [22]. Wildcats are kept and bred in many European zoos and captive individuals have already been reintroduced in three regions in Germany [23]. However, to date no coordinated breeding programme or studbook exists for the European wildcat. Hence, there is no information available on the captive stock concerning founder size, founder origin, captive population size and relatedness. As it is planned to establish a studbook for the European wildcat in the near future (A. Sliwa pers. comm.), a genetic analysis of the breeding stock can provide important basic information for this studbook. Another challenge in the conservation of the European wildcat is hybridization with domestic cats, which is a serious threat in some wild populations [24–26]. Thus, an additional advantage of a genetic analysis prior to coordinated breeding is the exclusion of potential hybrids.

Here, we present the first systematic genetic analysis of the European captive population of *Felis s. silvestris*. We used both microsatellites and mitochondrial DNA (mtDNA) sequencing in order to assess the genetic variability, structure and integrity of the *ex situ* population. We also analysed samples from a wild population to evaluate if the genetic variability in the captive population is comparable to the diversity found in the wild.

Methods

Sampling and DNA extraction

We contacted 124 zoos, 64 of which confirmed keeping European wildcats. We obtained 80 samples (12 buccal swabs, 8 tissue samples from deceased individuals, 10 blood samples and 50 hair samples) from 30 zoos spanning six European countries (see Text S2 in File S1 and Tab. S4 in File S1 for an overview). All ex situ samples were taken during routine veterinary treatments (like vaccination, or the placement of transponders for juveniles) or whenever a veterinary treatment was necessary. The collection of hair samples does not pose a severe stress or hurt to the cats, the same is true for buccal swaps from young individuals. Blood samples were only taken from anaesthetized individuals.

The in situ sample consisted of 89 individuals (26 tissue samples and 63 hair samples) from a natural wild population in the Harz mountains (Germany). The hair samples were collected in the context of a radio telemetry study (M. Götz). The tissue samples were from a monitoring program for roadside casualties. Both studies were carried out with the permit of the Landesamt für Umweltschutz Sachsen-Anhalt which included the collection of samples from European Wildcats (see Text S1 in File S1 for further information). Additionally, we received six samples (3 tissue and 3 hair samples) from road kills and live traps collected in the wild populations in Rhineland-Palatinate and Saarland.

Furthermore, we sampled 33 domestic cats with the permission of their owners, in order to detect hybrids. 15 of the domestic samples came from a private animal shelter which was located near the in situ population. We chose this shelter, as it would provide a realistic insight into the potential influence local feral cats might have on the wild population. For the remaining samples we contacted private owners. The private owners tore out a small bunch of hair, so that the sample included the follicles. This bunch was then sent or given to us in an envelope.

DNA was extracted from buccal swabs, tissue and blood using the DNeasy blood and tissue kit (Qiagen) following the manufacturer's instructions (with special protocols for blood and buccal swabs as provided by the manufacturer). For the hair samples, we extracted DNA with a modified Chelex 100 protocol, using a 10% Chelex solution with addition of 7 µl Proteinase K (18 mg/ml) per sample and overnight lysis [27,28]. No DNA could be extracted from seven of the 208 samples (5 from wild populations and 2 from the captive population). The respective samples from the wild population were hair samples which had been taken several years ago and had then been stored at room temperature. The captive samples had probably been conserved in denatured ethanol.

Genotyping

The six samples from the wild caught individuals in Saarland and Rhineland-Palatinate were not genotyped, as the samples size from these populations was too small for population genetic analysis. They served as reference samples for the mtDNA analysis (see below). Therefore only 194 of the 201 samples were genotyped at ten di-nucleotide repeat microsatellite loci (FCA08, FCA23; FCA43; FCA58, FCA77, FCA78, FCA90, FCA126, FCA132 and FCA149) characterized in the domestic cat [29,30]. Amplification was performed in a Mastercycler (Eppendorf) using the 2.5×5PRIME HotMasterMix (5PRIME). For each PCR we used 5 µl reaction mix containing 1.2 µl genomic DNA, 2.2 µl HotMasterMix, 2.2 µl water and 0.1 µl forward and reverse primers. The PCR conditions were as recommended by the manufacturer, with an annealing temperature of 55°C for most primers (exceptions: FCA90: 60°C; FCA78: 50°C and addition of 2.5 mM Mg^{2+}). The 5′-end of each forward primer was labelled with a fluorescent dye, either 5-FAM, TAMRA or JOE. The products were genotyped on a MegaBACE 1000 automated DNA sequencer (GE Healthcare). Fragment lengths were determined using Fragment Profiler 1.2 (Amersham Biosciences). To minimize genotyping errors due to low DNA concentrations (especially in the hair samples), we applied a multiple tube approach as recommended by Taberlet et al. [31]. Samples were only included into further analyses, if they yielded unambiguous results in three independent replications (190 samples met this criterion, Tab. S4 in File S1). Two captive samples had identical genotypes in all three replications and it turned out that this individual had been transferred to another zoo and sampled twice. Therefore, the duplicate sample was excluded from further statistical analysis and this left 189 samples for analysis.

Sequencing

For 70 wild, 66 captive and 33 domestic individuals we sequenced the mitochondrial DNA fragment NADH dehydrogenase subunit 5 (ND5) based on the study by Driscoll et al. [32]. Most juveniles of sampled mothers and siblings were excluded and later assigned to the same mtDNA haplotype as all sequenced juveniles had the same haplotype as their dam. We finally obtained 88 sequences of wild, 77 of captive and 33 of domestic individuals (Tab. S4 in File S1). The primers CD-ND56-F1C and CD-ND56-R4 [32] were used for amplification in a Multigene Gradient Thermal Cycler (Labnet). We used the 5PRIME HotMasterMix (5PRIME) for amplification. The PCR product was purified using the High pure PCR product purification kit (Roche) according to the manufacturer's protocol. Sequencing was performed with the DYEnamic ET terminator cycle sequencing kit (GE Healthcare) for sequencing reactions run on a MEGABACE 1000 automated sequencer (GE Healthcare). Base-calling was performed in Sequence Analyzer 4.0 (Amersham Biosciences).

Data analysis

Genotyping. As the integration of hybrids and genetic subgroupings were unknown, we first searched for genetic structure in our samples using STRUCTURE 2.3.4 [33]. We assumed admixture between groups and used the correlated allele frequency model with a burn-in period of 100,000 simulations, followed by one million Markov chain Monte Carlo simulations. Tests were run for K = 1-15 with ten iterations for each K. In order to detect potential hybrids between wildcats and domestic cats, we used the STRUCTURE results at K = 2 and assigned all individuals either to the cluster "wildcats", "domestic cats" or "hybrid" [34]. The "hybrid" cluster was chosen, if Q values varied between 0.2 and 0.8 as recommended by Randi and Schulte et al. [35,36]. The optimal values for K were assessed using both the method described by Pritchard et al. [33], and the method suggested by Evanno et al. [37]. The method described by Evanno et al. [37] tends to result in low K values [38,39] and generally works better for scenarios with strong genetic differentiation [40] as it detects the highest level of differentiation. The method described by Pritchard et al. [33] might lead to inflated K-values if closely related individuals exist in the sample. Therefore, the clusters might more likely represent family groups and lineages [41]. In our case the detection of family groups and breeding lines is a desired outcome, which provides deeper insight into the genetic structure of the captive population. We, therefore, present the results obtained by the method described in Pritchard et al. [33]. However, we stopped increasing K when Q values for the next cluster dropped below 0.9 in all individuals as proposed by Schulte et al. [35]. The individuals were assigned to genetic clusters using the highest assignment probability. The dataset was checked for null alleles using MICRO-CHECKER 2.2.3 [42]. We checked all loci for potential linkage using the linkage maps provided by Menotti-Raymond et al. [30,43–45]. A test for linkage disequilibrium was performed in Fstat 2.9.3.2 using a log-likelihood ratio G-statistic with Bonferroni corrections [46]. For the latter test, we removed juveniles from the dataset, as these might erroneously suggest the presence of a linkage disequilibrium due to their close relationships [47].

Measures of genetic differentiation have lately been subject to a broad discussion [48,49]. Thus, we estimated genetic differentiation between the genetic clusters with three different estimates, D_{est}, F_{ST} and R_{ST} using GenAlEx 6.501 [50,51]. F_{ST} is still a useful measure if the population split is rather recent (e.g. for comparing differentiation among captive populations), while R_{ST} is more useful if the split is deep and if it is likely that mutations have already contributed to population differentiation (e.g. for comparing wildcats and domestic cats) [52,53]. When we tested D_{est} estimates for correlation with R_{ST} and F_{ST}, we found a strong positive correlation between D_{est} and F_{ST} ($R^2 = 0.47$, $P = 0.004$) but a negative correlation with R_{ST} ($R^2 = 0.41$, $P = 0.01$). Therefore, only the results of F_{ST} and R_{ST} are provided. For the analyses of population differentiation, we excluded all individuals which could not clearly be assigned to a cluster at K = 2 based on qi>0.8 and where thus assigned as hybrids (see above; all excluded individuals are marked with * in Table S5 in File S1). Nested AMOVAs based on F_{ST} and R_{ST} were performed in GenAlEx with the sample categories "captive", "domestic" and "wild" as regions and the six genetic clusters as populations.

To analyse the degree of hybridization between domestic cats and wildcats, we performed an analysis in NewHybrids 1.1 Beta 3 [54]. We used Jeffreys-type priors for pi and theta, a burnin of 100,000 sweeps and run 1,000,000 sweeps afterwards. Moreover, we calculated a Principal Component Analysis (PCA) using the adegenet package [55] for R 3.0.3 [56] replacing missing data by the mean frequency of the corresponding allele. For this analysis, we assigned all individuals to their most likely STRUCTURE clusters. Wild-caught individuals that clustered with domestic cats in STRUCTURE as well as captive individuals that clustered with wildcats in STRUCTURE were assigned to a unique cluster. We then simulated 100 first generation hybrids as well as 100 backcrosses with either domestic cats or wildcats and performed a second PCA to inspect the degree of overlap of captive individuals with the simulated hybrids.

For comparing genetic diversity, we calculated the mean number of alleles, allelic richness and inbreeding coefficients (F_{IS}) for the captive, domestic and wild samples as well as for the genetic clusters in Fstat. GenAlEx was used to determine the expected (H_e) and observed (H_o) heterozygosity for each locus and each population and to test for deviations from Hardy-Weinberg equilibrium (HWE). Furthermore, we tested the number and frequency of private alleles for domestic cats and wildcats (excluding potential hybrids) and analysed the number of alleles shared with captive cats as well as their mean frequencies.

Effective population sizes (N_e) for each the captive population, domestic population and the wild population (excluding the hybrids) were assessed in ONeSAMP [57], which uses a Bayesian approach. The upper and lower bounds of the prior distribution for N_e were 2 and 500, respectively. The captive and wild samples were tested for genetic signatures of population bottlenecks using the Wilcoxon signed rank statistic implemented in BOTTLENECK 1.2.02 [58]. We examined three different mutational models; the infinite alleles model (IAM), the stepwise mutational model (SMM) and the two-phase mutational model (TPM). However, we mainly considered the TPM, as this is the most likely mutation model for microsatellites [59].

The mean relatedness between individuals within the wild and captive population was calculated using Coancestry [60]. We assessed the performance of all seven relatedness estimators by comparing the results with known relationships and chose the estimator based on Wang [61], which showed the smallest deviation from known relationships and the smallest variance. Additionally, we tested the relatedness between the 19 known breeding pairs of the ex situ population. To measure individual F, four estimators are available in Coancestry. We used the TrioML estimator based on Wang [62] as it fitted our data best based on the relatedness data for the whole captive population (see above).

Sequencing. DNA sequences were corrected and aligned by eye as no indels occurred. We excluded ambiguous data from the beginnings and ends of the fragments in the analyses. The final alignment contained 828 bp (positions 12642–13469 based upon GenBank entry FCU20753). Sequences were deposited in GenBank under the accession numbers KM246612-KM246625. The identification of haplotypes was carried out in DnaSP 5 [63]. For calculating a haplotype network we included 198 sequences (Table S4 & S5 in File S1). To verify the assignment of haplotypes to wildcats and domestic cats, we also performed a second analysis including reference samples for European wildcats and domestic cats published by Driscoll et al. [32] (identical haplotypes are given in Table S5 in File S1). The program TCS 1.18 [64] was used to construct a parsimony-based network representation of the mitochondrial haplotypes with a connection limit of 95%.

Results

Genetic structure

Evidence for null alleles was detected only for single markers in single genetic clusters. All pairwise tests for linkage disequilibrium were non-significant (p>0.05). In the STRUCTURE analysis, we found a clear distinction between individuals from the captive (red cluster in Fig. 1a) and wild (green cluster in Fig. 1a) populations at K = 2. All domestic cats and all wild-caught domestic cats were assigned to the captive cluster (Fig. 1a). At K = 3 (optimal K value according to ΔK; Fig. S3 in File S1) the three groups "wild" (green cluster in Fig. 1b), "captive" (blue cluster in Fig. 1b) and "domestic" (red cluster in Fig. 1b) were clearly distinguished. The most likely number of genetic clusters (K) revealed by STRUCTURE was six (Fig. 1c), reflecting geographical groups of breeding lines (Table 1). For example wildcats assigned to cluster 6 (light blue in Fig. 1c) all originated from Scandinavian zoos or had ancestors from these, cluster 3 (yellow in Fig. 1c) comprised mainly ex situ wildcats stemming from one German zoo, and cluster 4 (dark blue in Fig. 1c) mainly included individuals from eastern European zoos. The wild population (green in Fig. 1c) represented a single cluster together with six captive samples. These six samples came mostly from zoos within the range of the wild populations and one of these individuals was a foundling. However not all of them had wildcat mtDNA haplotypes. Another cluster (red in Fig. 1c) included all domestic samples together with the wild-caught domestic cats, some of which were morphologically suspicious.

Genetic differentiation based upon F_{ST} was significant for all pairs of genetic clusters (p = 0.001, Table 2), but not for R_{ST}. The wild cluster (cluster 2) showed a higher differentiation to the captive clusters (mean $R_{ST} = 0.032$, mean $F_{ST} = 0.187$) than the domestic cluster (mean $R_{ST} = 0.015$, mean $F_{ST} = 0.163$). For the domestic cluster, the highest differentiation occurred to the wild cluster ($R_{ST} = 0.059$; $F_{ST} = 0.254$). The nested AMOVA based on F_{ST} revealed that most of the genetic variation occurred within individuals (80%). Yet, a statistically significant portion was explained by the difference between the six genetic clusters (12%) and between the captive, domestic and wild samples (7%). In the R_{ST}-based AMOVA, the population level (i.e. the clusters) explained no variation at all, but the three categories (i.e. domestic, wild, captive) explained 7% of the variation. With R_{ST} only 14% of the genetic variation occurred within individuals. The private alleles analysis revealed that captive cats shared 30 private alleles only with domestic cats (mean allele frequency of 21.0%±3.1% SE) and six only with wildcats (8.6%±4.2% SE).

When analysing the data in NewHybrids, only two captive individuals were assigned as purebred wildcats, whereas most captive cats were either assigned as F2 hybrids (62 individuals) or backcrosses with domestic cats (13 individuals). The projected inertia of the PCA was 15.7% for the first axis and 6.9% for the second axis. The first axis separated the domestic cats (negatively loaded, cluster 1 in Fig. 2a) from the wildcats (positively loaded, cluster 2 in Fig. 2a). The captive clusters (3–6) had intermediate positions with negative loadings on the first axis. Captive individuals that clustered with the wildcats in the STRUCTURE analysis (cluster 7 in Fig. 2a) had a strong overlap with the wildcat cluster (except for two individuals). Wild-caught individuals that had been assigned to the domestic/hybrid cluster in STRUCTURE had a strong overlap with the domestic cluster in the PCA as well (cluster 8 in Fig. 2a). When performing the same analysis including the simulated hybrids and back-crosses, there was a substantial overlap of the captive clusters with first generation hybrids (clusters 3–6 in Fig. 2b) and of the wild-caught, morphologically suspicious individuals with second generation backcrosses with domestic cats (cluster 8 in Fig 2b).

We found 14 mtDNA haplotypes in our samples (Fig. 3), two of which (WC1, WC2) were found in 31 in situ wildcats, 15 captive individuals as well as the European wildcat reference samples published by Driscoll et al. [32]. Further ten haplotypes (DC1-DC10) were found in domestic cats (including the domestic cat reference samples), six known or suspected in situ hybrids and 45 zoo samples (as well as GenBank sequences from domestic cats). One haplotype found in four captive individuals (WC3) was identical to a haplotype found in an Iberian wildcat (Fsi257) by Driscoll et al. (2007). The remaining haplotype (WC4*) was found in 36 wild individuals from the Harz population as well as two captive individuals. It was closely related to the most common domestic haplotype DC1 (two substitutions). Altogether 68% (n = 45) of the captive individuals had domestic haplotypes, whereas 32% (n = 21) had haplotypes found in wild populations of *F. s. silvestris* (Figs. 1d, 3). Only one captive cluster (STRUCTURE cluster 6) was little affected by mitochondrial introgression with only a single individual possessing a domestic haplotype. However, all individuals belonging to this cluster were assigned either as F2 hybrids or backcrosses with domestic cats in the NewHybrids analysis.

Genetic diversity in captive and wild population

The effective population sizes of the captive, wild and domestic populations were 70±1.74, 32±1.11 and 53±3.02 (mean ± SE), respectively. The captive clusters had a higher allelic richness and higher expected and observed heterozygosities than the wild cluster (Table 3). There was little variation in genetic diversity among the captive clusters.

The captive population showed significant evidence of a recent bottleneck under both IAM and TPM, but not under SMM. For the domestic population there was only significant evidence of a bottleneck under the IAM, whereas the wild population showed no signs of a bottleneck (Table 4). When the genetic clusters were analysed separately, two captive clusters showed no evidence of a bottleneck, whereas all other clusters showed evidence of a bottleneck under the IAM.

The mean relatedness between dyads was $r = -0.044 \pm 0.007$ (mean ± SE) in the wild population and $r = -0.042 \pm 0.004$ in the captive population. The mean individual inbreeding coefficients were 0.149±0.019 in the wild and 0.139±0.019 in the captive population. In both populations most individuals (61%) had an individual inbreeding coefficient <0.125. The number of individuals with an inbreeding coefficient between 0.125 and 0.25 was 18% and 14% for the wild and captive population respectively and the number of individuals with values above 0.25 was 23% and 25%.

Within the breeding pairs we found a mean r of 0.068±0.059. Most breeding pairs (47%) had a low relatedness $r \leq 0.05$. However, there were five breeding pairs which showed relatedness values above the half-sib level (0.25) with two of them almost reaching the full sib level (0.5).

Discussion

Genetic structure of the European wildcat populations

We found four distinct genetic clusters within the captive population of *F. s. silvestris* (Fig. 1c), which reflected their origin from certain zoos or regional groupings of zoos. The differentiation of the clusters is probably caused by the separation of breeding lines of groups of zoos that often exchanged individuals or zoos that were successful breeders. The existence of isolated breeding lines confirms a lack of knowledge in zoos concerning the stocks of other zoos. Therefore, an exchange of individuals is usually restricted to known partners. When testing the data in STRUCTURE, only six captive individuals were assigned to the wild cluster, some of which were foundlings from the in situ populations that had recently been integrated into the captive

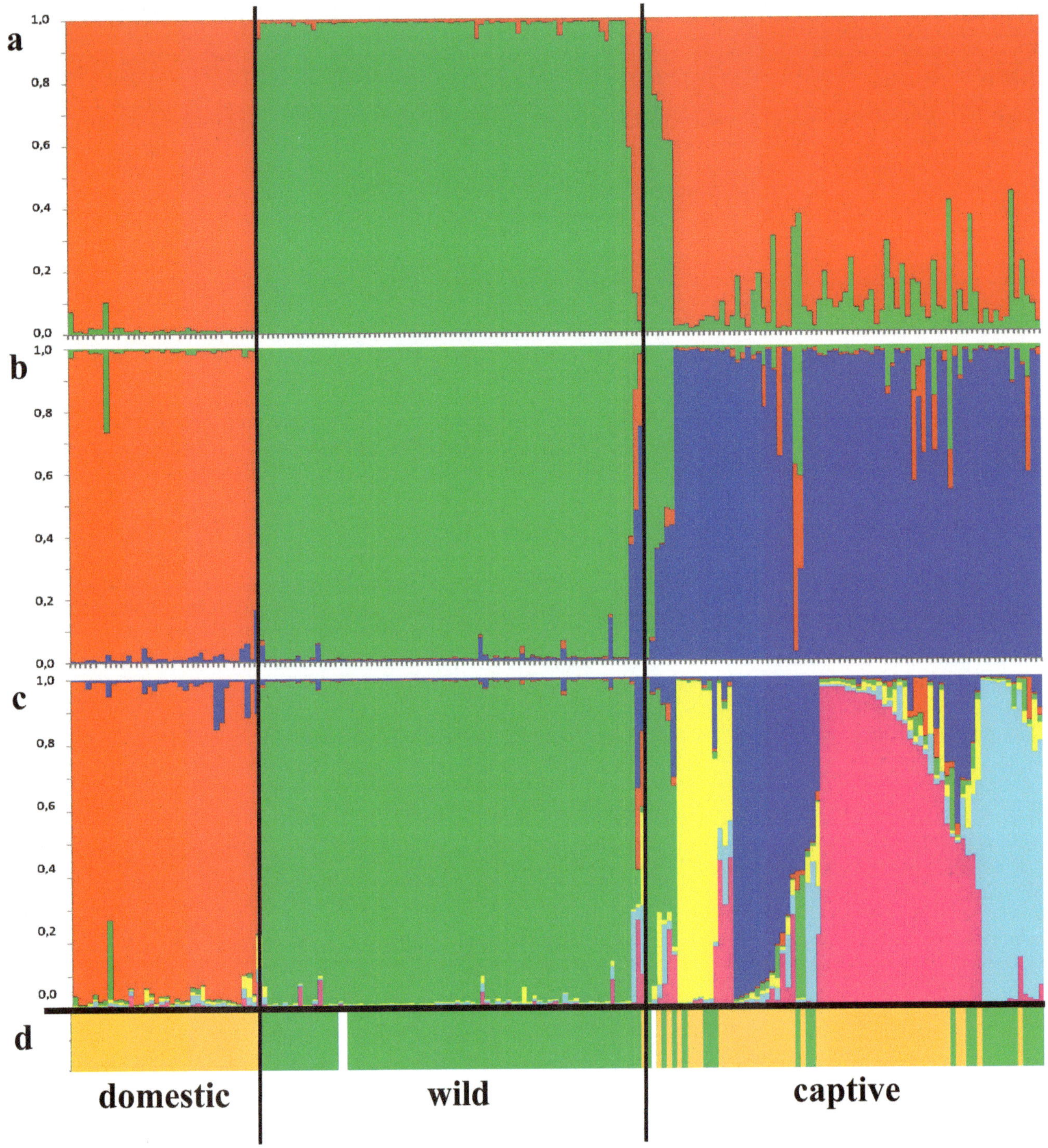

Figure 1. Genetic clusters obtained from the Structure analysis (n = 189). Each individual is represented by a single vertical line, divided into K colours. The coloured segment shows the individual's estimated proportion of membership to that genetic cluster; (a) assignment for K = 2 (red = domestic/captive, green = wild); (b) assignment for K = 3 (red = domestic, green = wild, blue = captive); (c) assignment for K = 6 (cluster 1: domestic; cluster 2: wild, clusters 3–6: captive). (d) Analogue to the cluster assignments the haplotype affiliation for each individual is indicated by a vertical bar (orange = domestic haplotype, green = wildcat haplotype, white = missing data). For the definition of clusters (geographic origin and ancestry of individuals) see Table 1.

population. However, only four of them carried wild mtDNA haplotypes. At K = 2 we found a clear distinction between wild individuals on the one hand and captive individuals together with domestic cats on the other hand. When analysing the data in NewHybrids, only two captive individuals were assigned to purebred wildcats and in the PCA there was a strong overlap of the captive clusters with simulated hybrids and backcrosses. Furthermore, the level of differentiation based upon R_{ST} values was much lower between captive clusters and domestic cats than between captive clusters and wildcats. This suggests that the

Table 1. STRUCTURE based clusters combined with available data on origin and ancestry of the included individuals.

cluster	grouping	individuals lived in or originated from zoos in
cluster 1	domestic/hybrid	all domestic samples and some wild-caught (morphologically suspicious) individuals
cluster 2	wild	wild-caught individuals from the Harz population and single ex situ samples from some German zoos
cluster 3	single German zoo	almost exclusively individuals originating from one German zoo; all four individuals with the Iberian haplotype (WC3) were assigned to this cluster
cluster 4	east-european zoos	cats originating from eastern European zoos or with ancestors from eastern European zoos
cluster 5	various zoos	various zoos
cluster 6	skand. zoos	cats originating from Scandinavian zoos or with ancestors from Scandinavian zoos

captive population is strongly influenced by gene flow with domestic cats and that the captive cats are not representative for the wild population.

In the wild population we found two distinct mtDNA clades, which is in agreement with other genetic studies [65,66]. One mtDNA lineage (consisting of the haplotypes WC1, WC2 and WC3) also included wildcat samples from Rhineland-Palatinate and the Saarland as well as wildcat reference samples of Driscoll et al. [32] (see also table S5 in File S1). This lineage ("wildcat", Fig. 3) was well separated from the other clades and probably represents the autochthonous mitochondrial lineage of the European wildcat (including the substantially differentiated Iberian clade WC3). The second in situ lineage ("wildcat introgressed", WC4*, consisting exclusively of individuals from the Harz population and two captive individuals) was closely related to a common domestic cat mtDNA haplotype (DC1). We, therefore, suspect that this lineage originated by ancient introgression from domestic cats. As this mtDNA haplotype is already derived from the domestic cat haplotype and no domestic cat shared exactly the same haplotype, we suppose that introgression probably occurred hundreds to thousands of years ago (due to the relatively small genetic distances a molecular clock is not applicable). This scenario is a reasonable explanation as the wild population in the Harz region has had a long history of geographic isolation due to the extinction of most adjacent populations [26,65,66]. During this period of isolation the population has most likely declined due to persecution and some of the remaining wild individuals might well have mated with domestic cats. On the other hand, it remains unknown if this lineage is even more widespread in eastern or south-eastern Europe as samples from this region are scarce. Due to the differences to pure domestic cat haplotypes, we treat this introgressed in situ lineage as a second wildcat lineage, which is also representative for a natural population.

In the wild population we found only few individuals (most of which were roadkills that were known or suspected to be hybrids) who shared haplotypes with domestic cats, confirming that hybridization still occurs in the wild but at a low frequency. Hybridization with domestic cats has been detected in several free-ranging populations of the European wildcat [25,26,32,65–69]. In contrast to the past introgression event in the wild population, the captive population seems to be strongly affected by recent introgression, as two thirds of the individuals shared haplotypes with domestic cats (Fig. 1 and 3). We infer, based on the data presented here, that some founders of the captive population were hybrids. Zoos and wildlife sanctuaries are often confronted with stray kittens, which are suspected to be wildcats and are integrated into the captive population (as was the case with some of our sampled individuals). However, there is a high chance that many of these individuals are of hybrid origin. Humans are more likely to encounter litters or stray kittens of domestic cats, as these are tamer and have a higher probability of living close to human settlements [70]. Hybrids are difficult to identify based on their morphology and even the discrimination of some domestic cats from pure wildcats remains difficult [24,32,67,71]. Without genetic screening there remains a high risk of integrating hybrids or backcrosses into the captive population.

Genetic integrity of the captive population

Our results show that the captive population of the European wildcat represents a hybrid swarm. Only six captive individuals (7.7%) were assigned as wildcats in the STRUCTURE analysis (at K = 2), two of which had domestic cat mtDNA haplotypes, and only two individuals were clearly assigned in the analysis with NewHybrids (one with a domestic cat mtDNA haplotype).

Table 2. F_{ST} values (above diagonal) and R_{ST} values (below diagonal) for genetic differentiation between the genetic clusters.

	Cluster 1 (domestic)	Cluster2 (*wild*)	Cluster 3	Cluster 4	Cluster 5	Cluster 6
Cluster 1	---	0.254*	0.185*	0.150*	0.147*	0.169*
Cluster 2	0.059*	---	0.234*	0.183*	0.147*	0.184*
Cluster 3	0.000	0.033	---	0.122*	0.132*	0.181*
Cluster 4	0.023	0.019	0.100*	---	0.108*	0.099*
Cluster 5	0.036*	0.050*	0.114*	0.161*	---	0.114*
Cluster 6	0.000	0.024	0.068*	0.121*	0.074*	---

The in situ samples were assigned to cluster 2, whereas the domestic samples were assigned to cluster 1. Significant values are marked with * (P<0.05).

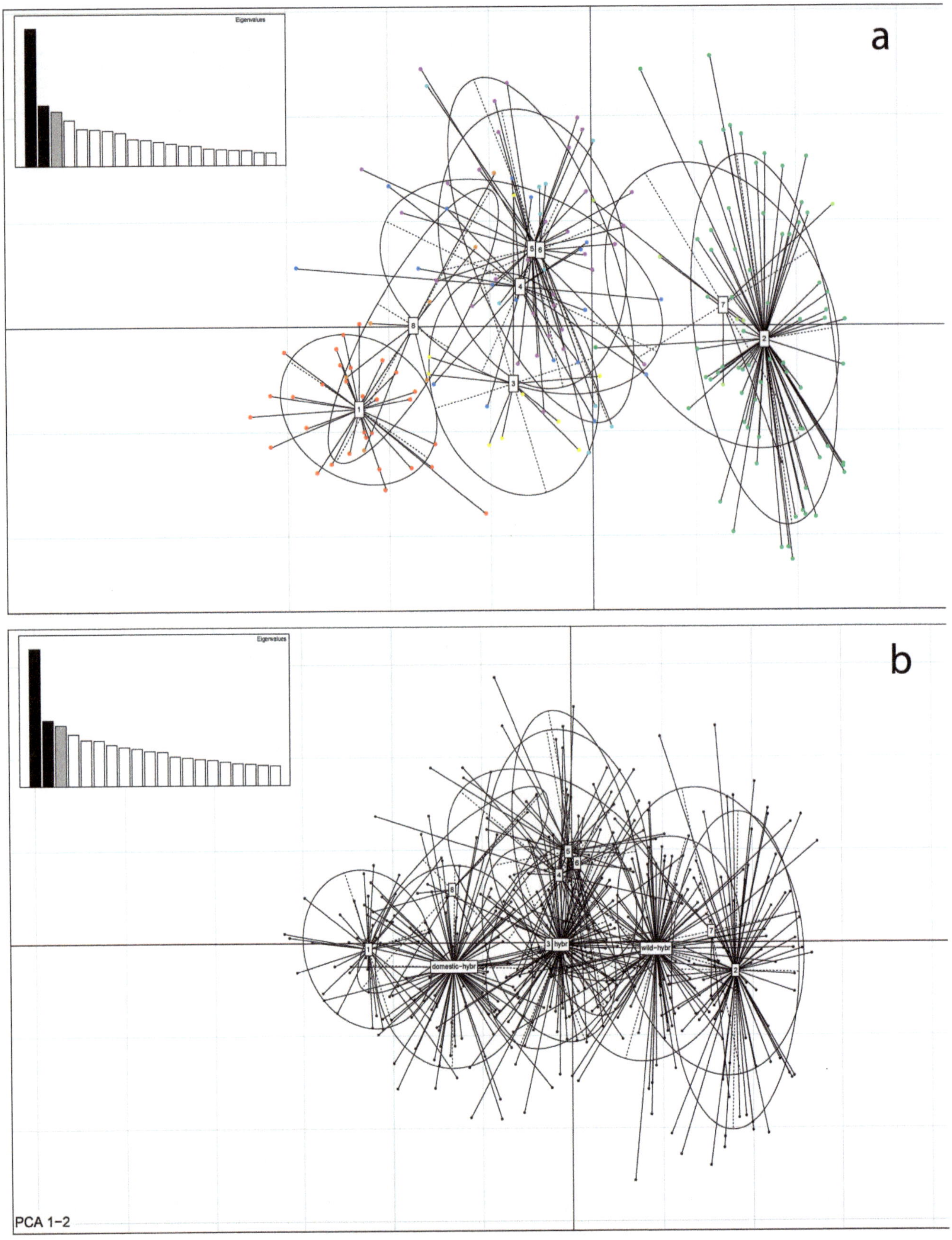

Figure 2. Plot of the first two axes of the Principal component analyses (a) including all obtained genotypes, (b) including simulated first generation hybrids as well as backcrosses with either domestic cats or wildcats. Numbers represent the cluster numbers obtained in STRUCTURE; colours in (a) correspond to the clusters in Figure 1. Captive individuals that were assigned to the wildcat cluster were assigned to a unique group (7) as well as wild-caught individuals that were assigned to the domestic/hybrid cluster (8). The inserted graph shows the distribution of Eigenvalues.

Altogether, 68% of the captive population possess domestic cat mtDNA haplotypes. Therefore, the captive population cannot be recommended for further breeding and reintroduction. As mtDNA is maternally inherited, the presence of domestic cat haplotypes in two thirds of the ex situ population does not allow any conclusions on the total amount of domestic cat nuclear introgression into the captive wildcat gene pool. Furthermore, the amount of introgression from male domestic cats remains unknown, as no Y-chromosome markers have been studied. However, the European wildcat is still widespread and it is possible to obtain new founders from the wild. Particularly the large populations in the Hunsrueck Mountains (Rhineland-Palatinate/Saarland) currently show no indication of past introgression, but more detailed studies are required for any final conclusions. Individuals of hybrid origin cannot be recommended for breeding. This leaves many zoos with the difficult decision of what to do with the individuals which do not qualify as "pure" wildcats. One might question whether the introgression affecting the captive population occurred recently or if it reflects another trace of an ancient introgression event. Hertwig et al. [65] suggested that "pure" wildcats without any traces of past introgression do not exist in Central Europe. They hypothesized that repeated introgression from domestic cats, followed by a spread and diversification of the wildcat populations might have played a major role in the establishment of European wildcat populations. However, the domestic cat haplotypes found in the captive population were only found in few wild individuals, all of which were declared to be morphologically suspicious and most probably were wild-caught domestic cats. Although we did not sample the captive individuals ourselves, we occasionally came across some individuals, which were also morphologically suspect. This supports our hypothesis that recent hybrids were integrated into the captive population as

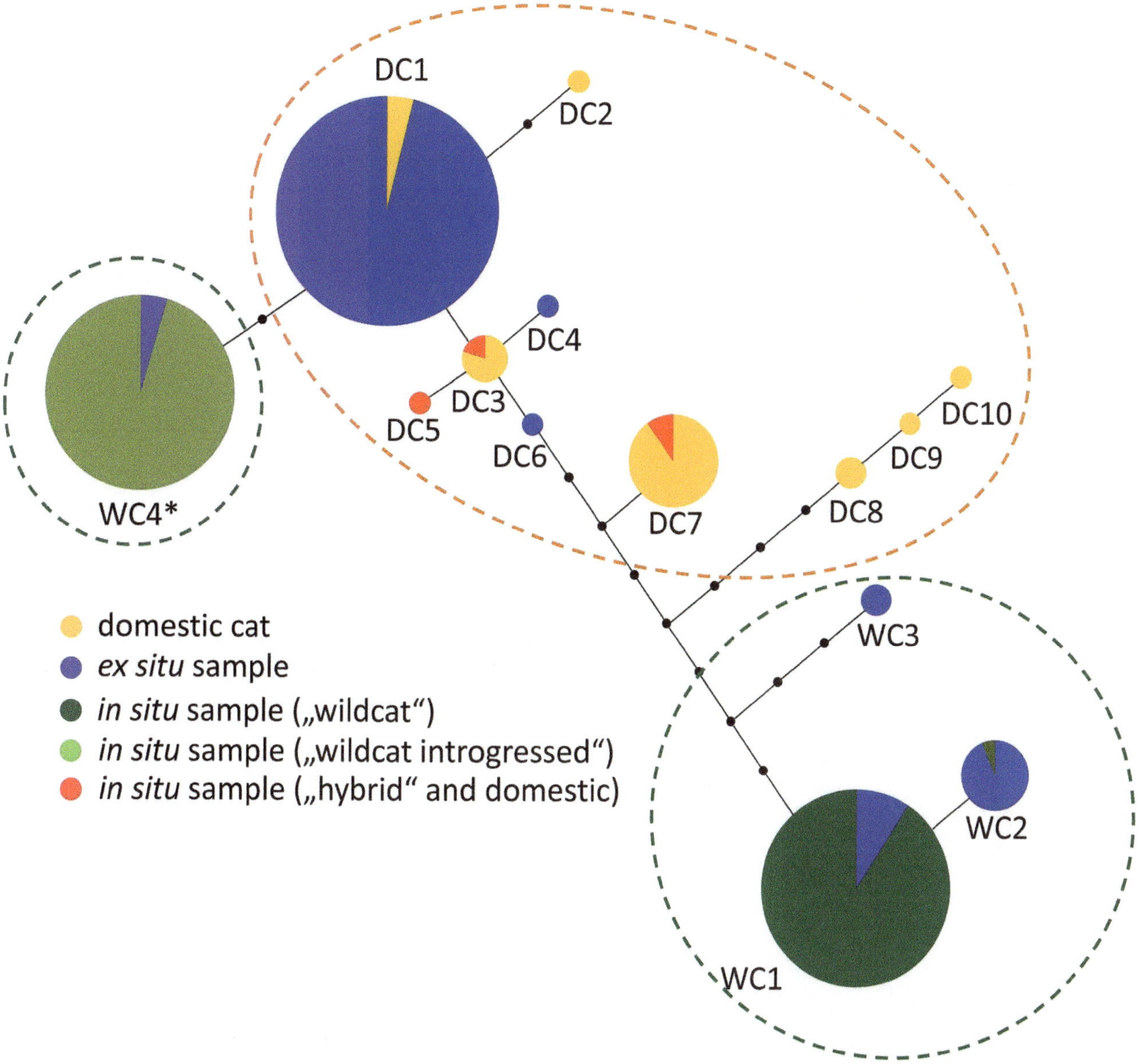

Figure 3. Unrooted parsimony network of mtDNA haplotypes based on an 828 bp sequence of ND5. The sizes of the circles roughly reflect haplotype frequencies in our samples.

Table 3. Genetic diversity and inbreeding in captive, domestic and wild populations and the STRUCTURE based clusters.

population/cluster	n	n_a	A_r	H_o	H_e	F_{IS}
cluster 1 (domestic)	32	8.2	5.9	0.731	0.741	0.005
cluster 2 (wild)	79	4.7	3.3	0.539	0.544	−0.012
cluster 3 (captive)	9	3.8	3.8	0.711	0.603	−0.170
cluster 4 (captive)	17	5.4	4.6	0.600	0.691	0.108
cluster 5 (captive)	33	6.3	4.6	0.709	0.697	−0.033
cluster 6 (captive)	12	4.1	3.9	0.717	0.664	−0.138
mean captive clusters	**71**	**4.9**	**5.7**	**0.684**	**0.649**	**−0.040**
captive	77	7.7	6.9	0.681	0.738	0.085
domestic	32	8.2	8.1	0.741	0.741	0.064
wild	73	4.1	3.7	0.532	0.533	0.021

n = sample size; n_a = number of alleles; A_r = allelic richness based on the lowest samples size (n = 9 for the cluster-based analysis; n = 32 for the three sampled populations); H_o = observed heterozygosity; H_e = expected heterozygosity; F_{IS} = inbreeding coefficient.

founders. Based upon these conclusions we recommend to abstain from any further reintroductions of captive European wildcats from the existing captive stock.

It is very likely that the reintroduced populations of the European wildcat in parts of Germany (e.g. Bavaria, Hesse) represent hybrid swarms as well as they stem from the captive population studied. Removing the reintroduced populations from the wild is not feasible as the administrative effort and costs for such a project would be far too high and it is very unlikely that all individuals could be caught and screened genetically. Moreover, it remains unknown to what extent the reintroduced wildcat populations originated from the captive population or immigrated from wild populations. A genetic survey of the reintroduced populations is therefore needed to assess their status. As the European wildcat currently expands its range (for example in Germany) [22,65,72], further reintroduction programmes are not needed and also not cost-efficient. Nevertheless, regional conservation administrations still aim at reintroduction in areas which are currently not colonized naturally.

Recommendations for a future breeding strategy

Outbreeding is a major threat for the success of *ex situ* conservation programmes [5,7,73]. Some principle decisions on the fate of the captive wildcat population have to be made by the holders and zoo associations. The first question concerns the priority of keeping the European wildcat in zoos. Space is a limited resource in zoos and it might be a wise decision to focus on breeding highly endangered species rather than non-threatened ones. However, zoos also have an educational function, which legitimates the display of native wild fauna. Therefore, if the strategic plan of zoo associations comprises a permanent husbandry of European wildcats, large parts of the current stock need to be replaced with pure wildcats. A future captive population should reflect the genetic structure and diversity found in the wild, particularly in populations which are little affected by hybridization with domestic cats. A sporadic integration of wild individuals into the captive breeding stock has been recommended by several authors [74,75]. In the case of the European wildcat, zoos are in the lucky position that they can obtain individuals from healthy wild populations and, therefore, are able to increase the genetic variability of their breeding stock. Nevertheless, each new individual should be screened genetically before it is integrated into the population.

The current plan to establish a studbook for the European wildcat also requires a coordinated breeding strategy. The most widely accepted management strategy is minimizing kinship [76],

Table 4. One-tailed p-values of the Wilcoxon test for heterozygosity excess under three mutational models for captive population, wild population and the genetic clusters.

population/cluster	IAM	TPM	SMM
captive	0.00049*	0.00244*	0.98389
domestic	0.00488*	0.53906	0.98779
wild	0.04199	0.31250	0.98779
cluster1	0.00684*	0.57715	0.99316
cluster2	0.00684*	0.24609	0.90332
cluster3	0.06543	0.24609	0.57715
cluster4	0.06543	0.34766	0.91992
cluster5	0.00146*	0.27832	0.98389
cluster6	0.00244*	0.00488*	0.11621

IAM: infinite alleles model; TPM: two-phase mutational model; SMM: stepwise mutational model. Significant values are marked with * ($P<0.005$).

which aims at minimizing the overall level of relationship in the population and maximizing the retention of genetic diversity [77,78]. However, for this approach, detailed pedigree data is needed to find optimal breeding pairs [77,79]. Genetic analysis during the establishment of a new captive population can help to minimize the risk of inbreeding and outbreeding. An artificial fragmentation of captive populations into several more or less independent subpopulations has been recommended by several authors to retain a maximum of genetic diversity [74,80]. Although such a fragmentation has been achieved in the current *ex situ* population of *F. s. silvestris* just by chance (due to the missing network of holders), and might have possibly helped to maintain the strong genetic diversity, it is likely that this diversity has mainly been caused by admixture with domestic cats. Increased genetic diversity caused by admixture of several evolutionary lineages has also been reported from invasive non-native populations [35].

Genetic monitoring of *ex situ* conservation projects

The importance of genetic assessments of captive populations has already been recognized decades ago [81]. Molecular genetic studies of captive populations have provided valuable information for the improvement of conservation breeding programmes [7]. These include the detection of hybrids [82], the detection of genetically valuable individuals that had been excluded from breeding [83], the validation of studbook data and detection of errors [11,12], and finally the detection of insufficient representation of genetic diversity found in the wild [10].

The present study shows that genetic analyses provide valuable insight into the status of wild and captive populations. Using two different marker systems (microsatellites, mtDNA) helps to unravel introgression patterns. Future studies should also integrate Y-chromosome markers to detect potential male introgression. The integration of hybrids into ex situ populations can have drastic effects on the genetic composition of captive populations. Hybridisation with closely related species or domesticated congeners represents a severe threat to the conservation of endangered species both in situ and ex situ [35,82,84–87]. Until now, many individuals from the captive population of the European wildcat have been reintroduced to the wild without any test of their genetic integrity. Such reintroductions may strongly deteriorate the genetic integrity of wild populations and promote cryptic extinctions [85]. In some parts of Europe (e.g. Scotland) the wild populations of the wildcat are severely threatened by hybridisation [22]. Therefore, measures to assess and constrict any further admixture of wild and domestic cats are of great importance [22]. Although introgression must have occurred also in nature (perhaps even repeatedly), the European wildcat can still be regarded as a distinct taxon that exists in sympatry with the domestic cat and needs to be managed as a separate unit.

Supporting Information

File S1 Text S1, Permits for sampling. **Text S2**, List of participating institutions for genetic analyses. **Figure S3**, Plots of LnP(D) and delta K (Evanno et al. 2005) for the microsatellite data obtained with STRUCTURE Harvester. **Table S4**, Sampling regime. 1) Most juveniles of sampled mothers and siblings were excluded from sequencing. Their haplotypes were inferred using their maternal relatives (number behind slash). 2) One captive individual was sampled twice and thus the duplicate sample was excluded. 3) The DNA quality of the sample from a preserved captive specimen was not sufficient for sequencing. **Table S5**, Information on the samples used in this study. The table contains all samples which were used in this study, together with their origin, the detected mtDNA haplotypes, the assignments to the STRUCTURE clusters at $K = 2$, $K = 3$ and $K = 6$ as well as the most probable assignment by NewHybrids. At $K = 2$ all individuals which qualify as hybrids based on $qi < 0.8$ are marked with an *. These individuals were excluded for the analyses of population differentiation. For reasons of data protection the origin of the ex situ samples can only be given as encrypted information. Remarks: 1) DNA quality was too low for genotyping; 2) individual was classified as morphologically suspicious; 3) known hybrid; 4) domestic cat (this individual had a wildcat number as it was included in the wildcat samples as a control by the provider of in situ samples); 5) single samples from wild populations as reference for sequencing, but not suitable for population genetic approaches; 6) zoo located next to wild population in the Harz mountains; 7) foundling within region of the wild population Rhineland Palatinate; 8) the cat was transferred to a new zoo which also took a sample, the duplicate sample was excluded from analysis.

Acknowledgments

We would like to thank Carlos Driscoll for supplying us with unpublished data on wildcat phylogeny and for helpful advice and background information on informative sequences. Malte Götz from the Technische Universität Dresden provided the samples from the wild population. Samples from other wild populations were provided by the Centre for Biodocumentation, Katharina Diergarten, Daniel Hoffmann and Sebastian Kirchhoff. We also would like to thank all zoos (Text S2 in File S1) for contributing samples from the captive population. Domestic cat samples were obtained from Jana Deppermann, Winfried Gerlach, Jessica Hillen, Sabine Kettermann, Katrin Scharafin, Heiko Strunk and Kornelia Ociepka. Katrin Scharafin and Petra Willems helped us in the lab with sequencing mtDNA.

Author Contributions

Conceived and designed the experiments: AH KW. Performed the experiments: KW. Analyzed the data: KW AH. Contributed reagents/materials/analysis tools: KW AH. Wrote the paper: KW AH.

References

1. Butchart S, Walpole M, Collen B, van Strien A, Scharlemann J, et al. (2010) Global biodiversity: indicators of recent declines. Science 328: 1164–1168.
2. Fischer J, Lindenmayer DB (2000) An assessment of the published results of animal relocations. Biol Conserv 96: 1–11.
3. Storfer A (1999) Gene flow and endangered species translocations: a topic revisited. Biol Conserv 87: 173–180.
4. WAZA (2005) Building a future for wildlife - The world zoo and aquarium conservation strategy. In: P. . Dollinger, editor editors. Bern: WAZA.
5. Boakes EH, Wang JL (2005) A simulation study on detecting purging of inbreeding depression in captive populations. Genet Res 86: 139–148.
6. Frankham R, Ballou JD, Briscoe DA (2010) Introduction to conservation genetics. Cambridge: University Press. 618 p.
7. Witzenberger KA, Hochkirch A (2011) Ex situ conservation genetics: a review of molecular studies on the genetic consequences of captive breeding programmes for endangered animal species. Biodivers Conserv 20: 1843–1861.
8. Cuaron AD (2005) Further role of zoos in conservation: Monitoring wildlife use and the dilemma of receiving donated and confiscated animals. Zoo Biol 24: 115–124.
9. Marshall TC, Spalton JA (2000) Simultaneous inbreeding and outbreeding depression in reintroduced Arabian oryx. Anim Conserv 3: 241–248.
10. Tzika AC, Remy C, Gibson R, Milinkovitch MC (2009) Molecular genetic analysis of a captive-breeding program: the vulnerable endemic Jamaican yellow boa. Conserv Genet 10: 69–77.

11. Bowling AT, Zimmermann W, Ryder O, Penado C, Peto S, et al. (2003) Genetic variation in Przewalski's horses, with special focus on the last wild caught mare, 231 Orlitza III. Cytogenet Genome Res 102: 226–234.
12. Witzenberger KA, Hochkirch A (2013) Evaluating ex situ conservation projects: Genetic structure of the captive population of the Arabian sand cat. Mamm Biol 78: 379–382.
13. O'Brien SJ (2006) Animal conservation genetics – an overview with relevance to captive breeding programmes. EAZA News 57: 26–35.
14. Leberg PL, Firmin BD (2008) Role of inbreeding depression and purging in captive breeding and restoration programmes. Mol Ecol 17: 334–343.
15. Hedrick PW, Miller PS, Geffen E, Wayne R (1997) Genetic evaluation of the three captive Mexican wolf lineages. Zoo Biol 16: 47–69.
16. Henry P, Miquelle D, Sugimoto T, McCullough DR, Caccone A, et al. (2009) *In situ* population structure and *ex situ* representation of the endangered Amur tiger. Mol Ecol 18: 3173–3184.
17. Boakes EH, Wang JL, Amos W (2007) An investigation of inbreeding depression and purging in captive pedigreed populations. Heredity 98: 172–182.
18. Kraaijeveld-Smit FJL, Griffiths RA, Moore RD, Beebee TJC (2006) Captive breeding and the fitness of reintroduced species: a test of the responses to predators in a threatened amphibian. J Appl Ecol 43: 360–365.
19. Lau J, Alberts AC, Chemnick LG, Gerber GP, Jones KC, et al. (2009) Isolation and characterization of 23 polymorphic microsatellite loci for a West Indian iguana (*Cyclura pinguis*) from the British Virgin Islands. Mol Ecol Res 9: 1412–1414.
20. Roldán VA, Navarro JL, Gardenal CN, Martella MB (2010) May captive populations of Greater rhea (*Rhea americana*) act as genetic reservoirs in Argentina? Zoo Biol 29: 1–6.
21. Ramirez O, Altet L, Ensenat C, Vilà C, Sanchez A, et al. (2006) Genetic assessment of the Iberian wolf *Canis lupus signatus* captive breeding program. Conserv Genet 7: 861–878.
22. Driscoll C, Nowell K (2010) *Felis silvestris*. 2013 IUCN Red List of Threatened Species. IUCN
23. Grabe H, Worel G (2001) Die Wildkatze. Zurück auf leisen Pfoten. Amberg: Buch und Kunstverlag Oberpfalz.
24. Macdonald DW, Yamaguchi N, Kitchener AC, Daniels M, Kilshwa K, et al. (2010) Reversing cryptic extinction: the history, present, and future ot the Scottish wildcat. In: D. W. . Macdonald and A. . Loveridge, editors. Biology and conservation of wild felids. Oxford, UK: Oxford University Press. pp.471–491.
25. Lecis R, Pierpaoli M, Birò ZS, Szementhy L, Ragni B, et al. (2006) Bayesian analyses of admixture in wild and domestic cats (*Felis silvestris*) using linked microsatellite loci. Mol Ecol 15: 119–131.
26. Pierpaoli M, Birò ZS, Herrmann M, Hupe K, Fernandes M, et al. (2003) Genetic distinction of wildcat (*Felis silvestris*) populations in Europe, and hybridization with domestic cats in Hungary. Mol Ecol 12: 2585–2598.
27. Walsh PS, Metzger DA, Higuchi R (1991) Chelex® 100 as a medium for simple extraction of DNA for PCR-based typing from forensic material. BioTechniques 10: 506–513.
28. Estoup A, Largiader CR, Perrot E, Chourrout D (1996) Rapid one-tube DNA extraction for reliable PCR detection of fish polymorphic markers and transgenes. Mol Mar Biol Biotechnol 5: 295–298.
29. Menotti-Raymond M, O'Brien SJ (1995) Evolutionary conservation of ten microsatellites loci in four species of felidae. The Journal of Heredity 86: 319–322.
30. Menotti-Raymond M, David VA, Lyons LA, Schäffer AA, Tomlin JF, et al. (1999) A genetic linkage map of microsatellites in the domestic cat (*Felis catus*). Genomics 57: 9–23.
31. Taberlet P, Griffin S, Goossens B, Questiau S, Manceau V, et al. (1996) Reliable genotyping of samples with very low DNA quantities using PCR. Nucleic Acids Res 24: 3189–3194.
32. Driscoll CA, Menotti-Raymond M, Roca AL, Hupe K, Johnson WE, et al. (2007) The near eastern origin of cat domestication. Science 317: 519–523.
33. Pritchard JK, Stephens M, Donnelly P (2000) Inference of population structure using multilocus genotype data. Genetics 155: 945–959.
34. Vähä J-P, Primmer C (2006) Efficiency of model-based Bayesian methods for detecting hybrid individuals under different hybridization scenarios and with different numbers of loci. Mol Ecol 15: 63–72.
35. Schulte U, Veith M, Hochkirch A (2012) Rapid genetic assimilation of native wall lizard populations (Podarcis muralis) through extensive hybridization with introduced lineages. Mol Ecol 21: 4313–4326.
36. Randi E (2007) Detecting hybridization between wild species and their domesticated relatives. Mol Ecol 17: 285–293
37. Evanno G, Regnaut S, Goudet J (2005) Detecting the number of clusters of individuals using the software STRUCTURE: a simulation study. Mol Ecol 14: 2611–2620.
38. Campana MG, Hunt HV, Jones H, White J (2010) CorrSieve: software for summarizing and evaluating Structure output. Mol Ecol Res.
39. Hausdorf B, Hennig C (2010) Species delimitation using dominant and codominant multilocus markers. Syst Biol 59: 491–503.
40. Waples R, Gaggiotti O (2006) What is a population? An empirical evaluation of some genetic methods for identifying the number of gene pools and their degree of connectivity. Mol Ecol 15: 1419–1439.
41. Rodrigues ST, Wang J (2012) The effect of close relatives on unsupervised Bayesian clustering algorithms in population genetic structure analysis. Mol Ecol Res 12: 873–884.
42. Van Oosterhout C, Hutchinson WF, Wills DPM, Shipley P (2004) MICRO-CHECKER: software for identifying and correcting genotyping errors in microsatellite data. Mol Ecol 4: 535–538.
43. Menotti-Raymond M, David VA, Chen ZQ, Menotti KA, Sun S, et al. (2003) Second-generation integrated genetic linkage/radiation hybrid maps of the domestic cat (*Felis catus*). J Hered 94: 95–106.
44. Menotti-Raymond M, David VA, Schaffer AA, Tomlin JF, Eizirik E, et al. (2009) An autosomal genetic linkage map of the domestic cat, *Felis silvestris catus*. Genomics 93: 305–313.
45. Menotti-Raymond M, David VA, Agarwala R, Schaffer AA, Stephens R, et al. (2003) Radiation hybrid mapping of 304 novel microsatellites in the domestic cat genome. Cytogenet Genome Res 102: 272–276.
46. Goudet J (1995) FSTAT (Version 1.2): A computer program to calculate F-statistics. J Hered 86: 485–486.
47. Falush D, Stephens M, Pritchard JK (2003) Inference of population structure using multilocus genotype data: Linked loci and correlated allele frequencies. Genetics 164: 1567–1587.
48. Jost L (2008) G_{ST} and its relatives do not measure differentiation. Mol Ecol 17: 4015–4026.
49. Gerlach G, Jueterbock A, Kraemer P, Deppermann J, Harmand P (2010) Calculations of population differentiation based on G_{ST} and D: forget G_{ST} but not all of statistics! Mol Ecol 19: 3845–3852.
50. Peakall R, Smouse PE (2006) GENALEX 6: genetic analysis in Excel. Population genetic software for teaching and research. Mol Ecol Res 6: 288–295.
51. Peakall R, Smouse PE (2012) GenAlEx 6.5: genetic analysis in Excel. Population genetic software for teaching and research—an update. Bioinformatics 28: 2537–2539.
52. Slatkin M (1995) A measure of population subdivision based on microsatellite allele frequencies. Genetics 139: 457–462.
53. Lugon-Moulin N, Brünner H, Wyttenbach A, Hausser J, Goudet J (1999) Hierarchical analyses of genetic differentiation in a hybrid zone of *Sorex araneus* (Insectivora: Soricidae). Mol Ecol 8: 419–431.
54. Anderson EC, Thompson EA (2002) A model-based method for identifying species hybrids using multilocus genetic data. Genetics 160: 1217–1229.
55. Jombart T (2008) adegenet: a R packae fort he multivariate analysis of genetic markers. Bioinformatics 24: 1403–1405.
56. R Development Core Team (2014) R: A Language and environment for statistical computing. Vienna: R Foundation for Statistical Computing.
57. Tallmon DA, Koyuk A, Luikart G, Beaumont MA (2008) ONeSAMP: a program to estimate effective population size using approximate Bayesian computation. Mol Ecol Res 8: 299–301.
58. Cornuet JM, Luikart G (1996) Description and power analysis of two tests for detecting recent population bottlenecks from allele frequency data. Genetis 144: 2001–2014.
59. Piry S, Luikart G, Cornuet J (1999) Computer note. BOTTLENECK: a computer program for detecting recent reductions in the effective size using allele frequency data. J Hered 90: 502–503.
60. Wang J (2010) COANCESTRY: a program for simulating, estimating and analysing relatedness and inbreeding coefficients. Mol Ecol Res 11: 141–145
61. Wang JL (2002) An estimator for pairwise relatedness using molecular markers. Genetics 160: 1203–1215.
62. Wang J (2007) Triadic IBD coefficients and applications to estimating pairwise relatedness. Genet Res 89: 135–153.
63. Librado P, Rozas J (2009) DnaSP v5: a software for comprehensive analysis of DNA polymorphism data. Bioinformatics 25: 1451–1452.
64. Clement M, Posada D, Crandall KA (2000) TCS: a computer program to estimate gene genealogies. Mol Ecol 9: 1657–1659.
65. Hertwig ST, Schweizer M, Stepanow S, Jungnickel A, Böhle U-R, et al. (2009) Regionally high rates of hybridization and introgression in German wildcat populations (*Felis silvestris*, Carnivora, Felidae). J Zool Syst Evol Res 47: 283–297.
66. Eckert I, Suchentrunk F, Markov G, Hartl GB (2010) Genetic diversity and integrity of German wildcat (*Felis silvestris*) populations as revealed by microsatellites, allozymes, and mitochondrial DNA sequences. Mamm Biol 75: 160–174.
67. Randi E, Pierpaoli M, Beaumont MA, Ragni B, Sforzi A (2001) Genetic identification of wild and domestic cats (*Felis silvestris*) and their hybrids using bayesian clustering methods. Mol Biol Evol 18: 1679–1693.
68. O'Brien J, Devillard S, Say L, Vanthomme H, Leger F, et al. (2009) Preserving genetic integrity in a hybridising world: are European Wildcats (*Felis silvestris silvestris*) in eastern France distinct from sympatric feral domestic cats? Biodivers Conserv 18: 2351–2360.
69. Oliveira R, Godinho R, Randi E, Alves PC (2008) Hybridization versus conservation: are domestic cats threatening the genetic integrity of wildcats (*Felis silvestris silvestris*) in Iberian Peninsula? Philosophical Transactions of the Royal Society B-Biological Sciences 363: 2953–2961.
70. Germain E, Benhamou S, Poulle ML (2008) Spatio-temporal sharing between the European wildcat, the domestic cat and their hybrids. J Zool 276: 195–203.
71. Hubbard AL, McOrist S, Jones TW, Boid R, Scott R, et al. (1992) Is survival of European wildcats Felis silvestris in Britain threatened by interbreeding with domestic cats? Biol Conserv 61: 203–208.

72. Steyer K, Simon O, Kraus RHS, Haase P, Nowak C (2013) Hair trapping with valerian-treated lure sticks as a tool for genetic wildcat monitoring in low-density habitats. Eur J Wildl Res 59: 39–46.
73. Frankham R (2010) Challenges and opportunities of genetic approaches to biological conservation. Biol Conserv 143: 1919–1927.
74. Frankham R (2008) Genetic adaptation to captivity in species conservation programs. Mol Ecol 17: 325–333.
75. Williams SE, Hoffman EA (2009) Minimizing genetic adaptation in captive breeding programs: A review. Biol Conserv 142: 2388–2400.
76. Falconer DS, Mackay TFC (1996) Introduction to quantitative genetics. Harlow, England: Longman.
77. Ballou JD, Lacy RC (1995) Identifying genetically important individuals for management of genetic diversity in pedigreed populations. In: J. D. Ballou, M. Gilpin and T. J. Foose, editors. Population Management for Survival & Recovery Analytical Methods and Strategies in Small Population Conservation. New York: Columbia University Press. pp. 76–111.
78. Saura M, Perez-Figueroa A, Fernandez J, Toro MA, Caballero A (2008) Preserving population allele frequencies in *ex situ* conservation programs. Conserv Biol 22: 1277–1287.
79. Caballero A, Toro MA (2000) Interrelations between effective population size and other pedigree tools for the management of conserved populations. Genet Res 75: 331–343.
80. Robert A (2009) Captive breeding genetics and reintroduction success. Biol Conserv 142: 2915–2922.
81. Templeton AR, Davis SK, Read B (1987) Genetic variability in a captive herd of Speke's gazelle (*Gazella spekei*). Zoo Biol 6: 305–313.
82. Luo S-J, Johnson WE, Martenson J, Antunes A, Martelli P, et al. (2008) Subspecies genetic assignments of worldwide captive tigers increase conservation value of captive populations. Curr Biol 18: 592–596.
83. Goncalves da Silva A, Lalonde DR, Quse V, Shoemaker A and Russello MA (2010) Genetic approaches refine *ex situ* Lowland Tapir (*Tapirus terrestris*) conservation. J Hered 101: 581–590.
84. Gottelli D, Sillerozubiri C, Applebaum GD, Roy MS, Girman DJ, et al. (1994) Molecular-genetics of the most endangered canid - the Ethiopian wolf (*Canis simensis*). Mol Ecol 3: 301–312.
85. Rhymer J, Simberloff D (1996) Extinction by hybridization and introgression. Annu Rev Ecol Syst 27: 83–109.
86. Allendorf FW, Leary RF, Spruell P, Wenburg JK (2001) The problems with hybrids: setting conservation guidelines. Trends Ecol Evol 16: 613–622.
87. Ely JJ, Dye B, Frels WI, Fritz J, Gagneux P, et al. (2005) Subspecies composition and founder contribution of the captive US chimpanzee (*Pan troglodytes*) population. Am J Primatol 67: 223–241.

12

DNA Barcode Identification of Podocarpaceae—The Second Largest Conifer Family

Damon P. Little[1]*, Patrick Knopf[2], Christian Schulz[2]

1 Lewis B. and Dorothy Cullman Program for Molecular Systematics, The New York Botanical Garden, Bronx, New York, United States of America, 2 Lehrstuhl für Evolution und Biodiversität der Pflanzen, Ruhr-Universität Bochum, Bochum, Nordrhein-Westfalen, Bundesrepublik Deutschland

Abstract

We have generated *matK*, *rbcL*, and nrITS2 DNA barcodes for 320 specimens representing all 18 extant genera of the conifer family Podocarpaceae. The sample includes 145 of the 198 recognized species. Comparative analyses of sequence quality and species discrimination were conducted on the 159 individuals from which all three markers were recovered (representing 15 genera and 97 species). The vast majority of sequences were of high quality ($B_{30} = 0.596–0.989$). Even the lowest quality sequences exceeded the minimum requirements of the BARCODE data standard. In the few instances that low quality sequences were generated, the responsible mechanism could not be discerned. There were no statistically significant differences in the discriminatory power of markers or marker combinations ($p = 0.05$). The discriminatory power of the barcode markers individually and in combination is low (56.7% of species at maximum). In some instances, species discrimination failed in spite of ostensibly useful variation being present (genotypes were shared among species), but in many cases there was simply an absence of sequence variation. Barcode gaps (maximum intraspecific p-distance > minimum interspecific p-distance) were observed in 50.5% of species when all three markers were considered simultaneously. The presence of a barcode gap was not predictive of discrimination success ($p = 0.02$) and there was no statistically significant difference in the frequency of barcode gaps among markers ($p = 0.05$). In addition, there was no correlation between number of individuals sampled per species and the presence of a barcode gap ($p = 0.27$).

Editor: Mehrdad Hajibabaei, University of Guelph, Canada

Funding: Funding from the Alfred P. Sloan Foundation (http://www.sloan.org/; 2010-6-02) to DPL is gratefully acknowledged. The funders had no role in study design, data collection and analysis, decision to publish, or preparation of the manuscript.

Competing Interests: The authors have declared that no competing interests exist.

* E-mail: dlittle@nybg.org

Introduction

Podocarpaceae is a family of evergreen trees and shrubs that are sometimes cultivated as ornamentals in suitably warm climates. In terms of number of species, Podocarpaceae is the second largest family of conifers [1]. Podocarpaceae are often a minor subcanopy component of angiosperm–dominated forests. They are most abundant in the mid– to high–elevation tropics where they thrive on nutrient–poor soils. In addition, Podocarpaceae are found in some unusual low–elevation forest types (e.g. kerangas of Borneo; [2]).

Accurate identification of tropical forest trees, such as Podocarpaceae, is often very difficult. The most easily accessed material is usually sterile. If fertile material is present, it is frequently either inaccessible or detached from the tree making it difficult to convincingly associate the fertile and sterile portions. Although sterile material of Podocarpaceae can usually be identified to genus using phyllotaxis and leaf form [3,4], accurate species identification often requires careful microscopic examination of internal [5–20] and external characteristics [21–27]. Proper use of the existing identification tools requires training in botanical terminology, skill in microtechnique, and familiarity with Podocarpaceae.

Species of Podocarpaceae are of conservation concern primarily as a result of small population sizes and limited available habitat. Twenty–seven Podocarpaceae species are included in the International Union for the Conservation of Nature (IUCN; [28]) red list under the categories of vulnerable (10 species), endangered (14 species), and critically endangered (three species). Two species are included in the appendices of the Convention on International Trade in Endangered Species (CITES; [29]): *Podocarpus parlatorei* is listed in Appendix I (trade is not allowed) and *Po. neriifolius* is listed in Appendix III (trade with, some limitations, is allowed).

Podocarpaceae have a minor role in commerce. *Nageia nagi*, when labeled as Asian bayberry, can legally be sold in the United States of America as an herbal dietary supplement [30]. The seeds are processed into an edible oil that is also used in manufacturing [31]. The young leaves are also edible, but not typically consumed [32]. The conspicuous fleshy reproductive structures (receptacles or epimatium) of *Afrocarpus falcatus*, *Dacrycarpus dacrydioides*, *Dacrydium cupressinum*, *Po. elatus*, *Po. macrophyllus*, *Po. totara*, and *Prumnopitys taxifolia* are eaten either raw or cooked [32].

Although their use is currently very limited, Podocarpaceae are known to have medicinal properties that benefit humans and animals [33,34]. The receptacles and leaves contain a variety of bio–active compounds such as antioxidants, nordi–terpenes, podocarpic acid, and tatarol [33,35,36]. Some of these compounds have antimicrobial, fungistatic, or bacteriostatic properties [33,37,38]. Other compounds have cytotoxic properties that may be useful in destroying cancer [39–43].

The rarity of large uniform stands coupled with a slow rate of growth for most species makes harvest of Podocarpaceae wood generally unsustainable [2]. The growth rate of *Po. totara* may however accommodate sustainable harvest [44]. Relative scarcity results in a meager international trade—primarily originating from New Zealand and South Africa [2,45–47]. Timber from Podocarpaceae, referred to as 'podo' (or 'yellow yew') in commerce, has straight even grain. The wood of some species is brittle when worked and not particularly durable outdoors [46,47]. Wood from *Po. totara* is durable and highly amenable to industrial machining [44]. Wood from *Lepidothamnus intermedius*, *Manoao colensoi*, and *Pr. taxifolia* is very rot resistant [48]. Timber of *Ma. colensoi* and *Le. intermedius* have long been used for railway ties [48]. In addition, the wood of some Podocarpaceae species is highly insect resistant (e.g. *Af. gracilior* [49], *Po. hallii* [50], *Po. macrophyllus* [51], and *Po. nivalis* [50]).

A reference library of Podocarpaceae DNA barcodes will allow researchers unfamiliar with the family's morphology and anatomy to make accurate identifications. We hope that DNA barcodes will permit foresters, ecologists, conservationists, customs authorities, etc. to make accurate biodiversity inventories and to monitor trade in threatened and endangered Podocarpaceae species so that future conservation and management decisions can be based on sound data.

We aim to generate and evaluate a DNA barcode reference library for Podocarpaceae. The library will be assessed both by the quality of the constituent sequences and the degree to which observed sequence variation unambiguously distinguishes Podocarpaceae species from one another.

Materials and Methods

We sampled 320 individuals representing all of the 18 extant genera of Podocarpaceae (including *Phyllocladus*). Our sample included 145 of the 198 recognized species (73.2%; [1]). Between 1 and 9 individuals per species were sequenced (median = 2; IQR = 1–3). All samples were expert-identified using a combination of morphology and leaf anatomy [20]. Voucher information is in Dataset S1.

Molecular techniques

DNA was extracted from herbarium specimens or silica-dried tissue using the Qiagen DNeasy96 kit. The manufacturer's protocol was modified for herbarium specimens: instead of the recommended incubation, homogenized tissue was digested with 30 µL (20 µg/µL stock) of proteinase K in 400 µL AP1 (supplied by the manufacturer) and 1 µL of DX (supplied by the manufacturer) at 42°C for 24 hours with slow mixing (60 rotations per minute).

The Polymerase Chain Reaction (PCR) was used to amplify *matK* in a 15 µL volume containing: 20 mM tris pH 8.8, 10 mM KCl, 10 mM $(NH_4)_2$ SO_4, 2 mM $MgSO_4$, 0.1% (v/v) Triton X-100, 5% (w/v) sucrose, 0.025% (w/v) cresol red, 0.025 µg/µL BSA, 0.2 mM dNTPs, 1 µM of Gym_F1A (5′-ATY GYR CTT TTA TGT TTA CAR GC-3′; [53]), 1 µM of Gym_R1A (5′-TCA YCC GGA RAT TTT GGT TCG-3′; [53]), 0.5 units *Taq*, and 0.5 µL genomic DNA. The reaction mixture was incubated for 150 sec at 95°C, cycled 35 times (30 sec at 95°C, 60 sec at 52°C, 40 sec at 72°C) and then incubated at 72°C for 10 minutes.

PCR amplification of nrITS2 was similar to that of *matK* except primer annealing was carried out for 30 sec at 58°C rather than 60 sec at 52°C and primers S2F (5′-ATG CGA TAC TTG GTG TGA AT-3′; [53]) and S3R (5′-GAC GCT TCT CCA GAC TAC AAT-3′ [53]) replaced Gym_F1A and Gym_R1A.

Unused primers and dNTPs were neutralized using ExoSAP-IT (USB). PCR products were bidirectionally sequenced, using the amplification primers, with BigDye v3.1 (Life Technologies) at the High-Throughput Genomics Unit (University of Washington). PCR amplification and sequencing of *rbcL* was described previously [54].

Data analysis

Bases were called and quality values (QV) assigned using KB 1.4 (Life Technologies). Sequencer 4.1 (Gene Codes) was used to construct sequence contigs, trim contigs to a uniform beginning/end (priming sites were excluded), and resolve differences between sequencing reads. Sequences of *matK* and *rbcL* were checked for stop codons and frameshift mutations. To identify potential contaminates, all sequences were queried against GenBank using BLAST 2.2.26 [55]. Only hits with an e-value of $10^{-20.0}$ or less were retained. Additional contaminates were identified by aligning each marker with MUSCLE 3.8 [56], coding the resulting indels using 'simple indel coding' [57,58], and resampling the resulting matrix 1000 times using the jackknife [59] procedure. For each resampled matrix, the search for optimal trees was conducted in TNT 1.1 [60]. The search consisted of ten random addition replicates with five trees held in memory per replicate and SPR followed by TBR (BB) branch swapping. The strict consensus tree from each resampled matrix was used to calculate the jackknife tree.

In order to make meaningful comparisons across makers, we only analyzed data from specimens for which *matK*, *rbcL*, and nrITS2 were able to be sequenced—referred to as 'complete samples' hereafter. Sequences from specimens that could not be definitively identified using morphology/anatomy were excluded.

Sequence quality was assessed using the barcode quality index (*B*; [61]) with the acceptable quality threshold (q) set to 30 (an average of one error per thousand sequenced bases). The expected coverage (x) was set to 2. The contig size (c) was set to the observed size. Linguistic complexity (*LC*; [62]), a measure of sequence repetitiveness, was calculated for each sequence, using COMPLEX 6.1.0 [63] with window size set to 100 bases, step size set to 1 base, minimum pattern size set to 3 bases, and maximum pattern size set to 6 bases. The threshold for significant increase in homopolymer (mononucleotide repeats) induced PCR artifacts has been empirically determined to be eight bases [64]—thus sequences with homopolymers, eight bases or longer, were identified. Statistical differences, in sequence quality, linguistic complexity, and homopolymer frequency among markers were evaluated with Scheffé's test [65–67] at $p = 0.05$ using the Gaussian distribution. Correlations between sequence quality and linguistic complexity as well as sequence quality and homopolymer frequency were measured by Spearman's rank correlation tests [67,68].

TNT 1.1 [60] was used to analyze phylogenetic relationships among complete samples. Each marker was aligned with MUSCLE 3.8 [56], indels were coded using 'simple indel coding' [57,59], and markers were combined by concatenation [69]. The resulting matrix was searched for optimal trees using 1000 random addition replicates: for each replicate two trees were held in memory, SPR branch swapping was followed by TBR (BB) branch swapping and a 200 iteration ratchet [70] perturbing 8% of the characters per iteration (4% up weighted, 4% down weighted). Clade support was assessed by 10,000 jackknife resamplings [59]. For each resampled matrix, the search for optimal trees consisted of ten random addition replicates with five trees held in memory per replicate and SPR followed by TBR (BB) branch swapping. The jackknife frequency of each clade in the strict consensus of the

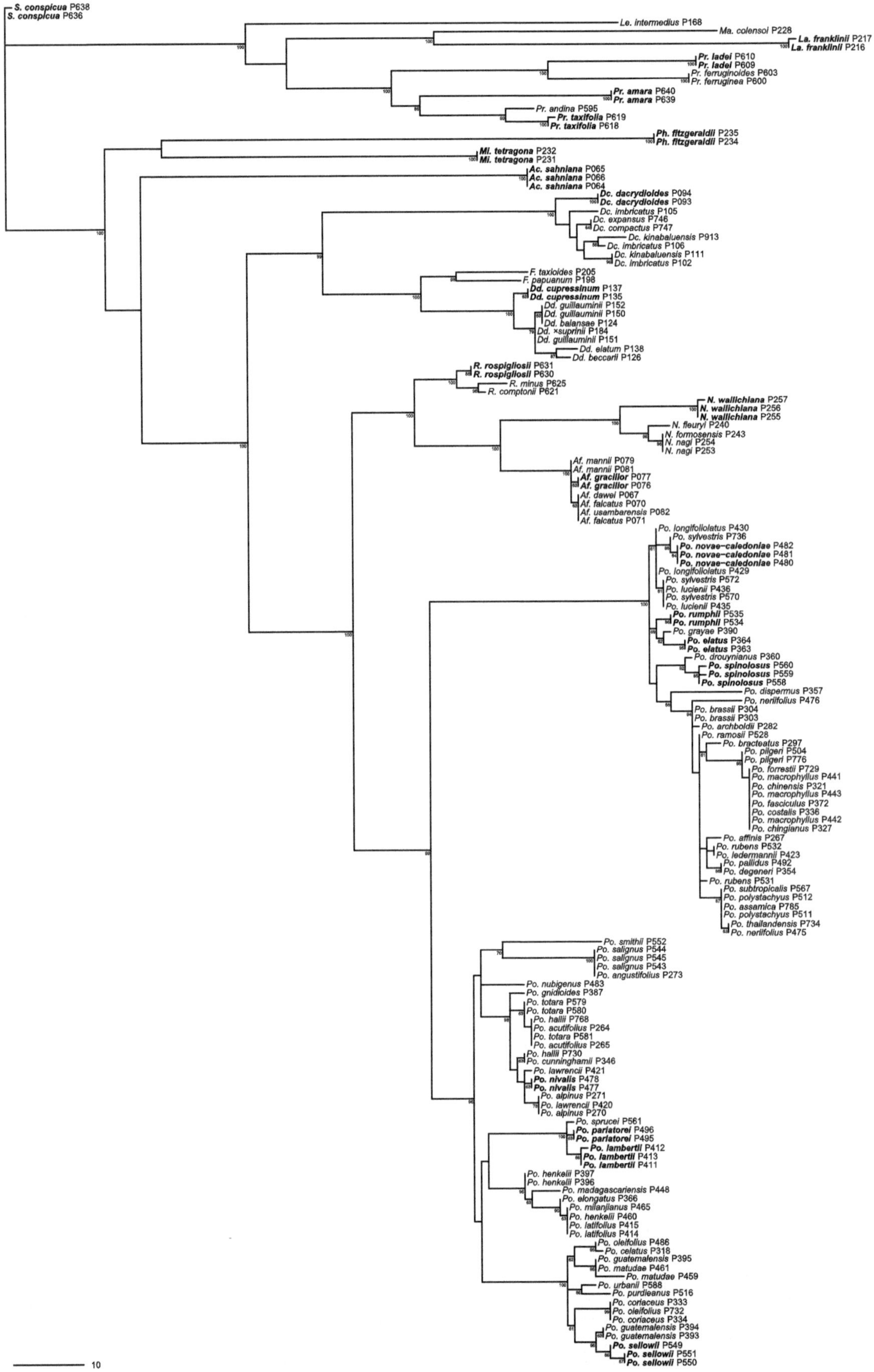
S. conspicua P638
S. conspicua P636
Le. intermedius P168
Ma. colensoi P228
La. franklinii P217
La. franklinii P216
Pr. ladei P610
Pr. ladei P609
Pr. ferruginoides P603
Pr. ferruginea P600
Pr. amara P640
Pr. amara P639
Pr. andina P595
Pr. taxifolia P619
Pr. taxifolia P618
Ph. fitzgeraldii P235
Ph. fitzgeraldii P234
Mi. tetragona P232
Mi. tetragona P231
Ac. sahniana P065
Ac. sahniana P066
Ac. sahniana P064
Dc. dacrydioides P094
Dc. dacrydioides P093
Dc. imbricatus P105
Dc. expansus P746
Dc. compactus P747
Dc. kinabaluensis P913
Dc. imbricatus P106
Dc. kinabaluensis P111
Dc. imbricatus P102
F. taxioides P205
F. papuanum P198
Dd. cupressinum P137
Dd. cupressinum P135
Dd. guillauminii P152
Dd. guillauminii P150
Dd. balansae P124
Dd. ×suprinii P184
Dd. guillauminii P151
Dd. elatum P138
Dd. beccarii P126
R. rospigliosii P631
R. rospigliosii P630
R. minus P625
R. comptonii P621
N. wallichiana P257
N. wallichiana P256
N. wallichiana P255
N. fleuryi P240
N. formosensis P243
N. nagi P254
N. nagi P253
Af. mannii P079
Af. mannii P081
Af. gracilior P077
Af. gracilior P076
Af. dawei P067
Af. falcatus P070
Af. usambarensis P082
Af. falcatus P071
Po. longifoliolatus P430
Po. sylvestris P736
Po. novae-caledoniae P482
Po. novae-caledoniae P481
Po. novae-caledoniae P480
Po. longifoliolatus P429
Po. sylvestris P572
Po. lucienii P436
Po. sylvestris P570
Po. lucienii P435
Po. rumphii P535
Po. rumphii P534
Po. grayae P390
Po. elatus P364
Po. elatus P363
Po. drouynianus P360
Po. spinolosus P560
Po. spinolosus P559
Po. spinolosus P558
Po. dispermus P357
Po. neriifolius P476
Po. brassii P304
Po. brassii P303
Po. archboldii P282
Po. ramosii P528
Po. bracteatus P297
Po. pilgeri P504
Po. pilgeri P776
Po. forrestii P729
Po. macrophyllus P441
Po. chinensis P321
Po. macrophyllus P443
Po. fasciculus P372
Po. costalis P336
Po. macrophyllus P442
Po. chingianus P327
Po. affinis P267
Po. rubens P532
Po. ledermannii P423
Po. pallidus P492
Po. degeneri P354
Po. rubens P531
Po. subtropicalis P567
Po. polystachyus P512
Po. assamica P785
Po. polystachyus P511
Po. thailandensis P734
Po. neriifolius P475
Po. smithii P552
Po. salignus P544
Po. salignus P545
Po. salignus P543
Po. angustifolius P273
Po. nubigenus P483
Po. gnidioides P387
Po. totara P579
Po. totara P580
Po. hallii P768
Po. acutifolius P264
Po. totara P581
Po. acutifolius P265
Po. hallii P730
Po. cunninghamii P346
Po. lawrencii P421
Po. nivalis P478
Po. nivalis P477
Po. alpinus P271
Po. lawrencii P420
Po. alpinus P270
Po. sprucei P561
Po. pariatorei P496
Po. pariatorei P495
Po. lambertii P412
Po. lambertii P413
Po. lambertii P411
Po. henkelii P397
Po. henkelii P396
Po. madagascariensis P448
Po. elongatus P366
Po. milanjianus P465
Po. henkelii P460
Po. latifolius P415
Po. latifolius P414
Po. oleifolius P486
Po. celatus P318
Po. guatemalensis P395
Po. matudae P461
Po. matudae P459
Po. urbanii P588
Po. purdieanus P516
Po. coriaceus P333
Po. oleifolius P732
Po. coriaceus P334
Po. guatemalensis P394
Po. guatemalensis P393
Po. sellowii P549
Po. sellowii P551
Po. sellowii P550
10

Figure 1. Phylogenetic relationships among complete samples. Strict consensus of 3600 most parsimonious trees (L = 1205; CI = 0.59; RI = 0.93; all tree statistics exclude uninformative nucleotide positions) obtained from the simultaneous analysis of *matK*, *rbcL*, and nrITS2 sequence data. Numbers at nodes indicated jackknife support above 50%. Species that can be distinguished from all other species using the 'least inclusive clade' method are in boldface (the least inclusive clade method cannot be applied to species with only one sample). Genera have been abbreviated: *Ac.* = *Acmopyle*, *Af.* = *Afrocarpus*, *Dc.* = *Dacrycarpus*, *Dd.* = *Dacrydium*, *F.* = *Falcatifolium*, *La.* = *Lagarostrobos*, *Le.* = *Lepidothamnus*, *Ma.* = *Manoao*, *Mi.* = *Microcachrys*, *N.* = *Nageia*, *Ph.* = *Pherosphaera*, *Po.* = *Podocarpus*, *Pr.* = *Prumnopitys*, *R.* = *Retrophyllum*, and *S.* = *Saxegothaea*. Sample codes correspond to those used in Dataset S1.

original matrix was calculated with SUMTREES 3.3.1 [71] using the strict consensus tree from each resampled matrix [67,72]. Trees were rooted following [54]. Tree–based species discrimination was assessed using the 'least inclusive clade' method [73].

Species discrimination was calculated using BRONX 2.0 [74,75]. Discrimination success would be overestimated if the reference database just included sequences in the complete sample—thus a BRONX reference database was constructed from all sequences for each marker and marker combination (Dataset S1). To calculate species discrimination, sequences of each complete sample were queried against the reference database. Species were considered distinct if all queries for a given species returned only sequences belonging to that species. The binomial distribution, with each species considered an independent test, was used to compute 95% confidence intervals [67,76,77]. Differences in species discrimination among markers and marker combinations were quantified using Scheffé's test [65–67] at $p = 0.05$. The binomial distribution was used for tests of species discrimination and the Gaussian distribution was used to test if the number of species conflated when identification failed varied among markers.

Relative variation within and among species—the 'barcode gap' [78]—was quantified by comparing pairwise distances for complete samples. Each pair of sequences in the complete sample was aligned separately with MUSCLE 3.8 [56] and the number of unambiguous nucleotide differences was divided by the total number of aligned positions to calculate the edit distance (uncorrected p–distance; [79]). To minimize sampling and analytic artifacts, the maximum intraspecific distance was compared to the minimum interspecific distance for each species [80]. For each marker, the frequency of barcode gap (maximum intraspecific > minimum interspecific) occurrence was assessed using the binomial distribution and Scheffé's test [65–67] at $p = 0.05$. The point–biserial correlation coefficient was used to examine the relationship between number of samples per species and the occurrence of a barcode gap [67,81]. Sequences from all three markers were used simultaneously with McNemar's test [67,82] to measure the correlation between the occurrence of a barcode gap and whether or not a species can be consistently distinguished from all other species using diagnostic nucleotide positions.

Results and Discussion

In total, 281 *matK*, 202 *rbcL*, and 212 nrITS2 finished sequences were generated (Dataset S1). BLAST [55] queries indicate that the newly generated sequences are consistent with other samples of Podocarpaceae deposited in GenBank (data not shown). Phylogenetic arrangement of genera and species is roughly consistent with previous molecular phylogenetic studies (Figure 1; [54,83–86]). Sequences derived from individuals of the same species are always in close phylogenetic proximity, but in some cases the sequences do not form a monophyletic group (sensu [87]). Sequences of some morphological/anatomical species are unambiguously polyphyletic (sensu [87]; e.g. *Podocarpus oleifolius*, Figure 1). Mismatches between morphological/anatomical species circumscription and barcode sequences warrant further investigation as they may indicate the presence of cryptic species, introgression, or ancestral polymorphism followed by incomplete lineage sorting. Together the BLAST and phylogenetic contaminate screens indicate that the sequences generated are indeed Podocarpaceae and that no PCR artifacts or errors in sample handling could be detected.

Newly generated sequences of *matK* vary from 760 to 775 bp (median = 769; IQR = 769–769), sequences of *rbcL* are uniformly 607 bp, and sequences of nrITS2 vary from 420 to 435 bp (median = 425; IQR = 425–425). The multiple sequence alignment of *matK* results in six indels—3 bp (two indels), 6 bp (three indels), and 9 bp (one indel), respectively. The 20 nrITS2 indels resulting from multiple sequence alignment range from 1 to 17 bp (median = 1; IQR = 1–2).

Of the 320 individuals sequenced, finished *matK*, *rbcL*, and nrITS2 sequences were generated for 159 individuals. These samples, representing 15 of the 18 extant genera (83.3%; [1]; *Halocarpus*, *Parasitaxus*, and *Phyllocladus* are not included in the complete sample) and 97 of the 198 recognized species (48.9%; [1]), were analyzed for sequence quality, linguistic complexity, species discrimination, and barcode gaps. The complete sample set contained between 1 and 3 individuals per species (median = 1; IQR = 1–2; Table 1). In total, there were 95 distinct *matK* sequence types, 70 *rbcL* sequence types, and 81 nrITS2 sequence types. The complete sample contained 71 (74.7%) *matK* sequence types, 61 (87.1%) *rbcL* sequence types, and 65 (80.2%) nrITS2 sequence types for a combined 90 distinct multilocus genotypes.

Sequence quality and complexity

Sequence quality, as measured by B_{30} [61], ranged from 0.775 to 0.989 for *matK* (median = 0.967; IQR = 0.960–0.975), 0.596 to 0.951 for *rbcL* (median = 0.938; IQR = 0.929–0.944), and 0.671 to 0.933 for nrITS2 (median = 0.924; IQR = 0.919–0.927; Figure 2). The vast majority of sequences were of high quality: across all markers, 93.5% of the positions in the median sequence were assigned a quality value of 30 or greater—indicating that few, if any, of the finished sequences contain erroneous base calls. Although differences in sequence quality among markers was statistically significant ($p = 0.05$; *matK* > *rbcL* > nrITS2), even the lowest quality sequences exceeded the minimum requirements of the BARCODE data standard (version 2.3; [88])—thus the statistical differences observed are not particularly meaningful in practice.

Published B_{30} values for sequences generated using different primer sets are not directly comparable to those reported here because the primer sets define (slightly) different marker regions. Although not comparable in all cases, the median Podocarpaceae sequence is of higher quality than the average sequence reported for angiosperms across all three markers [61,89,90]. The largest comparable difference between literature reports and newly generated Podocarpaceae sequences was observed in nrITS2 (0.829 versus 0.924; [89]). The high quality of Podocarpaceae *matK* sequences is notable, but the gymnosperm specific primers [52] used to generate the *matK* sequences make direct comparisons to published values for angiosperms tenuous.

Table 1. Diagnostic barcode variation for complete samples of Podocarpaceae.

species	n	discriminatory success				multilocus barcode gap	most frequently conflated species (ordered by frequency)
		matK	*rbcL*	nrITS2	combined		
Ac. sahniana	3	+	+	+	+	1	—
Af. dawei	1	–	–	–	–	0	*Af. gracilior, Af. mannii, Af. falcatus*, et al.
Af. falcatus	2	–	–	–	–	0	*Af. gracilior, Af. mannii, Af. usambarensis*, et al.
Af. gracilior	2	–	–	+	+	0	*Af. falcatus, Af. usambarensis, Af. dawei*, et al.
Af. mannii	2	–	–	–	+	0	*Af. falcatus, Af. usambarensis, Af. dawei*, et al.
Af. usambarensis	1	–	–	–	–	0	*Af. gracilior, Af. mannii, Af. falcatus*, et al.
Dc. compactus	1	–	–	–	–	0	*Dc. kinabaluensis, Dc. imbricatus, Dc. expansus*
Dc. dacrydioides	2	+	+	+	+	1	—
Dc. expansus	1	–	–	–	–	0	*Dc. kinabaluensis, Dc. imbricatus, Dc. compactus*
Dc. imbricatus	3	–	–	+	+	0	*Dc. expansus, Dc. compactus, Dc. kinabaluensis*
Dc. kinabaluensis	2	–	–	–	+	0	*Dc. expansus, Dc. compactus, Dc. imbricatus*
Dd. balansae	1	–	–	–	–	0	*Dd. araucarioides, Dd. cupressinum, Dd. nausoriense*, et al.
Dd. beccarii	1	–	+	+	+	1	*Dd. gracile, Dd. xanthandrum*
Dd. cupressinum	2	+	–	+	+	1	*Dd. guillauminii, Dd. araucarioides, Dd. balansae*, et al.
Dd. elatum	1	+	+	+	+	1	—
Dd. guillauminii	3	–	–	–	–	0	*Dd. araucarioides, Dd. cupressinum, Dd. nausoriense*, et al.
Dd. ×suprinii	1	–	–	–	–	0	*Dd. araucarioides, Dd. cupressinum, Dd. nausoriense*, et al.
F. papuanum	1	+	+	+	+	1	—
F. taxioides	1	+	+	+	+	1	—
La. franklinii	2	+	+	+	+	1	—
Le. intermedius	1	+	+	+	+	1	—
Ma. colensoi	1	+	–	+	+	1	*Le. intermedius*
Mi. tetragona	2	+	+	+	+	1	—
N. fleuryi	1	+	+	+	+	1	—
N. formosensis	1	–	–	–	–	0	*N. motleyi, N. nagi*
N. nagi	2	–	–	–	–	0	*N. motleyi, N. formosensis*
N. wallichiana	3	+	–	+	+	1	*N. motleyi*
Ph. fitzgeraldii	2	+	+	+	+	1	—
Po. acutifolius	2	–	–	–	–	0	*Po. cunninghamii, Po. totara, Po. hallii*
Po. affinis	1	–	+	+	+	1	*Po. degeneri, Po. rubens, Po. insularis*, et al.
Po. alpinus	2	–	–	–	–	0	*Po. lawrencei, Po. gnidioides, Po. nivalis*, et al.
Po. angustifolius	1	–	–	–	–	0	*Po. salignus*
Po. archboldii	1	+	–	–	+	1	*Po. drouynianus, Po. polystachyus, Po. thailandensis*, et al.
Po. assamica	1	–	–	–	–	0	*Po. drouynianus, Po. deflexus, Po. insularis*, et al.
Po. bracteatus	1	–	–	–	+	1	*Po. neriifolius, Po. pseudobracteatus*
Po. brassii	2	–	–	–	+	1	*Po. drouynianus, Po. thailandensis, Po. deflexus*, et al.
Po. celatus	1	–	–	–	+	1	*Po. guatemalensis, Po. coriaceus, Po. sellowii*, et al.
Po. chinensis	1	–	–	–	–	0	*Po. nakaii, Po. pilgeri, Po. annamiensis*, et al.
Po. chingianus	1	–	–	–	–	0	*Po. nakaii, Po. pilgeri, Po. annamiensis*, et al.
Po. coriaceus	2	–	–	–	–	0	*Po. guatemalensis, Po. trinitensis, Po. celatus*, et al.
Po. costalis	1	–	–	–	–	0	*Po. nakaii, Po. pilgeri, Po. annamiensis*, et al.
Po. cunninghamii	1	–	–	–	–	0	*Po. totara, Po. acutifolius, Po. hallii*
Po. degeneri	1	–	–	–	–	0	*Po. affinis, Po. insularis, Po. ledermannii*, et al.
Po. dispermus	1	+	+	+	+	1	—
Po. drouynianus	1	–	+	–	+	1	*Po. polystachyus, Po. thailandensis, Po. insularis*, et al.
Po. elatus	2	+	+	+	+	1	—
Po. elongatus	1	+	+	–	+	1	*Po. polystachyus, Po. milanjianus, Po. henkelii*, et al.
Po. fasciculus	1	–	–	–	–	0	*Po. nakaii, Po. pilgeri, Po. annamiensis*, et al.

Table 1. Cont.

species	n	discriminatory success				multilocus barcode gap	most frequently conflated species (ordered by frequency)
		matK	rbcL	nrITS2	combined		
Po. forrestii	1	–	–	–	–	0	*Po. nakaii, Po. pilgeri, Po. annamiensis*, et al.
Po. gnidioides	1	+	–	+	+	1	*Po. alpinus, Po. lawrencii*
Po. grayae	1	+	–	+	+	1	*Po. lucienii, Po. rumphii, Po. sylvestris*, et al.
Po. guatemalensis	3	–	–	–	–	0	*Po. tepuiensis, Po. brasiliensis, Po. coriaceus*, et al.
Po. hallii	2	–	–	–	–	0	*Po. totara, Po. acutifolius, Po. cunninghamii*
Po. henkelii	3	–	–	–	–	0	*Po. polystachyus, Po. elongatus, Po. milanjianus*, et al.
Po. lambertii	3	+	+	+	+	1	—
Po. latifolius	2	–	–	–	–	0	*Po. polystachyus, Po. elongatus, Po. milanjianus*, et al.
Po. lawrencii	2	–	–	–	–	0	*Po. lawrencei, Po. gnidioides, Po. nivalis*, et al.
Po. ledermannii	1	–	–	–	–	0	*Po. polystachyus, Po. thailandensis, Po. decipiens*, et al.
Po. longifoliolatus	2	–	–	–	+	1	*Po. rumphii, Po. grayae, Po. lucienii*, et al.
Po. lucienii	2	–	–	–	–	0	*Po. rumphii, Po. grayae, Po. decumbens*, et al.
Po. macrophyllus	3	–	–	–	–	0	*Po. nakaii, Po. pilgeri, Po. annamiensis*, et al.
Po. madagascariensis	1	+	+	+	+	1	—
Po. matudae	2	–	–	–	–	0	*Po. oleifolius, Po. guatemalensis*
Po. milanjianus	1	–	–	–	–	0	*Po. polystachyus, Po. elongatus, Po. henkelii*, et al.
Po. neriifolius	2	–	–	–	–	0	*Po. drouynianus, Po. insularis, Po. ledermannii*, et al.
Po. nivalis	2	–	+	–	+	1	*Po. lawrencei, Po. totara, Po. acutifolius*, et al.
Po. novae-caledoniae	3	–	–	–	+	1	*Po. sylvestris, Po. beecherae*
Po. nubigenus	1	–	+	+	+	1	*Po. atjehensis*
Po. oleifolius	2	–	–	–	–	0	*Po. trinitensis, Po. salicifolius, Po. rusbyi*, et al.
Po. pallidus	1	–	–	–	–	0	*Po. affinis, Po. insularis, Po. ledermannii*, et al.
Po. parlatorei	2	–	–	+	+	1	*Po. transiens, Po. sprucei*
Po. pilgeri	2	–	+	–	+	1	*Po. nakaii, Po. costalis, Po. fasciculus*, et al.
Po. polystachyus	2	–	–	–	+	0	*Po. drouynianus, Po. deflexus, Po. insularis*, et al.
Po. purdieanus	1	+	+	+	+	1	—
Po. ramosii	1	–	–	–	+	1	*Po. thailandensis, Po. insularis, Po. assamica*, et al.
Po. rubens	2	–	–	–	–	0	*Po. drouynianus, Po. polystachyus, Po. thailandensis*, et al.
Po. rumphii	2	–	–	+	+	1	*Po. lucienii, Po. sylvestris, Po. grayae*, et al.
Po. salignus	3	–	–	–	–	0	*Po. angustifolius*
Po. sellowii	3	+	–	–	+	1	*Po. tepuiensis, Po. celatus, Po. oleifolius*, et al.
Po. smithii	1	+	+	+	+	1	—
Po. spinolosus	3	–	–	+	+	1	*Po. dispermus, Po. drouynianus, Po. rostratus*, et al.
Po. sprucei	1	–	–	–	+	1	*Po. glomeratus, Po. lambertii, Po. transiens*, et al.
Po. subtropicalis	1	–	–	–	–	0	*Po. drouynianus, Po. deflexus, Po. insularis*, et al.
Po. sylvestris	3	–	–	–	–	0	*Po. rumphii, Po. grayae, Po. beecherae*, et al.
Po. thailandensis	1	–	–	–	–	0	*Po. drouynianus, Po. insularis, Po. ledermannii*, et al.
Po. totara	3	–	–	–	–	0	*Po. cunninghamii, Po. acutifolius, Po. hallii*
Po. urbanii	1	+	–	+	+	1	*Po. guatemalensis, Po. celatus, Po. oleifolius*, et al.
Pr. amara	2	+	+	+	+	1	—
Pr. andina	1	+	–	+	+	1	*Pr. exigua, Pr. montana*
Pr. ferruginea	1	–	–	–	–	0	*Pr. ferruginoides*
Pr. ferruginoides	1	–	–	+	+	0	*Pr. ferruginea*
Pr. ladei	2	+	+	+	+	1	—
Pr. taxifolia	2	+	+	+	+	1	—
R. comptonii	1	+	–	–	+	1	*R. minus, R. rospigliosii, R. vitiense*
R. minus	1	–	–	+	+	1	*R. comptonii*
R. rospigliosii	2	+	+	–	+	1	*R. comptonii, R. vitiense*

Table 1. Cont.

species	*n*	discriminatory success				multilocus barcode gap	most frequently conflated species (ordered by frequency)
		matK	*rbcL*	nrITS2	combined		
S. conspicua	2	+	+	+	+	1	—

Genera have been abbreviated: Ac. = Acmopyle, Af. = Afrocarpus, Dc. = Dacrycarpus, Dd. = Dacrydium, F. = Falcatifolium, La. = Lagarostrobos, Le. = Lepidothamnus, Ma. = Manoao, Mi. = Microcachrys, N. = Nageia, Ph. = Pherosphaera, Po. = Podocarpus, Pr. = Prumnopitys, R. = Retrophyllum, and S. = Saxegothaea. Species that can be consistently distinguished from all other species are indicated by a '+' while those that are conflated with other species are indicated with a '−'. The presence of a barcode gap is indicated with '1', absence of a barcode gap with '0'.

Linguistic complexity is a measure of the number of repeated 'words' in a sequence (words 3–6 bp were examined in this case; [62]). Sequences of *matK*, *rbcL*, and nrITS2 have statistically distinct linguistic complexity ($p = 0.05$) with *matK* being the simplest (median $= 0.443$; IQR $= 0.437$–0.449), followed by nrITS2 (median $= 0.527$; IQR $= 0.513$–0.566), and *rbcL* (median $= 0.584$; IQR $= 0.577$–0.590; Figure 2). The range of nrITS2 linguistic complexity is relatively broad especially in comparison to that of *matK* and *rbcL*—perhaps a result of different functional constraints on structural versus protein coding sequences.

One might expect that sequences with lower linguistic complexity (i.e. those with homopolymers and/or simple sequence repeats) will have lower sequence quality due to slip–strand mispairing at the site of repetitive sequence elements [64,91–93], however lower linguistic complexity is correlated with higher sequence quality in Podocarpaceae ($p < 2.2 \times 10^{-16}$). The sequences with the lowest linguistic complexity generally have sequence quality typical for the marker in question (Figure 2).

A homopolymer eight bases or longer was found in 35 sequences of the complete sample: 34 were *matK* sequences and one was an nrITS2 sequence. In the *matK* sequences from the complete sample, there is a single occurrence of A_8 and 33 occurrences of T_8. The T_8 homopolymers occupy alignment positions 465–472 (found in all samples of *Afrocarpus*, *Lepidothamnus*, *Nageia*, *Prumnopitys*, and *Retrophyllum*) and 720–727 (found in some *Podocarpus*). The frequency of homopolymer occurrence was significantly different among markers ($p = 0.05$). Counter to previous findings [92,93], high homopolymer frequency is correlated with high sequence quality in Podocarpaceae ($p = 2.1 \times 10^{-16}$). Previous investigations of the relationship between homopolymers and sequence quality focused on homopolymers ten bases or longer because they consistently result in low sequence quality [91–93], however homopolymers ten bases or longer are not found in any sequence of the complete sample. Thus we cannot determine if this length homopolymer has any effect on sequence quality.

The observed correlation between increased sequence quality and decreased linguistic complexity as well as the correlation between increased sequence quality and increased homopolymer frequency indicate that a mechanism other than slip–strand mispairing is responsible for the low quality sequences in the complete sample.

Species discrimination

For individual markers, BRONX [74,75] species discrimination ranged from 28.8% to 38.1% (Figure 3; Table 1). Discrimination for marker combinations was slightly better at 46.4% to 56.7%. Discriminatory power did not statistically differ ($p = 0.05$) among markers or marker combinations. When species identification failed, the number of conflated species ranged from a mean of 4.3 ($\sigma = 1.8$) to 5.6 ($\sigma = 4.1$) species for individual markers and a mean of 2.9 ($\sigma = 1.4$) to 3.6 ($\sigma = 2.1$) species for marker combinations. There were no unambiguous statistical differences ($p = 0.05$) in the number of conflated species among markers or marker combinations.

A synergistic effect was observed for marker combinations both in terms of an increase in discriminatory power and decrease in the number of conflated species (Table 1; e.g. the two specimens of *Af. mannii* examined cannot be consistently distinguished from *Af. dawei*, *Af. falcatus*, *Af. gracilior*, or *Af. usambarensis* by any single marker, but when *matK* and nrITS2 are combined, *Af. mannii* can be consistently distinguished from all other species). In no case did combining markers result in a loss of discriminatory power.

The core barcode markers (*matK* and *rbcL*) were able to consistently distinguish among 46.3% of the species in the complete sample (Figure 3). In comparison, studies that analyzed sequences of *matK*, *rbcL*, and nrITS2, individually and in combination, using comparable methods of species discrimination (the 'best match' procedure [94] or the 'simple pairwise matching' technique [95]) had a median success rate of 59.5% (range $= 35.7$–71.4) for core barcode markers (Table 2; [89,90,96–98]). In these same studies, species discrimination noticeably improved with the addition nrITS2 as a supplemental marker (median $= 92.6\%$; range $= 57.1$–99.3). Although species discrimination did improve in Podocarpaceae with the addition of nrITS2 (Figure 3), the rate of species discrimination (56.7%) is less than the lowest published value (Table 2; [97]).

Of the 49 species represented by two or more individuals in the complete sample, BRONX could distinguish 28 (57.1%) from all other species using a combination of three markers (Table 1). In contrast, the 'least inclusive clade' method could distinguish 21 (42.9%) species (Figure 1). This provides another example of the poor performance of tree–based algorithms for barcode sequence discrimination [73,74].

The complete sample was composed of 97 species represented by 90 distinct multilocus genotypes. Thus, if intraspecific variation is assumed to be near zero, one could plausibly expect that species discrimination would be close to 92.8%, however only 56.7% of species could be consistently distinguished using all three markers simultaneously.

In many cases, identification failed in spite of ostensibly useful variation being present—this most often occurred when genotypes were shared among species (e.g. *Po. guatemalensis* and *Po. matudae* are sister species [54] that have a total of three multilocus genotypes [Figure 1]: the first multilocus genotype is restricted to *Po. guatemalensis*, the second multilocus genotype is restricted to *Po. matudae*, and the third multilocus genotype is found in both species). In the cases where genotypes are shared across species boundaries, the data cannot definitively distinguish between the underlying causal mechanisms of recent introgression versus

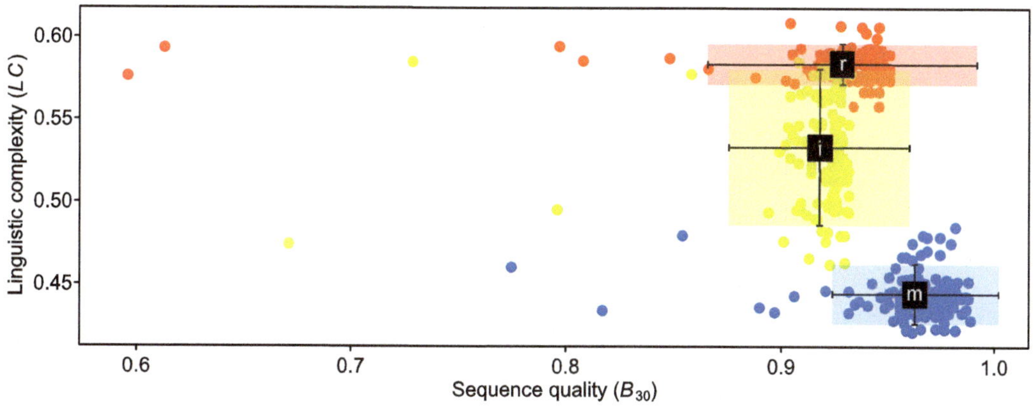

Figure 2. Barcode sequence quality (B_{30}) versus linguistic complexity (*LC*) for complete samples. Circles represent individual *matK* (blue; m), *rbcL* (red; r), and nrITS2 (yellow; i) sequences. Black squares indicate marker means. Error bars span three standard deviations.

ancestral polymorphism followed by incomplete lineage sorting. In these cases, it is unlikely that sequence data from additional markers will increase species discrimination.

In some cases, identification failure is the result of an absence of sequence variation (e.g. *Dc. compactus* and *Dc. expansus* are sister species [54] that have identical sequences for all three markers [Figure 1]). Sequence data from additional markers may improve species discrimination in these cases. Although we did not test the utility of supplementary plastid markers, it seems unlikely that better discrimination will be provided by additional plastid data given the small difference in species discrimination between *matK* and *rbcL* (4.1%; Figure 3)—discriminatory power for plastid markers usually plateaus at two markers [95]. Rather than sequencing more plastid markers, effort would be better invested in variable unlinked markers that are easily recovered from Podocarpaceae (e.g. *NEEDLY* intron 2 [54,99]).

Discrimination success was mixed for the two CITES-listed Podocarpaceae species (Table 1): *Po. parlatorei* (CITES Appendix I) can be distinguished from all other species using nrITS2 (*matK* and *rbcL* cannot distinguish *Po. parlatorei* from *Po. sprucei*; *rbcL* cannot distinguish *Po. parlatorei* from *Po. transiens*); *Po. neriifolius* (CITES Appendix III) cannot be distinguished from *Po. thailandensis* using all three markers (using single markers, *Po. neriifolius* can also be conflated with *Po. archboldii*, *Po. assamica*, *Po. brassii*, *Po. crassigemmis*, *Po. drouynianus*, *Po. gibbsiae*, *Po. insularis*, *Po. ledermannii*, *Po. philippinensis*, *Po. polystachyus*, *Po. ramosii*, *Po. rubens*, and/or *Po. subtropicalis*). The herbal dietary supplement, *N. nagi* (Asian bayberry), cannot be distinguished from *N. formosensis* using all three markers (*matK* also cannot distinguish *N. nagi* from *N. motleyi*).

Barcode gap

The barcode gap is a measure of the relative variation within and among species [78]. In the complete sample, 39.1% of species had a barcode gap for *matK*, 34.0% for *rbcL*, 38.1% for nrITS2, and 50.5% for all markers simultaneously (Figure 4; Table 1). There is no statistical difference ($p = 0.05$) in the frequency of barcode gaps among markers. The presence of a barcode gap is not correlated with sample size in Podocarpaceae ($r_{pb} = 0.06$; $p = 0.27$).

Barcode gaps quantify species distinctness at the barcode locus and thereby provide a crude measure of identification reliability (i.e. a species without a barcode gap may be more likely to be misidentified since it is not particularly distinctive; [71]). In this data set, whether a species can be consistently distinguished from all other species is unrelated to the presence or absence of a barcode gap ($p = 0.02$). For *matK* and *rbcL*, all of the species that can be consistently diagnosed have a barcode gap, but there are six

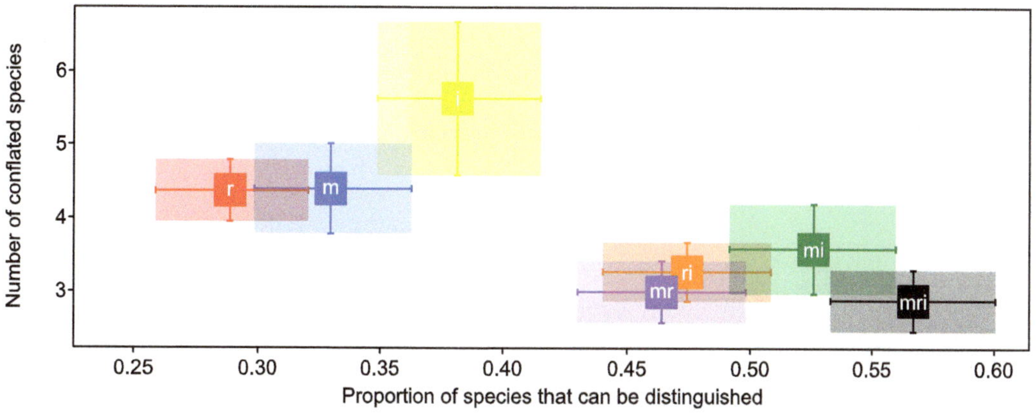

Figure 3. Species discrimination by barcode marker for complete samples. Squares indicate means for *matK* (blue; m), *rbcL* (red; r), nrITS2 (yellow; i), *matK* combined with *rbcL* (purple; mr), *matK* combined with nrITS2 (green; mi), *rbcL* combined with nrITS2 (orange, ri), and all markers combined (black; mri). Error bars indicate 95% confidence intervals.

Table 2. Rates of discriminatory success for barcoding studies that analyzed *matK*, *rbcL*, and nrITS2 sequences using algorithms comparable to BRONX.

study focus	*matK*+*rbcL*	*matK*+*rbcL*+nrITS2
Parnassia (Parnassiaceae) [98]	71.42%	96.19%
Chinese *Primula* sect. *Proliferae* (Primulaceae) [96]	68.75%	87.50%
Malagasy *Euphorbia* (Euphorbiaceae) [90]	**59.45%**	99.32%
Caryoteae (Arecaceae) [89]	51.85%	**92.59%**
Actaea (Ranunculaceae) [97]	35.71%	57.14%

Median rates of discriminatory success are in boldface.

species with barcode gaps that cannot be consistently differentiated from all other species (*matK*: *Dd. beccarii*, *Po. bracteatus*, *Po. novae-caledoniae*, *Po. nubigenus*, *Po. rumphii*, and *R. minus*; *rbcL*: *Manoao colensoi*, *N. wallichiana*, *Po. bracteatus*, *Po. pilgeri*, *Po. spinolosus*, and *Pr. andina*). In contrast, there are four species that do not have nrITS2 barcode gaps, but can be consistently diagnosed with nrITS2 (*Af. gracilior*, *Dc. imbricatus*, *Po. lambertii*, and *Pr. ferruginoides*). There are also four species that have nrITS2 barcode gaps that cannot be consistently differentiated from all other species using nrITS2 (*Po. bracteatus*, *Po. celatus*, *Po. longifoliolatus*, and *Po. sprucei*). There are no species with multilocus barcode gaps that cannot be consistently diagnosed using all three markers simultaneously, but there are six species that do not have multilocus barcode gaps that can be consistently diagnosed (*Af. gracilior*, *Af. mannii*, *Dc. imbricatus*, *Dc. kinabaluensis*, *Po. polystachyus*, and *Pr. ferruginoides*; Table 1).

The absence of a barcode gap coupled with discrimination success serves to contrast algorithmic approaches that use diagnostic nucleotide positions (i.e. those positions that consistently distinguish one species from all others) with distance-based methods. The presence of a barcode gap, does not guarantee that a species will be distinct. For example, a species may have a large amount of intraspecific variation combined with a small, but consistent, amount of interspecific variation rendering the species without a barcode gap, but consistently diagnosable—one nucleotide difference that consistently differentiates the species in question from all other species is all that is required. Thus, the absence of a barcode gap is a poor predictor of discrimination success.

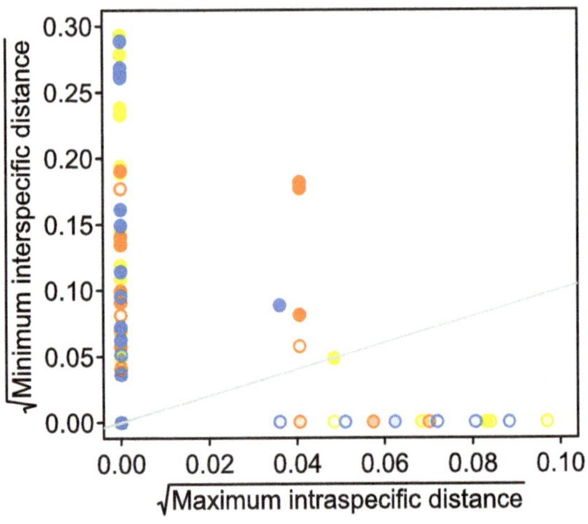

Figure 4. Barcode variation within and among species for complete samples. Circles represent the set of *matK* (blue), *rbcL* (red), and nrITS2 (yellow) sequences for each species. Opaque filled circles denote diagnostic sequence sets. Non-diagnostic sequence sets are indicated with semi-transparent filled circles. Equal intra- and inter-specific variation is marked by the gray line. Points above the gray line indicate species with 'barcode gaps'.

The presence of a barcode gap coupled with discrimination failure is an artifact of the analysis conducted: barcode gaps were computed using only sequences in the complete sample whereas discrimination was calculated with a reference database composed of all sequences. Thus, the barcode gap calculation did not necessarily include samples with zero interspecific distance that were included in the discrimination calculation. Restricting the discrimination calculation to sequences in the complete sample would have overestimated discrimination success for Podocarpaceae. At the same time, calculating the barcode gap using all sequences would have resulted incomparable values.

Sampling of additional individuals cannot decrease the maximum intraspecific distance, nor can it increase the minimum interspecific distance. Thus, new sequence data for *matK*, *rbcL*, and nrITS2 will either maintain or decrease the number of species with barcode gaps. Likewise, the rate of species discrimination cannot improve, and will most likely deteriorate, with additional sampling of individuals. New sequences of unlinked markers may however increase the number of species with barcode gaps and/or improve the rate of species discrimination.

Conclusions

The vast majority of barcode sequences generated for this study were of high quality (Figure 2). Even the lowest quality sequences exceeded the minimum requirements of the BARCODE data standard. In the few instances that low quality sequences were generated, the responsible mechanism could not be discerned: slip-strand mispairing at the site of repetitive sequence elements cannot adequately explain the low quality sequences observed.

The power of *matK*, *rbcL*, and nrITS2, individually and in combination, to discriminate among Podocarpaceae species is relatively low (56.7% of species at maximum; Table 1; Figure 3). There were no statistically significant differences in the discriminatory power of markers or marker combinations. Although the discrimination rate for Podocarpaceae is below the rate reported for comparably analyzed studies (Table 2), it is not markedly lower. Plant DNA barcoding studies that heavily sample within taxonomic groups usually report low rates of species discrimination.

Discrimination success was mixed for Podocarpaceae species important in commerce and of conservation concern (Table 1). The CITES Appendix I species, *Po. parlatorei*, can be distinguished from all other species using nrITS2. Unfortunately, the CITES

Appendix III species, *Po. neriifolius*, and the herbal dietary supplement, *N. nagi*, cannot be unambiguously distinguished from all other Podocarpaceae using all three markers.

The presence of a barcode gap was not predictive of discrimination success. There was no statistically significant difference in the frequency of barcode gaps among markers in Podocarpaceae (Figure 4). In addition, there was no correlation between number of individuals sampled per species and the presence of a barcode gap.

Sequences of additional variable unlinked markers that are easily recovered from Podocarpaceae (e.g. *NEEDLY* intron 2) may increase the rate of species discrimination.

Acknowledgments

We thank Joan Deutsch for providing excellent technical assistance. License for the use of TNT was generously provided by the Willi Hennig Society.

Author Contributions

Conceived and designed the experiments: DPL PK CS. Performed the experiments: PK CS. Analyzed the data: DPL CS. Contributed reagents/ materials/analysis tools: DPL PK CS. Wrote the paper: DPL PK CS.

References

1. Farjon A (2001) World checklist and bibliography of conifers. Richmond: Royal Botanic Gardens, Kew, 2 edition.
2. Cernusak LA, Adie H, Bellingham PJ, Biffin E, Brodribb TJ, et al. (2011) Podocarpaceae in tropical forests: a synthesis. In: Turner BL, Cernusak LA, editors, Ecology of the Podocarpaceae in tropical forests, Washington, D.C.: Smithsonian Insitiution Scholarly Press, volume 95. pp. 189–195.
3. de Laubenfels DJ (1969) A revision of the Malesian and pacific rainforest conifers, i. Podocarpaceae, in part. Journal of the Arnold Arboretum 50: 315–369.
4. de Laubenfels DJ (1988) Coniferales. Flora Malesiana 10: 337–453.
5. Buchholz JT, Gray NE (1948) A taxonomic revision of *Podocarpus* I: the sections of the genus and their subdivisions with special reference to leaf anatomy. Journal of the Arnold Arboretum 29: 46–63.
6. Buchholz JT, Gray NE (1948) A taxonomic revision of *Podocarpus* II: the American species of Podocarpus section Stachycarpus. Journal of the Arnold Arboretum 29: 64–76.
7. Buchholz JT, Gray NE (1948) A taxonomic revision of *Podocarpus* IV: the American species of section *Eupodocarpus* subsections C and D. Journal of the Arnold Arboretum 29: 123–151.
8. Gray NE, Buchholz JT (1948) A taxonomic revision of *Podocarpus* III: the American species of *Podocarpus* section *Polypodiopsis*. Journal of the Arnold Arboretum 29: 117–122.
9. Gray NE, Buchholz JT (1951) A taxonomic revision of *Podocarpus* V: the south Pacific species of *Podocarpus* section *Stachycarpus*. Journal of the Arnold Arboretum 32: 82–92.
10. Gray NE, Buchholz JT (1951) A taxonomic revision of *Podocarpus* VI: the south Pacific species of *Podocarpus* section *Sundacarpus*. Journal of the Arnold Arboretum 32: 93–98.
11. Gray NE (1953) A taxonomic revision of *Podocarpus* VII: the African species of *Podocarpus* section *Afrocarpus*. Journal of the Arnold Arboretum 34: 67–76.
12. Gray NE (1953) A taxonomic revision of *Podocarpus* VIII: the African species of section *Eupodocarpus*, subsections A and E. Journal of the Arnold Arboretum 34: 163–175.
13. Gray NE (1955) A taxonomic revision of *Podocarpus* IX: the south Pacific species of section *Eupodocarpus*, subsection F. Journal of the Arnold Arboretum 36: 199–206.
14. Gray NE (1956) A taxonomic revision of *Podocarpus* X. Journal of the Arnold Arboretum 37: 160–172.
15. Gray NE (1958) A taxonomic revision of *Podocarpus* XI. Journal of the Arnold Arboretum 39: 424–477.
16. Gray NE (1960) A taxonomic revision of *Podocarpus* XII: section *Microcarpus*. Journal of the Arnold Arboretum 41: 36–39.
17. Gray NE (1962) A taxonomic revision of *Podocarpus* XIII: section *Polypodiopsis* in the south Pacific. Journal of the Arnold Arboretum 43: 67–79.
18. Schoonraad E, Vanderschijff HP (1974) Anatomy of leaves of genus *Podocarpus* in South Africa. Phytomorphology 24: 75–85.
19. Knopf P, Nimsch H, Stützel T (2007) *Dacrydium* × *suprinii*, sp. Nova—a natural hybrid of *Dacrydium araucarioides* × *D. guillauminii*. Feddes Repertorium 118: 51–59.
20. Knopf P (2011) Differential diagnosis and evolution within the Podocarpaceae s. l. Ph.D. thesis, Ruhr-University Bochum.
21. Stockey RA, Ko H (1990) Cuticle micromorphology of *Dacrydium* (Podocarpaceae) from New Caledonia. Botanical Gazette 151: 138–149.
22. Stockey RA, Ko H, Woltz P (1992) Cuticle micromorphology of *Falcatifolium* de Laubenfels (Podocarpaceae). International Journal of Plant Sciences 153: 589–601.
23. Stockey RA, Ko H, Woltz P (1995) Cuticle micromorphology of *Parasitaxus* de Laubenfels (Podocarpaceae). International Journal of Plant Sciences 156: 723–730.
24. Stockey RA, Frevel BJ (1997) Cuticle micromorphology of *Prumnopitys Philippi* (Podocarpaceae). International Journal of Plant Sciences 158: 198–221.
25. Stockey RA, Frevel BJ,Woltz P (1998) Cuticle micromorphology of *Podocarpus*, subgenus *Podocarpus*, section Scytopodium (Podocarpaceae) of Madagascar and South Africa. International Journal of Plant Sciences 159: 923–940.
26. Mill RR, Schilling DM (2009) Cuticle micromorphology of *Saxegothaea* (Podocarpaceae). Botanical Journal of the Linnean Society 159: 58–67.
27. Schilling DMS, Mill RR (2011) Cuticle micromorphology of Caribbean and Central American species of *Podocarpus* (Podocarpaceae). International Journal of Plant Sciences 172: 601–631.
28. IUCN (2012) International union for the conservation of nature red list of threatened species, version 2012.2. Available: http://www.iucnredlist.org. Accessed 2013 March 5.
29. CITES (2012) Convention on international trade in endangered species, appendices I, II, and III. Available: http://www.cites.org. Accessed 2013 March 5.
30. McGuffin M, Kartesz JT, Leung AY, Tucker AO (2000) Herbs of Commerce. American Herbal Products Association, 2nd edition.
31. Fu L, Li Y, Mill RR (1994) Podocarpaceae. In: Wu Z, RavenPH, Hong D, editors, Flora of China, St. Louis: Missouri Botanical Garden, volume 4. pp. 78–84.
32. Facciola S (1990) Cornucopia: a source book of edible plants. Vista: Kampong Publications.
33. Abdillahi HS, Stafford GI, Finnie JF, Van Staden J (2010) Ethnobotany, phytochemistry and pharmacology of *Podocarpus* sensu latissimo (s.l.). South African Journal of Botany 76: 1–24.
34. Abdillahi HS, Verschaeve L, Finnie JF, Van Staden J (2012) Mutagenicity, antimutagenicity and cytotoxicity evaluation of South African *Podocarpus* species. Journal of Ethnopharmacology 139: 728–738.
35. Bauch J, Schmidt O, Hillis WE, Yazaki Y (1977) Deposits in heartshakes of *Dacrydium* species and their toxicity against fungi and bacteria. Holzforschung 31: 1–7.
36. Symonds EL, Konczak I, Fenech M (2012) The Australian fruit illawarra plum (*Podocarpus elatus* Endl., Podocarpaceae) inhibits telomerase, increases histone deacetylase activity and decreases proliferation of colon cancer cells. British Journal of Nutrition 15: 1–9.
37. Abdillahi HS, Finnie JF, Van Staden V (2008) Antibacterial activity of *Podocarpus* species. South African Journal of Botany 74: 359–360.
38. Abdillahi HS, Stafford GI, Finnie JF, Van Staden J (2008) Antimicrobial activity of South African *Podocarpus* species. Journal of Ethnopharmacology 119: 191–194.
39. Hayashi Y, Matsumoto T, Tashiro T (1979) Antitumor activity of norditerpenoid dilactones in *Podocarpus* plants—structure activity relationship on in vitro cytotoxicity against Yoshida sarcoma. Gann 70: 365–369.
40. Hembree JA, Chang CJ, McLaughlin JL, Cassady JM, Watts DJ, et al. (1979) Cytotoxic norditerpene dilactones of *Podocarpus milanjianus* and *Podocarpus seellowii*. Phytochemistry 18: 1691–1694.
41. Hembree JA, Chang C, Mclaughlin JL, Cassady J (1980) Milanjilactone A and milanjilactone B, 2 novel cytotoxic norditerpene dilactones from *Podocarpus milanjianus* Rendle. Experientia 36: 28–29.
42. Park HS, Takahashi Y, Fukaya H, Aoyagi Y, Takeya K (2003) S-R-podolactone D, a new sulfoxide-containing norditerpene dilactone from *Podocarpus macrophyllus* var. *maki*. Journal of Natural Products 66: 282–284.
43. Park HS, Takahashi Y, Fukaya H, Aoyagi Y, Takeya K (2004) New cytotoxic norditerpene dilactones from leaves of *Podocarpus macrophyllus* var. *maki*. Heterocycles 63: 347–357.
44. Bergin DO (2000) Current knowledge relevant to management of *Podocarpus totara* for timber. New Zealand Journal of Botany 38: 343–359.
45. Phillips EWJ (1941) The identification of coniferous woods by their microscopic structure. The Journal of the Linnean Society of London, Botany 52: 259–320.
46. Dallimore W, Jackson AB, Harrison SG (1967) A handbook of Coniferae and Ginkgoaceae. New York: St. Martin's Press.
47. Farjon A (2008) A natural history of conifers. Portland: Timber Press.

48. Cockayne L (1919) New Zealand plants and their story. Wellington: M. F. Marks, Goverment Printer.
49. Kubo I, Matsumoto T, Klocke JA (1984) Multichemical resistance of the conifer *Podocarpus gracilior* (Podocarpaceae) to insect attack. Journal of Chemical Ecology 10: 547–559.
50. Russell GB, Singh P, Fenemore PG (1972) Insect-control chemicals from plants: nagilactone c, a toxic substance from the leaves of *Podocarpus nivalis* and *P. hallii*. Australian Journal of Biological Sciences 25: 1025–1029.
51. Saeki I, Sumimoto M, Kondo T (1970) Termiticidal substances from wood of *Podocarpus macrophyllus* D. Don. Holzforschung 24: 83–86.
52. Li Y, Gao LM, Poudel RC, Li DZ, Forrest A (2011) High universality of *matK* primers for barcoding gymnosperms. Journal of Systematics and Evolution 49: 169–175.
53. Chiou SJ, Yen JH, Fang CL, Chen HL, Lin TY (2007) Authentication of medicinal herbs using PCR-amplified ITS2 with specific primers. Planta Medica 73: 1421–1426.
54. Knopf P, Schulz C, Little DP, Stützel T, Stevenson DW(2012) Relationships within Podocarpaceae based on DNA sequence, anatomical, morphological, and biogeographical data. Cladistics 28: 271–299.
55. Altschul SF, Gish W, Miller W, Myers EW, Lipman DJ (1990) Basic local alignment search tool. Journal of Molecular Biology 215: 403–410.
56. Edgar RC (2004) MUSCLE: multiple sequence alignment with high accuracy and high throughput. Nucleic Acids Research 32: 1792–1797.
57. Simmons MP, Ochoterena H (2000) Gaps as characters in sequence-based phylogenetic analysis. Systematic Biology 49: 369–381.
58. Little DP (2005) 2xread: a simple indel coding tool. Available: http://www.nybg.org/files/scientists/2xread.html. Accessed 2013 January 5.
59. Farris JS, Albert VA, Kallersjo M, Lipscomb D, Kluge AG (1996) Parsimony jackknifing outperforms neighbor-joining. Cladistics 12: 99–124.
60. Goloboff PA, Farris JS, Nixon KC (2008) TNT, a free program for phylogenetic analysis. Cladistics 24: 774–786.
61. Little DP (2010) A unified index of sequence quality and contig overlap for DNA barcoding. Bioinformatics 26: 2780–2781.
62. Pesole G, Attimonelli M, Saccone C (1996) Linguistic analysis of nucleotide sequences: algorithms for pattern recognition and analysis of codon strategy. Methods in Enzymology 266: 281–294.
63. Rice P, Longden I, Bleasby A (2000) EMBOSS: the European Molecular Biology Open Software Suite. Trends in Genetics 16: 276–277.
64. Shinde D, Lai Y, Sun F, Arnheim N (2003) *Taq* DNA polymerase slippage mutation rates measured by PCR and quasi-likelihood analysis: (CA/GT)n and (A/T)n microsatellites. Nucleic Acids Research 31: 974–980.
65. Scheffé H (1953) A method for judging all contrasts in the analysis of variance. Biometrika 40: 87–104.
66. de Mendiburu F (2012) Agricolae version 1.1-3. Available: http://cran.r-project.org/. Accessed 2013 January 5.
67. R development core team (2012) R: a language and environment for statistical computing (version 2.15.2). Vienna, Austria: R Foundation for Statistical Computing.
68. Spearman C (1904) The proof and measurement of association between two things. The American Journal of Psychology 15: 72–101.
69. Salinas NR, Little DP (2013) 2matrix: a utility for indel coding and phylogenetic matrix concatenation. Available: https://github.com/nrsalinas/2matrix. Accessed 2013 September 13.
70. Nixon KC (1999) The parsimony ratchet, a new method for rapid parsimony analysis. Cladistics 15: 407–414.
71. Sukumaran J, Holder MT (2010) DendroPy: a Python library for phylogenetic computing. Bioinformatics 26: 1569–1571.
72. Paradis E, Bolker B, Claude J, Cuong HS, Desper R, et al. (2013) The APE (Analyses of Phylogenetics and Evolution) package version 3.0-10. Available: http://cran.r-project.org/. Accessed 2013 January 5.
73. Little DP, Stevenson DW (2007) A comparison of algorithms for identification of specimens using DNA barcodes: examples from gymnosperms. Cladistics 23: 1–21.
74. Little DP (2011) DNA barcode sequence identification incorporating taxonomic hierarchy and within taxon variability. PLoS ONE 6: e20552.
75. Little DP (2012) BRONX: Barcode Recognition Obtained with Nucleotide eXposés version 2.0. Available: http://www.nybg.org/files/scientists/dlittle/BRONX2.html. Accessed 2013 January 5.
76. Wilson EB (1927) Probable inference, the law of succession, and statistical inference. Journal of the American Statistical Association 22: 209–212.
77. Harrell Jr FE (2012) The Hmisc package version 3.10-1. Available: http://cran.r-project.org/. Accessed 2013 January 5.
78. Meyer CP, Paulay G (2005) DNA barcoding: error rates based on comprehensive sampling. PLoS Biology 3: e422.
79. Levenshtein VI (1966) Binary codes capable of correcting deletions, insertions, and reversals. Soviet Physics|Doklady 10: 707–710.
80. Meier R, Zhang G, Ali F (2008) The use of mean instead of smallest interspecific distances exaggerates the size of the "barcoding gap" and leads to misidentification. Systematic Biology 57: 809–813.
81. Rizopoulos D (2013) Latent Trait Models under IRT version 0.9-9. Available: http://cran.r-project.org/. Accessed 2013 January 5.
82. McNemar Q (1947) Note on the sampling error of the difference between correlated proportions or percentages. Psychometrika 12: 153–157.
83. Conran JG, Wood GM, Martin PG, Dowd JM, Quinn CJ, et al. (2000) Generic relationships within and between the gymnosperm families Podocarpaceae and Phyllocladaceae based on an analysis of the chloroplast gene *rbcL*. Australian Journal of Botany 48: 715–724.
84. Kelch DG (2002) Phylogenetic assessment of the monotypic genera *Sundacarpus* and Manoao (Coniferales: Podocarpaceae) utilising evidence from 18S rDNA sequences. Australian Systematic Botany 15: 29–35.
85. Sinclair WT, Mill RR, Gardner MF,Woltz P, Jaffré T, et al. (2002) Evolutionary relationships of the New Caledonian heterotrophic conifer *Parasitaxus usta* (Podocarpaceae), inferred from chloroplast *trnL-F* intron/spacer and nuclear rDNA ITS2 sequences. Plant Systematics and Evolution 233: 79–104.
86. Biffin E, Conran JC, Lowe AJ (2011) Podocarp evolution: a molecular phylogenetic perspective. In: Turner BL, Cernusak LA, editors, Ecology of the Podocarpaceae in tropical forests, Washington, D.C.: Smithsonian Insitiution Scholarly Press, volume 95. pp. 1–20.
87. Farris JS (1974) Formal definitions of paraphyly and polyphyly. Systematic Zoology 23: 548–554.
88. Hanner R (2009) Proposed standards for BARCODE records in INSDC (BRIs). Technical report, Database Working Group, Consortium for the Barcode of Life, Available: http://barcoding.si.edu/PDF/DWG data standards-Final.pdf. Accessed 2013 January 5.
89. Jeanson ML, Labat JN, Little DP (2011) DNA barcoding: a new tool for palm taxonomists? Annals of Botany 108: 1445–1451.
90. Aubriot X, Lowry PP, Cruaud C, Couloux A, Haevermans T (2013) DNA barcoding in a biodiversity hot spot: potential value for the identification of Malagasy *Euphorbia L.* listed in CITES Appendices I and II. Molecular Ecology Resources 13: 57–65.
91. Devey DS, Chase MW, Clarkson JJ (2009) A stuttering start to plant DNA barcoding: microsatellites present a previously overlooked problem in non-coding plastid regions. Taxon 58: 7–15.
92. Fazekas AJ, Steeves R, Newmaster SG (2010) Improving sequencing quality from PCR products containing long mononucleotide repeats. BioTechniques 48: 277–285.
93. Fazekas AJ, Steeves R, Newmaster SG, Hollingsworth PM (2010) Stopping the stutter: improvements in sequence quality from regions with mononucleotide repeats can increase the usefulness of non-coding regions for DNA barcoding. Taxon 59: 694–697.
94. Meier R, Shiyang K, Vaidya G, Ng PKL (2006) DNA barcoding and taxonomy in Diptera: a tail of high intraspecific variability and low identification success. Systematic Biology 55: 715–728.
95. CBOL Plant Working Group (2009) A DNA barcode for land plants. Proceedings of the National Academy of Sciences 106: 12794–12797.
96. Yan HF, Hao G, Hu CM, Ge XJ (2011) DNA barcoding in closely related species: a case study of *Primula* L. sect. *Proliferae* Pax (Primulaceae) in China. Journal of Systematics and Evolution 49: 225–236.
97. Baker DA, Stevenson DW, Little DP (2012) DNA barcode identification of black cohosh herbal dietary supplements. Journal of AOAC International 95: 1023–1034.
98. Yang JB, Wang YP, Möller M, Gao LM, Wu D (2012) Applying plant DNA barcodes to identify species of *Parnassia* (Parnassiaceae). Molecular Ecology Resources 12: 267–275.
99. Little DP (2004) Documentation of hybridization between Californian cypresses: *Cupressus macnabiana* × *sargentii*. Systematic Botany 29: 825–833.

13

Mammalian Collection on Noah's Ark: The Effects of Beauty, Brain and Body Size

Daniel Frynta*, Olga Šimková, Silvie Lišková, Eva Landová

Department of Zoology, Faculty of Sciences, Charles University in Prague, Prague, Czech Republic

Abstract

The importance of today's zoological gardens as the so-called "Noah's Ark" grows as the natural habitat of many species quickly diminishes. Their potential to shelter a large amount of individuals from many species gives us the opportunity to reintroduce a species that disappeared in nature. However, the selection of animals to be kept in zoos worldwide is highly selective and depends on human decisions driven by both ecological criteria such as population size or vulnerability and audience-driven criteria such as aesthetic preferences. Thus we focused our study on the most commonly kept and bred animal class, the mammals, and we asked which factors affect various aspects of the mammalian collection of zoos. We analyzed the presence/absence, population size, and frequency per species of each of the 123 mammalian families kept in the worldwide zoo collection. Our aim was to explain these data using the human-perceived attractiveness of mammalian families, their body weight, relative brain size and species richness of the family. In agreement with various previous studies, we found that the body size and the attractiveness of mammals significantly affect all studied components of the mammalian collection of zoos. There is a higher probability of the large and attractive families to be kept. Once kept, these animals are presented in larger numbers in more zoos. On the contrary, the relative mean brain size only affects the primary selection whether to keep the family or not. It does not affect the zoo population size or the number of zoos that keep the family.

Editor: David L. Roberts, University of Kent, United Kingdom

Funding: The project was supported by the Grant Agency of the Czech Academy of Sciences (project number IAA 601410803). Personal costs of OŠ and SL were partially supported by the project number SVV-2013-267 201. The funders had no role in study design, data collection and analysis, decision to publish, or preparation of the manuscript.

Competing Interests: The authors have declared that no competing interests exist.

* E-mail: frynta@centrum.cz

Introduction

Nowadays, mankind covers about 83% of the Earth's land surface [1], causing global biodiversity to decline due to a quick loss of natural habitats of many species [2]. The proportion of potentially threatened species is rapidly increasing, leaving only very few species safe from a possible extinction [3]. It is therefore important not to miss any potential chance for animal conservation, including both in-situ and ex-situ conservational efforts. The world's zoos, aquaria, botanical gardens, and gene banks provide insurance for species and genetic diversity [4]. According to the ISIS (International Species Information System) online database, more than 7 million individual animals are kept in 872 zoos and aquariums (as recorded by the date of 12th January 2011). The high potential of zoos to serve as wildlife reservoirs, coupled with the rapid destruction of nature that we have faced in the last few decades, led [5] to frame the landmark Ark Hypothesis. The role of zoos as an ark proved viable in the case of amphibians (the "Amphibian Ark") that suffered a rapid population decline due to the chytridiomycosis disease. In response to this threat, The World Association of Zoos and Aquariums (WAZA), the IUCN/SSC (The International Union for Conservation of Nature/Species Survival Commission) Amphibian Specialist Group, and the IUCN/SSC Conservation Breeding Specialist Group (CBSG) worked together to collect a large number of species [6]. Moreover, in 1993 the EU recognized this conservation potential of zoos during the Convention on Biological Diversity (CBD), and obliged the zoos to manage the ex-situ and in-situ conservational role under the CBD's requirements [7], regardless of the lack of the government's systematic financial or other support [8]. EAZA (American Association of Zoos and Aquariums) and AZA (American Association of Zoos and Aquariums) supervise many specific ex-situ conservation programs such as Species Survival Plans and European Endangered Species Programmes. Moreover, they cooperate with CBSG, TAGs (Taxon Advisory Groups) and use various studbooks and data management systems, notably ISIS, to maintain the breeding of a variety of species which may also raise the effectiveness and possibility of species survival through captive breeding [9–11].

However, although the space to accommodate wild animals in zoos worldwide altogether is large, it is still very limited when compared to the list of all extant species. Only a small fraction of the world's animal population can board the Ark. In the year 2009, the Ark provided space for about 152 thousand individual mammals belonging to 990 species (18.5% of extant species) within a median worldwide zoo population size of 34.5 individuals. Empirical studies suggest that the minimum population size necessary for short-time captive maintenance of animal species/breeds under controlled conditions is about 50 [12], and populations over about 500 individuals are usually not affected by inbreeding depression [13]. Although these thresholds are only rough estimates because the effect depends on both effective population size and frequency of deleterious recessive mutations (cf. [14–16]). The zoos worldwide maintain 416 mammalian species represented by more than 50 individuals, which is 7.8% of all extant species. For the threshold of 500 individuals, the numbers count only 79 mammalian species, representing 1.5% of the extant mammalian diversity [17]. Similar numbers were confirmed in an independent study from April 2010 [18], finding that out of all 142 threatened mammalian species belonging to the IUCN categories Endangered, Critically Endangered, and Extinct in the wild, 68 species are being kept in zoos in more than 50 individuals, out of which 30 species are kept in more than 250 individuals.

Such small numbers point out that, even theoretically, if zoos tried to keep and breed endangered species, the space would be limited to hold only a tiny fraction of needful species at populations large enough to sustain a long-term captive breeding program of animals while avoiding an inbreeding depression. However, the presence of just a few unrelated individuals in zoos may occasionally save the species if the captive population is immediately expanded when necessary, e.g., after an unexpected crisis of the wild populations (but see [19] for negative effects of bottlenecking). This suggests that cooperative ex-situ conservation can help restore animal populations once the threat has diminished. Although many authors have questioned this assumption ever since (i.e., [20–24]), many researchers find the theoretical role of zoos in reintroduction programs as feasible (for a review of the limitations and solutions of availability of captive populations in reintroduction, see [25]; e.g., disease risk, behavioral competence of captive-reared individuals such as reduced ability to avoid predators or find food resources and attachment to humans; changes in genetic compositions, etc.), and a notable contribution of zoos to animals ex-situ breeding can be demonstrated by successful reintroductions which reduced the threat level of particular species. Species such as the Przewalski horse (*Equus ferus przewalskii*; [26]), the American bison (*Bison bison*; [27]), the European wisent (*Bison bonasus*; [28–30]), Pére David's deer (*Elaphurus davidianus*; [31]), or Arabian oryx [32] may serve as examples of successfully maintained zoo animal populations released back into the wild. Other notably successful reintroduced animals were the Asiatic wild ass (*Equus hemionus*; [33]), the golden lion tamarin (*Leontopithecus rosalia*; [34]), and the black-footed ferret (*Mustela nigripes*). The latter recovered from a very small population of only eighteen remaining individuals ([35]).

Moreover to the ex-situ breeding role of captive zoo populations, these groups might contribute to conservation purposes in other ways that do not necessarily demand high captive population densities. A very important role of today's zoological gardens presents education, especially promotion of increased public and political awareness of the need for in-situ conservation [36].

A single popular animal (or a small group of these) might serve as a flagship species and help its endangered relatives in the wild, or their natural habitat and its residents, to gain the necessary financial support from the public [37]. Another significant role of zoos resides in training specialists with the right knowledge about the breeding and care of the animals [38]. International studbooks running under the World Association of Zoos and Aquariums (WAZA) and collaboration with the ISIS database include husbandry and veterinary guidance for as many species as possible (www.waza.org). However, keeping a few individuals of a rare species, or their relatives, may help to retain the right specialists for future needs. The experience of staff and researchers working with living animals is irreplaceable by studbook information and guidelines, and their presence may help to save the species. Similarly, imagine a well-educated surgeon with no experience with real patients to perform operations on living humans.

The situation is complicated due to the fact that conservation activities are not the only purpose of the zoos. In fact, it is an issue discussed more intensively during the last several decades [38]. Zoos are vitally dependant on the funds gained from visitors [39]. Guests come to zoos mainly for recreational activities, expecting to see large, attractive, and active animals [40–42]. This may seemingly lead to the conclusion that there is a trade-off of which animals to keep in zoos to satisfy both conservational purposes and the visitors' recreational desires. The species' conservational status according to the IUCN was not the key factor for the selection of species to be included in the worldwide zoo collection (see [17,18] for terrestrial vertebrates; [43] for parrots, [44] for boid snakes). A question thus arises: is it the attractiveness and/or factors connected with the attractiveness of the animals that determines the composition of the collection of zoo animals around the world?

Balmford et al. [45] hypothesized that it is the size of the animal that determines its presence in a zoo and we confirmed this relationship in almost all examined taxa of terrestrial vertebrates in our previous papers [17,43,44]. Additionally, we found that attractiveness of the animals to human respondents also affects the world's zoo population numbers in some clades, namely snakes [44], parrots [43], terrestrial birds (grouping followed [46], excluding Accipitriformes, Strigiformes, Trogonidae and Coliidae; Psittaciformes, Opisthocomus and Cuculiformes were added), basal mammals (Monotremata, Marsupialia, Xenarthra and Afrotheria), and mammals of the carnivore-ungulate clade (Laurasiatheria, [17]).

On principle, larger animals are more conspicuous and visible to potential zoo visitors. This leads to the assumption that size can be difficult to separate from attractiveness, yet still there is more to beauty than size itself. In this paper, we measured separate data of attractiveness perceived by human respondents for a nearly complete set of mammalian families. We hypothesized that both size and attractiveness of mammals are good predictors of the size of their captive zoo populations (World Zoo Collection; further referred to as WZC).

Moreover to the above-mentioned physical properties, the brain size may modulate the attractiveness of animals to zoo curators and visitors. Adolf Portmann, the well-known Swiss biologist of the 20th century, hypothesized that humans categorize animals into the "higher" and "lower" ranked animal groups [47], and that this categorization, reflecting the brain size of the animal, affects the perceived attractiveness. Accordingly, we hypothesized that the brain size is a good predictor of behavioral attractiveness of mammalian species.

In this paper, we aimed to analyze the effect of body size, brain size, and attractiveness of mammalian species from almost all recent families to several variables explaining the WZC. We elaborate the previously-questioned number of individuals in the WZC in separate analyses to ask which factors determine whether the animal is present in any zoo or not, and to ask how many zoos actually keep those selected ones, and in how large or small numbers. There is a possibility that rather than species recognized by current taxonomists, taxa closer to genera or families represent the primary units of human spontaneous categorization [48]. Thus, the species within a family may compete with each other for the space available on the Ark (e.g., a zoo might select just one "mouse" or "rat" to keep in its collection instead of all 715 species of the family Muridae). Because of that, we included the species richness of the family as another factor to explain the analyzed WZC variables.

Materials and Methods

The dataset used for statistical analyses (see Appendix S1) includes 123 rows referred to as families. There were 119 families recognized by [49] and four infraorders (i.e., Microchiroptera, Megachiroptera, Mysticeti, Odontoceti). Cetaceans and chiropterans were pooled into infraorders because these specialized aerial and marine taxa deviate considerably from the typical mammalian body plan and their representation in zoos is rather poor.

Dependent variables

Information about the numbers of mammalian species and individuals kept in zoos worldwide was obtained from the ISIS (International Species Information System) online database (http://www.isis.org) covering more than 800 zoos and aquariums from 76 countries. We excluded hybrids, ambiguous genera, and domestic species/forms (see Appendix S2) from the dataset. We analyzed the following four variables derived from this dataset (accessed on 12th January 2011):

(1) Proportion of zoo species. The number of species kept in the WZC scaled to the total number of extant species of a given family. This binomial variable reflects the mean probability that a species is kept at least in one zoo.

(2) Mean population size. The world zoo population size (square-root transformed) per species present in the WZC (the families not represented in the WZC were excluded).

(3) Number of zoos. The natural log-transformed mean number of zoos keeping the species (species and families not represented in the WZC were excluded).

(4) Number of conspecifics per zoo. The mean number of conspecifics per zoo keeping the species (the families not represented in the WZC were excluded).

Explanatory variables

Species richness. The number of extant species of each family was log-transformed. In the case of infraorders, the number of extant species was divided by the number of families belonging to the infraorder.

Body size. We gathered body weight records (in grams) for representatives of most mammalian genera from literary sources (mostly from [50]). These values were naturally log-transformed and used for computation of family means.

Brain size. The relative brain size was substituted by an encephalization quotient. This variable, introduced by [51–53] and then repeatedly used as a measure of relative brain size (e.g. [54]), is a natural log-transformed ratio between observed brain weight and theoretical brain weight predicted by an allometric equation for mammalian species of the given body size. We performed an ordinary least square regression to calculate the allometric relationship between brain and body size. We gathered the primary data on brain and body weights for 1309 mammalian species from various literary sources (see Appendix S3). Because the individual families were unequally represented in the dataset, we employed a weighting by the variable inversely proportional to the number of included species belonging to the particular family (Figure 1). The resulting empirical allometric equation ($\ln B = -0.6601*\ln M-2.4100$; B = brain mass, M = body mass) was used for a calculation of the encephalization quotients.

Attractiveness. For the purpose of data collection, we defined four sets of 123 pictures depicting species from each family. Species representing individual families in each partial set were selected by a two-step (first genus then species) random choice process from the list of extant genera and species ([49]; domestic forms were excluded). Thus, duplicated presence of identical genera and species was avoided whenever possible. In monotypic families, the species was represented by different pictures. Sets coded as A, B and C consisted of illustrations while D consisted of photographs. The main sources of the pictures were [55–58]. In order to avoid possible effects of body size and background on rating, we adjusted the pictures with a white background and we resized them so that the pictured mammals were of a similar relative size (for illustration, see Figure 2).

The aesthetic attractiveness of the families was examined by presenting pictures of mammalian species to human volunteers (following [44]). The respondents were Czech citizens, mostly university students within the age range of 19–29 years. One can argue that as far as age, sex, and ethnic composition is concerned, our respondents did not properly represent the full scope of zoo visitors. The Czech Republic belongs among the least socially stratified countries in the world as demonstrated by the Gini index (GI) measuring the extent to which the distribution of income among individuals within an economy deviates from a perfectly equal distribution (0 corresponds to perfect equality while 100 to perfect inequality). The GI of the Czech Republic is 31 which corresponds to the 110 rank of 137 countries included in a comparison provided by the World Bank ([59]; compare, e.g., with GI = 45 for the USA which places them to the 41st position in the ranking; the higher the rank, the higher is the equality in society). Thus, we expected the Czech students to possess aesthetic preferences for animals comparable to the rest of the society, including zoo visitors (but see [60,61] who found socio-economic and educational differences in preferences, but their method did not focus purely on aesthetics). Since we preferred a homogenous example of respondents well motivated to performing the task, in which we could focus on the variables related to the tested stimuli rather than respondent characteristics, the students presented a good testing sample for the experiments evaluating human-perceived animal attractiveness.

Moreover, our previous studies revealed that the aesthetic ranking of animal species is highly stable with the factors of age, sex and ethnic composition having only a marginal effect ([43]). This is especially surprising in the case of cross-cultural comparisons, e.g., in such different cultures as are those in Europe and Papua New Guinea ([62,63] and new unpublished data; see also the agreement in physical attractiveness ratings of female faces across cultures [64], but see [65]). Note that the sample of respondents from non-European countries differed greatly in age and/or socioeconomic rank but their preferences for the examined animals still highly corresponded to those of the Czech students.

During the experiment, each respondent was exposed to one set of 123 pictures that were placed on a table in a random assemblage. Their objective was to "pack the photographs in an order corresponding to the beauty of the depicted species from the most beautiful to the least beautiful one", as we asked them. The order of the pictures in the pack was then coded by numerals from 1 (the most beautiful one) to 123, further referred to as ranks. Although no explicit time limit was given, all the respondents performed the task in about 30 minutes. Each set of pictures was evaluated by a comparable number of respondents: 77 (25 men), 85 (31), 77 (27) and 75 (25) for sets A, B, C and D, respectively. Altogether, we gathered data from 314 respondents; 206 of which were women and 108 were men.

All respondents agreed to participate in the project voluntarily. Each subject provided a written informed consent and additional information about gender, age and their affinity to mammals. The age and gender had no effect on the preferences (MANOVA, all

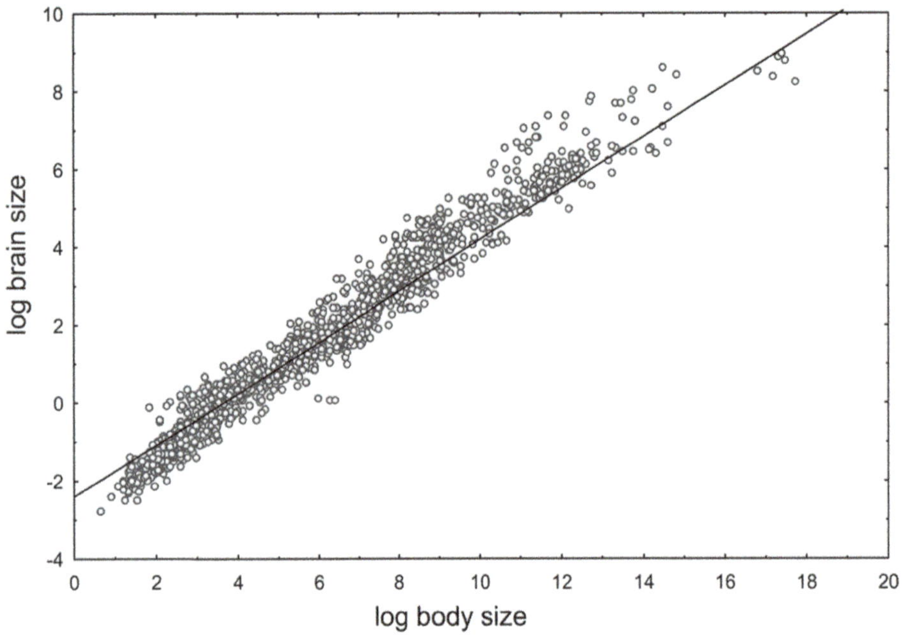

Figure 1. The allometric relationship between the brain and body sizes in 1309 mammalian species. Because the families were represented by an unequal number of data points (species), each family was given an equal weight when calculating the ordinary least-square regression. This correction for biased representation of the families resulted in a line seemingly unfitting the data points (representing species and not families). This adjusted regression line is considerably more reliable for further GLM analyses performed between family level, though. Allometric equation: ln(brain mass) = −0.6601 * ln(body mass)−2.4100.

Ps>0.05), which allowed us to pool the data. The affinity to mammals exhibited too low of a variance (91.85% respondents reported positive affinity to mammals, 8.15% neutral and 0 negative) to allow reliable testing. The experiments were performed in accordance with the European law and were approved by The Institutional Review Board of Charles University, Faculty of Science (No. 2009/2).

Figure 2. The pictures of the representatives of the family Felidae (the second most preferred family out of 123 examined ones). It illustrates the variation of body positions among four tested sets of pictures (for details see under Materials and Methods): clouded leopard *Neofelis nebulosa* (top left), serval *Leptailurus serval* (top right), jaguar *Panthera onca* (bottom left; pictures painted by Silvie Lišková) and caracal *Caracal caracal* (bottom right; photo from the archive of the Zoo Prague).

In order to quantify and test congruence in species ranking provided by different respondents and/or to compare sets composed of different species, we adopted Kendall's Coefficient of Concordance (W) as implemented in SPSS v.16.0 [66]. There was considerable congruence among the respondents in all four sets of pictures; W coefficients were 0.206, 0.264, 0.224 and 0.334 for sets A, B, C and D, respectively (all $p<0.001$).

Prior to further analyses, the raw ranks were transformed as follows: each value was divided by the number of evaluated families (123) and square-root arcsine transformed to improve its statistical distribution. Next, we computed the mean transformed rank for each set and family. Mean transformed ranks computed for individual sets were mutually significantly correlated ($r^2 = 0.334$, 0.450, 0.401, 0.324, 0.449 and 0.420 for A vs. B, A vs. C, A vs. D, B vs. C, B vs. D and C vs. D, respectively; all $p<0.0001$). This allowed us to compute family means from the mean preference ranks obtained for partial sets of pictures and to further use this variable as a simplified measure of aesthetic attractiveness of the family for humans. The numeric values of this variable were positivized (multiplied by −1) to make the explanation of the results more intuitive.

We included another explanatory variable in preliminary analyses: whether the family is mainly diurnal or nocturnal/fossorial. However, this variable correlates with the body size ($r = -0.51$), and when the influence of body size is removed, the factor itself explains neither of the analyzed dependant variables. As such, we removed it from further analyses.

Statistical treatment

In order to examine the effects of species richness, body size, attractiveness, and brain size on dependent variables, we generated General Linear Models (GLMs) in R 2.8.0 [67]. In the case of the proportion of zoo species, we adopted the binomial model with logit link function and Chi – square tests. We used the Gaussian distribution with identity link in remaining analyses. AIC criterion was used to reduce the original full models. The simplified model was also compared to the previous model by the ANOVA test to verify that the change in residual deviance was not significant ($P>0.05$).

Because species data are not independent as a result of shared phylogeny among more closely related taxa [68], we also carried out a phylogenetically controlled analysis using the independent contrast method [69]. For the purpose of this analysis, we used a phylogenetic tree of families compiled from recent studies dealing with molecular phylogenies. The main branching was adopted from Bininda-Emonds [70] and Arnason et al. [71], while specialized studies were used to improve the branching of partial crown taxa: Marsupialia [72], Xenarthra [73], Cetartiodactyla [74], Carnivora [75], Madagascar carnivores [76], Rodentia [77], Hystricognathi [78], Platacanthomyidae [79], Muridae [80], and Primates [81]. When phylogenetic information was equivocal, we resolved the tree in accordance with the conventional taxonomy. The independent contrasts of the arcsine-transformed mean preference rank, log-transformed WZPS, species number and body mass were computed using COMPARE, version 4.6b [82]. All branch lengths were set to 1 because the corresponding estimations were not available. Thus, for the contrasts analysis, we assumed the applicability of the punctuational model of evolution. The diagnostic proposed by [83] revealed that the contrasts were appropriately standardized. The multiple regression analyses based on independent contrast scores were performed in Statistica 6.0. [84] and constrained to pass through the origin [83].

Results

The analyzed zoos kept 179 868 mammals belonging to 1048 species and 103 families. The most represented families belonged predominantly to the carnivore-ungulates, primates and Xenarthra-Afrotheria clades (Appendix S1).

The highest aesthetic attractiveness was found in the large sized mammals (families Ailuridae, Felidae, Phascolarctidae, Ursidae, Giraffidae, Elephantidae, Equidae, Macropodidae, Mephitidae, and Cervidae). In contrast, the least preferred mammals (families Notoryctidae, Bathyergidae, Chrysochloridae, Spalacidae, Caenolestidae, Solenodontidae, Talpidae, Ctenomyidae, Geomyidae, and Dasypodidae) were predominantly small subterranean (fossorial) creatures with reduced eyes.

Proportion of zoo species

GLM revealed that the proportion of zoo species is associated positively with body size, attractiveness, and brain size, and negatively with species richness (all $Ps<0.0001$). This result was confirmed when the original variables were replaced by their independent contrasts and treated by a multiple regression through the origin (Table 1a, Figure 3).

Mean population size

Mean population size of zoo species was associated positively with body size ($P=0.0001$) and attractiveness ($P=0.0008$). Only the effect of attractiveness was confirmed by the independent contrast analysis ($P<0.0001$; Table 1b, Figure 4a,b).

Number of zoos

Both GLM of original data and multiple regression of independent contrasts agreed that the mean proportion of zoos keeping the species was significantly predicted by the body size ($P<0.0001$ and $P=0.0104$, respectively) and attractiveness ($P=0.0045$ and $P=0.0002$, respectively) of the animal (Table 1c, Figure 4c,d).

Number of conspecifics per zoo

GLM revealed negative effect of body size ($P=0.0016$) on the mean number of conspecifics kept in a zoo. This effect was confirmed by multiple regression of independent contrasts ($P=0.0002$). The latter analysis also revealed negative effect of species richness ($P=0.0318$; Table 1d, Figure 4e,f).

Discussion

We analyzed the representation of mammalian species in the WZC in the year 2011 and we found that it was very poor and highly selective, comparable to its state in 2009. [18] The list of currently recognized mammalian species [49] contains 5334 extant species, but only 1048 of them (16%) were actually present in the WZC. There were twenty families that were entirely absent in the WZC. Such selectivity may not only affect the putative conservation value of zoo populations, but it may also warn us about the existence of a large bias in species selection for conservation in general.

The selection of species into the WZC is determined by decisions made by humans, and although the selection criteria might be different, we may still assume that the psychological drives behind such selection are the same or similar to those for the selection of species for conservation and reintroduction. Such conclusion is supported by the fact that, between the years 1992–2009, out of the 12 evaluated reintroduced mammalian families, 11 of them are in the top-half when taking brain size (EQ) or attractiveness into account (for the list of reintroduced families with detailed published results, see Appendix S4). The "intelligent" and "beautiful" animals seem to be favored in human decision-makings. Moreover, many conservation programs strongly depend on financial support by the public, and it is most appropriate to assume that their decisions which species to support and which not to is driven by similar factors. Therefore, a deeper understanding of the factors that affect various aspects of the WZC is very important.

It was previously demonstrated that large species selectively attract human attention and conservation efforts [85,86]. We confirmed that the body size affects fundamentally all components of mammalian representation in the WZC. Mammalian species characterized by a large body size have a higher probability to be included in WZC. They tend to be represented by more numerous populations and they are also kept by more zoos. Because the material cost of keeping animals increases with the body size of the animal (Balmford 1996), the preferential representation of large mammals in the WZC is an interesting phenomenon. The metabolism (and thus the amounts of food and feces) and the required area of enclosure exponentially increase with the body size to about ¾ (0.72; [87,88]; for review see [89]) and 2/3 (i.e., the length of the suggested enclosure is roughly proportional to the body length of the animal in breeder's guidelines, e.g., [90]), respectively. The fact that keeping a large species is constrained by the available space and expenses is further illustrated by our results suggesting that the number of conspecifics per zoo tends to be smaller in species of a larger body size. However, it seems that the zoos are able to overcome these constraints and selectively keep large animals because these attract more visitors and thus secure countervailing income ([91]).

In our previous study, we performed a separate analysis of the WZC of selected mammalian clades and we demonstrated that human preferences affect the WZC positively in basal mammals

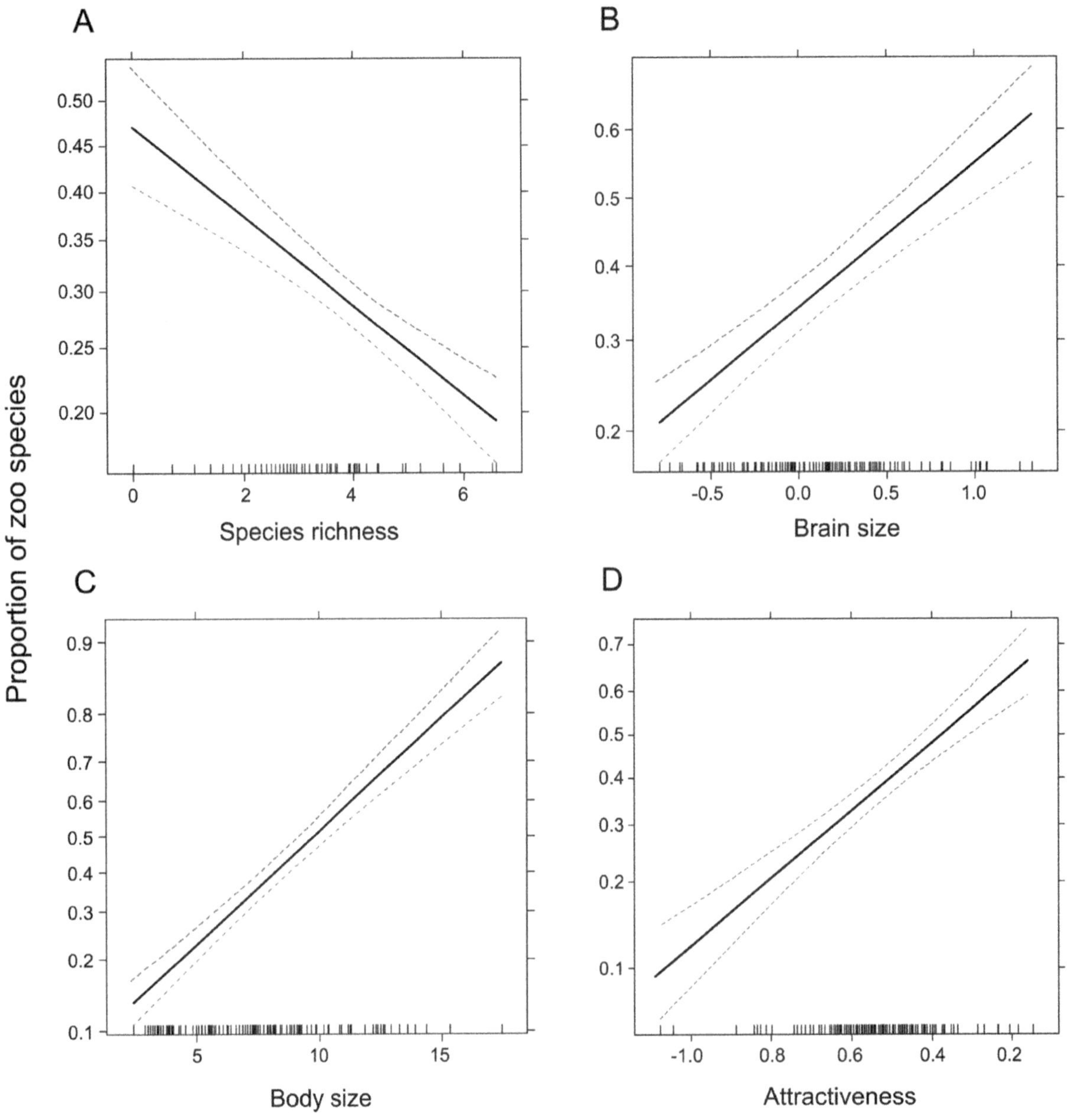

Figure 3. The proportion of zoo species as predicted by GLM. The effects of species richness (a), brain size (b), body size (c), and attractiveness (d). The dependent variable is the number of species kept in WZC scaled to the total number of extant species of a given family. This binomial variable reflects the mean probability that a species is kept in at least one zoo. For the definition and transformation of the explanatory variables see under the Material and methods section.

and Laurasiatheria [17]. The results of this study first show this relationship across all mammalian families; the species belonging to aesthetically attractive families have a higher probability to be included in the WZC, and they tend to be represented by more numerous populations as well as being kept by more zoos.

The positive effect of the encephalization index on the representation of mammalian species in the WZC predicted by [47] was only confirmed in the case of the proportion of zoo species. Apparently species of small-brained clades have a higher probability to be entirely omitted by the world zoos and their managers and curators. In agreement with Portman, the overall brain size (non-human primates: [92]), relative brain size (executive brain size: [93]; residuals: [94]) or the encephalization quotient, $EQ = E_a/E_e$, indicate the extent to which the brain size of a particular species E_a deviates from the expected brain size E_e, and are, to some extent, good predictors of so-called "intelligence" of a mammal (for a review, see [95]).

However, the exact definition of the term "intelligence", or higher cognitive abilities, is ambiguous. There are two main conflicting views: the adaptive specializations theory [96], saying that "intelligence" includes various learning and memory processes, which lead to adaptations for specific ecological task resolutions; and the general process view of "intelligence", described by the existence of general associative-learning abilities, which differ quantitatively among species [97]. A new, consensual theory describes "intelligence" as a behavioral flexibility [95] manifested in quick problem-solving task or number of innovations [94]. Fagen [98] assumed that the number of innovations and play

Table 1. The parameters of the reduced general linear models examining the effects of species richness, body size, brain size and attractiveness on dependent variables reflecting the representation of mammalian species in WZC.

A.	Anova					Coefficients	Independent contrasts		
	Df	**Deviance**	**Resid.Df**	**Resid.Dev**	**P**	**Estimate**	**Beta**	**B**	**p-level**
NULL	122	1855.22							
(Intercept)						−0.447			
Taxonomic uniqueness	1	333.04	121	1522.18	<0.0000	−0.199	−0.309	−0.063	0.0001
Body size	1	770.76	120	751.42	<0.0000	0.257	0.363	0.059	<0.0000
Brain size	1	81.73	119	669.69	<0.0000	0.861	0.176	0.165	0.0221
Attractiveness	1	71.33	118	598.36	<0.0000	3.178	0.242	0.582	0.0020
B.	**Anova**					**Coefficients**	**Independent contrasts**		
	Df	**Sum Sq**	**Mean Sq**	**F value**	**P**	**Estimate**	**Beta**	**B**	**p-level**
(Intercept)						11.152			
Body size	1	383.31	383.31	16.9380	0.0001	0.468			
Attractiveness	1	272.03	272.03	12.0210	0.0008	12.499	0.389	16.375	<0.0000
Residuals	100	2263	22.63						
C.	**Anova**					**Coefficients**	**Independent contrasts**		
	Df	**Sum Sq**	**Mean Sq**	**F value**	**P**	**Estimate**	**Beta**	**B**	**p-level**
(Intercept)						2.336			
Body size	1	44.23	44.23	38.2475	<0.0000	0.188	0.237	0.145	0.0104
Attractiveness	1	9.75	9.75	8.4308	0.0045	2.366	0.350	3.416	0.0002
Residuals	100	115.65	1.16						
D.	**Anova**					**Coefficients**	**Independent contrasts**		
	Df	**Sum Sq**	**Mean Sq**	**F value**	**P**	**Estimate**	**Beta**	**B**	**p-level**
(Intercept)						3.663			
Taxonomic uniqueness							−0.209	−0.152	0.0318
Body size	1	11.02	11.02	10.5723	0.0016	−0.129	−0.370	−0.193	0.0002
Attractiveness	1	2.17	2.17	2.0789	0.1525	−0.001			
Residuals	100	104.20	1.04						

a) Proportion of extant species kept in WZC.
b) World zoo population size per species in WZC.
c) Proportion of zoos keeping the species.
d) Individuals per Zoo Keeping the species.
The results of the phylogenetically adjusted analyses (multiple regression of independent contrasts performed through the origin) are also included. For the definition and transformation of the variables see under the Material and methods section.

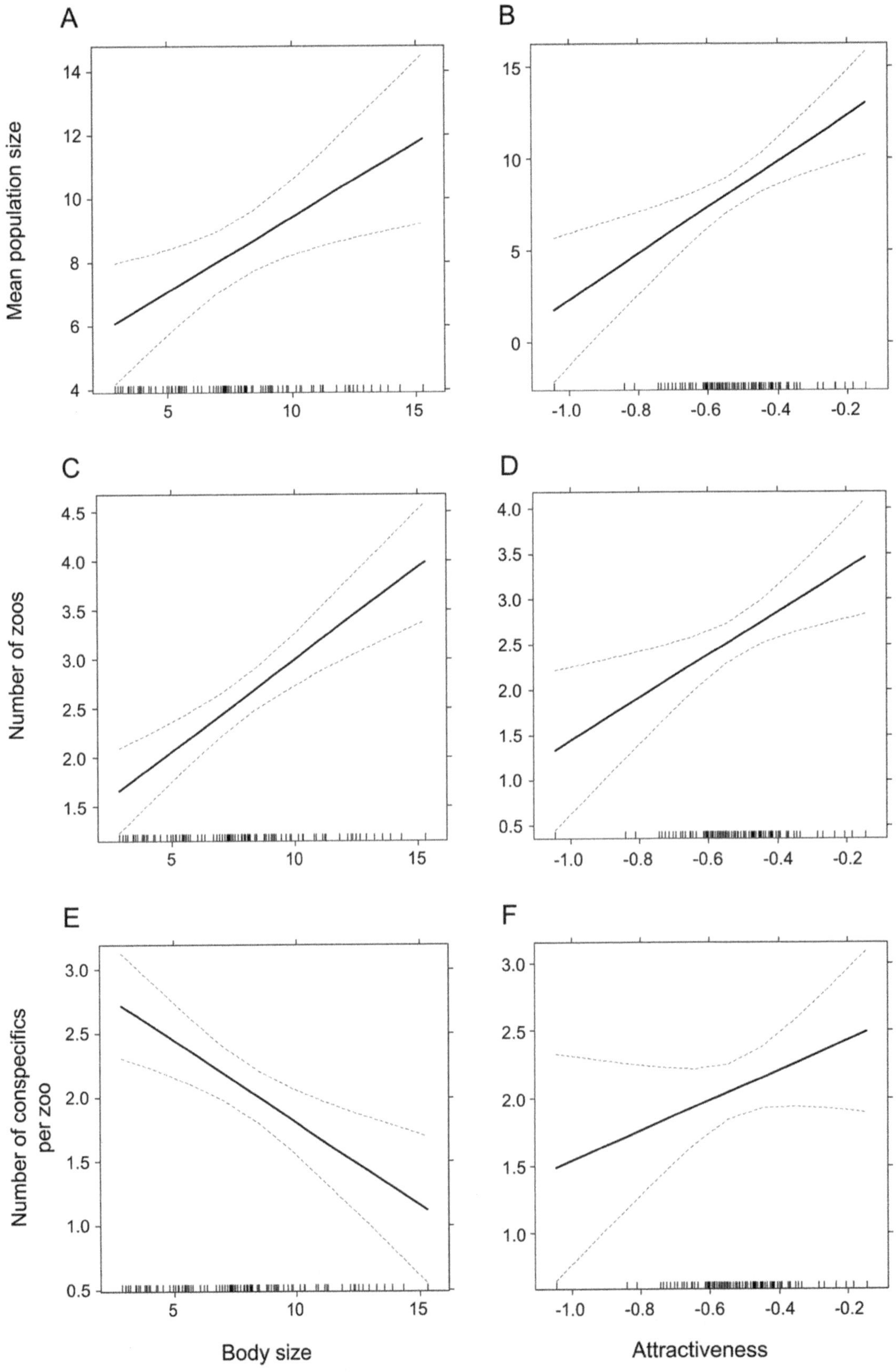

Figure 4. The effects of body size and attractiveness. The effects of body size and attractiveness on the mean population size (a,b), number of zoos (c,d) and number of conspecifics per zoo (e,f). The mean population size: The world zoo population size (square-root transformed) per species present in WZC. The number of zoos: The natural log-transformed mean number of zoos keeping the species. The number of conspecifics per zoo:

The mean number of conspecifics per zoo keeping the species. The families not represented in WZC were excluded from the analyses. For the definition and transformation of the explanatory variables see under the Material and methods section.

has close or causal relationship. In mammals, the large-brained taxa are more likely to contain species that play more often [99], and playful and active animals are more attractive to zoo visitors (e.g., Felidae: [100]). The attractiveness of animals with interesting behavior was also confirmed in [60], which reported higher willingness to support birds with special behavioral characteristics, e.g., courtship rituals.

Species richness of a family has a negative effect on the proportion of zoo species, the mean population size, and the number of conspecifics per zoo. Thus, the zoo curators tend to avoid simultaneous keeping of species belonging to the same family. This can be explained by human tendency to categorize mammals into primary cognitive categories frequently joining multiple scientific species into a single unit. Ethnobiologists repeatedly demonstrated that the primary units of human categorization of animals correspond to so-called "generic names" (for a review see [48]). These are one-word terms describing a typical species of a "genus"; additional species names are derived by adding the adjectives.

The knowledge of the factors affecting the selection of animals in the WZC might be applicable to a broad array of efforts influenced by human-induced selection of species. Once we become aware of it, we may adjust future planning of conservation projects to lead them into a better success, saving both time and finances along the way. Zoo curators may intentionally try to select unattractive, but needful, endangered species along with the attractive ones to be included in their collection to fulfill both the advisable ex-situ conservation role and the expectation of zoo visitors. For example, the least attractive mammalian family of the marsupial moles (Notoryctidae) includes only two species, both endangered according to the IUCN status. They are not kept in any zoo at present, and their future is very insecure unless selectively focused on, going against the unwanted bias. Another way to fulfill both of the roles is to select the most attractive species out of a list of species with a similar threat status. For example, there are both attractive and unattractive species within some animal families, as shown by Frynta et al. [43], on all 367 parrots of the family Psittacidae. From within such families, endangered yet attractive animals could be included in the WZC.

Whatever the conservation priorities are ([101]), if there is an existing tendency to prefer the conservation of some species over the others, the factors affecting this tendency should be known. Nowadays, the preservation of biodiversity is a widely accepted priority for species conservation (e.g., [102,103]). This priority is in conflict with human tendency to pay the proper attention selectively to large and attractive animals. However, the awareness of this selectivity may help the conservationists to improve their strategies. There are many small and/or unattractive species that are phylogenetically significant and thus key for biodiversity preservation. With only little or no support received, these species could be lost forever. A good example of possible application of species attractiveness assessment is in the case of the EDGE species (Evolutionarily Distinct and Globally Endangered [104]) – a selection of threatened species with high biodiversity value. Once known, the unattractive species putatively lacking public awareness and support could receive special attention by conservation specialists.

Moreover, knowing that the attractiveness of an animal itself plays a major role in a human's decision-making, they may be able to intentionally select the "beautiful" species not only to raise the zoo's popularity among visitors, but also to use such species in educational programs, or present them as flagship species for further in-situ protection of wildlife. A reasonable definition of flagship species was proposed by Verissimo et al. [105] (page 2): "A species used as the focus of a broader conservation marketing campaign based on its possession of one or more traits that appeal to the target audience". Smith et al. [106] found that large bodied mammalian species with forward-facing eyes are most frequently used as flagships by non-government conservational organizations (NGO) and, based on these characteristics, they suggested five critically endangered species with a strong potential to serve as good flagship species: the African wild ass, tamarau (dwarf buffalo), pygmy raccoon, Talaud bear cuscus and Pennant's red colobus. When compared with our results, four of these animals belong to attractive mammalian families positioned in the top-half rank of the attractiveness (up to the fifth position; Equidae 7, Bovidae 20, Procyonidae 38 and Phalangeridae 50). The high attractiveness of threatened species per se may further increase the potential of these or similarly selected species to serve as flagships. In contrast, the Pennant's red colobus belongs to the family Cercopithecidae which appeared to be rather unattractive (placing the 94th rank position). However, this family contains a large number of species the attractiveness of which may vary. A more detailed analysis on a species level could help to determine the actual attractiveness of the red colobus, or help find a species with similar attributes but higher attractiveness to be used as a flagship instead.

A properly selected flagship species may convince the public to donate more money for conservation, just as demonstrated by [107]. In their study, the respondents were willing to pay more for the conservation of an otter than that of a water vole. This is in accord with our finding that otters from the family Mustelidae are more attractive to humans, placing 34th in the preference ranking, than water voles from the family Muridae, which placed as far as 73rd (See Appendix S1). Also, when lumped together into one conservation program, these two animals received less support than otters alone [107]. This may be explained either by the sole presence of the unattractive animal in the program, which pushes the respondents back, or by the rising complexity of the message that was presented to the respondents. Either way, if conservationists select a single highly attractive animal to be presented to the public as a messenger for conservation planning, it may raise the success of the project. This may be caused both by raising the financial support of the project by people living far away from the place of question, or by local people who may re-think their view of the natural riches surrounding them [37]. Furthermore, we found that the attractiveness of snakes as perceived by humans is shared among such different cultures as Europeans and villagers from Papua New Guinea [62]. Another study confirmed these results on people from the five main inhabited continents [63]. If applicable to other animal taxa, the message from a single flagship species could touch people worldwide as well as people local to the conservation project. Although local communities face various problems with potential flagships that trigger conflicts as predatory animals, competitors or pests ([108,109]), the attractiveness may play its role when the flagship is selected from harmless, non-conflicting species. In case of highly attractive animals, the attractiveness may even outweigh the possible conflict. The family Equidae placed 7th rank of attractiveness in our study and the reintroduction of the Przewalski horse was well-accepted by local people in Mongolia (Hustain Nuru; Kůs E., Zoo Prague, personal communication) despite it being a competitor for domestic horses ([26]).

In conclusion, this study shows that the predictors associated with human attention, especially body size and aesthetic attractiveness, have a substantial effect on the composition of the WZC. In the 21st century, it is of an utmost importance to pay attention to the biodiversity preservation, and it might be up to the worldwide zoological gardens to play a significant role in this task. This is especially because zoos have the capacity to hold numerous species, a capacity larger than any other institution, together with the knowledge about the breeding of various species and properly managed studbooks. Whether they utilize this potential is a vision of the future. However, for the zoos and conservationists whose intentions are to conserve biodiversity, our study reveals one of many factors – the human factor – that may fundamentally affect the conservation efforts. Thus, conservation biologists should consider these psychological factors for proper management of the "Ark".

Supporting Information

Appendix S1 The dataset used for statistical analyses, sorted by attractiveness from the most attractive family to the least attractive one. (See under the Materials and methods section for the definition of the variables.)

Appendix S2 The forms and species listed in the ISIS database that were excluded from (or added to) the analysis.

Appendix S3 The literary sources of brain size data.

Appendix S4 Review of mammals reintroduced during the years 1992–2009. We collected journals and book sources of mammalian reintroductions within the years 1992–2009 with detailed information about the reintroduction events.

Acknowledgments

We thank Michaela Zelenková for help with the data collection, Pavel Němec and Zuzana Králová for providing a large portion of mammalian brain size data, Ludmila Konečná and Markéta Janovcová for the collection of papers dealing with reintroduced species, Jana Marešová and Barbora Eliášová for critical reading of the manuscript, Emily from Australia and Corey Stringer from North Carolina, USA, for a linguistic revision and Dr. Francisco Ceacero Herrador for a revision of the scientific styling of the manuscript, and all of the respondents for their kind participation in this project.

Author Contributions

Conceived and designed the experiments: DF EL. Performed the experiments: DF OS SL EL. Analyzed the data: DF OS SL EL. Wrote the paper: DF OS SL EL.

References

1. Sanderson EW, Jaiteh M, Levy MA, Redford KH, Wannebo AV, et al. (2002) The human footprint and the last of the wild. Bio Science 52: 891–904.
2. Brooks TM, Mittermeier RA, da Fonseca GAB, Gerlach J, Hoffmann M, et al. (2006) Global biodiversity conservation priorities. Science 313: 58–61.
3. Wilson EO (2002) The future of life. New York, NY: Alfred A. Knopf.
4. Marton-Lefèvre J (2010) Biodiversity is our life. Science 327: 1179.
5. Soulé M, Gilpin M, Conway W, Foose T (1986) The millenium Ark: How long a voyage, how many staterooms, how many passengers? Zoo Biol 5: 101–113.
6. McGregor Reid G, Zippel KC (2008) Can zoos and aquariums ensure the survival of amphibians in the 21st century? Int Zoo Yearb 42: 1–6.
7. Rees PA (2005) Will the EC Zoos Directive increase the conservation value of zoo research? Oryx 39: 128–131.
8. Gippoliti S (2012) Ex situ conservation programmes in European zoological gardens: Can we afford to lose them? Biodivers Conserv 21: 1359–1364.
9. Ebenhard T (1995) Conservation breeding as a tool for saving snímal species from extinction. Trends Ecol Evol 10: 438–443.
10. Stanley Price MR, Soorae PS (2003) Reintroductions: whence and whither? Int Zoo Yearb 38: 61–75.
11. Bowkett AE (2009) Recent captive-breeding proposals and the return of the ark concept to global species conservation. Conserv Biol 23: 773–776.
12. Soulé ME (1980) Thresholds for survival: maintaining fitness and evolutionary potential. In: Soulé ME, Wilcox BA, editors. Conservation Biology: An Evolutionary-ecological Perspective. Sunderland, MA: Sinauer. pp. 151–169.
13. Reed DH, Nicholas AC, Stratton GE (2007) Genetic quality of individuals impacts population dynamics. Anim Conserv 10: 275–283.
14. Kimura M (1983) The neutral theory of molecular evolution. Cambridge, MA: Cambridge University Press.
15. Lande R (1999) Extinction risks from anthropogenic, ecological, and genetics factors. In: Landweber LF, Dobson AP, editors. Genetics and the extinction of species: DNA and the conservation of biodiversity. Princeton, NJ: Princeton University Press. pp. 1–23.
16. Frankham R, Ballou JD, Briscoe DA (2002) Introduction to conservation genetics. Cambridge, UK: Cambridge University Press.
17. Frynta D, Marešová J, Landová E, Lišková S, Šimková O, et al. (2009) Are animals in zoos rather conspicuous than endangered? In: Columbus AM, Kuznetsov L, editors. Endangered species: new research. New York, NY: Nova Science Publishers. pp. 299–341.
18. Conde DA, Flesness N, Colchero F, Jones OR, Scheuerlein A (2011) An emerging role of zoos to conserve biodiversity. Science 331: 1390–1391.
19. Hale KA, Briskie JV (2007) Challenges to understanding the consequences of population bottlenecks for the conservation of endangered wildlife. Anim Conserv 10: 19–21.
20. Roberts L (1988) Beyond Noah's Ark: what do we need to know? Science 242: 1247.
21. Snyder NFR, Derrickson SR, Beissinger SR, Wiley JW, Smith TB, et al. (1996) Limitations of captive breeding in endangered species recovery. Conserv Biol 10: 338–348.
22. Fischer J, Lindenmayer DB (2000) An assessment of the published results of animal relocations. Biol Conserv 96: 1–11.
23. Mathews F, Orros M, McLaren G, Gelling M, Foster R (2005) Keeping fit on the ark: assessing the suitability of captive-bred animals for release. Biol Conserv 121: 569–577.
24. Lees CM, Wilcken J (2009) Sustaining the Ark: the challenges faced by zoos in maintaining viable populations. Int Zoo Yearb 43: 6–18.
25. Earnhardt JM (2010) The role of captive populations in reintroduction programs. In: Kleiman DG, Thompson KV, Baer CK, editors. Wild Mammals In Captivity. Chicago: University of Chicago Press. pp. 268–280.
26. Volf J (2009) Half a century of international cooperation in the preservation of the Przewalski Horse - direction: reintroduction. In: Anonymous, editor. Equus 2009. Prague, CZ: Zoo. pp. 39–56.
27. Freese CH, Aune KE, Boyd DP, Derr JN, Forrest SC, et al. (2007) Second chance for the plains bison. Biol Conserv 136: 175–184.
28. Ahrens TG (1921) The present status of the European bison or wisent. J Mammal 2: 58–62.
29. Pucek Z, editor (2002) European bison: current state of the species and an action plan for its conservation. Species Action Plan, LHI-WWF.
30. Tokarska M, Kawałko A, Wójcik JM, Pertoldi C (2009) Genetic variability in the European bison (*Bison bonasus*) population from Białowieża forest over 50 years. Bio J Linn Soc 97: 801–809.
31. Jiang Z, Yu C, Feng Z, Zhang L, Xia J, et al. (2000) Reintroduction and recovery of Père David's deer in China. Wildlife Soc B 28: 681–687.
32. Stanley Price MR (1989) Animal re-introductions, the Arabian Oryx in Oman. Cambridge, UK: Cambridge University.
33. Saltz D, Rubenstein DI (1995) Population dynamics of a reintroduced Asiatic wild ass (*Equus Hemionus*) herd. Ecol Appl 5: 327–335.
34. Stoinski TS, Beck BB (2004) Changes in locomotor and foraging skills in captive-born, reintroduced golden lion tamarins (*Leontopithecus rosalia rosalia*). Am J Primatol 62: 1–13.
35. Dobson A, Lyles A (2000) Black-footed ferret recovery. Science 288: 985–988.
36. Leader-Williams N, Balmford A, Linkie M, Mace GM, Smith RJ, et al. (2007) Beyond the ark: conservation biologists' views of the achievements of zoos in conservation. In: Zimmermann A, Hatchwell M, Dickie L, West C, editors. Zoos in the 21st Century: Catalysts for Conservation? Cambridge, UK: Cambridge University Press. pp. 236–254.
37. Dietz JM, Dietz LA, Nagagata EY (1994) The effective use of flagship species for conservation of biodiversity: the example of lion tamarins in Brazil. In: Olney PJS, Mace GM, Feistner ATC (eds) Creative conservation: interactive management of wild and captive animals. Chapman & Hall, London, pp 32–49.

38. Zimmermann A (2010) The role of zoos in contributing to in situ conservation. In: Kleiman DG, Thompson KV, Baer CK, editors. Wild mammals in captivity. Chicago: University of Chicago Press. pp. 281–287.
39. Mazur N, Clark TW (2000) Zoos and conservation: policy making and organizational challenges. Yale F&ES Bulletin 105: 1–17.
40. Bitgood S, Patterson D (1987) Principles of exhibit design. Visitor Behavior 2: 4–6.
41. Puan CL, Zakaria M (2007) Perception of visitors towards the role of zoos: a Malaysian perspective. Int Zoo Yearb, 41: 226–232.
42. Moss A, Esson M (2010) Visitor interest in zoo animals and the implications for collection planning and zoo education programmes. Zoo Biol 29: 715–731.
43. Frynta D, Lišková S, Bültmann S, Burda H (2010) Being attractive brings advantages: the case of parrot species in captivity. PLoS ONE 5: e12568.
44. Marešová J, Frynta D (2008) Noah's Ark is full of common species attractive to humans: The case of boid snakes in zoos. Ecol Econ 64: 554–558.
45. Balmford A, Mace GM, Leader-Williams N (1996) Designing the Ark: setting priorities for captive breeding. Conserv Biol 10: 719–727.
46. Ericson PG, Anderson CL, Britton T, Elzanowski A, Johansson US, et al. (2006) Diversification of Neoaves: integration of molecular sequence data and fossils. Biol Letters 2: 543–547.
47. Portmann A (1979) Nové cesty biologie. In: Fiala J, Neubauer Z, Pinc Z, editors. Scientia & Philosophia 7 (1997). Praha: Katedra matematické logiky a filosofie matematiky, Matematicko-fyzikální fakulta UK.
48. Berlin B (1992) Ethnobiological Classification: principles of categorization of plants and animals in traditional societies. Princeton, NJ: Princeton University Press.
49. Wilson DE, Reeder DAM, editors (2005) Mammal species of the world: a taxonomic and geographic reference (3rd ed). Baltimore, MD: Johns Hopkins, University Press. Checklist available on: http://nmnhgoph.si.edu/msw/.
50. Nowak RM (1999) Walker's mammals of the world. Baltimore, MD: The Johns Hopkins University Press.
51. Jerison HJ (1955) Brain to body ratios and the evolution of intelligence. Science 121: 447–449.
52. Jerison HJ (1963) Interpreting the evolution of the brain. Hum Biol 35: 263–291.
53. Jerison HJ (1973) Evolution of the brain and intelligence. New York: Academic Press.
54. Ashwell KWS (2008) Encephalization of Australian and New Guinean marsupials. Brain Behav Evol 71: 181–199.
55. Whitfield P, editor (1984) Longman illustrated animal encyclopedia. Harlow, Essex, England: Longman.
56. Anděra M (1997–2000) Svět zvířat I-III. Savci 1–3. Praha, Albatros.
57. Hutchins M, Kleiman DG, Geist V, McDade MC, editors (2004) Grzimek's animal life encyclopedia, 2nd edition. Volumes 12–16, Mammals I-V. Farmington Hills, MI: Gale Group.
58. Myers P, Espinosa R, Parr CS, Jones T, Hammond GS, et al. (2013) The Animal Diversity Web. Available: http://animaldiversity.org.
59. Central Intelligence Agency. "Country Comparison: Distribution of family income - Gini index".cia.gov. Central Intelligence Agency, n.d. Web. 27 Mar. 2013. Available: https://www.cia.gov/library/publications/the-world-factbook/rankorder/2172rank.html.
60. Veríssimo D, Fraser I, Groombridge J, Bristol R, MacMillan DC (2009) Birds as turism flagship species: a case study of tropical islands. Anim Conserv 12: 549–558.
61. Schlegel J, Rupf R (2010) Attitudes towards potential animal flagship species in nature conservation: A survey among students of different educational institutions. J Nat Conserv 18: 278–290.
62. Marešová J, Krása A, Frynta D (2009) We all appreciate the same animals: cross-cultural comparison of human aesthetic preferences for snake species in Papua New Guinea and Europe. Ethology 115: 297–300.
63. Frynta D, Petrů M, Šklíba J, Šumbera R, Krása A, et al. (2011) Crosscultural agreement in perception of animal beauty: Boid snakes viewed by people from three continents. Hum Ecol 39: 829–834.
64. Cunningham MR, Roberts AR, Barbee AP, Druen PB, Wu C (1995) "Their ideas of beauty are, on the whole, the same as ours": Consistency and variability in the cross-cultural perception of female physical attractiveness. J Pers Soc Psychol 68: 261–279.
65. Entwistle AC, Stephenson PJ (2000) Small mammals and the conservation agenda. In: Entwistle A, Dunstone N, editors. Priorities for theConservation of Mammalian Diversity. Has the Panda had its Day? Cambridge, UK: Cambridge University Press. pp. 119–139.
66. SPSS Inc. (2007) Spss, version 16.0. Available: http://www.winwrap.com.
67. R Development Core Team (2010) R: A language and environment for statistical computing. R foundation for statistical computing, Austria: Vienna.
68. Harvey PH, Pagel MD (1991) The comparative method in evolutionary biology. Oxford: Oxford University Press.
69. Felsenstein J (1985) Phylogenies and the comparative method. Am Nat 125: 1–15.
70. Bininda-Emonds ORP, Cardillo M, Jones KE, MacPhee RDE, Beck RMD, et al. (2007). The delayed rise of present-day mammals. Nature 446: 507–512.
71. Arnason U, Adegoke JA, Bodin K, Born EW, Esa YB, et al. (2002) Mammalian mitogenomic relationships and the root of the Eutherian tree. P Natl Acad Sci U S A 99: 8151–8156.
72. Meredith RW, Westerman M, Springer MS (2009) A phylogeny of Diprotodontia (Marsupialia) based on sequences for five nuclear genes. Mol Phylogenet Evol 51: 554–571.
73. Barros MC, Sampaio I, Schneider H (2008) Novel 12S mtDNA findings in sloths (Pilosa, Folivora) and anteaters (Pilosa, Vermilingua) suggest a true case of long branch attraction. Genet Mol Biol 31: 793–799.
74. Agnarsson I, May-Collado LJ (2008) The phylogeny of Cetartiodactyla: The importance of dense taxon sampling, missing data, and the remarkable promise of cytochrome b to provide reliable species-level phylogenies. Mol Phylogenet Evol 48: 964–985.
75. Dalerum F (2007) Phylogenetic reconstruction of carnivore social organizations. J Zool 273: 90–97.
76. Flynn JJ, Finarelli JA, Zehr S, Hsu J, Nedbal MA (2005) Assessing the impact of increased sampling on resolving enigmatic relationships. Syst Biol 54: 317–337.
77. Blanga-Kanfi S, Miranda H, Penn O, Pupko T, DeBry RW, et al. (2009) Rodent phylogeny revised: analysis of six nuclear genes from all major rodent clades. BMC Evol Biol 9: 71.
78. Huchon D, Douzery EJP (2001) From the Old World to the New World: a molecular chronicle of the phylogeny and biogeography of Hystricognath rodents. Mol Phylogenet Evol 20: 238–251.
79. Jansa SA, Giarlaa TC, Lim BK (2009) The phylogenetic position of the rodent genus Typhlomys and the geographic origin of Muroidea. J Mammal 90: 1083–1094.
80. Jansa SA, Weksler M (2004) Phylogeny of muroid rodents: relationships within and among major lineages as determined by IRBP gene sequences. Mol Phylogenet Evol 31: 256–276.
81. Chatterjee HJ, Ho SYW, Barnes I, Groves C (2009) Estimating the phylogeny and divergence times of primates using a supermatrix approach. BMC Evol Biol 9: 259.
82. Martins EP (2001) COMPARE Version 4.4. Computer Programs for Statistical Analysis of Comparative Data. Available: http://compare.bio.indiana.edu.
83. Garland T, Harvey PH, Ives AR (1992) Procedures for the analysis of comparative data using phylogenetically independent contrasts. Syst Biol 41: 18–32.
84. StatSoft (2001) Statistica, version 6.0. Available: http://www.statsoft.com.
85. Gunnthorsdottir A (2001) Physical attractiveness of an animal species as a decision factor for its preservation. Anthrozoos 14: 204–215.
86. Metrick A, Weitzman ML (1996) Patterns of behavior in endangered species preservation. Land Econ 72: 1.
87. Kleiber M (1932) Body size and metabolism. Hilgardia 6: 315–353.
88. Kleiber M (1961) The fire of life: an introduction to animal energetics. New York: Wiley.
89. Schmidt-Nielsen K (1984) Scaling: why is animal size so important? Cambridge, UK: Cambridge University Press.
90. Anonymous (1996). Gutachten über Mindestanforderungen an die Haltung von Säudetieren. BMVEL Bonn.
91. Balmford A (2000) Separating fact from artifact in analyses of zoo visitor preferences. Conserv Biol 14:1193–1195.
92. Deaner RO, Isler K, Burkart J, van Schaik C (2007) Overall brain size, and not encephalization quotient, best predicts cognitive ability across non-human primates. Brain Behav Evol 70: 115–124.
93. Reader SM, Laland KN (2002) Social intelligence, innovation, and enhanced brain size in primates. P Natl Acad Sci U S A 99: 4436–4441.
94. Overington SE, Morand-Ferronb J, Boogerta NJ, Lefebvrea L (2009) Technical innovations drive the relationship between innovativeness and residual brain size in birds. Anim Behav 78: 1001–1010.
95. Roth G, Dicke U (2005) Evolution of the brain and intelligence. Trends Cogn Sci 9: 250–257.
96. Shettleworth SJ (2003) Memory and hippocampal specialization in food-storing birds: challenges for research on comparative cognition. Brain Behav Evol 62: 108–116.
97. Macphail EM, Bolhuis JJ (2001) The evolution of intelligence: adaptive specializations versus general process. Biol rev Camb Philos 76: 341–364.
98. Fagen R (1974) Selective and evolutionary aspects of animal play. Am Nat 108: 850–858.
99. Iwaniuk AN, Nelson JE, Pellis SM (2001) Do big-brained animals play more? Comparative analyses of play and relative brain size in mammals. J Comp Psychol 115: 29–41.
100. Margulis SW, Hoyos C, Anderson M (2003) Effect of felid activity on zoo visitor interest. Zoo Biol 22: 587–599
101. Callicott JB (1990) Whither conservation ethics? Conserv Biol 4: 15–20.
102. Margules CR, Pressey RL (2000) Systematic conservation planning. Nature 405: 243–253.
103. Smith RJ, Veríssimo D, MacMillan DC (2011) Marketing and conservation: how to lose friends and in fluence people. In: Leader-Williams N, Adams W, Smith R, editors. Trade-Offs in conservation: deciding what to save. pp. 215–232. Oxford, UK: Wiley Blackwell.
104. Isaac NJ, Turvey ST, Collen B, Waterman C, Baillie JE (2007) Mammals on the EDGE: conservation priorities based on threat and phylogeny. PLoS ONE 2: e296.
105. Verissimo D, MacMillan DC, Smith RJ (2011) Toward a systematic approach for identifying conservation flagships. Conserv Lett 4: 1–8.

106. Smith RJ, Veríssimo D, Isaac NJB, Jones KE (2012) Identifying Cinderella species: uncovering mammals with conservation flagship appeal. Conserv Lett 5: 205–212.
107. White PCL, Gregory KW, Lindley PJ, Richards G (1997) Economic values of threatened mammals in Britain: A case study of the otter *Lutra lutra* and the watervole *Arvicola terrestris*. Biol Conserv 82: 345–354.
108. Kaltenborn BP, Bjerke T, Nyahongo JW, Williams DR (2006) Animal preferences and acceptability of wildlife management actions around Serengeti National Park, Tanzania. Biodivers Conserv 15: 4633–4649.
109. Bowen-Jones E, Entwistle A (2002) Identifying appropriate flagship species: the importance of culture and local contexts. Oryx 36: 189–195.

14

The Push and Pull of Land Use Policy: Reconstructing 150 Years of Development and Conservation Land Acquisition

Maria João Santos[1*¤], **Terry Watt**[2], **Stephanie Pincetl**[3]

1 Center for Spatial and Textual Analysis, Spatial History Project, and Bill Lane Center for the American West, Stanford University, Stanford, California, United States of America, **2** Terry Watt Planning Associates, San Francisco, California, United States of America, **3** California Center for Sustainable Communities, Institute of the Environment University of California Los Angeles, Los Angeles, California, United States of America

Abstract

The growth of human populations and their resource needs have stressed the conservation of natural land resources. Many policies and programs have been implemented to address the pressures on land resources and notwithstanding this pressure, significant acquisition of land for conservation has occurred throughout history in the U.S., and internationally. Here we assess the on-the-ground result of the evolution of land use policies in California as a pioneer forerunner, in the form of acquisition of land for conservation (i.e. Open Space), and its impact on the rest of the U.S. and beyond. To this end we describe the timeline and spatial representation of the growth of California's conservation network over the last 150 years, and link it to the history of land use policies. We then assess whether conservation land acquisition has consistently grown through time or occurred in specific decades. About ¼ of the state is now designated Open Space. Fewer and larger areas conserved and acquired at the beginning of the 20th century; the conservation network was complemented with a larger number of smaller sized properties. Despite acquisition of land in every decade, the process was uneven ($E = 0.3$ for California, $E = 0.14 \pm 0.08$ average for the state's counties), mostly due to the large acquisitions and land set asides in the 1900s, followed by 1930s and 1940s. This process was a result of a comprehensive set of legislation that evolved through time, and resulted from the competing needs for development and conservation. Even with the impressive 174,000 km^2 of public lands in California, the future of California's natural infrastructure and natural heritage cannot rely solely on these public lands, nor public agencies and their resources. Critically a future course of land preservation relying on the purchase of new lands – in California and beyond – for conservation is tremendously expensive.

Editor: Rob Slotow, University of KwaZulu-Natal, South Africa

Funding: MJS was funded by a postdoctoral fellowship from the Bill Lane Center for the American West and the Spatial History project at Stanford University. The funders had no role in study design, data collection and analysis, decision to publish, or preparation of the manuscript.

Competing Interests: The authors have declared that no competing interests exist.

* Email: mjsantos@stanford.edu

¤ Current address: Department of Innovation, Environmental and Energy Sciences, Utrecht University, The Netherlands

Introduction

The growth of human populations and their resource needs have stressed the conservation of natural land resources. Conservation planning has focused on moving beyond opportunism [1], and is using strategic planning to identify the best spatial solutions to meet conservation goals, and implement such solutions on-the-ground [2]. This is evident in the recent approaches to conservation of the Nature Conservancy, for example, identifying the most valuable and endangered conservation hotspots across the globe, and investing funds in those places to preserve land. However, temporal aspects have been somewhat neglected (but see [3]), only currently coming to play as changes in land cover/land use and climate may threaten the persistence of the already existing conservation networks [4]. This, of course raises difficult scientific and ethical issues beyond the scope of this paper, about moving fauna and flora to locations where they will survive under climate change conditions and other types of interventions. At the same time, the creation of a conservation network can in itself induces change [5], and this change can be tied with historical legacies or periods of intense action [3]. These periods can be identified in the timeline of protected area establishment and in the creation of key conservation legislation, but integration of both timelines is yet to be conducted. While the former is easily assessed worldwide, the latter is affected by local, regional, national and international contexts. Here we tie both timelines, to understand the push-and-pull of land use policies over the last 150 years, using California as case study to demonstrate the importance of the integration of both timelines and to suggest such research be conducted in other places. We focus on California for its preeminence in this area, and its international significance. California is arguably the state that led the implementation of the concept of resource reserves since the 19th century, in the country and elsewhere [3].

Forty four percent of the Earth's plant species and 35% of the Earth's terrestrial vertebrates are endemic to 25 hotspots, despite the fact these hotspots make up just 1.4% of the world's land cover [6]. These biodiversity hotspots have immense implications for conservation strategies to protect large number of species living in relatively small regions of land. They also represent lands that have been and are highly threatened by land cover and climate changes. The California Floristic Province is one of the world's 25 most biodiverse areas on the planet [6]. For California itself, 44% of its plant and vertebrate species are endemic to the state [7]. Today, ¼ of California are properties designated as Open Space. Open Space refers to *"lands protected through title ownership by a public agency or non-profit land conservation organization"* [8]. Open Spaces includes a wide array of types of properties from City Parks to National Parks, and some heavily managed land, such as that managed by the federal Bureau of Land Management (BLM). Fee title ownership by the public exists for all of these Open Spaces. However, Open Space does not include private land in easement contracts. Moreover, lands in public ownership – aside from the National Park lands – are subject to the Federal Land Policy and Management Act, FLPMA (1976) that allows multiple-use of public lands and for the most part does not recognize conservation as an exclusive use except on lands with designations such as wilderness.

Already in the later decades of the nineteenth century, preservation of California's remarkable natural resources was underway. In 1864 Yosemite Valley and the Mariposa Grove of Giant Sequoia trees were protected by Abraham Lincoln, with a grant of 80.9km^2 of federal land. At the time, the federal government maintained ownership of vast lands, part of the public domain, that had not been sold through the Homestead Act (1865) and other acts that distributed the lands the federal government inherited with its new territories. Much of California's early preserved land was land that belonged to the federal government and was "withdrawn" from sale to the public, and instead, conserved as part of the public domain. This eventually became the backbone of the lands of the National Park Service, and the National Forest Service. Similar patterns of land reservation occurred in many states of the American West. Nevada, for example, is 80% public lands, or federal ownership. In southern California, the San Gabriel and San Bernardino Forest reserves were created in 1892 and 1893 respectively, from the public domain by President Harrison, some of the first land reservations from sale from the public lands in the nation. California is in many ways the state that led the implementation of the concept of resource reserves since the 19th century [9], lobbying Congress and the President about the public interest in the public lands, and that they should be conserved. State interest in land conservation was reflected in 1927 by the then recently created State Parks Commission that instructed Frederick Law Olmsted Jr. *"to make a survey to determine what lands are suitable and desirable for the ultimate development of a comprehensive, well-balanced state park system, and to define the relation of such a system to other means of conserving and utilizing the scenic and recreational resources of the state"*[10]. The survey was conducted, and set the vision, for example, for the creation of the state's Redwood Parks. Birth home of the Sierra Club, there is no question that interest in protecting the state's scenic beauty has been important for nearly a century and a half, an idea exported to other parts of the country and the world to establish their own conservation timelines [3,9].

In this paper we review the history of land conservation in California from a land use and governance perspective, and assess the on-the-ground result of the evolution of land use policies in the state in the form of acquisition of land for conservation. To do so we reconstruct the conservation history of California, that is, the growth of the conservation network per decade over the last 150 years, and link it to the timeline of land use policies as an example for further research of this type. We assess whether conservation land acquisition has taken place relatively evenly over time or has been concentrated in specific decades. We then review the challenges of current land use regulations, including fiscal constraints and regulatory requirements, which create a complex terrain for further land conservation. New tools are emerging for protection but structural constraints that we will discuss, remain in place as well, and they will differ from state to state and from nation to nation. In this discussion we develop a method that can be replicated to better help understanding both the ways in which lands have been conserved and the potential pathways for further conservation given the structural conditions that exist in California.

Materials and Methods

Open Space land

To obtain geospatial data for all Open Spaces in the state, we used the California Protected Area Database (CPAD) as a base database [8]. CPAD was developed by GreenInfo Network as a GIS inventory of all fee-protected Open Space properties in California. We selected this database and not the World Database on Protected Areas [3,11] because CPAD has a more comprehensive representation of the Open Space properties in the state, from City Parks to National Parks. It does not include all public land, nor private owners, or properties protected through the use of easements. Neither WDPA nor CPAD included information on the acquisition and establishment dates of each and every Open Space property in California or other protected lands across the globe [8,11]. To complement this information gap, and reconstruct a conservation history for the state, we contacted federal and state agencies and Non-Governmental Organizations that manage Open Space properties to request information. We asked for acquisition date, that is, when the fee title was purchased, and establishment date – when the property was open to the public. Below we present statistics for data acquisition for the great majority of the properties and area of the state designated as Open Space. In the remaining analysis we use date of acquisition for our timeline of conservation land acquisition. To understand the conditions under which land were protected, knowing the timelines of establishment is important, as it allows better insights into the reasons why such acquisitions might have occurred [3].

Land use policies

To complement, compare and contrast the Open space acquisition timeline we developed the land use history timeline. We reviewed the legal documents that detailed land use policies in California. These included primary source legislation and published literature on examples, applications and assessments of the impacts of such legislation (for example see http://ceres.ca.gov/planning/state.html). Key literature was searched for the date of the legislation, its description, intended targets, and when possible implications of the legislation for land use planning and Open Space acquisition.

Data analysis

The timeline of conservation land acquisition describes the cumulative area acquired per decade in the state. To assess whether each decade contribution to the conservation network was higher or lower than the state's average, we calculated the slope of the regression line of all the decades and compared it with the slope of the regression for consecutive decades. Significantly different decades were illustrated in the state's timeline. We then tested whether the acquisition of land was even throughout time for the state and how it compared across counties. We calculated the Simpson evenness index (E) which varies from 0 (uneven

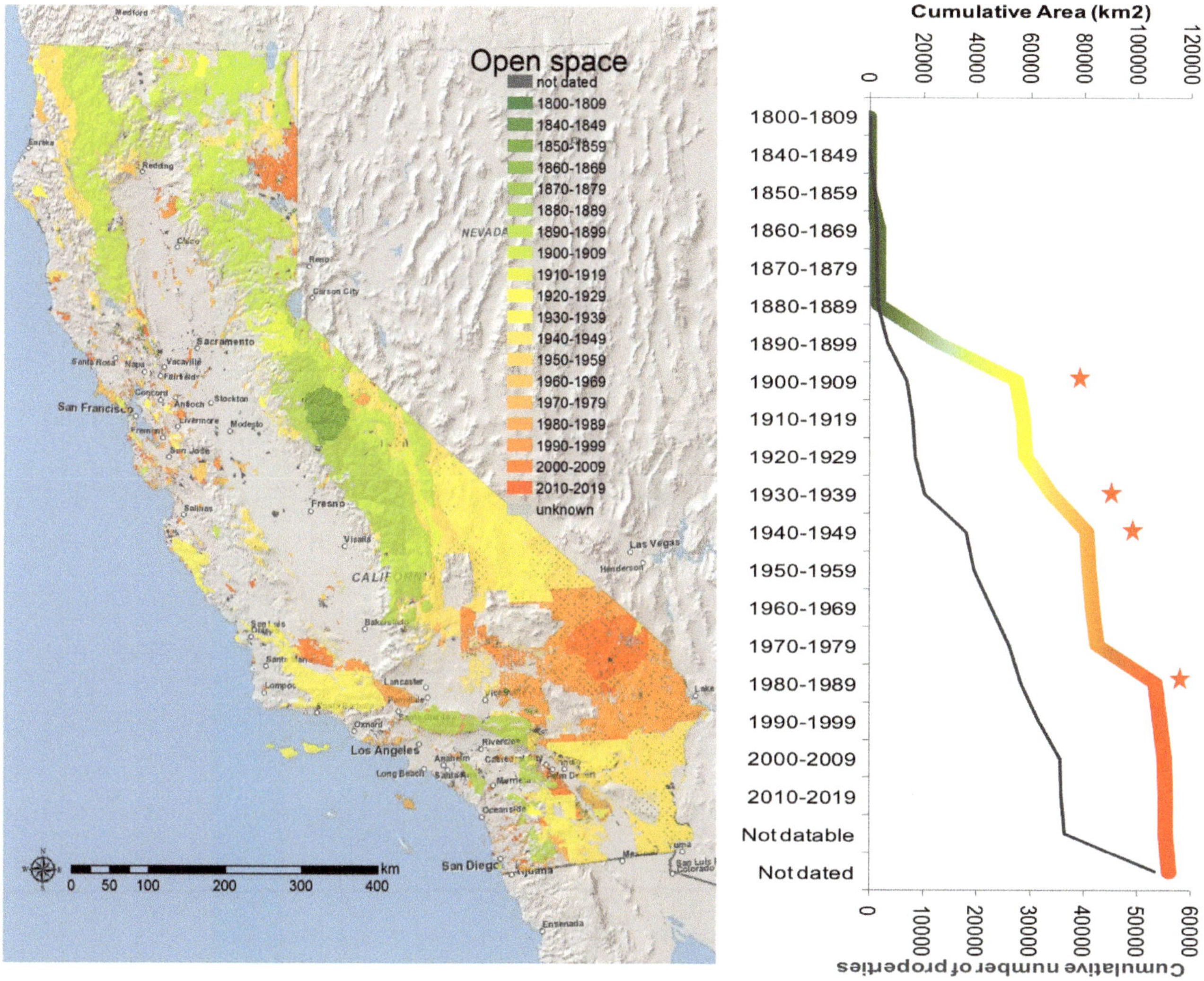

Figure 1. Timeline of Open Space acquisition and establishment in California over the last 150 years. Left panel – spatial representation, Right panel – timeline (red stars represent the decades where the conservation land acquisition was higher than the average over all the decades).

distribution of conservation action through the decades) to 1 (equal distribution of conservation action through the decades; [12]). We calculated evenness for the state and for each county, and determined whether counties differed from the state values. We then linked the timeline of conservation land acquisition with the history of land use policy in the state. Because of the link between funding for land acquisition and development, we analyzed in detail the most populous counties in the beginning of the 20th century.

Results

California has 16,000 Open Space properties in the form of 53,337 parcels that were acquired throughout the last 150 years. These properties cover ¼ of the state's area (112,156km^2). We were able to retrieve information on acquisition date for a total of 35,807 properties (67%), corresponding to 110,300km^2 (98.5%) of the state's Open Space. Merced, Orange, San Joaquin, Stanislaus, and Sutter were the counties from which it was harder to collect information. City Parks were the properties that had the least data about date acquired and/or created.

History of conservation land acquisition

Large tracts of land were set aside from the remaining public domain at the end of the 19th century. These were the mountain forest lands to be managed by the subsequently created United States Forest Service (Figure 1). Land acquisitions in the 1880s and 1890s contributed with an area of 30,000km^2, 30% of the current Open Space land. With the close of the public domain and into the twentieth century, additional land had to then be acquired for parks (mostly by federal agencies) as it was privately owned. Geographically, earlier land preservation focused on mountain regions, mostly in the Central and Southern Sierras, and the mountains surrounding Los Angeles (Figure 1). Land set asides and early acquisitions were predominantly in the northern California mountain ranges until 1920s (Figure 1). In the 1930's and 1940's, large tracts in southern California were purchased, as for example the lands for Death Valley and Joshua Tree National Parks. At this time the California State Parks further contributed to land acquisition, a pattern that continued into the 1950's (Figure 1). These acquisitions were also greatly aided by regulations passed in the 1940's that allowed Special Districts (a type of government unit that is formed to provide a specific service, such as schools, sewers, street lighting, parks or other) to acquire

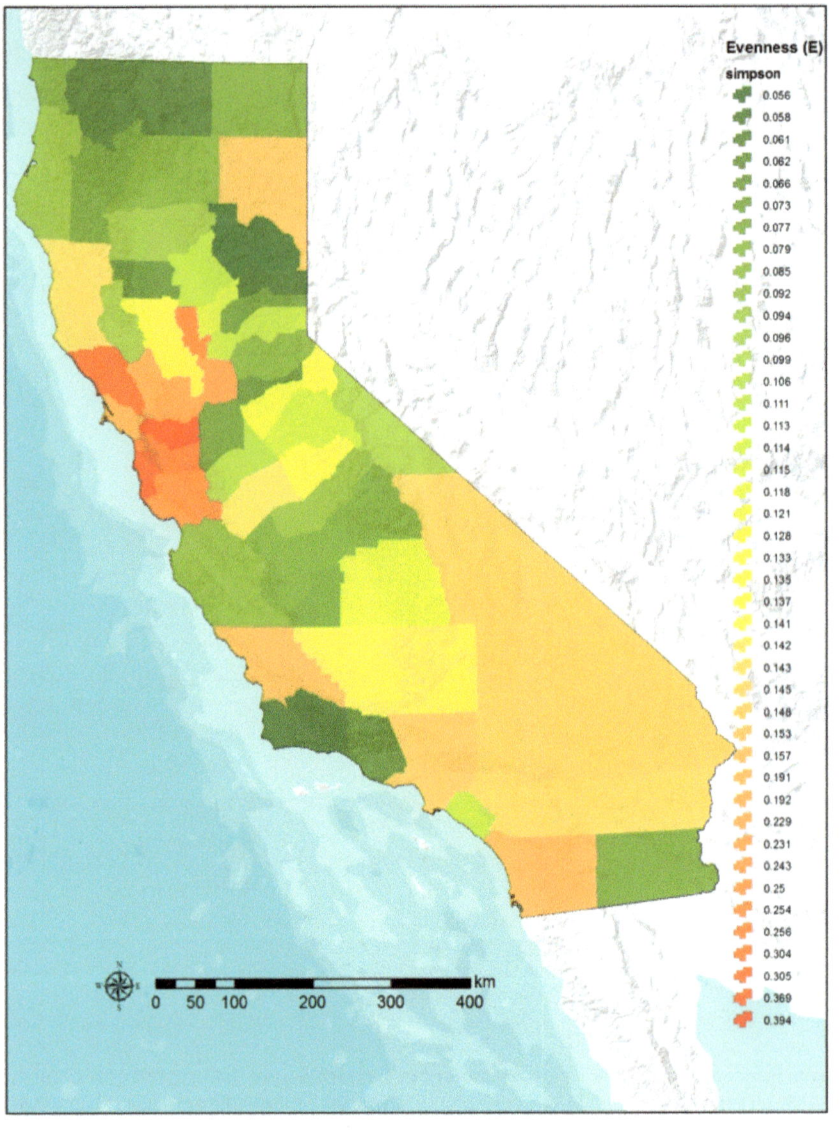

Figure 2. Per county evenness of the conservation timeline. Lighter colors represent higher evenness. The higher evenness is achieved around the San Francisco Bay area, and most of the counties show an evenness half of that of the state ($E_{California} = 0.3$).

conservation lands. From 1950 to 1980 more scattered and smaller sized properties filled in the conservation network, as the larger tracts of land were already public. However, this pattern shifted from the 1980's to today with the acquisition of large tracts of desert lands (Figure 1).

The rate of land acquisition (slope of the regression line for the state conservation acquisition) was 6874. 8 km^2 per decade (Figure 1). This rate was significantly higher in five decades (1890–1900, 1900–1910, 1920–1930, 1930–1940 and 1970–1980, indicated in Figure 1 with red stars), with acquisition rates ranging from 9697km^2 in 1930–1930, to 29799km^2 in 1900–1910, a very steep acquisition slope 5 times higher than the state's rate (Figure 1). The rate of conservation land acquisition in California over the last 150 years was also spatially uneven. The state's evenness value was 0.3, and 0.15 as the counties average (Figure 2). Plumas, Siskiyou and Santa Barbara counties were very uneven in their conservation land acquisition (Figure 2), while the Bay Area counties had a more even acquisition rate over time, and greater or equal to that of state. This is because of the greater amount of Open Space area acquired or set aside in the 1900's, followed by the 1930's and 1940's, with rates up to 5-times greater than the state's average rate of land acquisition (red stars in Figure 1).

Of the most populous counties in 1910, San Francisco engaged in some of the earliest acquisitions and then continued a constant rate of acquisition throughout the twentieth century (Figure 3). Los Angeles conservation land grew more slowly at the beginning of the twentieth century, showed a stepwise pattern. High federal involvement at the turn of the twentieth century and again in 1970–1990, was supplemented by land acquisitions by nonprofit organizations. San Diego shows a big increase in land acquisition in 1920's followed by a smooth rate thereafter; this is a similar pattern to Alameda, San Mateo and Santa Clara, but the peak of Open Space acquisition in these counties starts in 1950's. The most populous counties show, however, a very different proportion of conservation land; the largest being in San Diego and the smallest in San Francisco and Sacramento (Table 1).

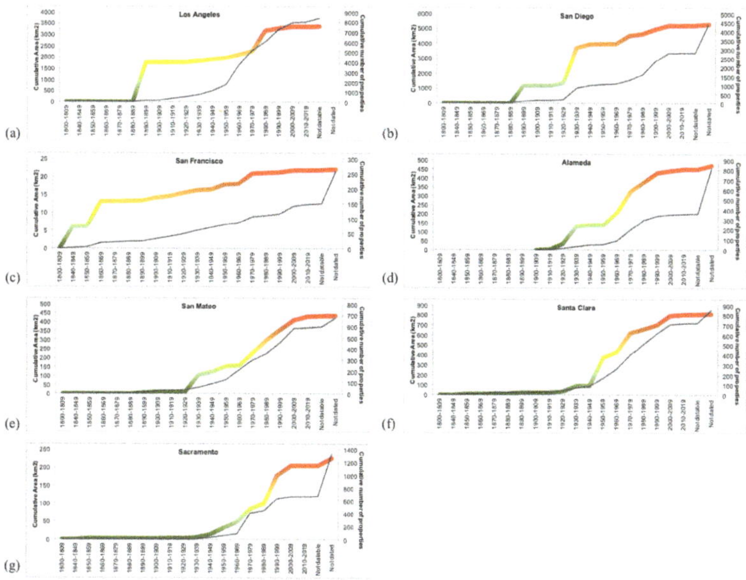

Figure 3. Timeline of conservation for the most populous counties in California in 1910.

History of land use policies

Local control over land use was instituted very early on to ensure autonomy of localities from state interference. This local independence was staunchly solidified by 1914 state legislation authorizing charter cities to make and enforce all laws and regulations in respect to municipal affairs ([13]; Table 2). This includes by 1937, the requirement to adopt comprehensive plans, which are guiding documents that set out a general vision for how the land in the jurisdiction of the city or unincorporated in the county should be utilized. Extensive land development patterns were facilitated by urban and state policies including uniform subdivision regulations in 1929 [14,15,16,17,18], and attempts to establish and enforce General Plans to guide development were overwhelmed by the tenacity of local developers and the proliferation of new municipalities, each with its own planning authority [14]. Most of the growth between 1950 and 1960 was located in suburban communities [19], aided by the recently built highways. By the early 1960s, state policy makers and Governor Pat Brown were concerned about growing fragmentation at the local level, including the increasing numbers of special districts formed to service urban growth. The state legislature created a state Planning Office in 1959. Governor Pat Brown appointed the Coordinating Council on Urban Policy to come up with policy proposals to improve planning and coordination of growth. The Coordinating Council's report urged the creation of one multi-purpose district in each of the state's metropolitan areas to deal with regional issues such as air pollution, water supply, sewage, parks and more. These proposals were roundly resisted by the Chamber of Commerce, cities and counties. In 1963, the state legislature, instead, created watered down county-level Local Agency Formation Commissions who were empowered to approve or disapprove any petition for incorporation, special-district formation or dissolution, and annexation in each county.

Frustrated by the lack of leadership at the state level, movements to attempt to curb growth and to protect the state's natural resources mobilized (Table 2). The attempt by the state Highway Commission to improve Highway 101 by cutting through portions of the Prairie Creek Redwoods and Jedediah Smith Redwoods State Parks, was defeated by the Sierra Club that then became a watch dog over projects proposed by the Highway Commission that often targeted land in state ownership, thus a less costly strategy that the purchasing of privately held lands. Another important example was the creation of California Tomorrow in 1962, a nonprofit educational institution that developed influential proposals for reforming state government and for the creation of

Table 1. Proportion of county land in Open Space.

County	County area (km^2)	Open Space area (km^2)	Area of county in Open Space (%)
Los Angeles	10242.41	3375.847	32.96
San Diego	10973.57	5286.147	48.17
San Francisco	277.02	22.12	7.98
Alameda	2126.87	469.39	22.07
San Mateo	1430.55	435.12	30.42
Santa Clara	3378.21	811.63	24.03
Sacramento	2580.21	225.94	8.76

Selected counties represent the most populous counties as in 1910s.

Table 2. The push-and-pull of development and Open space land use policies in the state.

The push and pull of development and natural resources legislation		
Natural resources		*For development*
	1914 – Charter cities	Authorizes the making and enforcement of all laws and regulations for municipal affairs
	1937 – Comprehensive (General) Plans	Sets out a general vision for the land in the jurisdiction of the city or county, and how it should be utilized
	1947 – State Highway Act	Authorize and promote the development of highways
	1956 and 1962 – Federal Highway Act	Authorize and promote the development of highways
Sets out a general vision for the water in the jurisdiction(s) and how it should be utilized	**1957 – California Water Plan**	
	1962 – California Tomorrow	Regional government to manage growth in the state
Stop the infill of the San Francisco Bay	**1965 – Bay Area Conservation and Development Commission**	
Allows contracts between the state and farmers to preserve agriculture lands	**1965 – Williamson Act**	
Developers to set aside a portion of their subdivisions as Parks or Open Space, or pay fees for parkland acquisition	**1965 – Quimby Act**	
	1970 - Open Space in General Plans	Requires that General Plans include Open Space elements
Protocol of analysis and public disclosure of environmental impacts of proposed projects, and measures for mitigation	**1970 – California Environmental Quality Act**	
	1971 – General Plans	Become required, not voluntary
For water protection	**1972 – California Wild and Scenic Rivers Act**	
For coastal area protection	**1972 – Coastal Protection Act**	
For protection of endangered species and their habitat	**1972 – Endangered Species Act**	
	1978 – Proposition 13	Property tax reduction initiative
	1978 – AB 857	Curb urban sprawl and directing new urban growth to existing cities and suburbs – infill
For protection of endangered species habitat	**1982 – Habitat Conservation Plans**	
For protection of natural communities (endangered or not)	**1991 – Natural Communities Conservation Planning**	
Infrastructure Planning	**2001 – AB857 Farmland Protection Act**	
Plan and manage natural communities for global climate change	**2006 – AB32 The Global Warming Solutions Act**	
Greenhouse Gas emissions regulation	**2007 – SB97 CEQA and Greenhouse Gas Emissions**	
	2008 – SB The Sustainable Communities and climate Protection Act	Adaptation and vulnerability for human populations

regional government to manage growth in the state. These environmental concerns contributed to the passage of the 1965 California Land Conservation Act (Williamson Act). The Williamson Act allowed farmers and ranchers to qualify for lower property tax rates if they entered into contracts keeping their lands in agriculture for a minimum of ten years. The 1965 Quimby Act was also passed. It allowed local governments to require developers to set aside a portion of their subdivisions as parks or Open Space, or pay fees for parkland acquisition and maintenance. The 1969 Santa Barbara oil spill had a jarring effect on the state, prompting a number of legislative initiatives, including the passage of the California Environmental Quality Act (CEQA) in 1970. Modeled on the federal National Environmental Policy Act (NEPA), it requires state and local agencies within California to follow a protocol of analysis and public disclosure of environmental impacts of proposed projects, and to adopt feasible measures for mitigating those impacts. It is a mandatory part of every California state and local agency process.

In 1972 state voters created the Coastal Commission, an entity appointed by the Governor, in charge of protecting California's coast from over development. This was one of the strongest measures to emerge from the first slow growth era [20], and a clear reaction to the increased pace of urbanization that was occurring. Other counteracting effect was that developments were being held up by the federal Endangered Species Act (ESA), passed in 1972 with the aim to preserve charismatic species. Environmental organizations used the ESA to challenge development with some success and developers found themselves stymied and slowed down. ESA challenges were based on a species by species endangerment threat, and it became clear that policies based on protecting individual species would scarcely achieve the goal of preserving species as they relied on habitats that needed to be preserved as well. Habitat conservation plans (HCP's) were added to the ESA by Congress in 1982 in an effort to address the need to conserve the habitat for imperiled species. Congress viewed HCPs as a win-win situation for imperiled species; HCPs took habitat into consideration, allowing a more encompassing approach, and they created Incidental Take Permits (ITPs), which allowed – for the first time – the taking of a limited amount of habitat in exchange for a commitment to an HCP to protect and manage other habitat areas, ensuring the species' overall recovery chances [20]. But HCPs still were created to protect one species at time. In response, California, under Governor Wilson, created Multiple Species Habitat Conservation Planning (MSHCP) and the 1991 Natural Communities Conservation Planning (NCCP) in an effort to encompass entire ecosystems and their processes. The NCCP process was intended to bring all stakeholders to the table in order to set aside coherent, regional habitat preserves [18,21]. NCCPs were endorsed by Secretary of the Interior, Bruce Babbitt as a legitimate implementation of the ESA that would allow the protection of endangered species, but not halt development. This is perhaps the impetus behind the increase in conservation land acquisition in San Diego (Figure 3), one of the precursor counties to develop a NCCP.

All in all, the 1970s were a period of active initiatives to preserve land at the local level with concurrent initiatives aimed to restrict or slow growth throughout the state, but in 1978 state voters passed Proposition 13, the property tax reduction initiative. Local government revenues plummeted from about $10 billion just prior to the passage of Proposition 13 to approximately half of that shortly thereafter. Growth management efforts were hampered by the need for revenues and most counties and cities adopted land use policies most likely to replenish their depleted budgets [14,18,22,23,24]. The impacts of Proposition 13 likely explain the asymptote of land acquisition rate from 1980s onwards (Figure 1). Some exceptions, however, emerge for counties like Sacramento, where the cumulative rate of conservation land acquisition still increased after 1980s (Figure 3). This is perhaps due to the emergence of creative solutions for raising funds for the acquisition of conservation land. Yet, in addition to Proposition 13 and Proposition 4, voters also passed Propositions 62 and 218 in 1996 extending the supermajority approval requirement of Proposition 13 to virtually all types of assessments, fees or taxes used by local government. This further constricted local government's flexibility to raise funds for necessary services, let alone to purchase additional lands for conservation purposes (Table 2).

With the decline in local government's ability to raise money for land preservation through tax dollars, new mechanisms began to be invented. One such new strategy was developed successfully by the Planning and Conservation League (PCL), a statewide environmental organization, which developed a set of inventive campaign funding tactics and ballot initiatives to finance the acquisition and development of park and recreation areas [25]. Nongovernmental organizations and quasi-governmental agencies emerged from this period as new stewards of public spaces, including habitat. State conservancies with budgets funded by grants, donations and foundation funding were created. Recent innovations include the Regional Advanced Mitigation Plan (RAMP), which is an effort to develop a more comprehensive and strategic approach to mitigating biological resource impacts caused by major infrastructure projects, before projects are constructed. RAMPs typically build on a conservation plan that validates and matches the mitigation actions and cumulative impacts. Another emerging concept is the inclusion of Regional Greenprints as an integral part of local General Plans and Sustainable Community Strategies. Greenprints would offer a new way to improve conservation planning by providing a process to map a region's important Open Space for a full range of ecosystem services including habitat, farmland, recreation, water resources and more. Conservation planning could provide a baseline on which land use plans are developed, and could be linked to CO_2 sequestration, a new and popular idea. For example, Preservation Ranch, a 7890ha property in Sonoma County, was just recently purchased with carbon credits funds (http://baynature.org/articles/preservation-ranch-big-conservation-thanks-to-carbon-credits/).

With the advent of climate change, priorities have expanded to include renewable energy and land use that reduces production of greenhouse gas emissions. All of these concerns are embodied in adopted state law and policy. Examples include the 1978 Urban Strategy, AB 857 (Wiggins codified at Section 65041.1 of the Government Code), the 2006 Assembly Bill 32 (California Global Warming Solutions Act), and the 2008 SB 97 (CEQA and Greenhouse Gas Emissions), and Senate Bill 375 (Sustainable Communities and Climate Protection Act).

Discussion

It can be argued that over the past 150 years a great deal of California's most scenic landscapes has been preserved. The fundamental dilemma of conserving land is the necessity for land to be purchased. Under the Fifth Amendment of the U.S. Constitution, government must purchase land at its market value, it cannot take land away from land owners, nor force them to sell below the land's market value. Thus for additional conservation to occur, as the state's population and cities grow into previously undeveloped areas that have important ecosystems, this land must be purchased from its owners. Moreover, since land development provides localities their revenues, there is a push to continue to zone land for urbanization. Such dynamics are typical throughout the U.S. but in addition, in California, for new developments, fees are assessed on units to create funds to acquire the conservation lands, near or adjacent to the new development – mitigation land. This leads to a paradox – for there to be funds to acquire conservation lands, there needs to be development [26]. In turn this quest for growth limits the land available for conservation – one of the original goals for which growth was planned for.

California's voluntary approach to meet conservation needs has been a reactive, and accretionary approach that creates obvious mixed messages between local power and regional planning efforts, and mismatches between policy goals and instruments for their implementation. Mixed messages are due to the autonomy of land use designations of cities while conservation needs and habitats do not observe political boundaries [27]. Land use planning is by definition based on the parcelizing of land for development, which is done by cities. This parcelizing then creates sections of land that can be bought and sold individually, allows the land market to work at the municipal and county scale, but this approach fragments the landscape and its governance. While there have been numerous attempts to create regional governments with regulatory authority, and to encourage greater collaboration among local governments [28]; they still have narrow single issue mandates (for example, regional "special districts" for air quality management and flood control). In contrast, conservation needs and habitats do not observe political boundaries, they are regional and even statewide. Preservation of ecosystem requires a landscape or regional approach, keeping parcels together to provide connectivity and or enough space for ecosystems to function well, and to preserve watersheds. Further, conservation lands require revenue to purchase them. Funding streams are often predicated on growth with development fees and various impact fees to finance conservation acquisitions [26]. Thus the state has a paradox: to conserve, there must be development.

There has been a century-long series of efforts to create effective institutions to help guide metropolitan growth and development in California [14,19,27]. While early conservation land acquisition tended to occur in the mountain regions, and away from the rapidly growing cities; the push-and-pull of land use policies lead to striking differences in the timeline of the 1910s most populous counties. The early most populous counties took different interpretations of the regulations, and the growth of the conservation network took different shapes. The most steady growth rate is found for San Francisco. San Francisco, along with Alameda, San Mateo, Santa Clara and 5 other counties form the Bay Area, the region that showed the higher evenness in conservation land acquisition. This has been a key region in the state and even an example of conservation success in a growing metropolitan area [28,29]. In 1965 the Bay Area Conservation and Development Commission was created, a single purpose regional planning agency, to supervise development around San Francisco Bay, and to stop Bay infill projects. The southern California counties show a stepwise growth of the conservation network, associated with the protection of the mountain areas around Los Angeles and desert areas around San Diego.

There are also mismatches between policy goals and instruments for their implementation. California has long recognized the costs of sprawl, and prioritized specific land use outcomes including orderly and efficient land use, infill, farmland protection and conservation of resource lands, but has developed weak planning tools and rules to strongly protect habitat. With the advent of climate change, priorities have expanded to include renewable energy and land use that reduces production of greenhouse gas emissions. These measures aim to curb urban sprawl and directing new urban growth to existing cities and suburbs – infill, revitalizing central cities and neighborhoods, and protecting resource lands, in alignment with the proposals put forward by California Tomorrow and others. While the most recent SB 375 statute is possibly the closest California has come to regional planning and encourages urban and suburban infill, clustered development, mixed land uses, transit oriented development to reduce GHGs [30], it does not change land use law and local control over local land use.

At the same time, other activities are occurring that will also challenge conservation. Paradoxically, concern about GHG emissions is fueling the Governor Jerry Brown's drive to have a high speed rail line built to link Southern California to the Bay Area and Sacramento. Its route, as currently designed, goes through some of the state's prime agricultural land, and is argued that it will support even more sub-urbanization in the already urbanizing Central Valley. Governor Jerry Brown is also pursuing more energy resources, including natural gas fracking. A great deal of the state's natural gas and petroleum resources are also found the Central Valley and will also compete for water, adding another demand in addition to water for agriculture, urban areas and for places of origin. Protected federal lands are not exempt either in California with large scale electricity generation projects in the high desert of the Mohave having been given the green light by the Department of the Interior, despite documented impacts on the desert tortoise among other species.

California has been fortunate to have had magnificent habitats preserved starting in the late 19th century. This is quite different from what has happened in the country and in the world [3,31]. At the same time, California has had tremendous population growth and urbanization as well. This has led to the push-and-pull in land use policies, creating conflicts between habitat preservation and development. Urban growth is still seen as the key to the prosperity of localities, so there is little incentive to build denser more contained cities and to collaborate across jurisdictions over infrastructure and/or land use. The state legislature, and voters through the ballot initiative process, have passed a number of laws, and created new agencies, to attempt to improve conservation planning and to protect conservation lands. But with the reduction of funding since the early 1980s, conserving lands has been more difficult. Any land purchased by government must be bought at market value, and since land in California is expensive, it means conserving more lands is costly. Innovative approaches to gather conservation acquisition funds are now being proposed and implemented.

Lost perhaps is the bigger vision of how we live on the land over time. Planning law, open space protection processes, finding funding for conservation through development and CO2 markets, are instrumental approaches to an issue that involves much a longer term vision about how people live on the land. Protected areas are not one more land use type – they are a result of a (land use) policy. While protected areas have distinct geographic boundaries they are part of a conservation network, which grows through a dynamic process [32,33]. U.S. land conservation policies, enabled by the vast area in public ownership, was seen as a model to emulate in many parts of the globe. For example, the International Union for the Conservation of Nature has been at the forefront of land conservation for habitat protection in Africa and increasingly in other parts of the world. Initially this was to simply preserve wildlife, now the mission includes attempting to anticipate the conservation needs due to climate change impacts. In Europe too there are now National Parks, based on the U.S. template, but often include working landscapes [11]. The reconstruction of the history of conservation can then describe the process of growth of a conservation network. This process is yet not well documented [3]; however, the creation of such baseline knowledge can help strategize the maintenance and future additions to the conservation network [33]. Further it can allow us to understand the feedback between land use policies and conservation action on meeting and maintaining conservation goals in California and elsewhere in the world [5,30,34].

Perhaps the most important value of historic analyses is what they indicate about the future. Policies could redirect future growth and conservation. As we observed, multiple funding sources contributed to the highest rates of land acquisition in different times in history (federal in 1890–1910, state in 1920–1940, and non-governmental organizations in 1970–1980). We also observed that reduction of development fee funding lead to two outcomes: creativity on methods to raise funds, and strategic acquisitions, perhaps because of the concurrent emergence of systematic conservation planning [1,2]. Perhaps the reduction of the funding stream imposed by Proposition 13 sparked innovative funding strategies, arguably more sustainable and less paradoxical. As future housing growth depends heavily on economic conditions [35], so do development fees. Innovative funding approaches that do not depend on development fees may be more sustainable in the long run. Linking histories of policies and conservation land acquisition allows reflections on the results of such policies but also on their perpetuity. Over the last 150 years we have seen a push-and-pull between development and conservation policies, some policies remained and some got replaced. The same applies to conservation lands, especially those that are not conserved in perpetuity. The future of these lands holds new challenges, with the onset of climate induced changes. Dynamic reserve systems are being proposed [4], which may take advantage of demoting some of these currently non-in-perpetuity conserved lands, or others, and it also presents challenges on funding sources. Perhaps a RAMP-like strategy can be applied to meet the conservation network dynamics needed to meet climate change-suited conservation. While solutions for the future of conservation are contextual and geographically restricted, the issues of fund raising, land acquisition, development versus conservation, and persistence of the conservation network operate at regional, national and global scales. Only through merging information at such scales can guarantee that the benefits of conservation can be enjoyed by the future generations.

Acknowledgments

We would like to thank Harold Mooney for connecting the authors in the paper.

Author Contributions

Conceived and designed the experiments: MJS TW SP. Performed the experiments: MJS TW SP. Analyzed the data: MJS TW SP. Contributed reagents/materials/analysis tools: MJS TW SP. Wrote the paper: MJS TW SP. Collected and analyzed the data on California conservation history: MJS. Contributed the history of land use policy: SP. Contributed the legislative timeline and provided oversight over the legislative components: TW.

References

1. Pressey RL, Humphries CJ, Margules CR, Vane-Wright RI, Williams PH (1993) Beyond opportunism: key principles for systematic reserve selection. Trends in Ecology and Evolution 8: 124–128.
2. Balmford A (2003) Conservation planning in the real world: South Africa shows the way. Trends in Ecology and Evolution 18: 435–439.
3. Radeloff VC, Beaudry F, Brooks TM, Butsic V, Dubinin M, et al (2013) Hot moments for biodiversity conservation. Conservation Letters 6: 58–66.
4. Pressey RL, Cabeza M, Watts ME, Cowling RM, Wilson KA (2007) Conservation planning in a changing world. Trends in Ecology and Evolution 22: 583–592.
5. Lovejoy TE (2006) Protected areas: a prism for a changing world. Trends in Ecology and Evolution 21: 329–333.
6. Brooks TM, Mittermeier RA, Mittermeier CG, da Fonseca GAB, Rylands AB, et al (2002) Habitat loss and extinction in the hotspots of biodiversity. Conservation Biology 16: 909–923.
7. Calsbeek R, Thompson JN, Richardson JE (2003) Patterns of molecular evolution and diversification in a biodiversity hotspot: the California Floristic Province. Molecular Ecology 12: 1021–1029.
8. GreenInfo Network (2013) California Protected Areas Database – CPAD. http://www.openspacecouncil.org/programs/index.php?program = 6. Accessed 2013 October 11.
9. Barton GA (2000) Empire forestry and American Environmentalism. Environment and History 6: 187–203.
10. Olmsted FL (1929) Report of State Parks Survey of California. California State Park Commission 106pp.
11. IUCN and UNEP-WCMC (2013) The World Database on Protected Areas (WDPA) [On-line]. Cambridge, UK: UNEP-WCMC. Available at: www.protectedplanet.net. Accessed 2013 October 11.
12. Simpson EH (1949) Measurement of diversity. Nature 163: 688.
13. Silva FJ, Barbour E (1999) The state-local fiscal relationship in California: A changing balance of power. Public Policy Institute of California, San Francisco.
14. Pincetl S (1999a) Transforming California, the Political History of Land Use in the State. Johns Hopkins University Press, Baltimore.
15. Fogelson RM (1967) The Fragmented Metropolis Los Angeles, 1850–1930. University of California Press, Berkeley.
16. Dear M (1996) In the City, Time Becomes Visible: Intentionality and Urbanism in Los Angeles, 1781–1991, in Scott AJ, Soja EW, eds., The City: Los Angeles and Urban Theory at the End of the Twentieth Century, University of California Press, Berkeley.
17. Hise G (1997) Magnetic Los Angeles; Planning the Twentieth Century Metropolis. Johns Hopkins University Press, Baltimore.
18. Pincetl S (1999b) The politics of influence: Democracy and the growth machine in Orange County, U.S. in Jonas A, Wilson D (eds.). The Urban Growth Machine: Critical Perspectives, Two Decades Later. State University of New York Press, Buffalo NY.
19. Barbour E (2002) Metropolitan Growth Planning in California 1900–2000. Public Policy Institute of California, San Francisco.
20. Press D (2002) Saving Open Space, the Politics of Local Preservation in California. University of California Press, Berkeley.
21. Fulton W (1997) The Reluctant Metropolis: The Politics of Urban Growth in Los Angeles. Solano Press Books, Point Arena.
22. Pincetl S (2004) The preservation of nature at the urban fringe. In Wolch J, Pastor M, Drier P (Eds.), Up Against the Sprawl, Public Policy and the Making of Southern California. University of Minnesota Press, Minneapolis, MN.
23. Pincetl S (2006) Conservation planning in the West, Problems, new strategies and entrenched obstacles. Geoforum 37: 246–255.
24. Wolch J, Pastor M, Drier P (2004) Up Against the Sprawl, Public Policy and the Making of Southern California. Minneapolis, MN: University of Minnesota Press.
25. Pincetl S (2003) Nonprofits and park provision in Los Angeles: an exploration of the rise of governance approaches to the provision of local services. Social Science Quarterly 84: 989–1001.
26. Pincetl S, Jonas AE, Sullivan J (2011) Political ecology and habitat conservation for endangered species planning in Southern California: Region, places and ecological governance. GeoForum 42: 427–438.
27. Schrag P (1998) Paradise Lost, California's Experience, America's Future. University of California Press, Berkeley.
28. Walker R (2007) The country in the city: the greening of the San Francisco Bay. University of Washington Press, Seattle, USA. 430pp.
29. Thorne JH, Santos MJ, Bjorkman J (2013) Regional assessment of urban impacts on landcover and Open space finds a smart urban growth policy performs little better than business as usual. PlosONE. DOI:10.1371/journal.pone.0065258.
30. California Transportation Commission, Regional Transportation Guidelines (2010) http://www.catc.ca.gov/programs/rtp/2010_RTP_Guidelines.pdf. Accessed 2013 March 12.
31. Moilanen A, Arponen A (2011) Administrative regions in conservation: balancing local priorities with regional to global preferences in spatial planning. Biological Conservation 144: 1719–1726.
32. Bishop K, Phillips A, Warren L (1995) Protected for ever? Factors shaping the future of protected areas policy. Land Use Policy 4: 291–305.
33. Wätzold F, Schwerdtner K (2005) Why be wasteful when preserving a valuable resource: review article on the cost-effectiveness of European biodiversity conservation policy. Biological Conservation 123: 327–338.
34. McDonald RI, Boucher TM (2011) Global development and the future of protected area strategy. Biological Conservation 144: 383–392.
35. Radeloff VC, Stewart SI, Hawbaker TJ, Gimmi U, Pidgeon AM, et al (2010) Housing growth in and near United States protected areas limits their conservation value. Proceedings of the National Academy of Sciences 107: 940–945.

15

Managing Conservation Reliant Species: Hawai'i's Endangered Endemic Waterbirds

Jared G. Underwood[1]*, Mike Silbernagle[2], Mike Nishimoto[3], Kim Uyehara[4]

1 Pacific Reefs National Wildlife Refuge Complex, U.S. Fish and Wildlife Service, Honolulu, Hawai'i, United States of America, 2 Oahu National Wildlife Refuge Complex, U.S. Fish and Wildlife Service, Haleiwa, Hawai'i, United States of America, 3 Maui National Wildlife Refuge Complex, U.S. Fish and Wildlife Service, Kihei, Hawai'i, United States of America, 4 Kauai National Wildlife Refuge Complex, U.S. Fish and Wildlife Service, Kilauea, Hawai'i, United States of America

Abstract

Hawai'I's coastal plain wetlands are inhabited by five endangered endemic waterbird species. These include the Hawaiian Coot ('alae ke'oke'o), Hawaiian Duck (koloa maoli), Hawaiian Stilt (ae'o), Hawaiian Gallinule (Moorhen) ('alae 'ula), and Hawaiian Goose (nēnē). All five species are categorized as being "conservation reliant." The current strategy to recover these endangered birds includes land protection and active management of wetlands. To assess the effectiveness of the current management paradigm, we compared species population trends across the state to those on six actively managed wetland national wildlife refuges (Refuges) thought to be critical for the survival of these endangered species. To perform the evaluation we relied on systematic semiannual population counts that have been conducted across most wetlands in the state and monthly population counts that have occurred on Refuges during the same time period. We found that statewide and Refuge populations of the Hawaiian Coot, Stilt and Gallinule have rebounded from historic lows and over the last 20 years have slowly increased or remained stable. We also documented that Refuges are important to each species year-round and that a disproportionately larger percentage of the population for each species is found on them. Understanding of why Refuges successfully house a disproportionate percentage of these "conservation reliant" species can inform current and future conservation efforts as well as ensure long-term population viability for these species.

Editor: David L. Roberts, University of Kent, United Kingdom

Funding: The authors have no support or funding to report.

Competing Interests: The authors have declared that no competing interests exist.

* E-mail: Jared_Underwood@fws.gov

Introduction

Hawai'i is home to more endangered species than any other state in the United States [1]. A quarter of all species listed under the United States Endangered Species Act of 1973, including one-third of all birds listed, are native to Hawai'i [1]. Listed species in Hawai'i include five endangered endemic waterbirds, the Hawaiian Coot ('alae ke'oke'o; *Fulica alai*), Hawaiian Duck (koloa maoli; *Anas wyvilliana*), Hawaiian Stilt (ae'o; *Himantopus mexicanus knudseni*), Hawaiian Gallinule (Moorhen) ('alae 'ula; *Gallinula galeata sandvicensis*), and Hawaiian Goose (nēnē; *Branta sandvicensis*). Endangerment of these five species has been attributed to a loss or modification of wetland habitats, introduced plants and predators, avian diseases, altered hydrology, hybridization, and historic overhunting [2,3]. Faced by this suite of cumulative threats, these endangered Hawaiian birds have recently been categorized as "conservation reliant", which implies that they will require active management into perpetuity [4,5,6].

These endangered waterbirds use a variety of wetland and open water environments in Hawai'i including freshwater/saltwater marshes and ponds, estuaries and mudflats, artificial reservoirs, agricultural lands (primarily taro-*Colocasia esculenta*- and stock watering ponds), irrigation and flood control ditches, sewage treatment ponds, and in the case of the Hawaiian Duck, montane streams and bogs [2]. How each species utilizes this system of wetlands and the interconnectedness of the wetland landscape is poorly understood. Today, approximately 6,190 ha of lowland coastal plain wetlands, the wetland type most commonly used by the endangered waterbirds, persist in the state [2,7]. This represents a 31% reduction from the total estimated in 1780, with the majority of remaining coastal plain wetland areas degraded by altered hydrology, invasive species, and contaminants [2,7]. It has generally been proposed that birds disperse across these wetlands during the rainy season (October- March) and congregate in key wetlands during the dry season (April-September) [2,3].

For the past several decades, the accepted active management strategy for these species has included three major components [2,8]. First is the control of invasive introduced plant species. Hawai'i experiences a year-round growing season and therefore management of invasive wetland plants must be constant. When wetlands are not actively managed, waterbird habitat can rapidly degrade as weedy invaders tend to fill open water and mudflat areas important to the birds. When combined with altered hydrology and sedimentation, wetlands can rapidly become completely unsuitable for endangered waterbirds. Control methods for invasive plants include herbicides, mechanical treatment, manipulation of water levels, and prescribed fire. Second is the control of introduced predators. Non-native cats, rats, mongooses, dogs, and to a lesser extent wild pigs, Barn Owls (*Tyto alba*), Cattle Egrets (*Bubulcus ibis*), predatory fish, and bullfrogs all directly depredate either eggs, young, or adult birds. To combat the mammalian predators that have the greatest negative impact, wetlands must have a barrier of fences, rodenticide baited stations, and live or kill traps around the areas used by waterbirds. Depredation rates and population-level effects of the non-mammalian predators mentioned, as well as effective control methods, are still being investigated. Third is the manipulation of wetland water levels to mimic natural hydrological processes that have been highly altered or lost due to anthropogenic impacts. Water control in wetlands is facilitated by a system of dikes and levees with water pumped from wells or procured from permanent free-flowing water sources (i.e., rivers, streams). This allows wetlands to maintain water levels in times of drought and to adjust water levels to best suit the needs of breeding birds. Although all three of these management actions may not occur at the same time/location, it is generally accepted by wetland managers in Hawai'i that all three management actions in concert are required to begin to restore the functionality of wetlands to meet the life-history requirements of these birds [2,8]. While the core components of the current active management strategy have remained the same, methods and tools have changed and will continue to change as managers face new threats and refine techniques for Hawai'i's wetland systems.

Recovery (downlisting or delisting from the endangered species list) for the Hawaiian Stilt, Coot, Gallinule and Duck is linked to the permanent protection and active management of key remaining coastal plain wetlands [2]. Six coastal plain wetland national wildlife refuges (Refuges) on the main Hawaiian Islands have all been identified as core wetlands that will contribute toward recovery [2,9]. These Refuges have some of the most actively managed of Hawai'i's wetlands. In this study we use Refuges to evaluate the effectiveness of the current management paradigm for these conservation reliant species.

To assess the importance of Refuges and their associated management practices to waterbird populations, we evaluated three metrics: long-term population trends, percentage of the population for each species found on Refuges over time, and seasonal use of Refuges. To assess long-term population trends, we analyzed and compared statewide population trends (all wetlands), Refuge specific trends, and non-Refuge wetland trends. In our analysis, non-Refuge wetlands refer to all surveyed wetlands other than the aforementioned six Refuges. These non-Refuge wetlands include some wetlands that are currently actively managed for waterbirds and others that have no active management. The analyses we performed, lumped managed and unmanaged non-Refuge wetlands. If the active management program on Refuge wetlands is an effective strategy, we might expect to document an enhanced population trend on Refuges compared to non-Refuge wetlands. Since we did not include all actively managed wetlands in our Refuge category, this expectation may not hold true as active management on non-Refuge wetlands may also have benefited these bird species. Second, we calculated the percentage of the population for each species that is found on the Refuges. This allowed for an evaluation of the relative importance of Refuge wetlands using bird population densities. If the active management program on Refuge wetlands is an effective strategy, we would expect to find proportionally higher bird populations on Refuges compared to wetland areas off Refuge. Lastly, we assessed the seasonal use of Refuges by each species to determine if these wetlands were important year-round for the birds. If the active management program on Refuge wetlands is an effective strategy, we would expect year-round use of Refuges rather than the seasonal use observed in non-Refuge wetlands.

Methods

Study Area

The endangered waterbird species are currently found on various "main Hawaiian Islands" (18°54′N to 22°12′N, 154°54′W to 160°12′W; Fig. 1). The term "main Hawaiian Islands" refers to the following eight islands: Ni'ihau, Kaua'i, O'ahu, Maui, Moloka'i, Lāna'i, Kaho'olawe, and Hawai'i. Semiannual waterbird surveys currently occur on all main Hawaiian Islands except for Ni'ihau and Kaho'olawe. The six coastal plain wetland Refuges occur across the main Hawaiian Islands (see Figure 1).

Coastal Plain Wetland Extent

The extent of coastal plain wetlands in the main Hawaiian Islands is approximately 6,190 ha. It should be noted that this estimate of wetlands does not include all areas used by the birds as not all wetlands such as reservoirs, irrigation and flood control ditches, golf course water features, streams, ephemeral springs and basins, or sewer treatment ponds are included. Of the coastal plain wetlands, 3,500 ha are identified as important for the recovery of waterbirds [2]. The six Refuges contain approximately 900 ha or 15% of all coastal plain wetlands and 26% of those identified as important to waterbirds. The 900 ha of wetlands on Refuges are as follows: Hanalei National Wildlife Refuge (NWR) (371 ha), Kealia Pond NWR (280 ha), James Campbell NWR (107 ha), Huleia NWR (98 ha), Pearl Harbor NWR (25 ha), and Kakahai'a NWR (18 ha) [2].

Species

For our analysis we focused on the Hawaiian Coot, Gallinule, and Stilt. We did not include the Hawaiian Duck in our analysis because population count results are complicated by the presence of feral Mallard (*Anas platyrhynchos*) and Hawaiian Duck hybrids. It can be difficult if not impossible to distinguish a pure Hawaiian Duck from a Mallard-Hawaiian Duck hybrid in the field, especially females. In addition, streams are not included in the surveys, which is where the majority of Hawaiian Ducks are believed to reside [2]. These factors make use of the state semiannual population counts for this species unreliable. We also did not incorporate the Hawaiian Goose as their primary habitats are in dry uplands and not coastal wetlands, although wetlands and other water bodies may be used during their annual cycle [10].

Data- Semiannual State Counts & Monthly Refuge Counts

Semiannual waterbird population surveys have been conducted statewide since the mid-1950s, but coverage was somewhat inconsistent until the 1980s. Currently, each census consists of simultaneous visits to wetlands on all islands during a single day or two-day time period in January and August [3,11]. This is done to reduce the possibility of double counting birds. These surveys include the majority of wetlands on each island, but do not cover all locations that support waterbirds such as streams, flood control ditches, and some private lands. During a survey, all waterbirds observed or heard on each wetland are counted. The counts resulting from these surveys are considered a minimum population estimate and certainly underestimate the actual population. Population estimates resulting from these surveys have not been corrected for detectability or observer differences/abilities. For the Hawaiian Coot and Stilt this probably does not have a significant impact as both species often use open water and mudflat areas and are considered to have detectability approaching 100% [2]. The Hawaiian Gallinule however, is secretive and often hides in densely vegetated areas resulting in detectability problems that

Figure 1. Map of "Main Hawaiian Islands" with the six coastal plain national wildlife refuges identified. Hawai'i, USA.

lead to undercounting [2,12,13]. Even with these limitations, since the methodology has remained consistent over time, we can still detect changes in relative abundance. During each census approximately 400 locations are surveyed, which include the remaining natural coastal plain wetlands as well as some anthropogenic wetlands (i.e., golf course ponds, waste treatment ponds, reservoirs). Monthly or bimonthly waterbird censuses have also been conducted on the six Refuges for varying lengths of time. To evaluate seasonal use patterns, we used data from five of the six Refuges. Data collected on Kakahai'a NWR only existed outside the timeframe analyzed. The methods employed during these censuses are similar to the semiannual state count methods in which all waterbirds observed or heard during the survey are counted. The data collected during Refuge surveys provide an index of year-round Refuge occupancy by the waterbirds that is not attainable from the semiannual state counts.

Data Analysis

To establish long-term population trends and seasonal use, we used data from the semiannual waterbird counts for 1986–2007 (dataset available from State of Hawai'i Division of Forestry and Wildlife upon request) [14] and from the monthly Refuge counts for approximately the same time period (1989–2013; dataset available from authors upon request). This time period was chosen because it represents the most complete set of data and corresponds to the time when active management was implemented on Refuges. An initial evaluation of the semiannual count dataset identified some discrepancies/errors during certain sampling periods. Errors included asynchronous data collection or missing data for multiple or key wetlands (see also Reed et al. [15] for an expanded discussion of this database and its limitations). We removed sampling periods with these errors from our analysis. Table 1 identifies the final set of data used to complete both the long-term trend and seasonal use analyses.

We analyzed the semiannual waterbird count data from 1986–2007 to identify the statistically significant long-term population trends across the state (all Refuge and non-Refuge wetlands), and on Refuge and non-Refuge wetlands independently. For the Hawaiian Coot, Gallinule, and Stilt we assessed long-term population trends and indices using Program TRIM Version 3.53 [16], which computes indices and trends for time series by means of log-linear Poisson regression and accounts for missing data points. Serial correlations were considered in TRIM with a Generalized Estimating Equations approach. We calculated the annual index and long-term population trend with standard errors for each species using the entire statewide database, data for Refuges only, and data excluding Refuges during the time period spanning 1986–2007. We estimated trends using the multiplicative trend, which reflects the average percent change per year [17]. The analysis started with a model incorporating change at each time point, and used a stepwise selection procedure to identify significant changes in slope based on Wald tests with a significance-level threshold value of $p<0.05$ [16]. All models were run with serial correlation considered, and a stepwise process was used to identify the best model.

We evaluated waterbird seasonal use patterns on Refuges by performing an analysis of variance and two multiple comparisons procedures on monthly Refuge count data spanning 1989–2013.

Table 1. Waterbird data included in analyses.

Location	Years with Usable Data for Population Trend Analyses[a]
Statewide Summer Data	1986,1987,1989–1993,1995,1998–2001,2003–2006
Statewide Winter Data	1987,1990,1993,1995,1999–2002,2004–2007
	Years with Data for Seasonal Occupancy Analysis[b]
James Campbell NWR	1989–2008 ($n = 328$)
	Monthly:1989–1992
	Bi-Monthly: 1992–2008
Pearl Harbor NWR	1989–2008 ($n = 344$)
	Monthly:1989–1992
	Bi-Monthly: 1992–2008
Kealia Pond NWR	1994–2007 ($n = 305$)
	Bi-Monthly: 1994–2007
Hanalei NWR	2010–2013 (n = 40)
	Monthly: 2010–2013
Huleia NWR	2010–2012 (n = 30)
	Monthly: 2010–2012
Kakahai'a NWR	None

[a]Data for the long-term population trend analyses includes data collected during semiannual waterbird counts across the state (Refuge and non-Refuge wetlands) for years spanning 1986–2007.
[b]Data for the seasonal occupancy analysis includes data collected only on Refuges for years spanning 1989–2013.

We used Tukey's honest significant differences and pairwise t-tests with a holm adjustment in R Stats version 2.12.1 [18] to detect monthly differences in use of selected Refuges for each species. We identified significant differences as those that had a p-value of $p<0.05$.

Finally, to understand the relative importance of the actively managed Refuges for each of the endangered waterbird species, we averaged the totals of the two semiannual statewide counts by species and year to obtain an annual population estimate. We then averaged the corresponding data for Refuges and calculated what percent of the total species' population occurred on Refuges for each species by year.

Results

We found that populations of the Hawaiian Coot, Stilt, and Gallinule were either stable or increasing across the state, and on both Refuges and non-Refuge wetlands when analyzed independently (Table 2). For Hawaiian Coots, the statewide population trend indicated an average population increase of 0.18% per year between 1986 and 2007.The population growth rate was −0.56% per year on Refuge wetlands and 0.83% for non-Refuge wetlands. The population trend, following criteria in TRIM, was identified as 'stable' (no significant increase or decline, and it is certain that trends are less than 5% per year) statewide and for both Refuge and non-Refuge wetlands. The statewide estimated average Hawaiian Coot population size was $1{,}777 \pm 310$ for the subset of data used spanning 1986–2007. State population estimates for Hawaiian Coots ranged from>2,500 in 1998 to~1,200 coots in 2002.

For Hawaiian Stilts, the statewide population trend indicated an average population increase of 1.98% per year between 1986 and 2007. The population growth rate was 3.11% per year on Refuge wetlands and 1.25% on non-Refuge wetlands. The population trend, following criteria in TRIM, was identified as 'moderate increase' (any significant trend increase<20% per year) statewide and for both Refuge and non-Refuge wetlands. State population estimates peaked at~2,200 stilts in our most recent estimate (2007).

For Hawaiian Gallinules, the statewide population trend indicated an average population increase of 6.86% per year between 1986 and 2007. The population growth rate was 5.46% per year on Refuge wetlands and 7.79% on non-Refuge wetlands. The population trend, following criteria in TRIM, was identified as 'moderate increase' (any significant trend increase<20% per year) statewide and for both Refuge and non-Refuge wetlands. State population counts exceeded~400 Gallinules in 2005. Results showing stable to increasing long-term trends for all three species

Table 2. Annual population trends for the Hawaiian Coot, Stilt, and Gallinule from count data spanning 1986–2007 in Hawai'i, USA.

Location[a]	Trend	SE	Interpretation[b]
Refuge Gallinule	1.0546	0.0202	Moderate Increase
Refuge Coot	0.9944	0.0046	Stable
Refuge Stilt	1.0311	0.0040	Moderate Increase
Non-Refuge Gallinule	1.0779	0.0049	Moderate Increase
Non-Refuge Coot	1.0083	0.0049	Stable
Non-Refuge Stilt	1.0125	0.0013	Moderate Increase
State Gallinule	1.0686	0.0097	Moderate Increase
State Coot	1.0018	0.0037	Stable
State Stilt	1.0198	0.0021	Moderate Increase

[a]Estimates by species for the entire state population (Refuge and non-Refuge lands combined); population on Refuges; and population on non-Refuge lands.
[b]As defined by program TRIM [16,17].

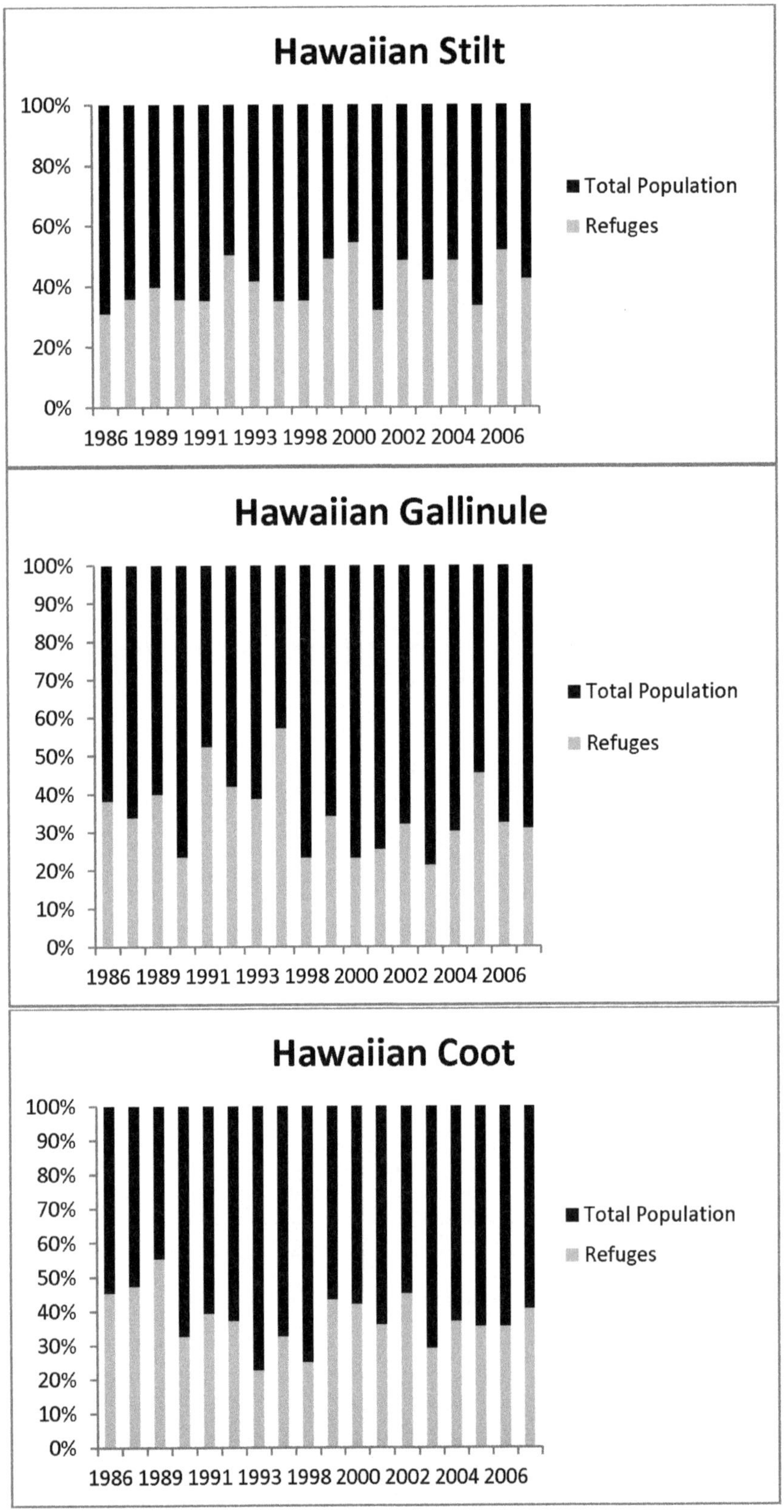

Figure 2. Percentage of endangered waterbird populations counted on six national wildlife refuges in Hawai'i, USA for the time period spanning 1986–2007. Refuges account for only 15% of available wetland habitat.

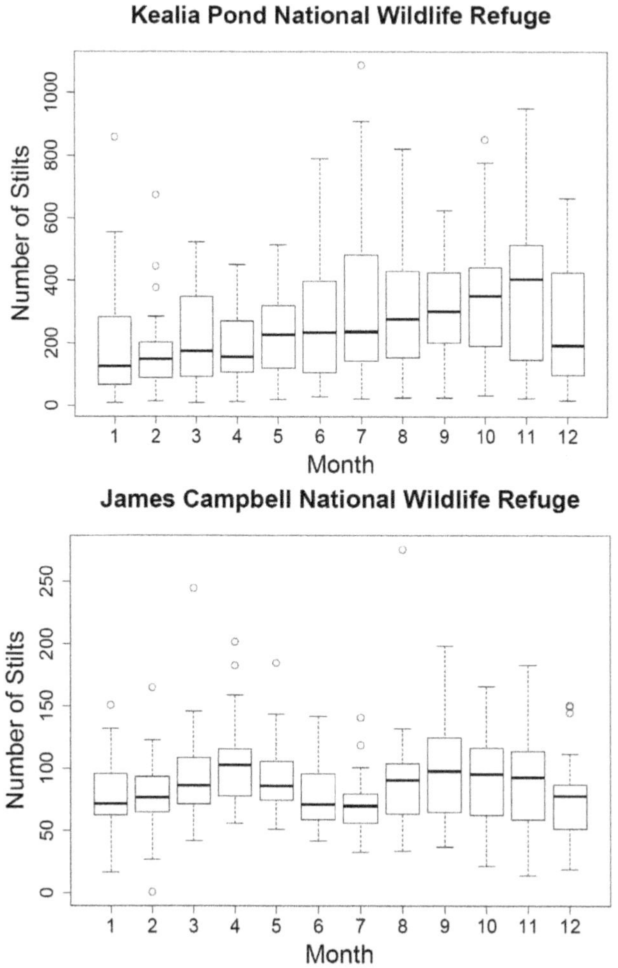

Figure 3. Seasonal use by Hawaiian Stilts on two national wildlife refuges in Hawai'i, USA for the time period spanning 1989–2008. The box represents the upper (75%) and lower (25%) quartiles, the horizontal line in the box shows the median, and vertical dashed lines (whiskers) indicate either the maximum and minimum values or 1.5 times the interquartile range (2 standard deviations). Data beyond the end of the whiskers are outliers and plotted as hollow circles.

are similar to those found by other recent analyses of similar data [15,19].

Although Refuges contain 15% of the total coastal plain wetland acreage in the state they support a disproportionate percentage of each waterbird species' population (see Figure 2). For Hawaiian Coots, the estimated average percentage of the statewide population found on Refuges for the subset of data used spanning 1986–2007 was 35.7%±4.7%. For Hawaiian Stilts, the estimated average percentage of the statewide population found on Refuges for the subset of data used spanning 1986–2007 was 41.5%±5.0%. For Hawaiian Gallinule, the estimated average percentage of the statewide population found on Refuges for the subset of data used spanning 1986–2007 was 33.7%±6.6%. When we assessed the trend for the relative contribution of the Refuges over time, we found that the pattern has remained stable.

We found that Refuge wetlands are used year-round by all three species. Some patterns of seasonal use were detected that differed between Refuges. Significantly fewer Hawaiian Stilts were found to be using James Campbell NWR during July when compared to April ($p<0.05$; see Figure 3). At Kealia Pond NWR, Hawaiian Stilts occurred in significantly less numbers in February than in October and November ($p<0.04$, $p<0.05$; see Figure 3). Hawaiian Coots (see Figure 4) and Gallinules did not exhibit any significant pattern of seasonal use.

Discussion

The status of the endangered Hawaiian Coot, Gallinule, and Stilt appears to be improving. All three species have increased in relative abundance over the last several decades and have rebounded from historic lows. We found that populations statewide and on Refuges have remained stable or have increased. Based on our metric of relative importance, we would conclude that the current active management strategy is successful as populations of all three species are increasing or stable, and a disproportionate number of birds are found on Refuges. That said, the relative importance of Refuges stayed the same over the time period analyzed and populations of two of the species are actually growing at a faster rate off Refuge lands. This is potentially the result of many factors, but we believe that this also corroborates the finding that the current active management strategy is successful because over the last two decades there has been an incremental increase in active management of wetlands for waterbirds both on Refuges and by other federal, state, local, private, and non-governmental conservation partners with wetland management responsibilities.

Populations off Refuge might also have increased at a faster rate as active management capacity on some Refuge wetlands remained fairly constant during the time period examined. Without active management of large additional areas of wetland on Refuges, the number of quality territories likely remained similar from year to year and young birds may have dispersed to non-Refuge wetlands. Limited band resights have documented Refuges as important source populations which contribute to populations in wetlands across the main Hawaiian Islands [20]. However, while the number of birds documented occupying non-Refuge wetlands may be increasing at a faster rate for the Hawaiian Coot and Gallinule, this does not mean birds in these wetlands are contributing to the long-term species viability in an equal manner. Although reproductive success is not routinely monitored at many of the unmanaged or minimally managed wetlands, occasional assessments of reproductive success on these wetlands have documented little to no reproductive success even when population counts are in the hundreds [21,22]. This finding highlights the importance of the active management strategy implemented on Refuge and other wetlands for the long-term viability of these endangered waterbird species.

Current Management Strategy

The current active management strategy appears to be successful at maintaining and even increasing populations for several of these endangered species. To meet the recovery goals for these species an active management strategy would need to be implemented on all key wetlands in the state identified in the Recovery Plan of Hawaiian Waterbirds [2]. This would result in expansion of the active management strategy to an additional 1,000–2,000 ha of wetlands [2]. Implementing such an action would require a sizeable commitment of both time and money, estimated in the Recovery Plan of Hawaiian Waterbirds [2] to be at least 2–3 million USD annually. Since current active management efforts are often time consuming and expensive, use of adaptive management frameworks should be expanded so

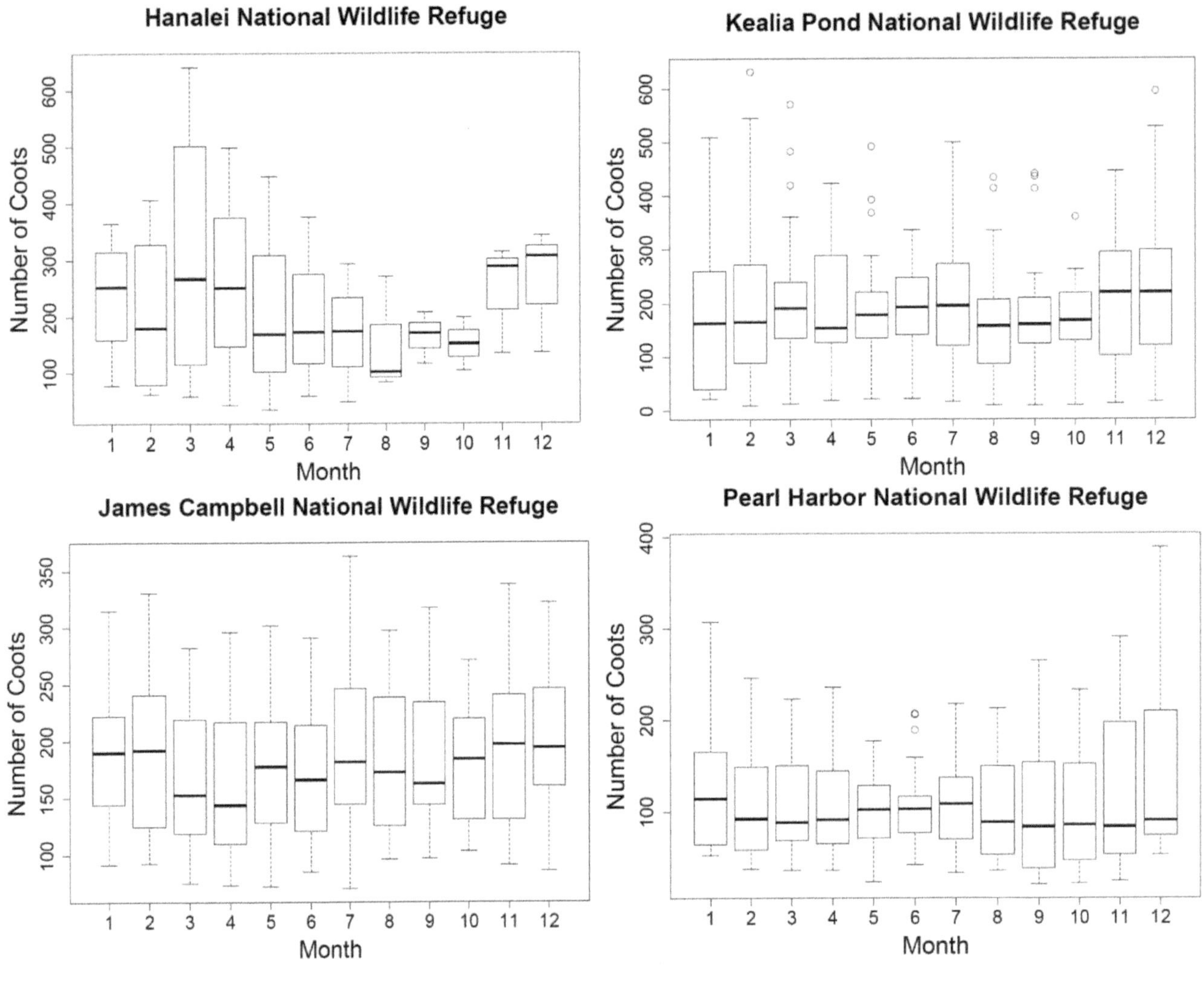

Figure 4. Seasonal use by Hawaiian Coots on four national wildlife refuges in Hawai'i, USA for the time period spanning 1989–2013. The box represents the upper (75%) and lower (25%) quartiles, the horizontal line in the box shows the median, and vertical dashed lines (whiskers) indicate either the maximum and minimum values or 1.5 times the interquartile range (2 standard deviations). Data beyond the end of the whiskers are outliers and plotted as hollow circles.

that managers can identify which active management actions provide the greatest benefit to endangered waterbirds.

It is our belief that without perpetual active management of wetlands in Hawai'i, the endangered waterbird species likely would experience population declines. As an example of what can happen when a wetland is not actively managed, Kawainui Marsh on the island of Oahu is one of the largest remaining natural wetlands in Hawai'i, totaling approximately 400 ha. It was named as a Wetland of International Importance under the RAMSAR convention in 2005 [23]. Historically it was used by hundreds of endangered waterbirds and migratory waterfowl [24,25,26]. However, growth of invasive plants and sedimentation resulting from altered hydrology had resulted in almost a complete loss of wetland functionality for waterbirds. As a result, one of the largest natural wetlands in the state provided habitat for only a handful of birds. Fortunately, in 2012 the State of Hawai'i and the U.S. Army Corp of Engineers began projects to restore a portion of this wetland to a functional state. Post restoration, this portion of the wetland will be actively managed for endangered waterbirds. When such restoration actions occur, we have observed on Refuges that populations of endangered waterbirds in the restored areas have increased dramatically (unpublished data). We believe that wetland restoration efforts both on and off Refuges are a major contributor to the increasing populations of endangered waterbirds.

In addition to the current threats of weedy invaders, disease, predation, and hybridization faced by the waterbirds, most wetlands used by these endangered waterbirds are also threatened with the impacts of sea level rise. Estimates of sea level rise range from 20 to 190 cm in Hawai'i by the end of the 21st century [27,28]. Sea level rise is projected to result in the loss of some wetland habitat and is expected to negatively impact the habitat quality of other wetlands by increasing salinity levels [29,30,31]. Increased salinity would have a negative impact on species that prefer freshwater such as the Hawaiian Duck and Gallinule [2]. Strategies to mitigate for the loss of wetlands and wetland functionality due to sea level rise include the restoration of wetlands located above the new projected sea level or the creation of additional wetlands in similar locations. To effectively manage these species into the future, the current active management paradigm must be expanded to consider the impacts of sea level rise.

Caveats. We acknowledge that the data used for this analysis may have errors for a variety of reasons. First, bird detectability is somewhat contingent on vegetative cover that can differ among wetlands. As surveys are reliant mainly on visual counts to estimate population size, locations with dense vegetative cover can obscure the true number of birds leading to underestimation. Second, the effort expended (amount of time and number of census stations) also varies between wetlands and can lead to differences in estimation. Third, observer experience or familiarity with the wetland being counted may have caused some differences. For example, Refuge biologists often have a more intimate familiarity with the wetlands they manage compared to surveyors that visit wetlands only for the semiannual counts. This may allow Refuge biologists to have a better idea of how/where to find the birds, resulting in more comprehensive counts. Lastly, the general survey methodology needs to be improved through greater standardization and consistency among islands in the identification of Hawaiian Ducks, feral Mallards, and their hybrids and more consistent coverage of wetlands each year to increase comparability over time. Implementation of more accurate methods of surveying for Hawaiian Gallinules (call counts), and inclusion of montane stream habitats used by Hawaiian Ducks would provide more accurate population estimates for theses species [2,8,11]. Finally, the feasibility of alternative methods such as point counts should be explored.

Conclusion

To maintain the recent increases in endangered Hawaiian waterbird populations, we will need to continue to improve both the quality and quantity of available wetland habitat. It is likely that the fate of these species will be perpetually tied to the amount of habitat that is actively managed for them. In addition to all currently required active management activities, it will also likely be necessary to mitigate for the loss of wetland habitat as a result of sea level rise. While the costs of active wetland management are not trivial, the endangered waterbirds of Hawai'i currently represent some of the only endangered Hawaiian species for which we have been able to reverse the population trajectory on a statewide scale. As a result, they represent one of the best returns on investments and conservation successes demonstrable in Hawai'i.

Acknowledgments

We thank and acknowledge all the volunteers and staff who have assisted with the Refuge and state waterbird counts over the last several decades. We also want to acknowledge N. Creps and J. Ho of the State of Hawai'i for providing the statewide waterbird database. We want to thank L. Cruz for her assistance in developing the figures and A Nadig, K. Kilbride, M. Reed and two anonymous reviewers for their reviews and helpful comments. The findings and conclusions in this article are those of the authors and do not necessarily represent the views of the U.S. Fish and Wildlife Service.

Author Contributions

Conceived and designed the experiments: JGU MS MN KU. Performed the experiments: JGU MS MN KU. Analyzed the data: JGU MS MN KU. Contributed reagents/materials/analysis tools: JGU MS MN KU. Wrote the paper: JGU MS MN KU.

References

1. U.S. Fish and Wildlife Service (USFWS) (2012) Endangered Species Program Website. Available: http://www.fws.gov/endangered/.Accessed 2012 Aug 12.
2. U.S. Fish and Wildlife Service (USFWS) (2011) Recovery Plan for Hawaiian Waterbirds, Second Revision. U.S. Fish and Wildlife Service, Portland, Oregon. xx+233 p.
3. Engilis A, Pratt TK (1993) Status and population trends of Hawai'i's native waterbirds, 1977–1987. Wilson Bull 105: 142–158.
4. Scott JM, Goble DD, Wiens JA, Wilcove DS, Bean M, et al. (2005) Recovery of imperiled species under the Endangered Species Act: The need for a new approach. Front Ecol Environ 3: 383–389.
5. Scott JM, Goble DD, Haines AM, Wiens JA, Neel MC (2010) Conservation-reliant species and the future of conservation. Conserv Lett 3: 91–97.
6. Reed JM, DesRochers DW, Vanderwerf EA, Scott JM (2012) Long-Term Persistence of Hawaii's Endangered Avifauna Through Conservation-Reliant Management. BioScience 62: 881–892.
7. State of Hawai'i (1996) Hawai'i Coastal Nonpoint Pollution Control Program Management Plan, Part III: Management Measures. Hawai'i Office of State Planning: Honolulu, HI. 249 p.
8. Griffin CR, Shallenberger RJ, Fefer SI (1989) Hawai'i's endangered waterbirds: a resource management challenge. In: Shwartz RR, Gibbons JW, editors. Freshwater Wetlands and Wildlife. Department of Energy symposium No. 61, U.S. Department of Energy, Oakridge, TN. 1165–1175.
9. North American Bird Conservation Initiative (NBCI), U.S. Committee (2011) The State of the Birds 2011 Report on Public Lands and Waters. U.S. Washington, DC: Department of Interior. 48 p.
10. U.S. Fish and Wildlife Service (USFWS) (2004). Draft Revised Recovery Plan for the Nene or Hawaiian Goose (*Branta sandvicensis*). U.S. Fish and Wildlife Service, Portland, OR. 148+xi p.
11. Hawai'i Division of Forestry and Wildlife (DOFAW) (2005) Hawai'i Waterbird Survey Instructions and Identification Guide. Hawai'i Natural Heritage Program and Pacific Islands Fish and Wildlife Office, Honolulu, HI.
12. DesRochers DW, Gee HKW, Reed JM (2008) Response of Hawaiian Moorhens to broadcast of conspecific calls and a comparison with other survey methods. J Field Ornithol 70: 448–457.
13. Chang PR (1990) Strategies for managing endangered waterbirds on Hawaiian National Wildlife Refuges. M.S. thesis, University of Massachusetts. 87 p.
14. Hawai'i Division of Forestry and Wildlife (1980–2008) Semiannual Hawaiian waterbird survey data. Summarized by Hawai'i Natural Heritage Program and Pacific Islands Fish and Wildlife Office, Honolulu, HI.
15. Reed JM, Elphick CS, Zuur AF, Ieno EN (2011) Long-term population trends of endangered Hawaiian waterbirds. Pop Ecol 53: 473–481.
16. Pannekoek J, van Strien A (2001) TRIM 3.0 for Windows (Trends and Indices for Monitoring data). Statistics Netherlands, Voorburg.
17. Pannekoek J, van Strien A (2006) TRIM 3.53 Manual (Trends and Indices for Monitoring data). Statistics Netherlands, Voorburg. 57 p.
18. R Development Core Team (2010). R: A language and environment for statistical computing. R Foundation for Statistical Computing, Vienna, Austria. ISBN 3-900051-07-0, URL http://www.R-project.org/.
19. Reed JM, Elphick CS, Zuur AF, Ieno EN (2007) Time series analysis of Hawaiian waterbirds. In: Zuur AF, Ieno EN, Smith GM, editors. Analysing Ecological Data. Springer. 615–631.
20. Reed JM, Silbernagle MD, Evans K, Engilis A, Oring L (1998) Subadult movement patterns of the endangered Hawaiian stilt (Himantopus mexicanus knudseni) as revealed by banding evidence. Auk 115: 791–797.
21. Ducks Unlimited, Inc. (2001) A Conservation Plan for the Hawaiian Stilt at Cyanotech aquaculture facility Keahole Point, Hawaii. Sacramento, CA. 136 p.
22. Robinson JA, Reed JM, Skorupa JP, Oring LW (1999) Black-necked Stilt (Himantopus mexicanus). In: Poole A, editor. The Birds of North America Online. Cornell Lab of Ornithology, Ithaca, NY. Retrieved from the Birds of North America Online. Available: http://bna.birds.cornell.edu/bna/species/449. Accessed 2013 May 15 May.
23. Hawai'i Audubon Society (2005) Kawainui and Hamakua Marsh Complex added to List of Wetlands of International Importance under Ramsar Convention. 'Elepaio 65: 41–42.
24. Hatch G, Harpham P (1950) Hawaii Audubon Society Bird Count 1949. The Elepaio 10: 45–51.
25. Peterson HD (1951) Hawaii Audubon Society Bird Count 1950. The Elepaio 11: 42–45.
26. Kaigler CG (1970) Honolulu Christmas Count 1969. The Elepaio 30: 70–74.
27. Intergovernmental Panel on Climate Cahnge (2007) Contribution of Working Group I to the Fourth Assessment Report of the Intergovernmental Panel on Climate Change. Section 10.6.5: Projections of Global Average Sea Level Change for the 21st Century. Intergovernmental Panel on Climate Change (World Meteorological Organization/United Nations Environment Programme), Geneva, Switzerland. Available: http://www.ipcc.ch. Accessed 2012 Aug 8.
28. Vermeer M, Rahmstorf S (2009) Global sea level linked to global temperature. Proceedings of the National Academy of Sciences, 2009; DOI: 10.1073/pnas.0907765106.
29. Clough JS, Larson EC (2010) Application of the Sea-Level Affecting Marshes Model (SLAMM 6) to Kakahai'a NWR. Warren Pinnacle Consulting, Warren VT. 33 p.

30. Clough JS, Larson EC (2010) Application of the Sea-Level Affecting Marshes Model (SLAMM 6) to Kealia Pond NWR. Warren Pinnacle Consulting, Warren VT. 44 p.

31. Clough JS, Larson EC (2010) Application of the Sea-Level Affecting Marshes Model (SLAMM 6) to Pearl Harbor NWR. Warren Pinnacle Consulting, Warren VT. 121 p.

16

Growth Dynamics of the Threatened Caribbean Staghorn Coral *Acropora cervicornis*: Influence of Host Genotype, Symbiont Identity, Colony Size, and Environmental Setting

Diego Lirman[1]*, Stephanie Schopmeyer[1], Victor Galvan[2], Crawford Drury[1], Andrew C. Baker[1], Iliana B. Baums[3]

1 Rosenstiel School of Marine and Atmospheric Science, University of Miami, Miami, Florida, United States of America, **2** Punta Cana Ecological Foundation, Punta Cana, Dominican Republic, **3** Pennsylvania State University, State College, Pennsylvania, United States of America

Abstract

Background: The drastic decline in the abundance of Caribbean acroporid corals (*Acropora cervicornis*, *A. palmata*) has prompted the listing of this genus as threatened as well as the development of a regional propagation and restoration program. Using *in situ* underwater nurseries, we documented the influence of coral genotype and symbiont identity, colony size, and propagation method on the growth and branching patterns of staghorn corals in Florida and the Dominican Republic.

Methodology/Principal Findings: Individual tracking of> 1700 nursery-grown staghorn fragments and colonies from 37 distinct genotypes (identified using microsatellites) in Florida and the Dominican Republic revealed a significant positive relationship between size and growth, but a decreasing rate of productivity with increasing size. Pruning vigor (enhanced growth after fragmentation) was documented even in colonies that lost 95% of their coral tissue/skeleton, indicating that high productivity can be maintained within nurseries by sequentially fragmenting corals. A significant effect of coral genotype was documented for corals grown in a common-garden setting, with fast-growing genotypes growing up to an order of magnitude faster than slow-growing genotypes. Algal-symbiont identity established using qPCR techniques showed that clade A (likely *Symbiodinium* A3) was the dominant symbiont type for all coral genotypes, except for one coral genotype in the DR and two in Florida that were dominated by clade C, with A- and C-dominated genotypes having similar growth rates.

Conclusion/Significance: The threatened Caribbean staghorn coral is capable of extremely fast growth, with annual productivity rates exceeding 5 cm of new coral produced for every cm of existing coral. This species benefits from high fragment survivorship coupled by the pruning vigor experienced by the parent colonies after fragmentation. These life-history characteristics make *A. cervicornis* a successful candidate nursery species and provide optimism for the potential role that active propagation can play in the recovery of this keystone species.

Editor: Peter Alan Todd, National University of Singapore, United States of America

Funding: This publication was made possible with support from NOAA's Coral Reef Conservation Program, NOAA's American Recovery and Reinvestment Act, The Nature Conservancy, the US ARRA Program, Counterpart International, and the Frohring Foundation. The funders had no role in study design, data collection and analysis, decision to publish, or preparation of the manuscript.

Competing Interests: The authors have declared that no competing interests exist.

* Email: dlirman@rsmas.miami.edu

Introduction

One of the Caribbean's predominant reef-building coral genera, *Acropora*, has suffered significant population declines caused by multiple natural and anthropogenic stressors over the past three decades [1,2,3]. The loss of *Acropora* has resulted in decreases in reef function and structure as acroporid corals are critically important for reef growth, island formation, fisheries habitat, coastal buffering, and biodiversity [4]. The significant decline in abundance of this keystone taxon has prompted a number of conservation measures aimed at protecting remaining populations and accelerating its recovery trajectory. These efforts include: 1) the recent listing of both species of Caribbean *Acropora* as threatened under the US Endangered Species Act [5], and 2) the development of regional propagation and restoration programs [6].

The active restoration projects developed for the Caribbean staghorn coral (*A. cervicornis*) have provided an opportunity to explore the early growth of staghorn fragments and colonies at an unprecedented level of detail. In this study, we report on the growth dynamics of >1700 staghorn individuals representing 37 distinct genotypes from Florida and the Dominican Republic that were followed from the time of collection/fragment creation to the development of complex branching morphology. While the growth patterns of *Acropora palmata* and *A. cervicornis* were documented in the past, growth was only reported based on linear extension of a subset of branches within colonies [7,8,9], and no information on the role of coral genotype was previously examined, except for the recent study by Griffin et al. [10] that reported on the growth of six staghorn genotypes in Puerto Rico.

Coral growth is the ultimate indicator of coral health, integrating the influence of internal factors such as coral genotype and symbiont identity, physiology, and external environmental factors. However, growth is often difficult to estimate for corals with complex or irregular morphologies. Colony sizes, in such cases, are often approximated to regular geometric shapes [11] that are then used to estimate surface area and colony volume, metrics commonly related to habitat value [12]. While these approximations are adequate to document higher-level metrics like coral cover or colony size-frequencies, they are not adequate, due to measurement error, to quantify colony growth accurately, especially for branching species like staghorn corals.

Coral growth was measured here as Total Linear Extension (TLE), a painstaking procedure that quantifies all of the tissue/skeleton on each individual. Using this unique dataset, we explore the role of fragment size, coral genotype, and symbiont identity on growth. The data presented here not only help elucidate the factors that influence growth at early stages of the asexual propagation process but also provide important guidelines for the propagation of *Acropora* within nursery settings to maximize the productivity of this important and threatened Caribbean taxon.

Materials and Methods

Coral Growth and Productivity

Total Linear extension (TLE, [13]) was estimated for each individual using a flexible ruler at the time of deployment and again after 12 months. Measurements included all branches and were calculated to the closest cm (Fig. S1). In addition to TLE, the number of terminal branches (> 1 cm in length) was assessed. When the measurements for a specific batch of fragments deviated > 1 week from the scheduled interval due to unfavorable field conditions, TLE values were extrapolated linearly to calculate 12-month values. Change in TLE over time (growth) was estimated as the total amount of tissue/skeleton produced between measurement intervals. In addition to growth, annual productivity was estimated as the amount of coral produced relative to the tissue/skeleton present at the start of the study (annual productivity = growth/TLE at start of study). Only fragments that were alive for the 12-month period and did not undergo partial tissue mortality or fragmentation were included in the calculations. This 1-year period was selected to have a uniform growth interval to compare among the different nursery programs that conducted fragmentation and transplantation at different times of the year or in different years, and to be able to track fragments from the time they were created as "fingerlings" with a simple morphology of 1–2 branches, to the time when complex branching morphology developed, marking the transition from fragments to colonies [14,15]. The measurement of TLE becomes increasingly difficult as colonies develop multiple branches [11] and fragmentation becomes more frequent due to physical impacts of fishing gear and storms, affecting growth measurements.

Each coral genotype was represented by multiple fragments that were labeled with either a colored cable tie or with numeric tags and measured individually. Growth and productivity data were compared: 1) among coral genotypes; 2) between fragments growing in an underwater nursery and natural reefs; and 3) among fragments growing on different nursery propagation platforms (i.e., frames, floating ropes). The growth and productivity data were log-transformed ($\log_{10}(x+1)$) due to lack of normality of some of the fragment groups. The assumptions of normality and homoscedasticity were tested for each comparison and, when the transformed data were either not normal or heteroscedastic, non-parametric tests were used instead of parametric tests.

Previous work by Lirman et al. [16] showed that the removal of small fragments (3–5 cm) appears to stimulate the growth of donor branches on parent colonies (after a brief initial hiatus in growth as the skeletal lesions heal and new apical polyps develop). However, this conclusion was based on the removal of <10% of the parent tissue. In the present study, we evaluated the impacts of high levels of fragmentation (up to 95% of the tissue removed) to determine whether the vigor associated with minor fragmentation is still observed when larger portions of colonies are removed. This was achieved by comparing the growth of colonies before and after pruning.

Study Sites

Sites in Florida included two underwater nurseries (25.488° N, 80.109° W; 25.362° N, 80.166° W) as well as 5 reefs where nursery-grown corals were outplanted and monitored (Site 1: 25.465° N, 80.149° W; Site 2: 25.462° N, 80.142° W; Site 3: 25.361° N, 80.177° W; Site 4: 25.357° N, 80.169° W; Site 5: 25.365° N, 80.171° W). The nurseries were established on sandy substrate at a depth of 5–6 m. Fragments in the two Florida nurseries were attached to cinder blocks containing PVC pedestals using epoxy (Fig. 1A). The blocks (20×40×10 cm) contained 10 fragments of a single genotype each and were spaced at a minimum distance of 1 m between blocks. Nurseries in Florida contained 40–50 blocks and 400–500 fragments, occupying an area on the bottom of roughly 20 m×20 m. Fragments within blocks were separated by 8–10 cm at the start of the study. Nursery-grown fragments were outplanted to 5 Florida reefs (4–5 m) using masonry nails pounded onto the reef substrate and attached to the nails using plastic cable ties (Fig. 1B). Underwater epoxy was used to further cement the nails and the bases of fragments onto the reef. The reef sites were separated by a minimum distance of 700 m. At each site, replicates (10–15 from each genotype) from multiple genotypes collected from the nurseries were deployed in a regular grid containing 50–100 corals at a spacing of 50–100 cm between fragments, with each grid covering an area of 50–100 m^2. Replicates of each coral genotype were haphazardly located within each plot.

The staghorn nursery and 9 outplant sites at the Dominican Republic (DR) are located at Punta Cana (68.347° W, 18.539° N). The nursery is located on sandy substrate (4–5 m of depth) and covers an area of approximately 30 m×30 m. Fragments are grown on metal frames secured to the bottom or on floating ropes (Fig. 1C, D). Multiple genotypes and numerous replicate fragments of each genotype are grown on each frame and rope. The Punta Cana nursery contained 14 frames and 3 ropes, holding> 1300 fragments. Initially (when fragments are small), frames and ropes can hold up to 100 fragments. Initial spacing of fragments within frames and ropes is approximately 20 cm. Corals from the nursery were outplanted to 9 sites located along a continuous reef in the Punta Cana region (depth = 4–6 m). Sites were separated by a minimum of 200 m. At each site, corals were

Figure 1. Staghorn fragments propagated within *in situ* nurseries. A) Cinder-block platform used to propagate corals in Florida, B) coral outplant attached to a reef using a masonry nail, epoxy, and plastic ties, C) A-frame used to propagate corals in the Dominican Republic, D) rope nursery.

attached to the bottom using similar methods to those used in Florida [13]. Each site contained a single plot of 50–80 fragments but only a subset of those corals was monitored for growth (15–20 fragments per site). Corals were deployed in a regular grid at a spacing of 50–100 cm between fragments, with each grid covering an area of 50–100 m^2. Between 15–20 replicates of each coral genotype were deployed at each site and fragments were haphazardly located within each plot.

Activities in Florida were conducted under permit BISC-2010-SCI-0008 provided by Biscayne National Park and Special Activity License SAL-10-1086-SCRP provided by the Florida Fish and Wildlife Conservation Commission. All activities conducted with *Acropora*, a threatened species listed under the US Endangered Species Act (ESA), were approved by a Section 7 Consultation Biological Opinion issued by The National Oceanic and Atmospheric Administration (NOAA) and the National Marine Fisheries Service (NMFS). Activities in the Dominican Republic were conducted under a permit provided by Ministerio de Medio Ambiente y Recursos Naturales (Ministry of the Environment and Natural Resources).

Coral Genotype Identity

For genetic analyses, a small section (1–2 cm) from a branch tip was removed from each target colony and preserved in 95% ethanol. DNA was extracted from each sample using the Qiagen DNEasy 96 blood and tissue kit (Qiagen, CA). Microsatellite markers were those used by Baums et al. [17,18] developed originally for *Acropora palmata*. Here, we used markers 166, 181, 182 and 207. Marker 192 does not amplify in *A. cervicornis*. PCR conditions were as in Baums et al. [19]. Thermal cycling was performed in an MJ Research PT200 or an Eppendorf Mastercycler Gradient cycler. Alleles were fluorescently labeled and then visualized and sized with internal standards on a PRISM 3100 Genetic Analyzer (Applied Biosystems). Electropherograms were visualized and allele sizes were called using GENEMAPPER 4.0 (Applied Biosystems). The microsatellite markers are highly heterozygous (mean observed heterozygosity = 0.77), thus there is a low probability of identifying two colonies or cells as clonemates when in fact they are distinct genets (this is called the Probability of Identity (PI) and equals 10^{-7} for *A. cervicornis*. These values were estimated for a dataset containing 278 samples from throughout the Caribbean [19]. We identified clonemates in the Florida dataset by comparing all new multilocus genotypes against the existing *A. cervicornis* genotypes (generated in the same lab under identical conditions, Baums et al., 2009) and against each other. Those colonies that shared identical alleles at all loci were deemed to be clonemates. Matching calculations were performed using-GenAlEx vers 6.4 [20].

Algal Symbiont Identity

Algal symbiont communities in genetic samples were characterized by extracting DNA following established protocols [21] and using an actin-based quantitative PCR (qPCR) assay to detect *Symbiodinium* in clades A, B, C and D. Primers and probes that targeted each clade, as well as the coral host, were modified from Mieog et al. [22] following the approach of Cunning and Baker [23]. Each sample was run in duplicate, with positive and no-template controls. This assay allowed us to identify the dominant *Symbiodinium* in the study corals at the clade level, as well as detect any symbionts present at low abundance (<10%).

Results

Effects of Colony Size

A total of 1715 fragments were measured in this study. Fragments grown and outplanted in the DR were generally larger than those used in Florida so, for comparison, growth and productivity data were calculated for four size classes: tips (<5 cm TLE), small (5 – 15 cm TLE), medium (>15 – 30 cm TLE), and large (> 30 cm TLE) (Table 1). While growth increased with increasing fragment size in both nursery and reef settings, productivity decreased with increasing fragment size in both Florida and the DR (Table 1). Growth was significantly and positively related to TLE recorded at the beginning of the growth interval across a wide range of initial sizes (1–175 cm TLE) (linear regression, $p<0.05$, $R^2 = 0.37$, $n = 1715$) (Fig. 2A). Moreover, a significant positive linear relationship was documented between TLE and number of branches after 12 months of growth (linear regression, $p<0.05$, $R^2 = 0.16$). Productivity declined exponentially with increasing initial colony size, with maximum annual productivity recorded for fragments <15 cm in TLE (non-linear regression, $p<0.05$, $R^2 = 0.34$) (Fig. 2B).

Effects of Coral Genotype and Algal Symbiont Identity

A total of 37 distinct coral genotypes were identified, 24 from Florida and 13 from the DR. The growth of individual coral genotypes was documented in nurseries in Florida (6–41 fragments per genotype) and the DR (7–31 fragments per genotype) over 1 year. Annual productivity values ranged from 0.9–9.8 cm of new coral produced by every cm of coral present at the start of the study for the 24 genotypes examined in Florida (Fig. 3A). In the DR, the range of annual productivity values was 2.2–16.2 cm for the genotypes examined (Fig. 3B). Significant differences in productivity were detected among genotypes in Florida and in the DR (Mann Whitney U test, $p<0.05$), but pair-wise comparisons revealed that only the fastest (ELKHORN in Florida, and B in the DR) and slowest genotypes (ACE 6 and STEPH in Florida, and W/B and W/Y in the DR) had significantly different productivity values (Tukey-Kramer test, $p<0.05$).

In both Florida and the DR, qPCR analysis revealed that *Symbiodinium* in clade A (likely *Symbiodinium* A3) was the dominant symbiont for all coral genotypes, except for one coral genotype in the DR (out of 13) and two coral genotypes in Florida (out of 24) that were dominated by clade C (Fig. 3A). In Florida, there were no significant differences in growth or productivity between corals dominated by *Symbiodinium* in clade C (2 coral genotypes) and corals dominated by clade A symbionts (22 coral genotypes) (t test, $p>0.05$ for all metrics). The coral genotype from the DR dominated by clade C was represented by only 3 surviving fragments and thus not included in the analyses. In the DR, qPCR analysis found that *Symbiodinium* in clades B and C were also commonly found at low abundance in colonies dominated by clade A, but clade D was never detected in any of these samples. In Florida, 8 out of 17 genotypes that were dominated by clade A also contained clade C at low abundance, with one genotype also containing clade D (likely D1a). Of the 2 genotypes that were dominated by clade C, one also contained low-abundance A and D, while the other contained low-abundance A and B.

Effects of Propagation Platform

A variety of propagation platforms are used in nursery programs to grow staghorn coral [13]. The platforms are classified into two types, fixed to the bottom (Fig. 1A, C) or floating (Fig. 1D). Here, we evaluate differences in growth metrics for small and medium fragments (the size categories represented in both ropes and frames) grown on wire frames ($n = 259$) and floating ropes ($n = 27$) in the DR. The same genotypes were grown on both platforms. Corals grown suspended on ropes had significantly higher annual productivity (6.6 cm (S.D. = 3.7) and number of branches per colony (5.7 (S.D. = 4.5)) than corals grown on frames fixed to the bottom (4.8 cm (S.D. = 5.1); 4.4 branches (S.D. = 4.7)) (t test, $p<0.05$ for each metric).

Effects of Habitat

Nursery-grown corals were outplanted to natural reefs in Florida, providing the opportunity to compare growth metrics

Table 1. Annual growth and productivity (± S.D.) of staghorn fragments in Florida and the DR.

Location	Type	Size Category	N fragments	Growth (cm per year)	Annual Productivity (growth/cm of initial coral)
DR	Nursery	Tips	*	*	*
DR	Nursery	S	130	59.3 (68.9)	5.6 (6.4)
DR	Nursery	M	156	97.2 (80.3)	4.5 (3.6)
DR	Nursery	L	94	173.1 (136.3)	3.3 (2.4)
DR	Reef	Tips	5	20.1 (15.9)	4.2 (3.1)
DR	Reef	S	72	29.7 (26.9)	3.0 (2.5)
DR	Reef	M	46	49.6 (43.2)	2.4 (1.9)
DR	Reef	L	*	*	*
Florida	Nursery	Tips	446	23.7 (16.1)	7.0 (5.1)
Florida	Nursery	S	140	48.5 (32.2)	6.0 (3.6)
Florida	Nursery	M	10	89.3 (29.5)	4.6 (1.4)
Florida	Nursery	L	*	*	*
Florida	Reef	Tips	75	28.1 (31.6)	5.9 (6.5)
Florida	Reef	S	349	36.5 (28.1)	4.2 (3.3)
Florida	Reef	M	25	71.1 (45.9)	4.0 (2.7)
Florida	Reef	L	*	*	*

* = <5 fragments of that size class were used.
Tips = <5 cm TLE, S = 5–15 cm TLE, M => 15 – 30 cm TLE, L => 30 cm TLE.

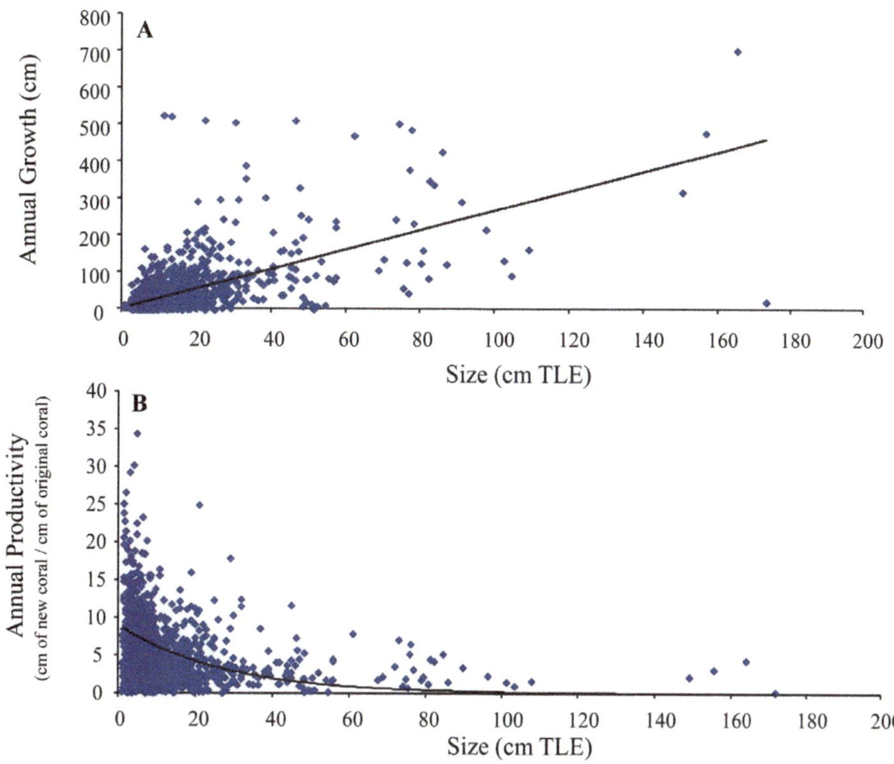

Figure 2. Growth and productivity of staghorn fragments. Annual growth (A) and annual productivity (B) of staghorn fragments.

between the nursery (Fig. 1A) and reefs (Fig. 1B). For fragments within the "small" category (the size category well represented in both the nursery and reef) (n = 140 nursery, n = 349 reef outplants),corals grown on blocks within the nursery had significantly higher annual productivity (6.0 cm (S.D. = 3.6)) than corals transplanted to reefs (4.2 cm (S.D. = 3.3)) (t test, $p<0.05$).

Effects of Pruning

The growth of fragments from 19 coral genotypes (n = 5–10 fragments per genotype) within the Florida nurseries was assessed for 1 year after collection (n = 402 fragments). After the first year of growth, the resulting colonies were pruned to produce a second generation of fragments that were used for outplanting activities. At this time, 40–95% of each colony was trimmed. The growth of the pruned colonies (n = 164 fragments) was then followed for an additional year to document the impacts of pruning on growth metrics (Fig. 4A). The average coral size at collection was 4.8 cm (S.D. = 2.0) compared to a lower average size immediately after pruning (3.2 cm (S.D. = 1.6)). The average number of branches per colony at collection and immediately after pruning was the same, 1.3 branches per colony (S.D. = 0.7). The mean size of colonies was 24.1 cm (S.D. = 17.5) 12 months after collection compared to more than double for the same colonies 12 months after pruning (58.4 cm (S.D. = 28.1)). Similarly, the colonies had 5.2 branches per colony (S.D. = 3.2) 12 months after initial collection compared to nearly double for the same colonies 12 months after pruning (9.5 branches per colony (S.D. = 3.5)). The yearly growth of colonies after collection was 21.9 cm yr^{-1} (S.D. = 15.6) compared to the yearly growth of the same colonies after pruning (55.6 cm yr^{-1} (S.D. = 27.0)). The annual productivity of colonies after initial collection was 4.2 cm (S.D. = 1.8) compared to the annual productivity of the same colonies after pruning (20.3 cm (S.D. = 8.8)). All of the growth metrics measured (i.e., growth, annual productivity, number of branches) were significantly higher for colonies after pruning (all genotypes grouped together, t test, $p<0.05$ for all metrics) (Fig. 4B). When the genotypes were examined individually, annual growth was significantly faster after pruning for 16 of 19 genotypes (Mann Whitney U test, $p<0.05$), while annual productivity was significantly greater after pruning for all 19 genotypes (Mann Whitney U test, $p<0.05$). The level of pruning (i.e., % of coral removed) was significantly and positively related to productivity (linear regression, $p<0.05$, $R^2=0.17$).

Discussion

Acropora cervicornis is an extremely fast-growing branching coral with annual productivity rates that often exceed 5 cm of new coral produced for every cm of existing coral in both Florida and the Dominican Republic. In addition to having prolific growth, this species benefits from high fragment survivorship coupled by the pruning vigor experienced by the parent colonies after fragmentation [16]. Here, we demonstrate that not only is the growth of donor colonies enhanced by pruning but that colonies can lose up to 95% of their tissue and skeleton through fragmentation and still have enhanced growth and recovery. The fast regrowth of damaged colonies after severe fragmentation has also been shown on wild populations of *Acropora palmata* after significant hurricane damage [24]. These life-history characteristics make acroporid corals highly successful nursery species and provide optimism for the potential role that active propagation can play in the their recovery.

Establishing clear relationships between coral colony size and growth has remained elusive due to the difficulties of measuring size and growth accurately, especially in natural conditions. The

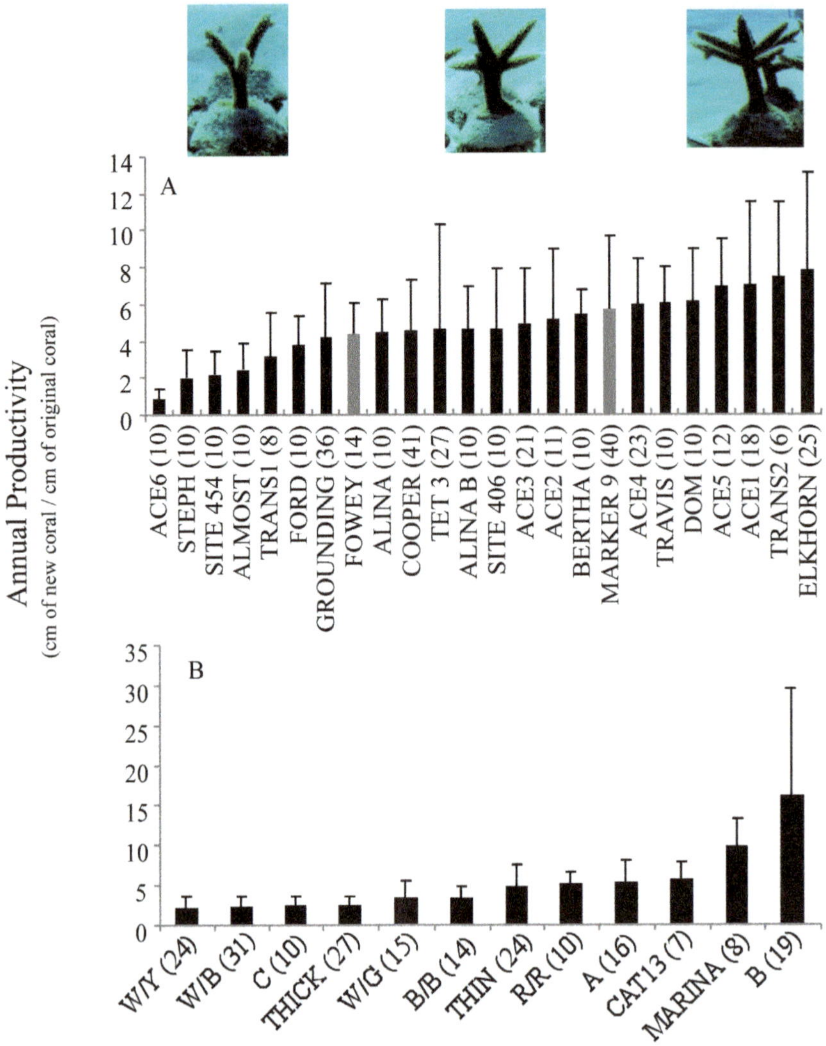

Figure 3. Productivity of staghorn fragments by coral genotype. Annual productivity of staghorn genotypes from Florida (A) and the Dominican Republic (B). Numbers in parentheses are sample sizes for each genotype. Grey bars in (A) are the two genotypes dominated by clade C algal symbionts. All other genotypes, including those in the DR were dominated by clade A. The coral genotype from the DR dominated by clade C was represented by only 3 surviving fragments and thus not included in the analyses. The images are representative of slow- (left), medium- (center), and fast-growing (right) genotypes after 1 year in the nursery. Pair-wise comparisons revealed that only ELKHORN and ACE 6/STEPH had significantly different productivity values in Florida, while only B and W/B/W/Y had significantly different productivity values in the DR (Tukey-Kramer test, $p < 0.05$).

majority of the information published on this key relationship is based on studies that examine nubbins or colonies of a limited size range [25], or estimate colony growth based on: 1) extension rates measured from skeletal cores [26] or a subset of marked branches [27]; 2) 2-dimensional estimation of colony surface area based on field measurements or photographs [28,29]; 3) 3-dimensional models [30,31]; or 4) buoyant weights [32]. This is one of the first studies to report genotype-specific, whole-colony growth for a large number of colonies of a wide range of colony sizes, expanding on the patterns reported by Griffin et al. [10] based on six staghorn genotypes.

In this study, we document a significant positive linear relationship between colony size (TLE) and growth and branch development that is not surprising considering that staghorn corals grow through the creation of new branches and subsequent linear extension from apical ends [8]. Clark and Edwards [33] also showed a significant positive linear relationship between colony size and growth for three species of Pacific branching *Acropora* spp. However, this positive relationship becomes a decreasing exponential function when size and productivity are related for *A. cervicornis*. While more coral is being produced by bigger colonies, the rate at which new coral is produced normalized by the amount of existing coral declines as colonies grow, with maximum productivity rates recorded for small fragments and colonies. A similar pattern, but documented over a very narrow size range (1–4 cm TLE), was observed for *Acropora pulchra* by Soong and Chen [34], where extension rates increased with size but productivity of larger fragments was lower than that of smaller ones. Similarly, reductions in net growth with increasing colony size were documented by Hughes and Connell [35] and Yap et al. [36]. In corals, the resources allocated initially to branch extension may be shifted later to support radial growth (our observations indicate that colonies get thicker at the base as they grow), reproduction, and recovery from fragmentation, thus resulting in

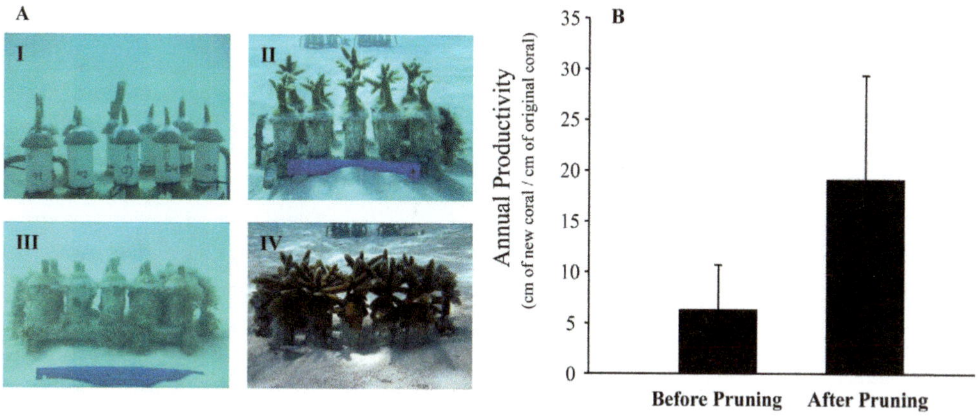

Figure 4. Effects of pruning on nursery corals. A) I: Fragments after initial collection, II: fragments 8 months after initial collection, III: fragments immediately after pruning (12 months after initial collection), IV: fragments 8 months after pruning. B) Annual productivity of fragments before pruning (after initial collection) and after pruning. Significant difference in productivity were found between groups (t test, $p<0.05$).

reduced productivity. An additional factor contributing to the reduction in productivity for this branching coral may be changes in branching patterns as colonies grow larger. If branching rates decrease over time due to physical limitation or within-colony crowding, linear growth rates would not be maintained. Within a coral propagation framework, results from the present study suggest that corals should be trimmed frequently to sustain productivity. While the optimal size to sustain high productivity will vary by species, the average annual productivity of staghorn colonies kept in nurseries in Florida and the DR dropped from> 5 cm for colonies <30 cm TLE to <3.2 cm for larger colonies. Lastly, considering the significant relationships demonstrated between colony size and growth and productivity in branching corals like *A. cervicornis*, caution should be exercised when comparing these metrics among treatments or even among different studies where initial fragment or colony size may be different.

The growth of *A. cervicornis* is influenced by the environmental setting where colonies are maintained. Coral nurseries are, as desired, locations of high productivity when compared to reef habitats where the productivity of transplanted, nursery-grown staghorn corals was, on average, 10% lower. Differences in productivity may be explained in part by considering that extensive resources are spent within nurseries to remove, manually, sediments deposited on colonies as well as coral predators and competitors like macroalgae and sponges [13]. Growth can also be influenced by propagation platform within the same habitat. Growing fragments on floating nursery structures that allow corals to develop branches in all directions was shown here to enhance branching and productivity and should be included by nursery managers as part of comprehensive propagation protocols. Indeed, some of the fastest growth rates reported for nursery-grown staghorn fragments are from floating nursery platforms in Puerto Rico where small fragments (4.4 cm TLE) grew an average of 52.5 cm in one year [10]. While the potential factors resulting in enhanced growth of corals suspended on ropes were not tested explicitly here, increased water motion has been shown to increase branching in *A. cervicornis* [37,38] and growth in *Pocillopora damicornis* and *P. meandrina* [32].

The unique framework provided by our propagation program, based on the collection of coral genotypes from a wide range of "home" habitats and subsequent growth in nursery environments, allowed us to explore the interactive role of holobiont genotype and environment on staghorn coral growth [39]. Within common gardens (nurseries), corals exhibited a wide range of productivity values, and significant differences in growth between the fastest and slowest-growing genotypes. Genotypic-based variation in coral growth within the same environment has been documented previously for *Favia* and *Diploastrea* by Todd et al. [39], *Porites* by Forsman et al. [25], and *Stylophora* by Rinkevich [40] and Shaish et al. [41]. In contrast, no differences in growth were documented for distinct genotypes of *Madracis* by Bruno and Edmunds [42].

Acropora cervicornis is known to host a variety of algal symbionts in *Symbiodinium* clades A, C and D, with dominance varying depending on environment [21,43,44,45]. The dominant symbiont in most shallow environments sampled to date appears to be A3 [43,44], but as our data show, members of other clades can occasionally dominate in shallow water (see also [45]). Our study found C-dominated *A. cervicornis* to be uncommon (3 of 37 genotypes, 8%) in shallow reefs of Florida and the DR, but these symbionts are likely to be much more common in deep water populations [43,44] which, to date, have been little studied. It is also possible that symbionts are more temporally and spatially variable than our results suggest. Additional within-colony sampling and long-term repeated sampling may reveal these associations to be more dynamic than our results suggest (but see [44]).

Corals containing different *Symbiodinium* can vary in their thermotolerance, especially corals dominated by certain *Symbiodinium* in clade D, including D1a in the Caribbean [46,47,48]. However, in our study, no colonies were found to be dominated by clade D *Symbiodinium* in Florida and no presence of D was found in colonies from the DR (assuming our limited sampling is representative of the symbiont communities found in each colony), and it is therefore unlikely that these symbionts would contribute significantly to the potential bleaching resistance of their hosts. However, corals with background, low abundance of various symbionts might recover more quickly from bleaching (i.e., be more resilient), especially if thermotolerant symbionts are present in their tissue. In this context, the presence of low-abundance clade D symbionts in Florida corals and the absence of D symbionts in DR corals might reflect differences in bleaching resilience, thermal regime, or disturbance (bleaching) history between these two populations. Monitoring the fate of symbionts during bleaching and recovery would help resolve the functional significance, if any, of background symbionts, which appear to be common in scleractinian corals [49]. C-dominated colonies might be expected to perform better under lower light conditions, compared with their A-dominated counterparts, but since these colonies co-existed at our sampling sites it is possible that, at the specific depths of our study, different *Symbiodinium* were equally beneficial to their coral hosts. Consequently, it may not be surprising to find that A- and C-dominated colonies have similar growth rates under these conditions, especially given our unbalanced statistical comparison (2 C-dominated genotypes vs. 22 A-dominated genotypes). A better test of physiological differences between these symbionts might be provided by transplanting colonies to deeper or shallower depths, or exposing them to combined heat and light stress to induce bleaching.

The realization that different coral genotypes can exhibit significantly different growth patterns within the same exact environment highlights the pressing need to incorporate, explicitly, genotype tracking into studies evaluating coral growth responses to environmental and experimental conditions. Without proper tracking, genotype influences on coral growth may mask or amplify a mean response if not taken into account in the sampling or experimental design. Moreover, with the possibility that colonies within patches or reefs can be clones produced by asexual propagation (prevalent in branching corals; [50]), the unbalanced inclusion of clone-mates in the estimation and comparison of growth patterns may raise issues related to lack of independence among samples and influence the statistical power to detect differences.

Our study is the first to document, in detail, the patterns of algal symbiont identity, growth, and productivity of asexually produced fragments of a large number of genotypes of the threatened Caribbean staghorn coral *Acropora cervicornis*. We show that even small fragments of this species are capable of extremely fast growth, but that productivity of developing colonies decline with colony size, indicating that regular pruning is required to maximize nursery productivity. In addition, large variability in growth was observed among coral genotypes maintained within a common environment, with an order-of-magnitude difference in growth between slow- and fast-growers. These findings have important implications for the propagation of corals and for the restoration of damaged reefs. While the identification of fast-growing coral genotypes may prompt practitioners to concentrate efforts on these "winners" to minimize time to ecosystem recovery, restoring genetic and genotypic diversity should also be considered as a key goal of restoration plans [51]. Moreover, further research is needed to fully understand the relationship between coral growth, algal symbiont identity, and resistance to stressors such as temperature anomalies and acidification. To date only a few coral genotypes have been tracked at both the nursery and outplanting phase of reef restoration and there still remains the possibility that coral growth by genotype is modulated by the environment and that "losers" within a nursery setting may become, over time, "winners" under different conditions experienced on reefs [39]. While documenting the reaction norm for corals remains a challenge, the expansion of *in situ* propagation programs [52] should be able to provide researchers with the ramets and genets required to conduct the challenging, large-scale transplantation experiments required to address this fundamental aspect of coral ecology and provide scientific support for the active recovery of this and other threatened species.

Supporting Information

Figure S1 Field measurements and growth of staghorn fragment. Photographs of a staghorn fragment taken after deployment at the Florida nursery (left image) and after 6 months of growth (right image). The yellow segments show the measurements taken. The sum of all of the segments represents the Total Linear Extension (TLE) of the fragment. The fragment on the left has a TLE of 12 cm and 2 branches, the fragment on the right has a TLE of 32 cm and 8 branches. The annual growth of this fragment was calculated at 40 cm per year. The annual productivity was calculated at 3.3 cm of new coral/cm of initial coral. The images were taken by J. Herlan. The figure was composed based on the guidelines for the estimation of TLE proposed by Johnson et al. [13].

Data S1 Size and number of branches of the staghorn corals included in this study.

Acknowledgments

We thank the many divers in Florida and the DR who helped collect coral growth data, especially J. Herlan, T. Thyberg, R. Santos, C. Vilmar, B. Huntington, C. Hill, and J. Cole. We are indebted to R. Silverstein for completing the symbiont identifications and X. Serrano for completing the identification of the coral genotypes from the DR, with help from other members of the Baker Lab. We also thank our collaborators J. Byrne, C Lustic, and M. Johnson (TNC), J. Kheel (PCEF), and A. Bowden-Kerby. Our manuscript was improved by insightful comments and suggestions made by the Academic Editor and two anonymous reviewers.

Author Contributions

Conceived and designed the experiments: DL SS VG. Performed the experiments: DL SS VG CD. Analyzed the data: DL SS AB IB. Wrote the paper: DL SS IB AB.

References

1. Jaap WC, Halas JC, Muller RG (1988) Community dynamics of stony corals (Milleporina and Scleractinia) at Key Largo National Marine Sanctuary, Florida, during 1981–1986. Proc 6th Int Coral Reef Symp 2: 237–243.
2. Porter JW, Meier OW (1992) Quantification of loss and change in Floridian reef coral populations. Am Zool 32: 625–640.
3. Precht W, Bruckner A, Aronson R, Bruckner R (2002) Endangered acroporid corals of the Caribbean. Coral Reefs 21: 41–42.
4. Bruckner AW (2002) *Acropora* workshop: Potential application of the US endangered species act as a conservation strategy. NOAA Technical Memorandum NMFS-OPR.
5. National Marine Fisheries Service (2006) Endangered and threatened species: final listing determinations for elkhorn and staghorn coral. Federal Registry 71: 26852–26861.
6. Young C, Schopmeyer S, Lirman D (2012) A review of reef restoration and coral propagation using the threatened genus *Acropora* in the Caribbean and western Atlantic. Bull Mar Sci 88: 1075–1098.
7. Shinn EA (1966) Coral growth-rate, an environmental indicator. J Paleontol 40: 233–240.
8. Gladfelter WB (1982) White-band disease in *Acropora palmata*: Implications for the structure and growth of shallow reefs. Bull Mar Sci 32: 639–643.
9. Bak RP, Nieuwland G, Meesters EH (2009) Coral growth rates revisited after 31 years: What is causing lower extension rates in *Acropora palmata*? Bull Mar Sci 84: 287–294.
10. Griffin S, Spathias H, Moore T, Baums I, Griffin B (2012) Scaling up *Acropora* nurseries in the Caribbean and improving techniques. Proc 12th Int Coral Reef Symp 20A.
11. Huntington BE, Miller MW (2013) Location-specific metrics for rapidly estimating the abundance and condition of the threatened coral *Acropora cervicornis*. Restor Ecol doi:10.1111/rec.12057.
12. Fisher WS, Fore LS, Hutchins A, Quarles RL, Campbell JG, et al. (2008) Evaluation of stony coral indicators for coral reef management. Mar Pollut Bull 56: 1737–1745.
13. Johnson ME, Lustic C, Bartels E, Baums IB, Gilliam DS, et al. (2011) Caribbean *Acropora* restoration guide: Best practices for propagation and population enhancement. The Nature Conservancy, Arlington, VA.

14. Lirman D (2003) A simulation model of the population dynamics of the branching coral *Acropora palmata*: Effects of storm intensity and frequency. Ecol Model 161: 167–180.
15. Herlan J, Lirman D (2008) Development of a coral nursery program for the threatened coral *Acropora cervicornis* in Florida. Proc 11th Int Coral Reef Symp 24: 1244–1247.
16. Lirman D, Thyberg T, Herlan J, Hill C, Young-Lahiff C, et al. (2010) Propagation of the threatened staghorn coral *Acropora cervicornis*: Methods to minimize the impacts of fragment collection and maximize production. Coral Reefs 29: 729–735.
17. Baums IB, Miller MW, Hellberg ME (2005a) Regionally isolated populations of an imperiled Caribbean coral, *Acropora palmata*. Mol Ecol 14: 1377–1390.
18. Baums IB, Devlin-Durante MK, Brown L, Pinzón JH (2009) Nine novel, polymorphic microsatellite markers for the study of threatened Caribbean acroporid corals. Mol Ecol Res 9: 1155–1158.
19. Baums IB, Hughes CR, Hellberg MH (2005b) Mendelian microsatellite loci for the Caribbean coral *Acropora palmata*. Mar Ecol Prog Ser 288: 115–127.
20. Peakall R, Smouse PE (2006) GENALEX 6: genetic analysis in Excel. Population genetic software for teaching and research. Mol Ecol Notes 6:288–295.
21. Baker AC, Rowan R, Knowlton N (1997) Symbiosis ecology of two Caribbean acroporid corals. Proc 8th Int Coral Reef Symp 2: 1295–1300.
22. Mieog JC, van Oppen MJH, Berkelmans R, Stam WT, Olsen JL (2009) Quantification of algal endosymbionts (*Symbiodinium*) in coral tissue using real-time PCR. Mol Ecol Resour 9: 74–82.
23. Cunning R, Baker AC (2013) Excess algal symbionts increase the susceptibility of reef corals to bleaching. Nature Climate Change 3: 259–262.
24. Lirman D (2000) Fragmentation in the branching coral *Acropora palmata* (Lamarck): growth, survivorship, and reproduction of colonies and fragments. J. Exp. Mar. Biol. Ecol. 251: 41–57.
25. Forsman ZH, Rinkevich B, Hunter CL (2006) Investigating fragment size for culturing reef-building corals (*Porites lobata* and *P. compressa*) in ex situ nurseries. Aquaculture 261: 89–97.
26. Hudson JH, Shinn EA, Halley RB, Lidy B (1976) Sclerochronology: A tool for interpreting past environments. Geology 1: 361–364.
27. Shinn EA (1976) Coral reef recovery in Florida and the Persian Gulf. Geo 1: 241–254.
28. Hughes TP (1984) Population dynamics based on individual size rather than age: a general model with a reef coral example. Am Nat 123:778–795.
29. Lirman D, Fong P (2007) Is proximity to land-based sources of coral stressors an appropriate measure of risk to coral reefs? An example from the Florida Reef Tract. Mar Pollut Bull 54: 779–791.
30. Courtney LA, Fisher WS, Raimondo S, Oliver LM, Davis WP (2007) Estimating 3-dimensional colony surface area of field corals. J Exp Mar Biol Ecol 351:234–242.
31. Enochs I, Manzello D, Carlton R, Schopmeyer S, van Hooidonk R, et al. (2014) Effects of light and elevated pCO2 on the growth and photochemical efficiency of *Acropora cervicornis*. Coral Reefs 33: 477–485.
32. Jokiel PL (1978) Effects of water motion on reef corals. J Exp Mar Biol Ecol 35: 87–97.
33. Clark S, Edwards AJ (1995) Coral transplantation as an aid to reef rehabilitation: Evaluation of a case study in the Maldive Islands. Coral Reefs 14: 201–213.
34. Soong K, Chen T (2003) Coral transplantation: Regeneration and growth of *Acropora* fragments in a nursery. Restor Ecol 11: 62–71.
35. Hughes TP, Connell JH (1987) Population dynamics based on size or age? A reef-coral analysis. Am Nat 129: 818–829.
36. Yap HT, Alvarez RM, Custodio HM III, Dizon RM (1998) Physiological and ecological aspects of coral transplantation. J Exp Mar Biol Ecol 229: 69–84.
37. Bottjer DJ (1980) Branching morphology of the reef coral *Acropora cervicornis* in different hydraulic regimes. J Paleontol 54: 1102–1107.
38. Becker LC, Mueller E (2001) The culture, transplantation and storage of *Montastraea faveolata*, *Acropora cervicornis* and *Acropora palmata*: What we have learned so far. Bull Mar Sci 69: 881–896.
39. Todd P, Ladle R, Lewin-Koh N, Chou L (2004) Genotype x environment interactions in transplanted clones of the massive corals *Favia speciosa* and *Diploastrea heliopora*. Mar Ecol Prog Ser 271: 167–182.
40. Rinkevich B (2000) Steps towards the evaluation of coral reef restoration by using small branch fragments. Mar Biol 136: 807–812.
41. Shaish L, Levy G, Gomez E, Rinkevich B (2008) Fixed and suspended coral nurseries in the Philippines: Establishing the first step in the "gardening concept" of reef restoration. J Exp Mar Biol Ecol 358: 86–97.
42. Bruno JF, Edmunds PJ (1997) Clonal variation for phenotypic plasticity in the coral *Madracis mirabilis*. Ecology 78: 2177–2190.
43. LaJeunesse T (2002) Diversity and community structure of symbiotic dinoflagellates from Caribbean coral reefs. Mar Biol 141: 387–400.
44. Thornhill DJ, LaJeunesse TC, Kemp DW, Fitt WK, Schmidt GW (2006) Multi-year, seasonal genotypic surveys of coral-algal symbioses reveal prevalent stability or post-bleaching reversion. Mar Biol 148: 711–722.
45. Bowden-Kerby A, Carne L (2012) Thermal tolerance as a factor in Caribbean *Acropora* restoration. Proc 12th Inter Coral Reef Symp 20A.
46. Glynn PW, Maté JL, Baker AC, Calderón MO (2001) Coral bleaching and mortality in Panama and Ecuador during the 1997–1998 El Niño-Southern Oscillation event: spatial/temporal patterns and comparisons with the 1982–1983 event. Bull Mar Sci 69: 79–109.
47. Berkelmans R, van Oppen MJ (2006) The role of zooxanthellae in the thermal tolerance of corals: A 'nugget of hope' for coral reefs in an era of climate change. Proc Biol Sci 273: 2305–2312.
48. LaJeunesse TC, Smith RT, Finney J, Oxenford H (2009) Outbreak and persistence of opportunistic symbiotic dinoflagellates during the 2005 Caribbean mass coral 'bleaching' event. Proc Biol Sci 276: 4139–4148.
49. Silverstein RN, Correa AM, Baker AC (2012) Specificity is rarely absolute in coral-algal symbiosis: Implications for coral response to climate change. Proc Biol Sci 279: 2609–2618.
50. Baums IB, Miller MW, Hellberg ME (2006) Geographic variation in clonal structure in a reef-building Caribbean coral, *Acropora palmata*. Ecol Monogr 76: 503–519.
51. Baums IB (2008) A restoration genetics guide for coral reef conservation. Mol Ecol 17: 2796–2811.
52. Rinkevich B (2014) Rebuilding coral reefs: Does active reef restoration lead to sustainable reefs? Curr Opin Environ Sustain 7: 28–36.

17

A Longitudinal Study of Steller Sea Lion Natality Rates in the Gulf of Alaska with Comparisons to Census Data

John M. Maniscalco[1,2]*, Alan M. Springer[2], Pamela Parker[1], Milo D. Adkison[2]

1 Department of Science, Alaska SeaLife Center, Seward, Alaska, United States of America, **2** Institute of Marine Science, University of Alaska Fairbanks, Fairbanks, Alaska, United States of America

Abstract

Steller sea lion (*Eumetopias jubatus*) numbers in the Western Distinct Population Segment are beginning to recover following the dramatic decline that began in the 1970s and ended around the turn of the century. Low female reproductive rates (natality) may have contributed to the decline and remain an issue of concern for this population. During the 2000s we found high natality among Steller sea lions in the Gulf of Alaska indicating a healthy population. This study extends these previous estimates over an additional three years and tests for interannual variations and long-term trends. We further examine the proportions of pups to adult females observed on the rookery and nearby haulouts during the birthing season to assess whether census data can be used to estimate natality. Open robust design multistate models were built and tested using Program MARK to estimate survival, resighting, and state transition probabilities in addition to other parameters dependent on whether or not a female gave birth in the previous year. Natality was estimated at 70% with some evidence of interannual variation but a long-term increasing or decreasing trend was not supported by the data. Bootstrap and regression comparisons of census data with natality estimates revealed no correlation between the two methods suggesting that census data are not an appropriate proxy for natality in this species. Longitudinal studies of individual animals are an appropriate method for estimating vital rates in species with variable detection over time such as the Steller sea lion. This work indicates that natality remains high in this region and is consistent with a population in recovery.

Editor: Daniel E. Crocker, Sonoma State University, United States of America

Funding: This study was funded by U.S. federal grant appropriations through the National Marine Fisheries Service - NOAA Award NA09NMF4390169 (https://grantsonline.rdc.noaa.gov). The funders had no role in study design, data collection and analysis, decision to publish, or preparation of the manuscript.

Competing Interests: The authors have declared that no competing interests exist.

* Email: johnm@alaskasealife.org

Introduction

Over the past 2 centuries, several fur seal and sea lion (Otariidae) populations have recovered to healthy numbers following catastrophic collapses [1]. Steller sea lions (*Eumetopias jubatus*) in Alaska, USA may be emulating this trend as numbers over a large portion of the western distinct population segment (WDPS) have begun to increase in recent years following a precipitous 30-year decline that started in the early 1970s [2,3]. Indeed, Steller sea lion populations may have undergone dramatic fluctuations in abundance several times during the past 4 millennia [4]. The most recent decline is primarily attributed to reduced juvenile survival during the 1980s in addition to minor reductions in reproductive rates and adult survival [5–7]. Changes in those vital rates could have resulted from nutritional stress, killer whale (*Orcinus orca*) predation, or a combination of these and other factors such as direct interactions with fisheries [8–11]. The latest upswing in Steller sea lion numbers was estimated to have begun in the early 2000s [12] and may be attributed to improved juvenile survival [13] and high natality rates that are reflective of a healthy population [14]. However, some work based on population demography and theoretic modeling has suggested natality reached an all-time low during the past decade [15].

Continued monitoring of vital rates is essential for species of concern such as the WDPS of Steller sea lions, which is classified as endangered under the Endangered Species Act. Of the two approaches to obtaining vital rate information, longitudinal, individual-based studies have many advantages over cross-sectional, population-level studies [16]. Yet, published estimates of vital rates based on longitudinal studies are rare for the WDPS of Steller sea lions even though hundreds of millions of dollars have been dedicated to researching these animals. Here we provide an update of our previous work [14] on natality rates for these animals in the Gulf of Alaska based on a longitudinal study of adult females and assess the potential for using census data (counts) to predict trends in natality over time.

Natality is defined as the birthrate or the proportion of births to some segment of the general population – generally mature females. Female Steller sea lions become reproductively mature at three to seven years of age and give birth to one pup per year but not necessarily every year [17]. The birthing season typically ranges between late May and early July in Alaska [18]. Twinning is extremely rare [19] and juveniles are often weaned during the spring months at age 1, age 2 or even older [13,20]. However, natality does not include neonatal or juvenile survival, and it is important to differentiate these life history variables because of the widely different factors that may affect them. Low rates of natality among pinnipeds have been related to disease, contaminants, and poor nutrition [6,21]; whereas neonatal or juvenile survival may be more influenced by weather conditions, predation, or food limitation [22–24]. By lumping life history stages together, information pertinent to understanding population dynamics can be hidden or lost entirely.

We previously outlined several reasons why it may be difficult to estimate natality based on proportions of pups on rookeries and haulouts late in the breeding season [14]. Primarily, these are uncertainties about the extent of neonatal mortality due to storm waves for example, and proportions of females foraging at sea which may change systematically with shifting oceanic regimes [14]. Assuming stability in these variables over time might lead to erroneous conclusions about the proportion of pups that are born to a population of sea lions. We further suggested that disparities in natality estimation techniques may be resolved by long-term comparisons of census data with direct estimates of natality based on longitudinal observations of adult females and mark-recapture modeling [14]. In this study, we added 3 additional years to our previous time series and extended our natality estimates over a 10-yr period, 2003–2012, using robust-design mark-recapture statistics. We further compared these results to census counts of the proportion of pups to adult females on rookeries and haulouts to determine if census data can be used to predict trends in natality. If proportions of pups to adult females, or non-pups, can be used to predict natality, we should expect positive correlations between these proportions and mark-recapture estimates of natality over time.

Materials and Methods

Ethics Statement

This research was conducted in accordance with Alaska SeaLife Center Institutional Animal Care and Use Committee Protocol No. R10-03-01 and National Marine Fisheries Service Permit No. 14324 for research on endangered Steller sea lions. The Chiswell Island group is part of the U. S. Fish and Wildlife Service National Maritime National Wildlife Refuge. Research was conducted on refuge lands under Right-of-Way Permit No. M-344-AM and Special Use Permit No. 74500-10-001 and earlier versions.

Study Site and Observational Methods

The focal area of this research included the Steller sea lion rookery on Chiswell Island (59° 59.18′ N, 149° 23.40′ W) and nearby haulouts in Kenai Fjords (Figure 1) which lie within the range of the endangered WDPS. The population decline at the Chiswell rookery was similar to that of other rookeries in the central Gulf of Alaska – that is, abundance fell by 90% from 1,106 adults in 1976 [25] to approximately 90 adults and 50–80 pups in the 2000s [26]. Beginning in 1999, up to six remotely operated video cameras were used to monitor Steller sea lions (see [26] for details). Video images, which provided complete spatial coverage of the Chiswell Island rookery, were viewable and controllable in real-time from the Alaska SeaLife Center 65 km away. Cameras were also installed and monitored at nearby haulouts beginning in 2000 (Figure 1).

Most adult Steller sea lions can be individually identified by unique scars, fungal patches, and/or flipper patterns, and longitudinal studies have been successfully conducted on animals identified by such means [23,24,26,27]. Over the course of this study, female sea lions with unique markings (n = 184) were tracked and digital photos of these animals and their distinguishing marks were taken on a regular basis (at least twice per month) at all remotely-monitored sites in Kenai Fjords. Some breeding females were identified by flipper tags (n = 5) or brands (n = 16), and age was known only for these animals. Females that did not have at least two distinguishing marks and could not be reliably resighted from one year to the next were not used in the analysis. Although pictures and data for some females were collected as early as 1999, the period 1999–2002 was not considered in the analysis of natality rates because there was a more focused effort on sighting females giving birth over females that did not. All females with unique markings were tracked from 2003 onward whether or not they gave birth.

Our observations were conducted year-round but intensified during the birthing and breeding season beginning in mid-May. Hour-long scan sampling for identifiable females and their pups was conducted four to ten times daily from 0600 h to 2200 h; earlier and later hours were added around the summer solstice (21 June) when light levels were sufficient for viewing sea lions. After 10 August, observations were recorded from approximately sunrise to sunset as diminishing daylight allowed. Events such as births and deaths were opportunistically recorded as they occurred or within 4 hrs of their known occurrence [24,26]. Births that happened overnight were recorded the following morning as having occurred at the half-way point of non-observation hours.

Complete census counts of all sea lions by age class (adult males, adult females, juveniles 1–4 yrs old, and pups 0–1 yrs old) on the rookery and at two nearby haulouts (Figure 1) were made at approximately 1100 h throughout the breeding season. Ratios of pups to adult females on the rookery and haulouts were summarized between 25 June and 15 July each year to determine if these counts could be used as a proxy for natality rates. This date range corresponds to when the National Marine Fisheries Service conducts their range-wide flight surveys for population counts [2,28] which are used in some models for vital rates estimation [7,15]. In some cases when census counts were not conducted (≤2 each yr; 14 out of 208 total), daily numbers of each age class were interpolated from previous and subsequent counts.

Steller sea lion mothers in the WDPS will normally remain with their newborn pups for 8 to 12 days following parturition [26,27]. Given the duration and detail of observations in this study (frequent scans and complete spatial coverage of the rookery), it is highly unlikely that any births went unnoticed. Furthermore, females that give birth to stillborns are not normally considered to be productive nor are stillbirths considered in the definition of natality. However, we include full-term stillbirths in our analysis (approximately 2%; [24]) so that our estimates are comparable to historic estimates during the 1970s when the population was presumed healthy, and in the 1980s during the height of the decline in this region. Those earlier estimates were based on late-term pregnancies of adult females collected in the field and could not account for stillbirths [6]. In addition, those previous studies only considered reproductively mature females whose status was known by examination of copora lutea in ovaries [6,17]. It was neither possible nor the intention of this study to verify reproductive maturity for females that were not observed to give birth even when age was known. Therefore, females of known age were included in this study beginning at 5 years of age to be consistent with the average age of sexual maturity of 4.6 yrs [17], which would indicate that age of first pupping would be at about 5.6 yrs. Females of unknown age were included in this study if they were judged to be at least 4 yrs of age as approximated by visual size comparisons to other known-age females lying nearby.

Non-breeding haulouts within the study area were also monitored during the birthing season to account for females that may have spent more time at those locations and to reduce sample

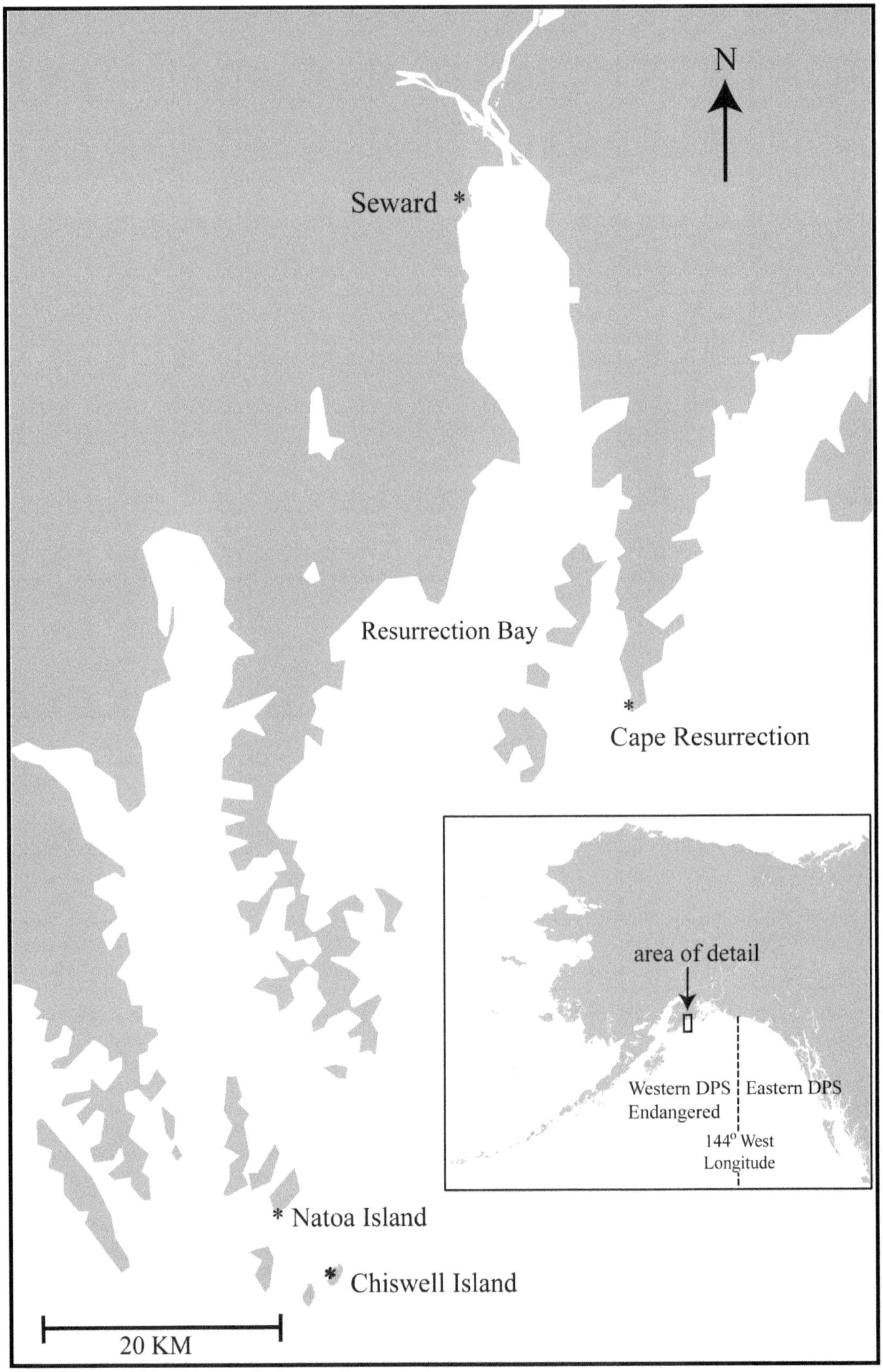

Figure 1. Location of the Chiswell Island Steller sea lion rookery and remotely monitored haulouts at Cape Resurrection and Natoa Island in Kenai Fjords, Gulf of Alaska.

bias toward more fecund females that may spend more time at the rookery seasonally and across years. Females at haulouts were included in the analyses if they met the aforementioned sighting and maturity criteria. Adult females that were observed without a pup during any birthing season were classified as not giving birth whether on the rookery or a haulout. However, some of the animals in the Chiswell Island population that were not giving birth on the rookery in any given year spent the summer elsewhere, presumably outside of the study area. Adult females that returned to the study area later in the year were typically without a pup but a few (<2%) did return with a suckling pup indicating that they gave birth elsewhere. Therefore, females that were without a pup when observed only during the non-pupping season could not be defined with certainty as not giving birth during the previous summer. We were able to account for this uncertainty by modeling the data as a hidden Markov process using a multi-state robust design [29].

Data Analysis

Open robust design (ORD) multi-state models with state uncertainty and seasonal effects were constructed and run using Program MARK [30]. Data were modeled across 10 primary periods (years 2003–2012) and 4 secondary periods – summer (Jun–Aug), autumn (Sep–Nov), winter (Dec–Feb), and spring (Mar–May) for a total of 40 sampling occasions. For each occasion, the states of every female were recorded as "b" – observed birth or with pup, "n" – observed but did not give birth, "0" – not observed, and "u" – observed during the non-birthing season without a pup, state uncertain. ORD models allow for entry and departure from the study area between primary periods and secondary periods, as often occurs among sea lions between seasons, but these models also greatly increase the number of parameters that can be estimated compared to models without a robust design. Nevertheless, robust design models allow estimation of temporary emigration from the study area and provide more precise estimates of survival and state transitions [29,31,32]. Furthermore, multistate ORD models do not necessarily require that the state of females be known with certainty or that females with and without a pup have equal sighting probabilities [31,33]. A goodness of fit test for multi-state ORD models when both states are observable has not been developed. However, a goodness of fit test using a simpler modeling approach for much of these data has shown only a slight and insignificant degree of overdispersion [14] and we have no reason to believe that data in the current model structure are much different.

The following parameters were estimated for this study with some initial constraints that were appropriate for the behavioral biology of Steller sea lions and to keep model run times reasonable:

- S_i^x = probability that a female in state x in year i survives until i+1.
- p_i^x = probability that a female is sighted in time period i in state x, given that she is present in the study area in period i. For secondary sampling periods a scaled covariate was added to reflect seasonal changes in observer effort which was highest during summer and lowest during winter.
- ψ_i^{xy} = transition probability or probability that a female in state x in year i is in state y in year i+1, given that she survived from year i to i+1.
- π_t^x = probability that a female is first observed in state x in year i for years 2–9 of the study.
- ω_t^x = probability that a female is in state x in year i, used as an estimate of natality in this study.
- δ_t^x = probability that a female is classified correctly to state. Constrained to be equal across all times periods for females with a pup. Constrained to be equal for females not seen with a pup across primary periods but allowed to vary between 1st secondary period (summer) and 2nd–4th secondary period combined.
- $pent_t^x$ = probability that a female in state x enters into the study area between a given secondary period. Constrained to be equal across primary periods for both states but allowed to vary between states and secondary periods with the mlogit link function to sum to ≤1.
- d_t^x = probability that a female in state x departs from the study area between secondary periods. Constrained to be equal across primary periods but allowed to vary between states and between secondary periods.
- α_t^x = probability that the attribute for state assignment has appeared (i.e., pup). Constrained to 1 for the 1st secondary period and to 0 for all other sampling periods because females only give birth during summer.
- c_t^x = probability that the attribute ceases – i.e., pups are weaned. Weaning often occurs during the spring months (Trites et al. 2006) and therefore was fixed to 0 for the 1st and 2nd secondary period intervals and allowed to vary during the 3rd interval.

These model parameters were combined across primary and secondary sampling periods into a joint multinomial likelihood using the mlogit link function in Program MARK. Additional constraints were placed on the models with regard to biological relevance in the search for parsimony. For example, survival (S) and state transitions (ψ^{xy}) were constrained to be equal over time in some models to assess whether the data more closely fit those model structures compared to time-varying model structures. The average ratio of pups to adult females on Chiswell Island and nearby haulouts was included as an annual covariate in some models to assess the effect on model fit. In the same manner, the total numbers of pups born on Chiswell each year was included as annual covariate for female state (ω) in addition to modeling linear trends. All models were compared with Akaike's Information Criteria (AIC; [34]), corrected for small sample bias (AICc; [35]). AICc weights, calculated from model differences in AICc values (ΔAICc), indicated relative support for the various models. Likelihood ratio tests (LRT) were also run to test for meaningful differences in models of interest. Finally, all models were averaged (multimodal inference; [36]) for estimates of parameter values relevant to this study.

We further assessed the possibility of using ratios of pups to females observed on rookeries and haulouts as a proxy for natality rates. Census data were log transformed to approximate normality and the ratios, *ln* (pups)/*ln* (adult females), were regressed against natality rates estimated from the ORD analysis for each year. This was conducted for each of 21 days between 25 June and 15 July each year to determine if any of these days provide a more accurate representation of natality over other days in that time period. We also conducted a bootstrap randomization test on standardized census data. The pup/adult female ratio was standardized to remove intra-seasonal patterns of haulout use by subtracting the daily means across years and adding back the overall mean to the daily count. We ran 1000 iterations to compare with estimated natality rate for the corresponding year. If count ratios are representative of natality rates across years, we would expect to find a significant correlation between these variables.

Results

A total of 16 models were run and ranked in order of lowest AICc score (Table 1). Models with no time differentiation in survival (S), sighting probability (p), and state transitions (ψ) showed more support for the data compared to those with variation across years (primary sampling periods). Time variation across seasons (secondary periods) was retained for sighting and departure (d) probabilities due to seasonal movements into and out of the study area (Alaska SeaLife Center, unpublished data), and the desire to model these changes. There was some evidence of differences in survival between females giving birth (89.0±1.4%) and those that did not (86.1±2.4%) as variation in this state was expressed in two of the four best models with no significant differences (Table 2; LRT: Model 1 vs 2: $\chi^2 = 1.602$; $P = 0.206$; Model 3 vs 4: $\chi^2 = 1.107$; $P = 0.293$). There was also little evidence of annual variation in the proportion of females giving birth (ω^b) as noted in a comparison of Models 1 and 3 (Table 1; LRT: $\chi^2 = 16.940$; $P = 0.050$) and Models 2 and 4 (Table 1; LRT: $\chi^2 = 16.444$; $P = 0.058$). See also Table 2 for time varying differences in ω^b for models 3 and 4.

An *a priori* decision was made to test for an increasing trend in natality, as the population in this region had been increasing over the study period [2]. However, fitting a linear increasing trend to natality did not improve model fit (Model 10 vs Model 4). Upon examination of natality estimates from the top ranked models, it was determined that a linear decrease in natality should also be tested for, but this also did not improve model fit (Model 11). We further tested for an improvement in model fit using either the total number of pups born or the average ratio of pups to adult females as linear covariates for ω^b each year. Both greatly reduced model fit compared to the model with full time dependence in ω^b (Models 5 & 6 vs Model 3 and Models 8 & 12 vs Model 4).

The four best models held 99.9% of the total weight (Table 1) and therefore, model averaged parameter estimates were based primarily on these structures. Survival to a subsequent year for females giving birth (S^b; 88.49±1.33% SE) was slightly higher than for those that did not give birth (S^n; 87.27±2.06%). Sighting probabilities were higher for females giving birth (p^b) compared to those without a pup (p^n) and seasonal sighting probabilities were especially low during winter (Table 2). Natality rates (ω^b) were high with an overall average of 70.5±1.6% that ranged narrowly from 67.6%±5.7% to 74.5%±7.3% (Figure 2). Also, females that gave birth in year i were nearly as likely to give birth in year i+1 ($\psi^{bb} = 0.708 \pm 0.021$) as females that did not give birth in year i ($\psi^{nb} = 0.736 \pm 0.033$).

Natality based on the ratio of pups to adult females on the rookery and nearby haulouts was higher than estimates from the mark-recapture modeling and had greater variation (91.3%±3.2%; Figure 2). Estimates for some years were unrealistically >100% because varying numbers of adult females were foraging at sea and unavailable to be counted. However, the intent of this analysis was not to obtain exact estimates of natality but to determine if natality rates based on census counts are correlated with direct estimates of natality and could therefore be used as a meaningful proxy for changes in natality. To test for a correlation, those rates were regressed against the natality rates (ω^b) estimated from Program MARK for each year. Slopes of the 21 regressions ranged from −0.250 to +0.223 and none were significant (*P*-value range: 0.080–0.925). The bootstrap analysis of the pup/adult female ratios also resulted in a non-significant relationship with natality ($r^2 = 0.048$; $P = 0.542$) and had slopes centered around zero (Figure 3).

Table 1. Comparison of models tested using a multi-state open robust design in Program Mark for estimation of survival (S), sighting probabilities (p), state transitions (ψ), and state occupation (ω) among Steller sea lions in Kenai Fjords.

Model Rank	Model Structure	No. Param.	AICc	ΔAICc	AICc Weights	Model Likelihood
1	$S_{.}\ p_{st.2°}\ \psi_{st.}\ \pi_{.}\ \omega_{.}\ d_{st.2°}$	30	6106.846	0.000	0.4067	1.0000
2	$S_{st.}\ p_{st.2°}\ \psi_{st.}\ \pi_{.}\ \omega_{.}\ d_{st.2°}$	31	6107.313	0.467	0.3220	0.7916
3	$S_{.}\ p_{st.2°}\ \psi_{st.}\ \pi_{.}\ \omega_{t}\ d_{st.2°}$	39	6108.617	1.771	0.1678	0.4125
4	$S_{st.}\ p_{st.2°}\ \psi_{st.}\ \pi_{.}\ \omega_{t}\ d_{st.2°}$	40	6109.601	2.755	0.1026	0.2522
5	$S_{.}\ p_{st.2°}\ \psi_{st.}\ \pi_{.}\ \omega_{p}\ d_{st.2°}$	39	6119.334	12.488	0.0008	0.0019
6	$S_{st.}\ p_{st.2°}\ \psi_{st.}\ \pi_{.}\ \omega_{r}\ d_{.2°}$	39	6124.224	17.378	0.0001	0.0002
7	$S_{st.}\ p_{st.2°}\ \psi_{st.}\ \pi_{t}\ \omega_{t}\ d_{st.2°}$	47	6124.259	17.413	0.0001	0.0002
8	$S_{st.}\ p_{st.2°}\ \psi_{st.}\ \pi_{.}\ \omega_{r}\ d_{st.2°}$	40	6125.533	18.687	0.0000	0.0001
9	$S_{.t}\ p_{st.2°}\ \psi_{st.}\ \pi_{t}\ \omega_{t}\ d_{st.2°}$	54	6129.711	22.865	0.0000	0.0000
10	$S_{st.}\ p_{st.2°}\ \psi_{st.}\ \pi_{.}\ \omega_{up}\ d_{st.2°}$	40	6133.919	27.073	0.0000	0.0000
11	$S_{st.}\ p_{st.2°}\ \psi_{st.}\ \pi_{.}\ \omega_{dn}\ d_{st.2°}$	40	6136.540	29.694	0.0000	0.0000
12	$S_{st.}\ p_{st.2°}\ \psi_{st.}\ \pi_{.}\ \omega_{p}\ d_{st.2°}$	40	6136.540	29.694	0.0000	0.0000
13	$S_{st.}\ p_{st.2°}\ \psi_{st.}\ \pi_{.}\ \omega_{t}\ d_{.2°}$	37	6138.479	31.634	0.0000	0.0000
14	$S_{st.t}\ p_{st.2°}\ \psi_{st.}\ \pi_{t}\ \omega_{t}\ d_{st.2°}$	63	6139.086	32.241	0.0000	0.0000
15	$S_{st.t}\ p_{st.2°}\ \psi_{st.t}\ \pi_{t}\ \omega_{t}\ d_{st.2°}$	79	6156.477	49.632	0.0000	0.0000
16	$S_{st.t}\ p_{st.1°.2°}\ \psi_{st.t}\ \pi_{t}\ \omega_{t}\ d_{st.2°}$	151	6190.485	83.639	0.0000	0.0000

Proportion of the population released in each state (π) and departure probabilities (d) were also manipulated to test their effects on model fit. Initial constraints on parameters not listed here are outlined in the Methods.

Notes: Subscripts indicate state variation (st; with vs without pup), time variation (t), and increasing (up) or decreasing (dn) linear trends across primary periods in addition to time variation specific to primary periods (1°) and secondary periods (2°). Additional subscripts identify annual covariates of total number of pups born (p) and the average ratio of pups to adult females (r) observed.

Table 2. Estimates (± SE) from the four top-ranked models in Table 1 for survival of females that gave birth (S^b) and those that did not (S^n), occupation of birthing state (ω^b) in any given year, and sighting probabilities by season for females that gave birth (p^b) and those that did not (p^n).

Parameter	Model 1	Model 2	Model 3	Model 4
S^b	88.1% (1.2%)	89.0% (1.4%)	88.1% (1.2%)	89.0% (1.4%)
S^n	88.1% (1.2%)	86.1% (2.4%)	88.1% (1.2%)	86.1% (2.4%)
ω^b-2003	70.5% (1.6%)	70.5% (1.6%)	85.5% (5.0%)	85.5% (5.0%)
ω^b-2004	70.5% (1.6%)	70.5% (1.6%)	75.3% (5.7%)	75.3% (5.7%)
ω^b-2005	70.5% (1.6%)	70.5% (1.6%)	60.0% (5.7%)	60.0% (5.7%)
ω^b-2006	70.5% (1.6%)	70.5% (1.6%)	67.6% (5.4%)	67.6% (5.4%)
ω^b-2007	70.5% (1.6%)	70.5% (1.6%)	75.5% (4.7%)	75.5% (4.7%)
ω^b-2008	70.5% (1.6%)	70.5% (1.6%)	68.2% (4.8%)	68.2% (4.8%)
ω^b-2009	70.5% (1.6%)	70.5% (1.6%)	71.8% (4.6%)	71.8% (4.6%)
ω^b-2010	70.5% (1.6%)	70.5% (1.6%)	64.7% (4.9%)	64.7% (4.9%)
ω^b-2011	70.5% (1.6%)	70.5% (1.6%)	76.6% (4.2%)	76.6% (4.2%)
ω^b-2012	70.5% (1.6%)	70.5% (1.6%)	66.3% (4.6%)	66.3% (4.6%)
p^b-summer	99.9% (<0.1%)	99.9% (<0.1%)	99.9% (<0.1%)	99.9% (<0.1%)
p^n-summer	98.3% (0.9%)	97.7% (1.2%)	97.7% (1.2%)	97.7% (1.3%)
p^b-fall	94.6% (1.8%)	94.6% (1.8%)	94.6% (1.8%)	94.6% (1.8%)
p^n-fall	74.3% (4.9%)	72.8% (4.4%)	72.9% (4.4%)	72.8% (4.4%)
p^b-winter	24.6% (3.6%)	24.6% (3.6%)	24.6% (3.6%)	24.6% (3.6%)
p^n-winter	12.7% (2.3%)	14.3% (4.0%)	14.3% (4.0%)	14.3% (4.0%)
p^b-spring	70.4% (4.1%)	70.4% (4.1%)	70.4% (4.1%)	70.4% (4.1%)
p^n-spring	39.4% (4.2%)	40.1% (4.7%)	40.1% (4.7%)	40.1% (4.7%)

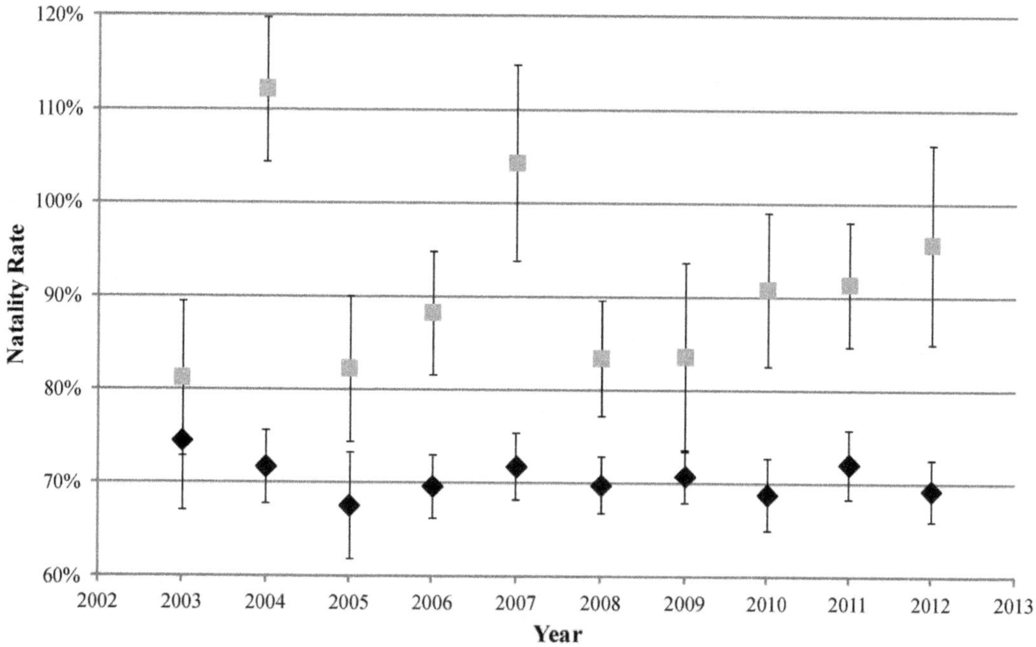

Figure 2. Mark-recapture estimates of natality rates (black diamonds) and natality estimates based on census data (gray squares) for adult female Steller sea lions at the Chiswell Island rookery from 2003 to 2012 with 95% confidence intervals. Note that the mark-recapture estimates are different from estimates expressed in Table 2 due to model averaging.

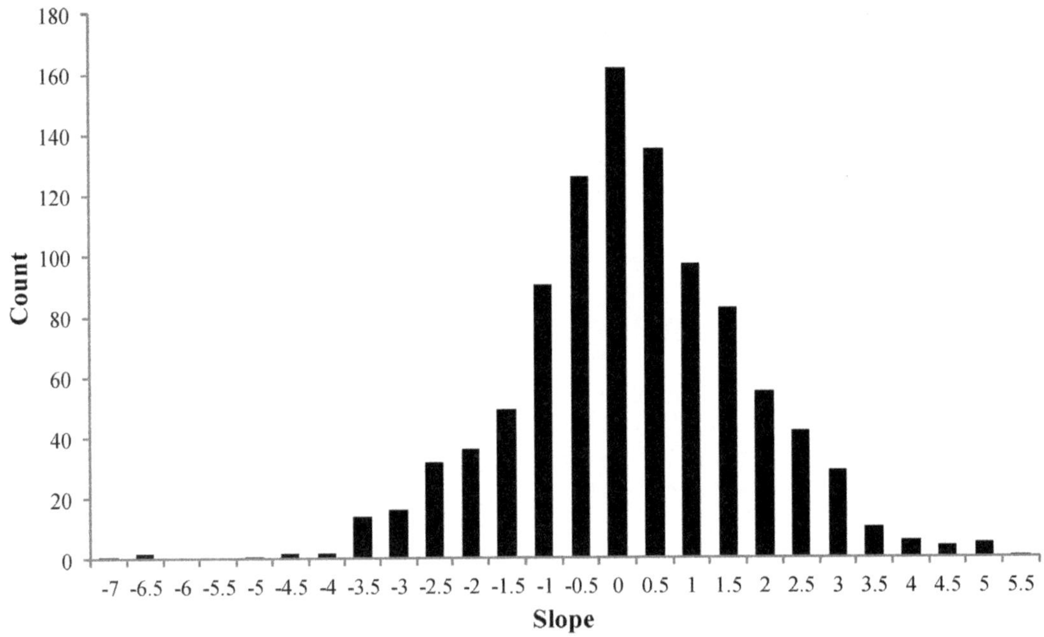

Figure 3. Bootstrap histogram of regression slopes based on pup to adult female ratios versus mark-recapture natality estimates. This figure shows that the regression slopes were strongly centered around zero with no correlative trend.

Discussion

Mark-Recapture Parameter Estimates

We previously examined natality of Steller sea lions in this region over seven years using a standard mark recapture multistate approach without robust design [14]. The current analysis was undertaken to improve our estimates using the robust design, to examine possible interannual and long-term changes in natality, and to assess the use of census data as a proxy for natality.

Not surprisingly, estimates of natality were similar to those previously found in this region at about 70% but with a smaller standard error (1.6 vs. 2.5%; [14]), which may be due partly to the larger sample size in this study. In addition, parameter estimates such as these are notably improved using the robust design as reported by others [29,32]. Sighting probabilities were high during the summer months for females giving birth and for those not giving birth in any given year (Table 2). Yet many non-breeders were not observed on the rookery during the birthing season (mid-May to Mid-July) and arrived later in the summer or autumn. If we only considered sightings during the birthing season much of the information about non-breeders would be lost and hence lead to greater uncertainty in our estimates.

Natality was not much different from pre-decline levels (67%) but better than estimated during the height of the population decline (55%) in the Gulf of Alaska during the 1980s [6]. It has previously been suggested that natality rates bottomed out at about 43% in the early 2000s, as estimated from an inferential population dynamic model [15]. A rapid change in natality from percentages in the mid-40s to nearly 70% would seem quite unusual, hence it was of interest to test for an increasing trend in this study. However, the model we tested with a increasing trend in natality did not fit the data, nor did one with a decreasing trend. In our analysis, natality ranged narrowly around 70%, with 2 of the 4 best models supporting interannual fluctuations but no trends. Therefore, we have little evidence to support unusually low natality rates during any year of this study.

Notwithstanding variation in survival, natality rates of 60% to 75% have been generally associated with stable or increasing populations of pinnipeds [37–40], including the Eastern Distinct Population Segment (EDPS) of Steller sea lions [41]. Natality of 55% or lower has been associated with declining populations of otariids and related to the adverse effects of density dependant factors such as intraspecific competition for food or breeding space [6,42] but see also [43]. Therefore, our estimate of natality from the Chiswell Island population of Steller sea lions is indicative of a population where adult females seemingly are not under resource limitation. Populations in this region and as far west as the eastern Aleutian Islands have been increasing since the early 2000s [2]. We cannot say with certainty that natality rates estimated here in the eastern Gulf of Alaska are representative of natality as far west as the eastern Aleutian Islands, but given the similar population trends [2], we currently have no reason to suppose otherwise.

During periods of food limitation long-lived mammals may exhibit a cost of reproduction in terms of reduced probability of survival or reproduction in successive years compared with times of food abundance [44,45]. Such a cost likely affected Steller sea lions during the 1980s when pregnancy was negatively correlated with lactation status [6]. Similar to our previous work [14], we found no evidence of a cost of reproduction among Steller sea lions during the period of this study (2003–2012) with survival and subsequent reproduction not being correlated with previous birthing status. This finding is consistent with a healthy population that is not exhibiting signs of broad scale food limitation [8,45,46].

Comparisons with Census Data

Long-term census-based studies provide important information regarding changes in demographics, distribution, densities, and trends of wildlife populations including the potential influences of climate change and human activities [16]. However, using census data to estimate vital rates in variably detectable species such as pinnipeds that spend a large proportion of their time foraging at sea may be difficult. Using several methods, we were unable to find a correlation between either pup counts or pup to female ratios and our natality estimates. As our natality estimates showed little year to year variation, this could have made such relationships difficult to detect. Nevertheless, if census counts can be used as a proxy for natality, we could expect at least weak positive correlations, but they were strongly centered around zero (Figure 3).

Using census counts for modeling vital rates is unusual for many large vertebrates and caution is warranted when modeling such data for vital rate estimates [47]. Census counts of Steller sea lions can vary with tide height, storm waves, time of day, and food availability [48–50]. Pup counts may or may not reflect the actual number of births because storm waves during the peak of birthing are variable across years and can cause huge losses of neonates [23,24]. These variations are typically assumed to be consistent over time [7,15] but they are not [14]. This is a primary reason why census data cannot be reliably used to estimate vital rates in species such as Steller sea lions unless adjustments are made to the data that appropriately reflect environmental stochasticity over time.

Accounting for variation in sighting probabilities is integral to longitudinal studies employing mark-recapture analysis. Long-term studies of a representative sample of individuals are necessary for accurate estimates of age-specific survival and natality, but are invariably more time consuming and costly than snapshot population counts [16,51]. Nevertheless, these mark-recapture methods provide critical information for understanding life histories and behavior of species of concern. In addition to natality, survival, and sighting probabilities, many other parameters can be estimated with the mark-recapture data analysis techniques currently available [29,32,33]. For instance, longitudinal studies of branded Steller sea lions are providing insights into temporary and permanent migration between the distinct population segments [52]. Without such work, we might mistakenly assume that populations receiving such immigrants were increasing solely due to improved survival or natality.

Census counts provide information on population trends over time, whereas individual-based longitudinal studies will more specifically inform researchers of why and how populations are changing. Based on census count data, we know that the EDPS of Steller sea lions has been increasing over the past 30 years [53]. Based on longitudinal studies we also recognize that the population increases likely result from a combination of good juvenile survival [54], good natality [41], and some immigration from the WDPS [51]. All of this information and more has lead to the recent delisting of the EDPS from the Endangered Species List [55]. The WDPS remains endangered, but with population trends continuing to increase, downlisting to threatened status is possible by 2015 [2]. Current population growth in the WDPS can be attributed to good natality and adult survival ([14]; this study) in addition to improved juvenile survival [13] since the height of the decline. Continued monitoring of vital rates is essential to detect changes that could threaten recovery of the species. Longitudinal studies of this nature can detect changes in vital rates that may happen quickly and drastically in response to increasing environmental stochasticity in the face of a shifting global climate [56].

Acknowledgments

Data collection for this study was accomplished by many hard-working technicians and interns at the ASLC including Melinda Fowler, Karin Harris, Juliana Kim, Carlene Miller, Emily Teate, and many other technicians and interns. We thank them and Lara Horstmann and Chuck Frost for discussions, commentaries, and much helpful advice on this material. Chris Oosthuizen also provided very constructive comments on the manuscript.

Author Contributions

Conceived and designed the experiments: JMM. Performed the experiments: JMM PP. Analyzed the data: JMM MDA. Contributed reagents/materials/analysis tools: JMM MDA. Contributed to the writing of the manuscript: JMM AMS PP MDA.

References

1. Gerber LR, Hilborn R (2001) Catastrophic events and recovery from low densities in populations of otariids: implications for risk of extinction. Mammal Review 31: 131–150.
2. Fritz L, Sweeney K, Johnson D, Lynn M, Gelatt T, et al. (2013) Aerial and ship-based surveys of Steller sea lions (*Eumetopias jubatus*) conducted in Alaska in June-July 2008 through 2012, and an update on the status and trend of the western distinct population segment in Alaska. US Dep Commer, NOAA Tech Memo, NMFS-AFSC-251, 91 p.
3. Loughlin TR (1998) The Steller sea lion: A declining species. Biosphere Conservation 1: 91–98.
4. Maschner HDG, Trites AW, Reedy-Maschner KL, Betts M (2013) The decline of Steller sea lions (*Eumetopias jubatus*) in the North Pacific: insights from indigenous people, ethnohistoric records and archaeological data. Fish and Fisheries doi:10.1111/faf.12038.
5. York AE (1994) The population dynamics of northern sea lions, 1975–1985. Marine Mammal Science 10: 38–51.
6. Pitcher KW, Calkins DG, Pendleton GW (1998) Reproductive performance of female Steller sea lions: an energetics-based reproductive strategy? Canadian Journal of Zoology 76: 2075–2083.
7. Holmes EE, York AE (2003) Using age structure to detect impacts on threatened populations: a case study with Steller sea lions. Conservation Biology 17: 1794–1806.
8. National Resource Council (2003) The decline of the Steller sea lion in Alaskan waters: Untangling food webs and fishing nets. Washington: National Academies Press, 179 p.
9. Springer AM, Estes JA, van Vliet GB, Williams TM, Doak DF, et al. (2003) Sequential megafaunal collapse in the North Pacific Ocean. Proceedings of the National Academy of Sciences 100: 12223–12228.
10. Trites AW, Donelly CP (2003) The decline of Steller sea lions *Eumetopias jubatus* in Alaska: a review of the nutritional stress hypothesis. Mammal Review 33: 3–28.
11. Wolf N, Mangel M (2008) Multiple hypothesis testing and the declining population paradigm in Steller sea lions. Ecological Applications 18: 1932–1955.
12. Eberhardt LL, Sease JL, DeMaster DP (2005) Projecting the trend of Steller sea lion populations in western Alaska. Marine Mammal Science 21: 728–738.
13. Maniscalco JM (2014) The effects of birth weight and maternal care on juvenile Steller sea lion (*Eumetopias jubatus*) survival. PLoS ONE 9(5): e96328.
14. Maniscalco JM, Springer AM, Parker P (2010) High natality rates of endangered Steller sea lions in Kenai Fjords, Alaska and perceptions of population status in the Gulf of Alaska. PLoS ONE 5(4): e10076.
15. Holmes EE, Fritz LW, York AE, Sweeney K (2007) Age-structured modeling reveals long-term declines in the natality of western Steller sea lions. Ecological Applications 17: 2214–2232.
16. Clutton-Brock T, Sheldon BC (2010) Individuals and populations: the role of long-term, individual-based studies of animals in ecology and evolutionary biology. Trends in Ecology and Evolution 25: 562–573.
17. Pitcher KW, Calkins DG (1981) Reproductive biology of Steller sea lions in the Gulf of Alaska. Journal of Mammalogy 62: 599–605.
18. Pitcher KW, Burkanov VN, Calkins DG, Le Boeuf BJ, Mamaev EG, et al. (2001) Spatial and temporal variation in the timing of births of Steller sea lions. Journal of Mammalogy 82: 1047–1053.
19. Maniscalco JM, Parker P (2009) A case of twinning and the care of two offspring of different age in Steller sea lions. Marine Mammal Science 25: 206–213.
20. Trites AW, Porter BP, Deecke VB, Coombs AP, Marcotte ML, et al. (2006) Insights into the timing of weaning and the attendance patterns of lactating Steller sea lions (*Eumetopias jubatus*) in Alaska during winter, spring, and summer. Aquatic Mammals 32: 85–97.
21. Gilmartin WG, DeLong RL, Smith AW, Sweeney JC, DeLapp BW, et al. (1976) Premature parturition in the California sea lion. Journal of Wildlife Diseases 12: 104–115.
22. Doidge DW, Croxall WP, Baker JR (1984) Density-dependent pup mortality in the Antarctic fur seal *Arctocephalus gazella* at South Georgia. Journal of Zoology, London 202: 449–460.
23. Kaplan CC, White GC, Noon BR (2008) Neonatal survival of Steller sea lions (*Eumetopias jubatus*). Marine Mammal Science 24: 443–461.

24. Maniscalco JM, Calkins DG, Parker P, Atkinson S (2008) Causes and extent of natural mortality among Steller sea lion (*Eumetopias jubatus*) pups. Aquatic Mammals 34: 277–287.
25. Calkins DG, Becker E, Spraker TR, Loughlin TR (1994) Impacts on Steller sea lions. 119–139 In: Loughlin TR, editor, Marine Mammals and the Exxon Valdez. Academic Press, San Diego.
26. Maniscalco JM, Parker P, Atkinson S (2006) Interseasonal and interannual measures of maternal care among individual Steller sea lions (*Eumetopias jubatus*). Journal of Mammalogy 87: 304–311.
27. Milette LL, Trites AW (2003) Maternal attendance patterns of Steller sea lions (*Eumetopias jubatus*) from stable and declining populations. Canadian Journal of Zoology 81: 340–348.
28. Fritz LW, Lynn M, Kunisch LE, Sweeney K (2008) Aerial, ship and land-based surveys of Steller sea lions (*Eumetopias jubatus*) in the Western Stock in Alaska, June and July 2005–2007. US Dep of Commer, NOAA Tech Memo, NMFS-AFSC-183, 70 p.
29. Kendall WL, Nichols JD (2002) Estimating state transition probabilities for unobservable states using capture-recapture/resighting data. Ecology 83: 3276–3284.
30. White GC, Burnham KP (1999) Program MARK: survival estimation from populations of marked animals. Bird Study 46: 120–139.
31. Kendall WL, Nichols JD, Hines JE (1997) Estimating temporary emigration using capture-recapture data with Pollock's robust design. Ecology 78: 563–578.
32. Kendall WL, White GC, Hines JE, Langtimm CA, Yoshizaka J (2012) Estimating parameters of hidden Markov models based on marked individuals: use of robust design data. Ecology 93: 913–920.
33. Kendall WL, Hines JE, Nichols JD (2003) Adjusting multistate capture-recapture models for misclassification bias: manatee breeding proportions. Ecology 84: 1058–1066.
34. Akaike H (1973) Information theory and an extension of the maximum likelihood principal. 267–281 In: Petran BN, Csaaki F, editors. International Symposium on Information Theory. Budapest, Hungary: Akadeemiai Kiado.
35. Hurvich CM, Tsai C-L (1989) Regression and time series selection in small samples. Biometrika 76: 297–307.
36. Burnham KP, Anderson DR, Huyvaert KP (2011) AIC model selection and multimodel inference in behavioral ecology: some background, observations, and comparisons. Behavioral Ecology and Sociobiology 62: 23–35.
37. Lunn NJ, Boyd IL, Croxall JP (1994) Reproductive performance of female Antarctic fur seals: the influence of age, breeding experience, environmental variation and individual quality. Journal of Animal Ecology 63: 827–840.
38. McKenzie J, Parry LJ, Page B, Goldsworthy SD (2005) Estimation of pregnancy rates and reproductive failure in New Zealand fur seals (*Arctocephalus forsteri*). Journal of Mammalogy 86: 1237–1246.
39. Melin SR, Laake JL, DeLong RL, Siniff DB (2012) Age-specific recruitment and natality of California sea lions at San Miguel Island, California. Marine Mammal Science 28: 751–776.
40. Testa JW (1987) Long-term reproductive patterns and sighting bias in Weddell seals (*Leptonychotes weddelli*). Canadian Journal of Zoology 65: 1091–1099.
41. Taylor RL, Boor GKH (2012) Beyond the robust design: accounting for changing, uncertain states and sparse, biased detection in a multistate mark-recapture model. Ecological Modelling 243: 73–80.
42. Dabin W, Beauplet G, Crespo EA, Guinet C (2005) Age structure, growth, and demographic parameters in breeding-age female subantarctic fur seals, *Arctocephalus tropicalis*. Canadian Journal of Zoology 82: 1043–1050.
43. Gibbens J, Parry LJ, Arnould JPY (2010) Influences on fecundity in Australian fur seals (*Arctocephalus pusillus doriferus*). Journal of Mammalogy 91: 510–518.
44. Bell G (1980) The costs of reproduction and their consequences. American Naturalist 116: 45–76.
45. Hamel S, Cote SD, Gaillard J-M, Festa-Bianchet M (2009) Individual variation in reproductive costs of reproduction: high-quality females always do better. Journal of Animal Ecology 78: 143–151.
46. DeMaster D, Atkinson S (2002) Steller sea lion decline: Is it food II. AK-SG-02-02. Fairbanks: University of Alaska Sea Grant College Program, 78 p.
47. Eberhardt LL (1985) Assessing the dynamics of wild populations. Journal of Wildlife Management 49: 997–1012.
48. Kastelein RA, Weltz FC (1991) Distribution and behavior of Steller sea lions (*Eumetopias jubatus*) in Prince William Sound, Alaska, June 1989. Aquatic Mammals 17: 91–97.
49. Kucey L (2005) Human disturbance and the hauling out behavior of Steller sea lions (*Eumetopias jubatus*). M. S. Thesis, University of British Columbia, Vancouver, 67 p.
50. Withrow DE (1982) Using aerial surveys, ground truth methodology, and haul out behavior to census Steller sea lions, *Eumetopias jubatus*. M. S. Thesis, University of Washington, Seattle, 102 p.
51. Clutton-Brock TH, Editor (1988) Reproductive success: Studies of individual variation in contrasting breeding systems. Chicago: The University of Chicago Press, 538 p.
52. Jemison LA, Pendleton GW, Fritz LW, Hastings KK, Maniscalco JM, et al. (2013) Inter-population movements of Steller sea lions in Alaska with implications for population separation. PLoS ONE 8(8): e70167.
53. Pitcher KW, Olesiuk PF, Brown RF, Lowry MS, Jeffries SJ, et al. (2007) Abundance and distribution of the eastern North Pacific Steller sea lion (*Eumetopias jubatus*) population. Fishery Bulletin 107: 102–115.
54. Hastings KK, Jemison LA, Gelatt TS, Laake JL, Pendleton GW, et al. (2011) Cohort effects and spatial variation in age-specific survival of Steller sea lions from southeastern Alaska. Ecosphere 2(10): 111.
55. U. S. Federal Register (2013) Endangered and threatened species; delisting of the Easter Distinct Population Segment of Steller sea lion under the Endangered Species Act. 50 CFR Parts 223 and 224. 78: 66140.
56. Boyce MS, Haridas CV, Lee CT, NCEAS Stochastic Demography Working Group (2006) Demography in an increasingly variable world. Trends in Ecology and Evolution 21: 141–148.

18

Genetic Diversity and Structure of *Sinopodophyllum hexandrum* (Royle) Ying in the Qinling Mountains, China

Wei Liu, Dongxue Yin, Jianjun Liu*, Na Li

College of Forestry, Northwest A & F University, Yangling, China

Abstract

Sinopodophyllum hexandrum is an important medicinal plant whose genetic diversity must be conserved because it is endangered. The Qinling Mts. are a *S. hexandrum* distribution area that has unique environmental features that highly affect the evolution of the species. To provide the reference data for evolutionary and conservation studies, the genetic diversity and population structure of *S. hexandrum* in its overall natural distribution areas in the Qinling Mts. were investigated through inter-simple sequence repeats analysis of 32 natural populations. The 11 selected primers generated a total of 135 polymorphic bands. *S. hexandrum* genetic diversity was low within populations (average $H_e = 0.0621$), but higher at the species level ($H_e = 0.1434$). Clear structure and high genetic differentiation among populations were detected by using the unweighted pair group method for arithmetic averages, principle coordinate analysis and Bayesian clustering. The clustering approaches supported a division of the 32 populations into three major groups, for which analysis of molecular variance confirmed significant variation (63.27%) among populations. The genetic differentiation may have been attributed to the limited gene flow ($N_m = 0.3587$) in the species. Isolation by distance among populations was determined by comparing genetic distance versus geographic distance by using the Mantel test. Result was insignificant ($r = 0.212$, $P = 0.287$) at 0.05, showing that their spatial pattern and geographic locations are not correlated. Given the low within-population genetic diversity, high differentiation among populations and the increasing anthropogenic pressure on the species, in situ conservation measures were recommended to preserve *S. hexandrum* in Qinling Mts., and other populations must be sampled to retain as much genetic diversity of the species to achieve ex situ preservation as a supplement to in situ conservation.

Editor: Raffaella Balestrini, Institute for Sustainable Plant Protection, C.N.R., Italy

Funding: This work was supported by the program from the Forestry Research Foundation for the Public Service Industry of China (200904004). The funders had no role in study design, data collection and analysis, decision to publish, or preparation of the manuscript.

Competing Interests: The authors have declared that no competing interests exist.

* Email: ljj@nwsuaf.edu.cn

Introduction

Sinopodophyllum hexandrum (Royle) Ying, family Berberidaceae, the only species of this genus in China, commonly known as Himalayan mayapple, is an endangered and medicinal perennial herb native to the Himalayan regions at elevations ranging from 2 700 m to 4 500 m [1–4]. Plants provide us with many important medicaments, including anticancer and antiinfective agents [5], and traditional Chinese medicine has contributed to identifying these substances [6]. *S. hexandrum* is a traditional Chinese medicine that has been used in folk medicine [7]. The roots and rhizomes of *S. hexandrum* contain large amounts of lignans. The most important lignan for human health is arguably the most active cytotoxic aryltetralin lignan, podophyllotoxin, with three times the podophyllotoxin levels compared to the American species *Podophyllum peltatum* [8,9,10], as a precursor for the semi-synthesis of the anticancer pharmaceuticals, such as etoposide (VP-16), teniposide (VM-26), GP-7, NK-611, etopophos, GL-331 and TOP-53 [9–15]. The destructive harvest of these plants added *S. hexandrum* to the endangered species list of the Convention on International Trade in Endangered Species of Wild Fauna and Flora [16]. *S. hexandrum* was classified as an endangered species (grade 3) in 1987 by the Chinese Plant Red Book [4].

Currently, with the enhanced awareness of its medicinal value and superior efficacy in clinical applications, the wild *S. hexandrum* populations in China has been noted to be very small and to be rapidly declining. The availability of podophyllotoxin from plants has become increasingly limited due to intense collection, habitat fragmentation, low natural regeneration rate, and the lack of organized cultivation. Wild *S. hexandrum* populations could become extinct without timely and effective protective measures, costing humans an ideal drug against cancer. Knowledge of the genetic diversity at intraspecific levels is an important prerequisite for species conservation and a rational exploitation program planning. However, previous studies have mainly focused on the identification and separation of the chemical components [17–25], biological properties [26–32], and micropropagation of *S. hexandrum* [33]. Only studies on genetic diversity in *S. hexandrum* from the Northwestern Himalayan region are available [34,35,36]. In particular, two areas where *S. hexandrum* is grown, namely, western Sichuan Province and Himalaya–Hengduan Mt. region in China, were investigated to identify genetic diversity of this species through inter-simple sequence repeats (ISSR) and amplified fragment length polymorphism (AFLP) markers [37,38,39], respectively. These reports with same results showed that *S. hexandrum* populations had relatively high genetic diversity ($H_e = 0.2944 \sim 0.3377$). However, there is a consensus that alpine plants are faced with pollinator restriction [40]. The unclear extent of the species dispersal mechanisms makes it interesting to study the relationships between populations. Furthermore, many reports about genetic diversity of other medicinal plants, such as *Calamagrostis porteri ssp. insperata* [41], *Aegiceras corniculatum* [42], *Sonneratia alba* [43], *Coscinium fenestratum* [44], *and Lilium pumilum* [45], have been published based on the ISSR approach.

Molecular markers are very useful tools for genetic diversity studies. ISSR markers are molecular markers especially suited to genetic polymorphisms analyses of species without available sequence information [46,47]. Studies on the population relationships, genetic diversity and conservation of *S. hexandrum* in Qinling Mts. are requisite because climate change and local overexploitation may cause unknown endangering mechanisms. The present study aims to establish management strategies for the conservation genetics of *S. hexandrum* by (1) examining the levels of genetic variability within and among *S. hexandrum* populations sampled from Qinling Mts. where historical records showed *S. hexandrum* having grown naturally, (2) assessing the possible factors that affect the genetic variation observed, and (3) comparing these within and among *S. hexandrum* populations with data published for itself or other plant taxa with similar characteristics.

Materials and Methods

Ethics statement

The endangered species were collected, and research activities were scientifically conducted under the permits issued by the local forestry department. A detailed description of the experimental material collection and procedures is provided. The experimental procedures were approved by the Ethics Committee for Plant Experiments of Northwest A & F University and the State Forestry Administration, P. R. China. The names of the authorities that issued the permit for each location were listed in Table S1 in Text S1.

Study area

This study was performed in the Qinling Mts. (32°41′ to 34°59′ N, 103°54′ to 110°34′ E), which are located in central of China (Figure 1). The Qinling Mts., a 1 500 Km mountain chain, run east–west and act as an important watershed divider between two great Chinese rivers, the Yangtze River and the Yellow River, which constitute a transitional zone between the northern subtropical zone and warm-temperate zone. The Qinling Mts. were considered to be a biodiversity hotspot in China [48]. As one of the distribution areas of *S. hexandrum*, Qinling Mts. have unique environmental features which have high impact on the evolution of the species.

Plant materials

S. hexandrum distribution pattern and extent of in Qinling Mts. were investigated from 2010 to 2011. The *S. hexandrum* populations are distributed in small and scattered patches. According to the field survey information, a total of 32 wild *S. hexandrum* populations were sampled for DNA analysis between July 19, 2012, and September 17, 2012, to ensure collection period consistency (Table 1). *S. hexandrum* has a wide geographic distribution throughout the Qinling Mts. (Figure 1). The altitude of the sample sites ranged from 1 013 m to 2 883 m (Table1). Geographical distances between populations ranged from 5.5 km to 276.8 km. 20 plants were sampled from each population. The horizontal and vertical distances between sampled plants within each population were over 20 and 5 m, respectively, to increase the likelihood of sampling inter-individual variation within each population [37,38]. About 2 g to 10 g of fresh young leaves per plant was immediately frozen in liquid nitrogen and then kept at −80 °C until DNA isolation. The key information on *S. hexandrum* populations in all sampling sites is summarized in Table 1.

DNA extraction

Total genomic DNA was extracted from frozen leaves by using a plant genomic DNA rapid extraction kit (Spin-column) (BioTek Corporation, Beijing, China; http://bioteke.biogo.net/). The extracted DNA was quantified by comparing with known DNA of standard quantity (Lambda DNA) through electrophoresis in ethidium bromide-stained 1.0% agarose gels (Gene Genius Bio Imaging System; Synegene), and the extracted DNA was diluted in TE buffer to a final concentration of 50 ng/μL and stored at−20°C before PCR amplification.

Primer screening and ISSR-PCR amplification

A total of 100 ISSR primers (synthesized by Shanghai Sheng Gong Biotechnology CO. LTD, China) were screened based on the primer set published by the Biotechnology Laboratory, University of British Columbia, Canada (UBC set No. 9) and the studies on Himalayan mayapple [34–39]. An optimum reaction system was obtained by screening DNA, Mg^{2+}, dNTP, primer (UBC900 was used for preliminary test), and Taq DNA polymerase concentrations and annealing temperature, and reaction conditions. The optimization showed that 20 μL of reaction system is ideal. Each 20 μL amplification reaction consisted of 1×PCR buffer (10 mM Tris-HCl at pH 8.3, 50 mM L^{-1} KCl, 0.001% gelatin, and 1.5 mmol L^{-1} $MgCl_2$), 1.6 mmol L^{-1} dNTP mix, 0.6 μmol L^{-1} primer (UBC900 was used for preliminary test), 15 ng of template DNA, and 1.0 U Taq DNA polymerase (TaKaRa Biotechnology, Dalian, China), using a cycling profile of initial 5 min at 94°C, followed by 45 cycles of 30 s at 94°C, 45 s annealing at 50°C, and 90 s extension at 72°C, ending with a final extension of 7 min at 72°C.

The optimized PCR experiment conditions were applied for primer screening in a PTC100TM Programmable Thermal Controller (MJ Research, Waltham, MA, USA). Six populations (TCG, LJG, BG, MDS, YMG and CTG) with observable variations (morphology, habitat, etc) were selected to initially screen 100 primers by using 10 samples for each population. Primers that generated scorable bands and high levels of polymorphisms were selected by genotyping all populations. The amplification products were electrophoresed on 1.0% agarose gels buffered with 1.0×TBE for 2.5 h at 100 V and were detected through ethidium bromide staining, and the gels were imaged in the Gene Genius Bioimaging System. Band size was estimated from a 0.1 kbp DNA ladder (TaKaRa Biotechnology, Dalian, China). Each primer was amplified in triplicate to confirm reliability and reproducibility. A reaction without DNA was used as negative control.

Data analyses

Amplification results were scored according to the positions of the DNA bands from electrophoresis, being labeled "1" for presence of the bands and "0" for absence of the bands in the data

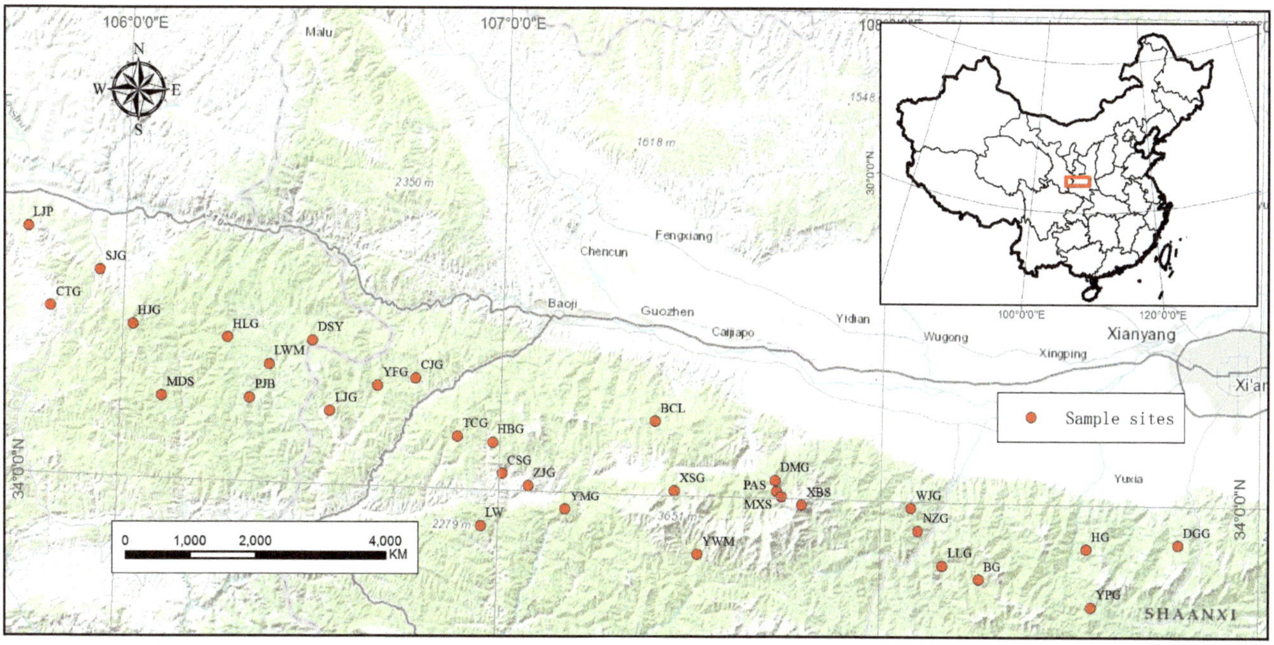

Figure 1. Locations of the 32 *S. hexandrum* populations in Qinling Mts. sampled for this study.

matrix. Only stable bands with repeatable differences were considered valid for polymorphism loci.

The resulting ISSR phenotype data matrix (binary matrix from 0 to 1) was analyzed using the POPGENE software (version 1.31) [49] to compute genetic diversity parameters, such as allele frequencies, percentage of polymorphic bands (PPB), number of alleles per locus (A_o), effective number of alleles per locus (A_e), total gene diversity (H_t), the level of gene flow (N_m), Nei's gene diversity (H_e), gene distance (GD), Shannon's information index (H_o), within-population diversity (H_s), and mean coefficient of gene differentiation (G_{st}). An unweighted pair groups mean arithmetic (UPGMA) mean dendrogram was constructed using PowerMarker 3.23 to examine the genetic relationship at the species level [50]. A bootstrap (resampling) test was performed 1 000 times to determine distances between the populations using PHYLIP version 3.69 (PHYLogeny Inference Package) programs [51]. Bayesian analysis of population structure was performed as implemented in STRUCTURE (version 2.2) to infer the most likely number of population genetic clusters (K) in the ISSR dataset [52]. K ranged from 1 to 10, with 10 replicate runs for each K, and a burn-in period of 2×10^5 and 5×10^4 iterations. The "no admixture model" and independent allele frequencies were chosen for this analysis. The most likely number of clusters was estimated according to the model values (ΔK) based on the second order rate of change, with respect to K, of the likelihood function [53]. To detect within-group structure, subsequent runs were performed for each obtained clusters using the same settings as previous. Population similarity was also explored and visualized through principle coordinate analysis (PCoA) using NTSYSpc 2.10e [54]. Analysis of molecular variance (AMOVA), which partitions total phenotypic variance within and among populations, was performed using WIN AMOVA (version 1.55), which was provided by the Genetics and Biometry Laboratory, University of Geneva, Geneva, Switzerland [55]. The AMOVA input files with the Euclidean distance matrix were created using AMOVA-PREP 1.01 [56]. Significance level was tested by comparing the frequency distributions from the original data and the data generated by a set of 1 000 computer simulations. A Mantel test was performed using Tools for Population Genetic Analysis (TFPGA) [56] for computing 5 000 permutations to test the isolation by distance (IBD) among populations by comparing genetic distance between all pairwise combinations of populations versus geographic distance. Geographical distance, L_{ab}, was computed using the following formula: $L_{ab} = \text{Arccos}\ [\cos(LAT_a)\text{-}COS(LONG_a)\cos(LAT_b)\cos(LONG_b) + \cos(LAT_b)\sin(LONG_b)\cos(LAT_b)\sin(LONG_b) + \sin(LAT_a)\sin(LAT_b)] \times R$ [57]. $LONG_a$, LAT_a and $LONG_b$, LAT_b are the longitudes and latitudes (in radians) of sampling sites a and b, respectively; R indicates the radius of the earth, which is 6 378 km; and L_{ab} represents the geographic distance between sampling sites a and b.

Results

Genomic DNA amplification results

For polymorphism testing of the 640 *S. hexandrum* individuals from 32 populations, forty-eight ISSR primers amplified visible bands and were chosen from the initial set of 100 primers to screen for reproducible markers. PCR amplification results (Table 2) show that 11 primers produced 241 clear and replicated bands (250 bp to 2 000 bp), of which 135 were polymorphic (56.02%) with 100% reproducibility. Individual primers detected between 19 (UBC825) and 29 (UBC900) loci can amplify clear bands, with an average of 21.91. The percentage of polymorphism ranged from 42.86% (UBC834) to 68.97% (UBC900), indicating that the selected primers are highly polymorphic across *S. hexandrum* populations. The H_t was 0.1434, whereas H_s was found to be 0.0599. The G_{st} value of 0.5823 indicated that 41.77% of the genetic diversity resided within the populations. The N_m among the sampled populations was calculated as 0.3587 using G_{st} through the formula $(0.5(1-G_{st})/G_{st})$.

Population genetic diversity

Detailed statistical analyses were performed on the ISSR amplification results (Table 3). The numbers of polymorphic

Table 1. Sample information of the 32 sampling sites.

No.	Population	Prefecture	Coordinates	N	Altitude/m	SO/SD/°	Vegetation type
1	Caotangou	Maiji	E105°48′N34°21′	20	1 899	SE70	Shrub-grass
2	Tancaogou	Fengxian	E106°53′N34°6′	20	1 728	SW45	Shrub-grass
3	Chunshugou	Fengxian	E107°1′N34°2′	20	1 512	SW65	Shrub-grass
4	Hougou	Fengxian	E108°33′N33°55′	20	1 439	N75	Shrub-grass
5	Xiaoshagou	Taibai	E107°27′N34°1′	20	2 332	NW60	Forest-edge, Shrub-grass
6	Doumugong	Meixian	E107°43′N34°2′	20	2 748	NW75	Forest-edge, Shrub-grass
7	Pinganssi	Meixian	E107°43′N34°1′	20	2 815	NW75	Forest-edge, Shrub-grass
8	Mingxingsi	Meixian	E107°44′N34°0′	20	2 637	E75	Forest-edge, Shrub-grass
9	Xiabansi	Meixian	E107°47′N33°59′	20	2 883	SE50	Forest-edge, Shrub-grass
10	Laojungou	Liangdang	E106°33′N34°9′	20	1 902	SE60	Cliff, edge of forest
11	Youfanggou	Fengxian	E106°40′N34°12′	20	1 545	SW70	Shrub-grass
12	Baicaoling	Taibai	E107°24′N34°9′	20	1 373	S60	Forest-edge, Shrub-grass
13	Nianzigou	Zhouzhi	E108°6′N33°56′	20	1 821	SW75	Meadow, edge of forest
14	Wenjiagou	Zhouzhi	E108°5′N33°59′	20	1 765	S80	Meadow, edge of forest
15	Liulingou	Zhouzhi	E108°10′N33°52′	20	1 013	NW75	Meadow, edge of forest
16	Beigou	Zhouzhi	E108°16′N33°51′	20	1 579	NW85	Meadow, edge of forest
17	Dagangou	Huxian	E108°47′N33°55′	20	1 335	NW65	Shrub-grass
18	Yaowangmiao	Taibai	E107°31′N33°53′	20	1 779	SE40	Forest-edge, Shrub-grass
19	Yingpangoukou	Huxian	E108°34′N33°47′	20	1 487	SE60	Shrub-grass
20	Maiduoshigou	Liangdang	E106°7′N34°10′	20	1 520	SE70	Cliff, edge of forest
21	Panjiaba	Liangdang	E106°20′N34°10′	20	1 976	SW45	Cliff, edge of forest
22	Zhangjiagou	Fengxian	E107°5′N34°1′	20	1 469	SW75	Shrub-grass
23	Longwangmiao	Fengxian	E106°57′N33°55′	20	1 475	SW75	Shrub-grass
24	Huangbaigou	Fengxian	E106°59′N34°6′	20	1 745	SW80	Shrub-grass
25	Dashuiyugou	Maiji	E106°30′N34°18′	20	1 878	W75	Shrub-grass
26	Chenjiagou	Fengxian	E106°46′N34°13′	20	1 515	SW45	Shrub-grass
27	Longwangmiao	Maiji	E106°23′N34°15′	20	1 736	NW60	Shrub-grass
28	Yinmagou	Taibai	E107°11′N33°58′	20	2 252	S60	Forest-edge, Shrub-grass
29	Hualingou	Maiji	E106°17′N34°18′	20	1 793	SW55	Shrub-grass
30	Huojigou	Maiji	E106°2′N34°19′	20	1 718	SE65	Shrub-grass
31	Shijiagoucun	Maiji	E105°56′N34°25′	20	1 366	SE60	Shrub-grass
32	Liujiaping	Maiji	E105°44′N34°30′	20	1 690	SW30	Shrub-grass

Note: N means sample size; SO/SD means slope orientation/degree.

Table 2. Description of the11 selected primers used for ISSR amplification.

No.	Primer	Sequences 5′→3′	N_t	N_p	P_r %	H_t	H_s	G_{st}
1	UBC825	$(AC)_8T$	19	12	63.16	0.1715	0.0835	0.5132
2	UBC834	$(AG)_8YT$	21	9	42.86	0.1035	0.0440	0.5748
3	UBC844	$(CT)_8AGC$	22	11	50.00	0.1160	0.0517	0.5543
4	UBC845	$(CT)_8AGG$	23	12	52.17	0.1192	0.0494	0.5855
5	UBC853	$(CT)_8AGT$	20	9	45.00	0.1121	0.0352	0.6858
6	UBC855	$(AC)_8YT$	21	11	52.38	0.1347	0.0645	0.5212
7	UBC857	$(AC)_8CTG$	20	13	65.00	0.1763	0.0763	0.5672
8	UBC867	$(GGC)_6$	19	10	52.36	0.1251	0.0564	0.5492
9	UBC873	$(GACA)_4$	22	11	50.00	0.1165	0.0441	0.6214
10	UBC895	AGAGTTGGTAGCTCTTGATC	25	17	68.00	0.1875	0.0765	0.5920
11	UBC900	ACTTCCCCACAGGTTAACACA	29	20	68.97	0.2150	0.0773	0.6404
Sum			241	135				
Mean			21.91	12.27	56.02	0.1434	0.0599	0.5823

Note: N_t-No. of total amplified bands; N_p-No. of polymorphic bands; P_r-Polymorphism rate; H_t- Total gene diversity among population; H_s-within population diversity; G_{st}-Mean coefficient of gene differentiation.

bands are different among populations. The highest PPB (39.17%) was observed in the DMG population. However, only 20.73% of the bands were polymorphic in the SJG population. The PPB was 56.02% at the species level, whereas those of the single populations ranged from 20.73% (SJG) to 39.17% (DMG), with an average of 27.33%. A_e ranged from 1.0182 (HJG) to 1.2643 (BG), with 1.1630 at the population level and 1.3732 at the species level. H_e within populations was lower (0.0621) than that of the species level (0.1434). Within each population, the H_e of most populations ranged from 0.0226 to 0.0971. Only in the population HG, a high H_e (0.1229, Table 3) was observed. H_o ranged from 0.0244 to 0.1038, with an average of 0.0637 at the population level and 0.2362 at the species level. Allele frequencies calculated using Popgene software were shown in Table S2 in Text S2.

Genetic structure and differentiation of the populations

UPGMA cluster analysis. UPGMA clustering analysis defined three major groups among 32 populations (Figure 2). Group 1 included 11 populations, which were further divided into three subgroups. Populations TCG, CSG and HG were in subgroup 1a; populations XSG, DMG, PAS, MXS and XBS were in subgroup 1b; and populations LJG, YFG, and BCL were in subgroup 1c. Populations NZG, WJG, LLG and BG (Zhouzhi County), population DGG (Huxian County), population YWM (Taibai County) and population YPG (Huxian County) were grouped in cluster 2. Group 3 contained 14 populations, which were sampled from two adjacent cities, Baoji and Tianshui, and were subdivided into two clusters. Populations MDS, PJB, ZJG, LW, and HBG (Baoji) and DSY (Tianshui) were in subgroup 3a. Populations LWM, HLG, HJG, SJG, CTG, and LJP (Tianshui) and CJG, YMG (Baoji) were in subgroup 3b. Seven populations including DSY, LWM, HLG, HJG, SJG, CTG, and LJP in Group 3 were from Gansu Province. The other populations clustered in Groups 1, 2, and 3 were from Shaanxi Province. The *S. hexandrum* population was not clustered on the UPGMA tree according to geographic distance, which may indicate no obvious correlation between geographic distribution and genetic distance.

Principle coordinate analysis. PCoA was used for ordination and exploration of the similarity between populations. All samples were clearly separated into three major groups on the first (PCo1) and second (PCo2) principal coordinates (Figure 3). PCo1 explained 13.6% of the total variance, and PCo2 explained 8.7% of the total variance. Groups 1 and 2 are clustered within each other's vicinity even though clearly separated, indicating higher similarity between these two groups. The clustering of the populations was in agreement with the UPGMA dendrogram.

Mantel test for IBD. A Mantel test for IBD was performed to assess the correlation between the genetic distance matrix and the corresponding geographic distance matrix of the wild *S. hexandrum* populations. The correlation coefficient r of genetic distance and geographic distance was 0.212 ($P = 0.287$), and the correlation analysis diagram (Figure 4) was comprised of many disordered and scattered points, indicating that the IBD of the wild *S. hexandrum* populations in Qinling Mts. was not significant at the level of 0.05. The Mantel test did not indicate correlation between genetic distance and geographical provenance.

Bayesian clustering. The genetic structure of the *S. hexandrum* samples were further analyzed using Bayesian clustering algorithm in the STRUCTURE software. The ΔK method indicated that the most likely K value was 3 (Figure 5). Sharp divisions were observed for the three clusters (Figure 6). The assignments of the populations to Groups 1, 2, and 3 were stable and consistent with UPGMA and PCoA clustering.

Table 3. Descriptive statistics summary of the *S. hexandrum* populations.

Population	Code	PPB (%)	A_o	A_e	H_e	H_o
Tancaogou	TCG	23.63	1.3546	1.1684	0.0559	0.0737
Chunshugou	CSG	22.62	1.2687	1.1556	0.0337	0.0689
Hougou	HG	25.66	1.5642	1.2357	0.1229	0.1038
Xiaoshagou	XSG	25.32	1.5721	1.1957	0.0777	0.0733
Doumugong	DMG	39.17	1.3905	1.1404	0.0296	0.0255
Pinganssi	PAS	33.95	1.4397	1.1833	0.0567	0.0442
Mingxingsi	MXS	26.94	1.4720	1.1667	0.0381	0.0667
Xiabansi	XBS	23.51	1.4986	1.231	0.0542	0.0614
Laojungou	LJG	28.52	1.2199	1.1937	0.0426	0.0566
Youfanggou	YFG	21.95	1.3688	1.1846	0.0533	0.0755
Baicaoling	BCL	28.84	1.5579	1.1795	0.0603	0.0715
Nianzigou	NZG	25.18	1.473	1.2166	0.0868	0.0508
Wenjiagou	WJG	31.99	1.4238	1.1737	0.0797	0.0321
Liulingou	LLG	26.85	1.5053	1.2000	0.0882	0.0733
Beigou	BG	29.01	1.5319	1.2643	0.0670	0.0680
Dagangou	DGG	27.07	1.5912	1.2128	0.0448	0.0781
Yaowangmiao	YWM	32.03	1.5309	1.2024	0.0728	0.0972
Yingpangoukou	YPG	24.93	1.6054	1.2290	0.0778	0.0799
Maiduoshigou	MDS	24.63	1.1576	1.0445	0.0226	0.0656
Panjiaba	PJB	32.95	1.1872	1.1293	0.0252	0.0277
Zhangjiagou	ZJG	26.17	1.2364	1.1722	0.0523	0.0464
Longwangmiao	LW	24.46	1.2953	1.2199	0.0498	0.0636
Huangbaigou	HBG	24.07	1.3276	1.1913	0.0684	0.0994
Dashuiyugou	DSY	31.74	1.2435	1.0573	0.0804	0.0704
Chenjiagou	CJG	28.75	1.4232	1.2048	0.0643	0.0544
Longwangmiao	LWM	33.62	1.2577	1.0735	0.0422	0.0722
Yinmagou	YMG	23.52	1.4565	1.2381	0.0971	0.061
Hualingou	HLG	23.34	1.1842	1.1088	0.0387	0.0603
Huojigou	HJG	30.96	1.0761	1.0182	0.0541	0.0244
Shijiagoucun	SJG	20.73	1.1253	1.0611	0.0412	0.0431
Caotangou	CTG	26.42	1.1088	1.0826	0.0315	0.0533
Liujiaping	LJP	25.93	1.2165	1.0802	0.0573	0.0961
Average		27.33	1.3645	1.1630	0.0621	0.0637
Total		56.02	1.7801	1.3732	0.1434	0.2362

Note: A_o, observed number of alleles per locus; A_e, effective number of alleles per locus; He, Nei's gene diversity; H_o, Shannon's information index; PPB, percentage of polymorphic bands.

Analysis of molecular variance. The pairwise GD values (Table 4) were small and ranged from 0.0467 (XSG and DMG) to 0.3997 (BG and LW), indicating low differentiation within populations. Low population differentiation indicated that gene flow within each clustered group may be high or that isolation time was recent.

AMOVA (Table 5) was also performed for population differentiation to further evaluate genetic structure. Highly significant ($P<0.000$ 2) genetic variance was expectedly observed among the populations and explained 63.27% of the total variance, supporting the results from the hierarchical and Bayesian clustering. Only 36.73% of the total genetic variance occurred within populations, indicating higher genetic differentiation between the populations than within each population and the emergence of genetic differentiation among populations. The G_{st} value (0.5823) (Table 2) showed there were more variation among populations than that within populations, confirming the AMOVA results.

Discussion

Genetic diversity of *S. hexandrum* in Qinling Mts

The ISSR markers developed in this study effectively revealed low genetic diversity within the *S. hexandrum* populations sampled in Qinling Mts. Populations are isolated given the population differentiation and clear clustering. The comparison of the average genetic diversity to that of *S. hexandrum* from Northwestern Himalayan region [34,36,38] and other Berberidaceae species [58,59] based on the ISSR approach showed that *S. hexandrum* populations of Qinling Mts. have low genetic diversity (average $H_e=0.0621$). The *S. hexandrum* in the Northwestern Himalayan

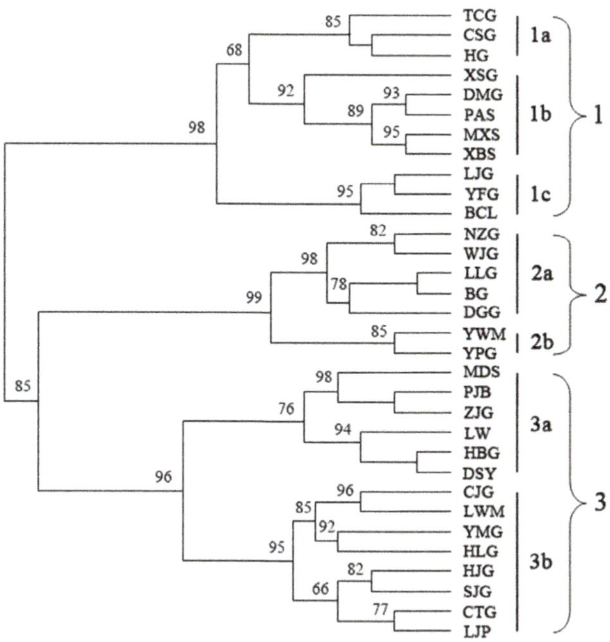

Figure 2. UPGMA clustering of *S. hexandrum* populations in Qinling Mts.

region [34,36,38] showed high genetic variation ($H_e = 0.2944$, 0.092, respectively). *S. hexandrum* is native to the Himalayan region, growing in valleys with secondary vegetation, or under shrubs or around trees [2]. Their habitat is significantly different from that of Qinling Mts. *S. hexandrum* reproduces through vegetative reproduction and seeds. Insects and birds are limited in the high altitude regions, implying that *S. hexandrum* pollination is easier than in the Qingling Mts. The sizes of wild populations of *S. hexandrum* are very small and declines each year in Qinling Mts. because of habitat fragmentation and deterioration caused by human disturbance (overcollection due to economic interests). The rapid decrease in individuals in wild populations may also cause loss of the genetic diversity of this endangered species. Another Berberidaceae species, *Dysosma versipellis*, is an endangered species endemic to China and has been listed as a key protected wild plant in China due to habitat fragmentation or destruction. Qiu et al. found this species had high level of genetic diversity in China ($H_e = 0.378$) [58]. *Dysosma pleiantha*, a threatened medicinal plant species distributed in southeastern China, sexually and asexually reproduces. High H_e (0.364) was observed in this species [59]. The relatively high level of genetic variation observed within two species suggested that the balance between vegetative reproduction and sexual reproduction was more in favor of sexual reproduction in the populations *D. versipellis* and *D. pleiantha* than in the *S. hexandrum* populations. The H_e found in this study was 0.1434 at the species level, lower than those of some strictly self-pollinating soybean species ($H_e = 0.1714$) [60] and the self-pollinating *Oryza granulata* ($H_e = 0.210$) [61], which also indicated that potential selfing system in these populations reduced genetic diversity of *S. hexandrum* populations. Historical events are also responsible for the variation in genetic diversity [62].

Genetic diversity is affected by multiple factors, such as geographical distribution, mating system, life form, pollen and seed dispersal [63,64]. Low genetic variation within populations could be attributed to seed dispersal and the predominant clonal reproduction in *S. hexandrum* in this high mountainous area. However, the genetic diversity of *S. hexandrum* is not much lower than other endangered species analyzed using ISSR markers, such as *Leontice microrhyncha* (Berberidaceae) ($H_e = 0.021$), which is a polycarpic perennial herb found in deciduous or coniferous forests

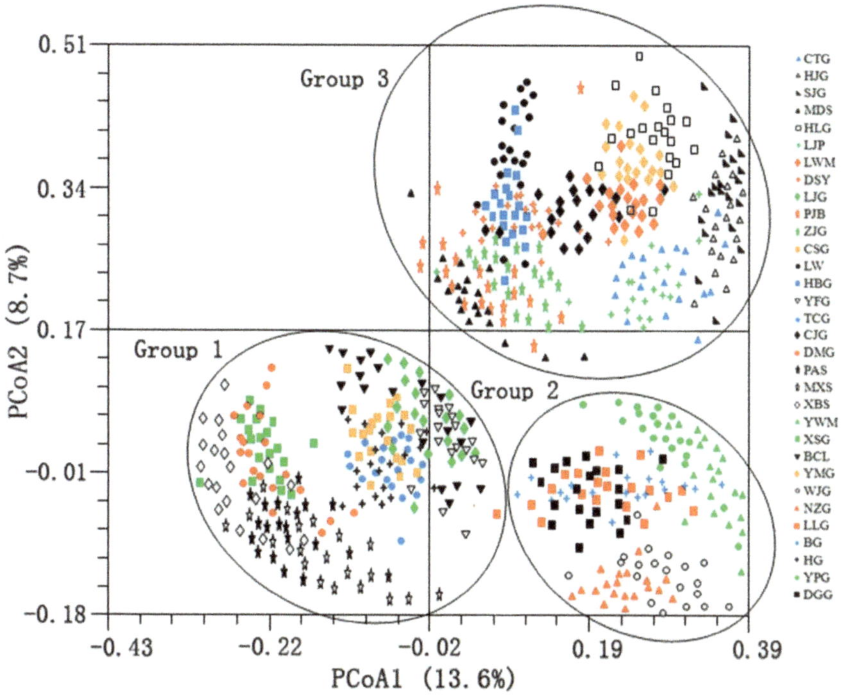

Figure 3. Distribution of individuals of the 32 *S. hexandrum* populations from Qinling Mts., according to the first (PCo1) and second (PCo2) principal coordinates. PCo1 and PCo2 account for 13.6 and 8.7% of the total variation, respectively.

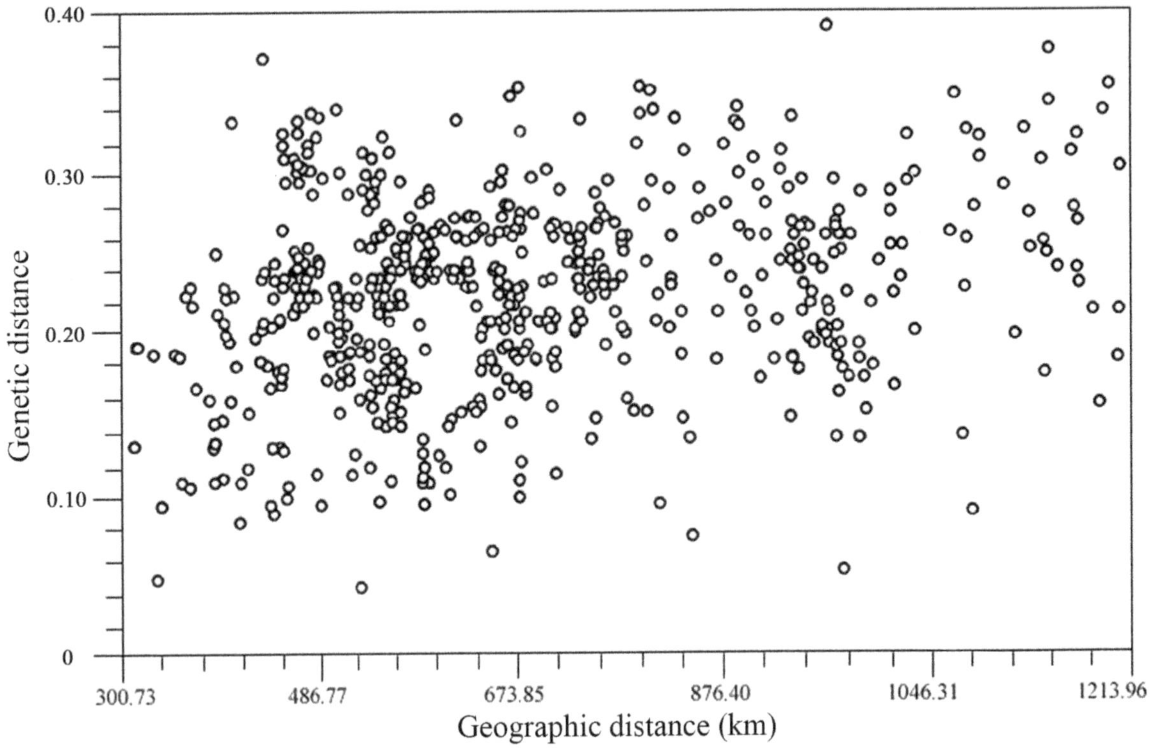

Figure 4. Mantel regression of the pairwise relationship between genetic and geographical distances for *S. hexandrum* populations.

in Korea and Northeast China [65]. For species *L. microrhyncha,* each pollinated flower produces an 8 mm berry and seed dispersal is restricted due to its heavy berry [66]. The genetic diversity of *S. hexandrum* was also higher than those of two other species, the endangered *Pinus squamata* ($H_e = 0.020$) [67] and the first-degree endangered species *Manglietia decidua* ($H_e = 0.0637$) [68]. *P. squamata* and *M. decidua* are extremely rare and endangered tree species in China. The extremely low genetic diversity of this two species could have resulted from the severe bottleneck effect during their evolutionary process. The gene drift and inbreeding may further decrease their genetic diversity in the shrinking populations. The weak competitive ability against broad - leaved trees and human activities may also accelerated the decrease of genetic variation.

S. hexandrum is reasonably long lived because its rhizomes easily reproduce, which could slow down the loss of genetic diversity. Pollen dispersal is generally restricted to a small region due to the large pollen size, which limits gene flow to increase or

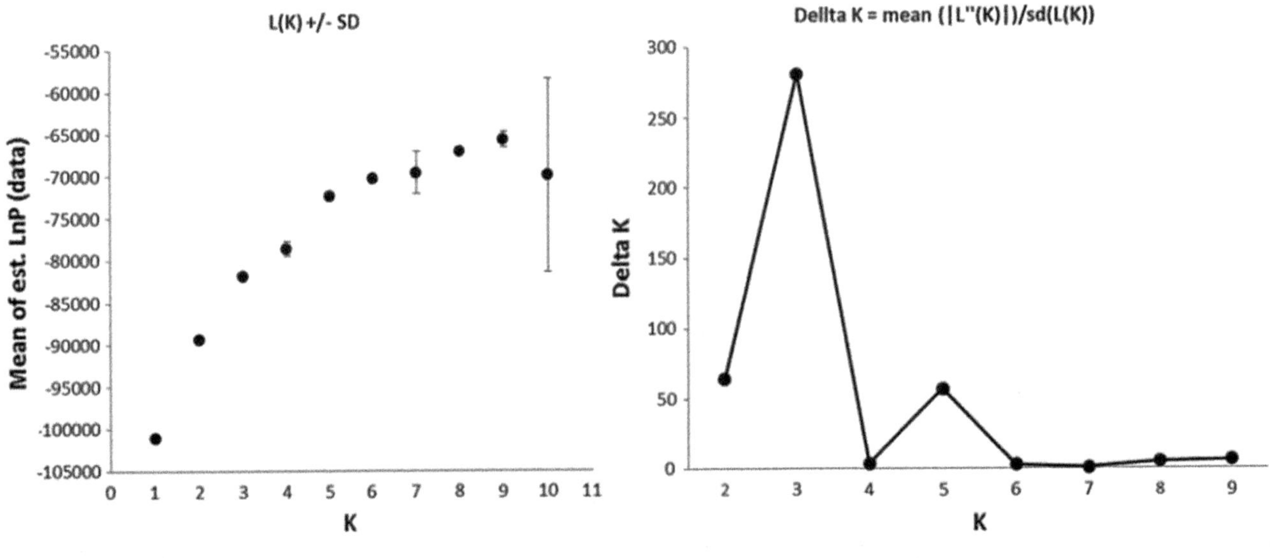

Figure 5. The probable K value estimated by likelihood of the probability of data L(K) and ad hoc quantity ΔK.

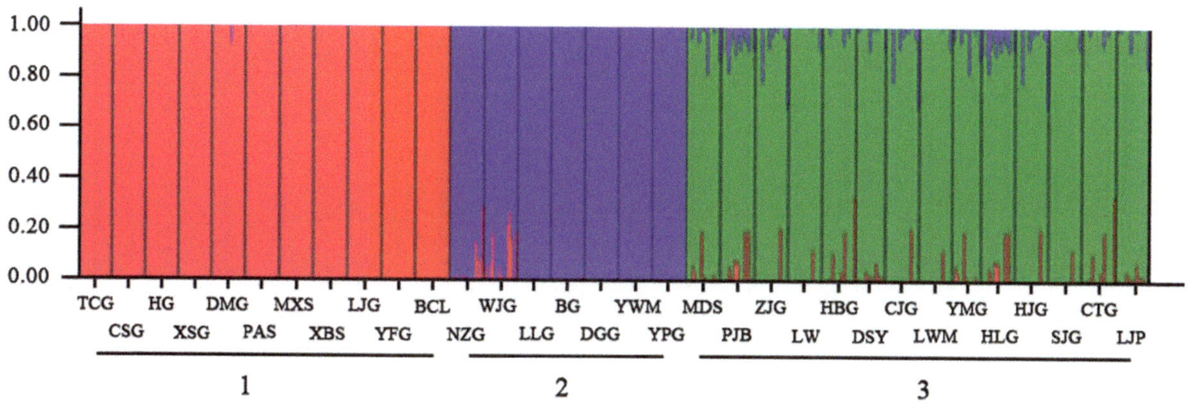

Figure 6. Bayesian clustering for infering population structure of *S. hexandrum* populations from Qinling Mts.

maintain genetic diversity. *S. hexandrum* is native to the Himalayan region, including China, India, Nepal, and Myanmar. No other genetic diversity studies of *S. hexandrum* exist in other locations. Further research should include more populations in other regions of China and countries.

High genetic differentiation and distinct genetic structure

High level of genetic differentiation and clear population structure was detected in this study. K = 3 in the Bayesian clustering as a meaningful value because group 1, 2 and 3 could be divided by detecting the within-group substructure. This result is also supported by distance-based clustering and PCoA. Groups 1 and 2 are genetically close but significantly different. Estimation of the number of clusters K should be treated with care because it is computationally difficult to obtain accurate estimates and the method merely provides an ad hoc approximation [52].

All sampled individuals were strongly assigned to their original populations, and all data strongly support the conclusion that the 32 *S. hexandrum* populations distributed in the Qinling Mts. are clustered into three major groups. These methods consistently showed that high genetic differentiation existed among *S. hexandrum* populations, which is consistent with genetic variation studies in certain selfing species [69]. This would mean that *S. hexandrum* should be a selfing species or a selfing predominant species, which is consistent with previous studies on *S. hexandrum* by Ma et al. [70]. Aside from the breeding system, the high genetic differentiation across populations may also be caused by genetic drift [71]. Wright [72] noted that genetic drift would lead a small population to emerge with a distinct genetic differentiation when the N_m value is lower than 1.0. The N_m of *S. hexandrum* (0.3587) determined using the POPGENE software was lower than 1.0 in the present study, which suggested that some genetic drift may have emerged among the populations of this species. The distribution of *S. hexandrum* populations obviously tend to fragment based on the field investigation, which is consistent with the possibility of genetic drift.

Migration of plant populations can occur through dispersal of pollen and seed [73]. But a number of factors such as fragmented geographical distribution, lack of pollinators or seed dispersers can be a barrier to gene flow between populations [74,75]. Limited gene flow among *S. hexandrum* populations may be related to inbreeding of the species and limited seed propagation distance. Some studies have found that seed dispersal is the primary factor influencing variation of gene flow and population structure [76]. Heavy mature berries of *S. hexandrum* usually drop to the ground because of rain or wind, settling some seeds in the soil, whereas others are dispersed by cattle, birds, or humans. Therefore, the short distance of seed dispersal of *S. hexandrum* probably resulted in limited gene flow among populations. Mountain ranges and rivers are possible barriers to either dispersal of pollen or rhizomes of *S. hexandrum*, reproductively isolating the populations. The restriction of gene flow associated with geographical distance is consistent with the results of previous studies on this species [34,35,36].

Implications for conservation

S. hexandrum is a rare and threatened species [4] in China. Assessment of genetic diversity is important for designing conservation strategies for threatened and endangered species [77,78]. The results of this study showed that there was low genetic diversity among *S. hexandrum* populations and genetic differentiation among populations was higher than within populations. Genetic diversity loss has deleterious effects on species fitness and threatens the population survival and could be the key reason that explains the endangerment of *S. hexandrum* in Qinling Mts. [79,80]. The estimation of the genetic diversity and population genetic structure could provide bases for *S. hexandrum* conservation and its reasonable utilization. The results will help determine what to conserve and where and how to conserve this species.

The field survey showed that the habitats of some populations have been destroyed by human disturbance for great medical value. Damage to natural habitats would led to a decrease in population sizes and probably a subsequent increase in inbreeding, decreasing its genetic diversity. In situ conservation effectively and sustainably prevents this problem. The establishment of *S. hexandrum* reserves should be the primary method because the Qinling Mts. are situated in state forest conservation areas, where cutting and hunting are restricted. The management for the conservation of genetic variability in this species should aim to preserve not only large populations but also as many of the small populations outside nature reserves as possible. Reduced levels of genetic variation, especially in the smaller populations, will affect the species' ability to adapt to changes in its habitat [81]. Positive correlations between population size, expected heterozygosity, and plant fitness were found in *Gentiana pneumonanthe* [82] and *Arnica Montana* [81]. Thus, policy plans should also be developed to stimulate seedling recruitment in the small populations (e.g., PAS and MXS). It may be dangerous to mix highly divergent populations because it could cause loss of adaptive diversity [45]. Therefore, it is necessary to improve gene flow among populations

Table 4. Pairwise genetic distance estimates from 241 ISSR markers in *S. hexandrum*.

Population	TCG	CSG	HG	XSG	DMG	PAS	MXS	XBS	LJG	YFG	BCL	NZG	WJG	LLG	BG	DGG	YWM	YPG	MDS	PJB	ZJG	LW	HBG	DSY	CJG	LWM	YMG	HLG	HJG	SJG	CTG	LJP
TCG	***																															
CSG	0.1362	***																														
HG	0.0712	0.1379	***																													
XSG	0.1756	0.1444	0.1064	***																												
DMG	0.1833	0.1433	0.1343	0.0467	***																											
PAS	0.1654	0.1373	0.1312	0.0513	0.1514	***																										
MXS	0.1532	0.1920	0.1255	0.0568	0.2222	0.0634	***																									
XBS	0.1743	0.1636	0.1168	0.0764	0.3867	0.0754	0.0822	***																								
LJG	0.1503	0.1520	0.1205	0.0608	0.1290	0.0879	0.1075	0.0732	***																							
YFG	0.1853	0.1585	0.1484	0.0654	0.1640	0.0819	0.1015	0.0797	0.1755	***																						
BCL	0.1897	0.1574	0.1453	0.0709	0.1684	0.1575	0.1262	0.0786	0.1799	0.2085	***																					
NZG	0.1974	0.1514	0.1396	0.0905	0.1461	0.1786	0.1278	0.3726	0.1876	0.1907	0.1586	***																				
WJG	0.2095	0.2061	0.1309	0.0680	0.3582	0.1546	0.1093	0.1273	0.1697	0.1790	0.1555	0.3824	***																			
LLG	0.1673	0.1777	0.2213	0.1457	0.3891	0.1896	0.1158	0.0989	0.1830	0.1994	0.1498	0.2020	0.1616	***																		
BG	0.2448	0.1592	0.2445	0.1568	0.3483	0.1940	0.1147	0.3804	0.1885	0.1907	0.1411	0.3772	0.2163	0.2304	***																	
DGG	0.2727	0.1657	0.2449	0.1368	0.1980	0.2017	0.1087	0.0869	0.2081	0.1829	0.1422	0.1818	0.1879	0.2020	0.2035	***																
YWM	0.2696	0.1646	0.2185	0.1356	0.2179	0.2138	0.1334	0.0858	0.1815	0.1772	0.1384	0.1873	0.1763	0.1835	0.1948	0.2076	***															
YPG	0.2639	0.1586	0.3831	0.1431	0.1844	0.1716	0.1350	0.0798	0.1861	0.1685	0.2015	0.2069	0.1828	0.1900	0.3893	0.2272	0.2040	***														
MDS	0.1552	0.2133	0.3086	0.0830	0.2136	0.2014	0.1142	0.1345	0.1624	0.1612	0.2136	0.1784	0.1817	0.1889	0.2172	0.2047	0.2251	0.2114	***													
PJB	0.2520	0.1849	0.3223	0.1026	0.2202	0.1462	0.1207	0.1061	0.1453	0.1891	0.1696	0.3726	0.2137	0.1829	0.2141	0.2093	0.1991	0.2318	0.2197	***												
ZJG	0.1799	0.1641	0.3050	0.0741	0.1530	0.1127	0.0896	0.3760	0.1674	0.1860	0.1694	0.0781	0.1853	0.2376	0.2084	0.1729	0.2341	0.2231	0.3793	0.1879	***											

Table 4. Cont.

Population	TCG	CSG	HG	XSG	DMG	PAS	MXS	XBS	LJG	YFG	BCL	NZG	WJG	LLG	BG	DGG	YWM	YPG	MDS	PJB	ZJG	LW	HBG	DSY	CJG	LWM	YMG	HLG	HJG	SJG	CTG	LJP
LW	0.2768	0.1706	0.3384	0.0787	0.1413	0.0843	0.1136	0.0971	0.1927	0.1803	0.1524	0.0977	0.1405	0.2092	0.3997	0.2079	0.2385	0.2474	0.2108	0.1822	0.2160	***										
HBG	0.1778	0.1695	0.1430	0.0842	0.1617	0.0753	0.1683	0.3744	0.1667	0.2150	0.1747	0.0729	0.2006	0.1884	0.1905	0.2123	0.2162	0.1946	0.2154	0.1735	0.2206	0.2416	***									
DSY	0.1574	0.1635	0.3868	0.1394	0.1530	0.0818	0.2399	0.0540	0.2017	0.1510	0.1711	0.0775	0.1307	0.1949	0.2184	0.2200	0.2283	0.2011	0.3709	0.1772	0.2261	0.2208	0.2214	***								
CJG	0.1925	0.2182	0.1617	0.1110	0.3773	0.0807	0.1154	0.3891	0.2061	0.1556	0.1624	0.1916	0.1769	0.1938	0.2153	0.2021	0.2161	0.2000	0.2405	0.2051	0.2457	0.2273	0.2461	0.0884	***							
LWM	0.1963	0.1898	0.1586	0.0865	0.1169	0.0747	0.1219	0.0929	0.1838	0.1611	0.1569	0.2112	0.1676	0.1878	0.2096	0.1899	0.2372	0.1940	0.2139	0.2020	0.2087	0.2262	0.2177	0.0624	0.0477	***						
YMG	0.2042	0.1653	0.1529	0.1514	0.3719	0.1294	0.0908	0.1008	0.1959	0.1807	0.1848	0.1836	0.1757	0.2425	0.2009	0.2110	0.2145	0.2487	0.3815	0.1963	0.2152	0.2202	0.1631	0.0974	0.0756	0.1038	***					
HLG	0.1864	0.1718	0.1442	0.1410	0.1563	0.1010	0.1148	0.0930	0.1837	0.1651	0.1817	0.1882	0.1880	0.2141	0.1936	0.1870	0.1941	0.2203	0.2240	0.1876	0.2141	0.2749	0.1910	0.1018	0.0725	0.0772	0.1090	***				
HJG	0.1747	0.1707	0.1369	0.1481	0.1640	0.0825	0.1395	0.0919	0.2048	0.1697	0.1760	0.3937	0.1935	0.1896	0.2215	0.2220	0.2292	0.2148	0.2436	0.1844	0.2081	0.2465	0.0774	0.0795	0.0668	0.0818	0.0874	0.1415	***			
SJG	0.0991	0.1647	0.3648	0.1470	0.1761	0.0890	0.1111	0.0859	0.1821	0.1752	0.1673	0.2133	0.2131	0.1961	0.2184	0.2264	0.2330	0.2344	0.2159	0.2123	0.2628	0.2220	0.0717	0.0916	0.0581	0.0873	0.0939	0.1131	0.0940	***		
CTG	0.1864	0.1894	0.1617	0.1722	0.1339	0.1366	0.1021	0.1106	0.1617	0.1948	0.1581	0.1622	0.1975	0.1950	0.2127	0.2341	0.2409	0.2096	0.3624	0.2092	0.2344	0.2285	0.0839	0.0794	0.0526	0.1069	0.0928	0.0886	0.0880	0.0853	***	
LJP	0.2107	0.1610	0.1560	0.1686	0.3655	0.1082	0.1086	0.3822	0.1968	0.1723	0.1860	0.1687	0.2021	0.1890	0.3834	0.2462	0.2231	0.2142	0.2213	0.2153	0.2700	0.2274	0.0894	0.1005	0.0805	0.0793	0.0868	0.0951	0.0847	0.0918	0.0907	***

Table 5. Analysis of molecular variance (AMOVA) for 640 individuals in 32 populations of *S. hexandrum* using 11 selected ISSR primers.

Source of variation	df	Sum of squares	Mean squares	Variance component	Total variance (%)	*P*-value
Among population	31	4,028.326	129.946	3.9128	63.27	<0.000 2
Within population	608	950.912	1.564	1.564	36.73	<0.000 2

within each group through some artificial means, such as transplanting individuals (by seed, rhizomes from one population to another). Furthermore, to avoid human overcollection, greater awareness for *S. hexandrum* protections must be emphasized, and related forest departments should be encouraged to undertake conservation through an integrated conservation strategy based on demographic, ecological, and genetic aspects.

As a supplement to in situ conservation, ex situ conservation would also be feasible as underlined by other studies on endangered species [83,84,85]. Populations may be partially preserved through seed banks or in vitro germplasm collections. *S. hexandrum* has favorable sexual reproduction. Each plant produces approximately 60 seeds, with a maximum of approximately 180 seeds [70]. Seed collection is easier for *S. hexandrum* than other endangered species. Thus, a strategy involving extensive collection to ensure full sampling of genetic diversity, subsequent cultivation in a garden at least 1000 m above sea level [7], and reintroduction into their original wild habitats seems feasible, although *S. hexandrum* mainly grows wild on high altitude mountain ranges. For *S. hexandrum* populations in Qinling Mts., there are some preserved forest farms which could be used for relocation. However, ex situ conservation has many drawbacks because it is impossible to recreate the habitat as a whole. The new environment may have important ecological differences compared with the original habitat, and the approach is technically challenging and is often expensive. Therefore, ex situ conservation is recommended only to supplement in situ conservation or as a last resort. In vitro techniques are also proven to be an effective alternative means of propagation that facilitates the recovery of the rare and endangered *S. hexandrum* [33]. At present, an effective protocol of in vitro propagation, involving multiple shoot formation from zygotic embryos and subsequent rooting, could be available for *S. hexandrum*. In vitro propagation may well be used as a means to rescue zygotic embryos for this species. In vitro techniques induce variability, but plants raised from tissue cultures may be screened for useful somaclonal variants and exploited to obtain plants or cultures with high podophyllotoxin contents, which possibly reduces the pressure on natural *S. hexandrum* populations.

Most of the genetic diversity of the important medicinal and endangered species *S. hexandrum* in Qinling Mts. must be guaranteed with these combined and sustained efforts.

Acknowledgments

The authors are also grateful to Genlu Bai, Guowei Xia and all the colleagues in the same laboratory for the assistance in the work.

Author Contributions

Conceived and designed the experiments: WL JJL. Performed the experiments: WL DXY NL. Analyzed the data: WL. Contributed reagents/materials/analysis tools: JJL. Contributed to the writing of the manuscript: WL.

References

1. Editorial committee, Chinese Academy of Sciences (2010) Vol. 29, Flora of China. Beijing: Science Press. 249–251 p. (In Chinese.).
2. Ying TS (1979) Study on Dysosma Woodson and Sinopodophyllum Ying (new genus) of the Berberidaceae. Acta Phytotax Sin 17: 17–23. (In Chinese with English abstract.).
3. Chatterjee R (1952) Indian podophyllum. Econ Bot 6: 342–354.
4. Fu LG (1992) Plant red book of China: Rare threatened plant. Beijing: Science Press. (In Chinese.).
5. Cragg GM, Newman DJ, Snader KM (1997) Natural products in drug discovery and development. J Nat Prod 60: 52–60.
6. Dong JE, Ma XH, Wei Q, Peng SB, Zhang SC (2011) Effects of growing location on the contents of secondary metabolites in the leaves of four selected superior clones of *Eucommia ulmoides*. Ind Crop Prod 34: 1607–1614.
7. Li GM (1975) Introduction a medicine plant: *Sinopodophyllum emodi Wall. var. Chinense Sprague*. Journal of Botony 2: 28. (In Chinese with English abstract.).
8. Fay DA, Ziegler HW (1985) Botanical source differentiation of Podophyllum resin by high performance liqid chromatography. J Liq Chromatogr 8: 1501–1506.
9. Giri A, Narasu ML (2000) Production of podophyllotoxin from *Podophyllum hexandrum*: a potential natural product for clinically useful anticancer drugs. Cytotechnology 34: 17–26.
10. Stahelin HF, Von Warhurg A (1991) The chemical and biological route from podophyllotoxin glucoside to etoposide: ninth cain memorial avard lecture. Cancer Res 51: 5–11.
11. Kamil WM, Dewick PM (1986) Biosynthetic relationship of aryltetralin lactone lignans to dibenzylbutyrolactone lignans. Phytochemistry 25: 2093–2102.
12. Holthuis JJM (1988) Etoposide and teniposide. Pharm Weekbl 10: 101–116.
13. Imbert TF (1998) Discovery of podophyllotoxin. Biochimie 80: 207–222.
14. Moraes RM, Lata H, Bedir E, Maqbool M, Cushman K (2002) The American mayapple and its potential for podophyllotoxin production. In: Janick J, Whipkey A, editors. Trends in new crops and new uses. Alexandria, VA: ASHS Press. pp. 527–532.
15. Yousefzadi M, Sharifi M, Behmanesh M, Moyano E, Bonfill M, et al. (2010) Podophyllotoxin: current approaches to its biotechnological production and future challenges. Eng Life Sci 4: 281–292.
16. Lata H, Moraes RM, Bertoni B, Pereira Ana MS (2010) In vitro germplasm conservation of *Podophyllum peltatum* L. under slow growth conditions. In Vitro Cell Dev Biol-Plant 46: 22–27.
17. Lin MC, Lin JH, Chen SK, Cheng YW, Cheng HW (2008) Simultaneous determination of podophyllotoxin, quercetin and kaempferol in podophyllin by liquid chromatography tandem mass spectrometry. J Food Drug Anal 16: 29–40.
18. Zhou Y, Jiang SY, Ding LS, Cheng SW, Xu HX, et al. (2008) Chemical fingerprinting of medicinal plants "Gui-jiu" by LC-ESI multiple-stage MS. Chromatographia 68: 781–789.
19. Purohit MC, Bahuguna R, Maithani UC, Rawat MSM (1999) Variation in podophylloresin and podophyllotoxin content in different populations of *Podophyllum hexeandrum*. Curr Sci India 77: 1078–1079.
20. Zhao CQ, Cao W, Nagatsu A, Ogihara Y (2001) Three new glycosides from *Sinopodophyllum emodi* (Wall.) Ying. Chem Pharm Bull 49: 1474–1476.
21. Kong Y, Xiao JJ, Meng SC, Dong XM, Ge YW, et al. (2010) A new cytotoxic flavonoid from the fruit of *Sinopodophyllum hexandrum*. Fitoterapia 81: 367–370.

22. Zhao CQ, Zhu YY, Chen SY, Ogihara Y (2011) Lignan glucoside from *Sinopodophyllum emodi* and its cytotoxic activity. Chinese Chem Lett 22: 181–184.
23. Sun YJ, Li ZL, Chen H, Liu XQ, Zhou W, et al. (2011) Three new cytotoxic aryltetralin lignans from *Sinopodophyllum emodi*. Bioorg Med Chem Lett 21: 3794–3797.
24. Qin Y, Gui MY, Yu LN, Ru H, Jin YR, et al. (2009) RP-HPLC determination of lignans in *Sinopodophyllum emodi* Wall. Chin J Pharm Anal 29: 1490–1493. (In Chinese with English abstract.).
25. Huang K, Jiang W, Zhao JF, Wang CH, Liu X, et al. (2012) Determination of podophyllotoxin and total lignans in *Sinopodophyllum emodi*. China journal of Chinese material medicia 37: 1360–1365. (In Chinese with English abstract.).
26. Inamori Y, Kubo M, Tsujibo H, Ogawa M, Baba K, et al. (1986) The biological activities of podophyllotoxin compounds. Chem pharm Bull 34: 3928–3932.
27. Goel HC, Prasad J, Sharma A, Singh B (1998) Antitumour and radioprotective action of *Podophyllum hexandrum*. Indian J Exp Biol 36: 583–587.
28. Chattopadhyay S, Bisaria VS, Panda AK, Srivastava AK (2004) Cytotoxicity of invitro produced podophyllotoxin from *Podophyllum hexandrum* on human cancer cell line. Nat Prod Res 18: 51–57.
29. Reddy PB, Paul DV, Agrawal SK, Saxena AK, Kumar HMS, et al. (2008) Design, synthesis, and biological testing of 4b-[(4-Substituted)-1,2,3-triazol-1-yl] podophyllotoxin analogues as antitumor agents. Arch Pharm Chem Life Sci 341: 126–131.
30. Zhu CG, Yang J, Xiong Y (2004) Research progress in natural antineoplastic podophyllotoxin and its derivatives. Drug Eval 1: 306.
31. Wang DW, Guo FX, Ma XY (1997) The antitumor activity of *Sinopodophyllum emodi*. Chinese medicine matria 20: 571–574. (In Chinese with English abstract.).
32. Li GY (2005) Effects of ethanol extracts from *Podophylum emodi var.* on proliferation and apoptosis of breast careinoma cell line MCF-7. Chinese Journal of New Drugs 26: 2185–2188. (In Chinese with English abstract.).
33. Nadeem M, Palni LMS, Purohit AN, Pandey H, Nandi SK (2000) Propagation and conservation of *Podophyllum hexandrum* Royle: an important medicinal herb. Bio Conserv 92: 121–129.
34. Alam MA, Naik PK, Gulati P, Gulati AK, Mishra GP (2008) Characterization of genetic structure of *Podophyllum hexandrum* populations, an endangered medicinal herb of Northwestern Himalaya, using ISSR-PCR markers and its relatedness with podophyllotoxin content. Afr J Biotechnol 7: 1028–1040.
35. Alam MA, Gulati P, Aswini KG, Gyan PM, Pradeep KN (2009) Assessment of genetic diversity among *Podophyllum hexandrum* genotypes of Northwestern Himalayan region for podophyllotoxin production. Indian J Biotechnol 8: 391–399.
36. Naik PK, Alam MA, Singh H, Goyal V, Parida S, et al. (2010) Assessment of genetic diversity through RAPD, ISSR and AFLP markers in *Podophyllum hexandrum*: a medicinal herb from the Northwestern Himalayan region. Physiol Mol Biol Plants 16: 135–148.
37. Xiao M, Li Q, Guo L, Luo T, Duan WX, et al. (2006) AFLP analysis of genetic diversity of the endangered species *Sinopodophyllum hexandrum* in the tibetan region of Sichuan province, China. Biochem Genet 44: 47–59.
38. Xiao M, Li Q, Wang L, Guo L, Li J, et al. (2006) ISSR analysis of the genetic diversity of the endangered species *Sinopodophyllum hexandrum* (Royle) Ying from Western Sichuan province, China. J Integr Plant Bio 48: 1140–1146.
39. Li Y, Zhai SN, Qiu YX, Guo YP, Ge XJ, et al. (2011) Glacial survival east and west of the 'Mekong–Salween Divide' in the Himalaya–Hengduan mountains region as revealed by AFLPs and cpDNA sequence variation in *Sinopodophyllum hexandrum* (Berberidaceae). Mole Phylogenet Evol 59: 412–424.
40. Totland Ø (1999) Effects of temperature on performance and phenotypic selection on plant traits in alpine *Ranunculus acris*. Oecologia 120: 242–251.
41. Esselman EJ, Li J, Crawford DJ, Winduss JL, Wolfe AD (1999) Clonal diversity in the rare *Calamagrostis porteri ssp. insperata* (Poaceae): Comparative results for allozymes and random amplified polymorphic DNA (RAPD) and inter-simple sequence repeat (ISSR) marker. Mol Ecol 8: 443–451.
42. Ge XJ, Sun M (1999) Reproductive biology and genetic diversity of a cryptoviviparous mangrove *Aegiceras corniculatum* (Myrsinaceae) using allozyme and inter-simple sequence repeat (ISSR) analysis. Mol Ecol 8: 2061–2069.
43. Li HS, Chen GZ (2004) Genetic diversity of *Sonneratia alba* in China detected by inter-simple sequence repeat (ISSR) analysis. Acta Bot Sin 46: 515–521.
44. Thriveni HN, Sumangala RC, Shivaprakash KN, Ravikanth G, Vasudeva R, et al. (2014) Genetic structure and diversity of *Coscinium fenestratum*: a critically endangered liana of Western Ghats, India. Plant Syst Evol 300: 403–413.
45. Tang N, Mo G, van Tuyl JM, Arens P, Liu JJ, et al. (2014) Genetic diversity and structure of *Lilium pumilum* DC. in southeast of Qinghai–Tibet plateau. Plant Syst Evol 300: 1453–1464.
46. Bornet B, Branchard M (2001) Nonanchored inter simple sequence repeat (ISSR) markers: reproducible and specific tools for genome fingerprinting. Plant Mol Biol Rep 19: 209–215.
47. Nagaoka T, Ogihara Y (1997) Applicability of inter-simple sequence repeat polymorphisms in wheat for use as DNA markers in comparison to RFLP and RAPD markers. Theor Appl Genet 94: 597–602.
48. Li JJ, Zhu ZC, Min ZL (1989) Comprehensive survey of the Taibai Mountain preserve (1st ed). Xi'an: Shaanxi Normal University Press. 1–9 p. (In Chinese.).
49. Yeh FC, Yang RC, Boyle T, Ye ZH, Mao JX (1997) POPGENE: The user friendly shareware or population genetic analysis. Molecular Biology and Biotechnology Center, University of Alberta, Edmonton.
50. Liu KJ, Muse SV (2005) PowerMarker: an integrated analysis environment for genetic marker analysis. Bioinformatics 21: 2128–2129.
51. Felsenstein J (1985) Confidence limits on phylogenies: an approach using the bootstrap. Evolution 39: 783–791.
52. Pritchard JK, Stephens M, Donnelly P (2000) Inference of population structure using multilocus genotype data. Genetics 155: 945–959.
53. Evanno G, Regnaut S, Goudet J (2005) Detecting the number of clusters of individuals using the software Structure: a simulation study. Mol Ecol 14: 2611–2620.
54. Jensen RJ (1989) Ntsys-Pc-numerical taxonomy and multivariate analysis system-version 1.40. Q Rev Biol 64: 250–252.
55. Exeoeffr L, Smouse PE, Quattro JM (1992) Analysis of molecular variance inferred from metric distances among DNA haplotypes: applications to human mitochondrial DNA restriction data. Genetics 131: 479–491.
56. Miller MP (1997) Tools for Population Genetic Analysis (TFPGA), Vesion 1.3. USA: Department of Biological Sciences, Northern Arizona University, Ariozna.
57. Larson SR, Palazzo AJ, Jensen KB (2003) Identification of western wheatgrass cultivars and accessions by DNA fingerprinting and geographic provenance. Crop Sci 43: 394–401.
58. Qiu YX, Li JH, Liu HL, Chen YY, Cheng XF (2006) Population structure and genetic diversity of *Dysosma versipellis* (Berberidaceae), a rare endemic from China. Biochem Syst Ecol 34: 745–752.
59. Zong M, Liu HL, Qiu YX, Yang SZ, Zhao MS, et al. (2008) Genetic diversity and geographic differentiation in the threatened species *Dysosma pleiantha* in China as revealed by ISSR analysis. Biochem genet 46: 180–196.
60. JinY, Zhang WJ, Fu D X, Lu BR (2003) Sampling strategy within a wild soybean population based on its genetic variation detected by ISSR markers. Acta Bot Sin 45: 995–1002. (In Chinese with English abstract.).
61. Wu CJ, Chen ZQ, Huang XQ, Yin SH, Cao KM, et al. (2004) Genetic diversity among and within populations of *Oryza granulata* from Yunnan of China revealed by RAPD and ISSR markers: implications for conservation of the endangered species. Plant Sci 167: 35–42.
62. Karron JD (1991) Patterns of genetic variation and breeding systems in rare plant species. In: Falk DA, Holsinger, KE, editors. Genetics and conservation of rare plants. New York: Oxford University Press. pp. 87–98.
63. Hamrick JL, Godt MJW, Sherman-Broyles SL (1992) Factors influencing levels of genetic diversity in woody plant species. New Forest 6: 95–124.
64. Ohsawa T, Saito Y, Sawada H, Ide Y (2008) Impact of altitude and topography on the genetic diversity of *Quercus serrata* populations in the Chichibu Mountains, central Japan. Flora 203: 187–196.
65. Lee TB (1980) Illustrated flora of Korea. Seoul: Hyang-mun Pub. Co. (In Korean.).
66. Chang CS, Kim H, Park TY, Maunder M (2004) Low levels of genetic variation among southern peripheral populations of the threatened herb, *Leontice microrhyncha* (Berberidaceae) in Korea. Biol Conserv 119: 387–396.
67. Zhang ZY, Li DZ (2003) Conservation genetics of extremely endangered pine, *Pinus squamata*. Acta Botanica Yunnanica 25: 544–55. (In Chinese with English abstract.).
68. Liao WF, Xia NH, Deng YF, Zheng QY (2004) Study on genetic diversity of *Manglietia decidua*. Acta Botanica Yunnanica 26: 58–64. (In Chinese with English abstract.).
69. Nybom H, Bartish IV (2000) Effects of life history traits and sampling strategies on genetic diversity estimates obtained with RAPD markers in plants. Plant Ecol Evol Syst 3: 93–114.
70. Ma SB, Xu ZY, Hu ZH (1997) A contribution to the reproductive biology of *Sinopodophyllum hexandrum* (Role) Ying (Berberidaceae). Acta Bot Boreal – Occident Sin 17: 49–55. (In Chinese with English abstract.).
71. Hamrick JL, Godt MJW (1989) Allozyme diversity in plant species. In: Brown AHD, Clegg MT, Khaler AL, Weir BS, editors. Plant population genetics, breeding and genetic resources. Sunderland: Sinauer Associates, Inc., Publisher. pp. 43–63.
72. Wright S (1965) The interpretation of population structure by F-statistics with special regard to systems of mating. Evolution 19: 395–420.
73. Ennos RA (1994) Estimating the relative rates of pollen and seed migration among plant-populations. Heredity 72: 250–259.
74. Slatkin M (1985) Gene flow in natural-populations. Annu Rev Ecol Syst 16: 393–430.
75. Zhou TH, Qian ZQ, Li S, Guo ZG, Huang ZH, et al. (2010) Genetic diversity of the endangered Chinese endemic herb *Saruma henryi Oliv.* (Aristolochiaceae) and its implications for conservation. Popul Ecol 52: 223–231.
76. Kalisz S, Hanzawa FM, Tonsor SJ, Thiede DA, Voigt S (1999) Ant-mediated seed dispersal alters pattern of relatedness in a population of *Trillium grandifloru*. Ecology 80: 2620–2634.
77. Hamrick JL (1983) The distribution of genetic variation within and among natural plant populations. In: Schonewald-Cox CM, Chambers SM, McBryde B, Thomas WL, editors. Genetics and conservation. Menlo Park: Benjamin Cummings Publishing Company. pp. 335–348.
78. Francisco-Ortega J, Santos-Guerra A, Kim SC, Crawford DJ (2000) Plant genetic diversity in the Canary Islands: a conservation perspective. Am J Bot 87: 909.
79. Malone CL, Knapp CR, Taylor JF, Davis SK (2003) Genetic consequences of pleistocene fragmentation: isolation, drift, and loss of diversity in rock iguanas (Cyclura). Conserv Genet 4: 1–15.

80. Reed DH (2003) Correlation between fitness and genetic diversity. Conserv Biol 17: 230–237.
81. Luijten SH, Dierick A, Oostermeijer JGB, Raijmann LEL, Den Nijs JCM (2000) Population size, genetic variation and reproductive success in the rapidly declining, self-incompatible perennial (*Arnica montana*) in the Netherlands. Conserv Biol 14: 1776–1787.
82. Oostermeijer JGB, Van Eijck MW, Van Leeuwen NC, Den Nijs JCM (1995) Analysis of the relationship between allozyme heterozygosity and fitness in the rare *Gentiana pneumonanthe* L. J Evol Biol 8: 739–757.
83. Bunn E (2005) Development of in vitro methods for ex situ conservation of *Eucalyptus impensa*, an endangered mallee from southwestern Western Australia. Plant Cell Tiss Org 83: 97–102.
84. Cochrane JA, Crawford AD, Monks LT (2007) The significance of ex situ seed conservation to reintroduction of threatened plants. Aust J Bot 55: 356–361.
85. Li QM, Xu ZF, He TH (2002) Ex situ genetic conservation of endangered *Vatica guangxiensis* (Dipterocarpaceae) in China. Biol Conserv 106: 151–156.

19

Great Apes and Biodiversity Offset Projects in Africa: The Case for National Offset Strategies

Rebecca Kormos[1]*, Cyril F. Kormos[2], Tatyana Humle[3], Annette Lanjouw[4], Helga Rainer[5], Ray Victurine[6], Russell A. Mittermeier[7], Mamadou S. Diallo[8], Anthony B. Rylands[9], Elizabeth A. Williamson[10]

1 Department of Integrative Biology, University of California, Berkeley, California, United States of America, **2** The WILD Foundation, Berkeley, California, United States of America, **3** Durrell Institute of Conservation and Ecology, School of Anthropology and Conservation, University of Kent, Canterbury, United Kingdom, **4** Strategic Initiatives and Great Apes Program, The Arcus Foundation, Cambridge, United Kingdom, **5** Conservation Program, The Arcus Foundation, Cambridge, United Kingdom, **6** Business Conservation Initiative and Conservation Finance, Wildlife Conservation Society, Bronx, New York, United States of America, **7** Conservation International, Arlington, Virginia, United States of America, **8** Guinée Écologie, Conakry, Republic of Guinea, **9** Conservation International, Arlington, Virginia, United States of America, **10** Scottish Primate Research Group, School of Natural Sciences, University of Stirling, Scotland, United Kingdom

Abstract

The development and private sectors are increasingly considering "biodiversity offsets" as a strategy to compensate for their negative impacts on biodiversity, including impacts on great apes and their habitats in Africa. In the absence of national offset policies in sub-Saharan Africa, offset design and implementation are guided by company internal standards, lending bank standards or international best practice principles. We examine four projects in Africa that are seeking to compensate for their negative impacts on great ape populations. Our assessment of these projects reveals that not all apply or implement best practices, and that there is little standardization in the methods used to measure losses and gains in species numbers. Even if they were to follow currently accepted best-practice principles, we find that these actions may still fail to contribute to conservation objectives over the long term. We advocate for an alternative approach in which biodiversity offset and compensation projects are designed and implemented as part of a National Offset Strategy that (1) takes into account the cumulative impacts of development in individual countries, (2) identifies priority offset sites, (3) promotes aggregated offsets, and (4) integrates biodiversity offset and compensation projects with national biodiversity conservation objectives. We also propose supplementary principles necessary for biodiversity offsets to contribute to great ape conservation in Africa. Caution should still be exercised, however, with regard to offsets until further field-based evidence of their effectiveness is available.

Editor: Clinton N. Jenkins, Instituto de Pesquisas Ecológicas, Brazil

Funding: Funding for part of this study was provided by the Arcus Foundation grant no. 1104-38 (http://www.arcusfoundation.org/). The funders had no role in study design, data collection and analysis, decision to publish. They did contribute to discussions of these ideas and preparation of the manuscript.

Competing Interests: The authors have declared that no competing interests exist.

* Email: rebeccakormos@yahoo.com

Introduction

Great apes–gorillas, chimpanzees, and bonobos–are distributed across 21 countries on the African continent [1]. Their conservation is important in several respects. Their geographic ranges are strongly associated with the tropical forests that harbor some of the richest biodiversity in the world and overlap extensively with those of many endemic species [2]. Great apes have large home ranges [3], and thus protection of their habitat will also bring many other species under protection. Great apes are keystone species, playing important roles in maintaining the health and diversity of their ecosystems through their seed dispersal [4–6]. In addition, they act as physical ecosystem engineers [7,8] shaping the forest structure by trampling, bending and breaking vegetation as they travel, forage and build nests [9,10]. Apes and ape habitat are important for people; protecting ape habitats protects important water catchment areas. In Rwanda, for example, the Volcanoes National Park provides much of that country's water [11], and the Fouta Djallon in Guinea is the source of a number of West Africa's major rivers, including the Niger. Tourism with great apes also provides significant income to local communities through revenue sharing and local businesses, thus providing livelihood opportunities for local people [12].

All great ape taxa are listed as either Endangered (EN) or Critically Endangered (CR) by the International Union for Conservation of Nature (IUCN) [13]. Threats to great apes include habitat loss, hunting and disease [13]. These threats are exacerbated by large-scale development activities such as hydro-electric projects, roads, and extractive industries [14], all of which result in the destruction of large areas of ape habitat and provide access to remote areas, facilitating bushmeat hunting [15]. Industrial development projects result in large influxes of people, exposing great apes to human diseases that can be fatal to them [16]. Human presence and activity can cause apes to leave their habitual ranges, which can result in competition, conflict and stress, with long-term consequences for the health and reproduction of the population [17,18]. Mortalities are likely to occur if chimpanzees are forced into an area that is already occupied by conspecifics because chimpanzees are highly territorial and often attack intruders [19]. The effects of development projects are intensified for great apes because of their reproductive biology and slow maturation [14]. Even low levels of disturbance to ape populations can result in declines that require decades for their subsequent recovery [20].

Industrial development is proliferating throughout Africa [21,22]. The interface between development projects and great ape conservation will, therefore, intensify in coming decades. Most countries in Africa where great apes occur rank high on the United Nations poverty index and are undergoing intensive infrastructure development [23]. A mining boom is occurring [24,25], which will result in the expansion of transportation infrastructure [25]. More than 50% of the range of chimpanzees and gorillas in Western Equatorial Africa has been allocated to logging concessions [26]. The extensive overlap between the distribution of commodities, biodiverse areas and great ape ranges means that companies will increasingly need to mitigate the negative impacts of their projects on great ape populations [14,27].

Options for mitigating impacts on great apes are limited. Relocation is risky [28] and can lead to mortalities [29,30], and restoring habitat is not feasible on a time-scale meaningful to great apes [31]. Unless great ape habitat is avoided entirely, in most cases mitigation is unlikely to prevent great ape losses and most projects will result in some population decline.

"Biodiversity offsets" are increasingly used worldwide to compensate for the negative impacts of development and private sector projects on biodiversity [31–40]. The Business and Biodiversity Offset Program (BBOP), a broad consortium including civil society and private sector organizations, financial institutions, governments, and intergovernmental organizations, defines biodiversity offsets as "measurable conservation outcomes resulting from actions designed to compensate for significant residual adverse biodiversity impacts arising from project development after appropriate prevention and mitigation measures have been taken" [36]. According to BBOP, the goal of biodiversity offsets is "to achieve no net loss and preferably a net gain of biodiversity on the ground with respect to species composition, habitat structure, ecosystem function and people's use and cultural values associated with biodiversity" [36]. Offsets are distinct from the broader category of biodiversity compensation projects, which mitigate impacts but do not follow a mitigation hierarchy or comply with other offset requirements [36].

Seventeen countries worldwide have national policies requiring biodiversity offsets, and more than 29 countries have national policies that suggest or enable the use of offsets [41]. No countries in the range of great apes in West and Central Africa, however, currently have policies guiding or requiring offsets [41]. Biodiversity offsets are therefore guided by private sector internal standards or those of lenders, rather than by government policy [42–45]. Several international organizations such as BBOP, the International Council on Mining and Metals (ICMM) and IUCN have proposed best practices for biodiversity offsets [36,46]. BBOP best practices include a list of 10 guiding "principles" for the design and implementation of biodiversity offsets. These principles are reflected in many national offset policies around the world [34], in the scientific literature [31–35,37–39,47], in private sector internal guidelines [42], and in other international best practices [46], and are reflected in the lending standards and principles of many of the largest international banks [45,46].

Despite this new emphasis on biodiversity offsets, there is little empirical evidence from the field to demonstrate that they are achieving conservation objectives over the long term. This is a result of the lack of standardized evaluation criteria, limited monitoring of projects, under-reporting of projects that are not working [31], and lack of access to information on projects due to confidentiality of reports. In addition, many offsets projects are still in the design or early implementation phase and do not yet have results to report. Biodiversity offset policies are still in their infancy [38], and more reviews of field projects are urgently needed before encouraging their wider use as a conservation tool [37,38].

To assess how well field projects adhere to international best practice principles, we examine four projects in Africa where private sector or development projects are impacting great apes and their habitat and are seeking to use offsets and compensation projects to counterbalance these impacts. We measure them against the six BBOP principles related to biological criteria: (1) limits to what can be offset, (2) adherence to the mitigation hierarchy, (3) additional conservation outcomes, (4) landscape context, (5) no net loss, and (6) long-term outcomes. We also assess whether full compliance to these principles would be sufficient to generate conservation benefits for great apes in light of their EN and CR status, their shrinking habitat, and their vulnerability to disturbance. Although this paper focuses on great apes, we believe the results of this study will also apply to many other EN and CR taxa.

Case Studies

We examine four projects in Africa that are investigating either biodiversity offsets or compensation for residual impacts to great apes and their habitat, or have attempted such projects (Fig. 1). They are: (1) the Simandou Project in the Republic of Guinea, (2) the Global Alumina Project (GAP) in the Republic of Guinea, (3) the Bumbuna Hydroelectric Project (BHP) in Sierra Leone, and (4) the Lom Pangar Dam in Cameroon. For each project, we researched all publicly available documents that included information on measuring and mitigating impacts on great apes and their habitat. Table 1 provides a list of those documents, available as of May 2014. We are also aware of other projects that have considered, or are considering, the use of offsets to compensate for residual damage to ape habitat in Africa. We focus on these four case studies because they are those of which the authors have the most direct experience. It will be important to eventually expand this type of analysis to include a larger set of projects. The current analysis, however, is an important first step. Table 2 provides a summary of predicted impacts on great apes and proposed mitigation measures for each project.

The Simandou Project, Republic of Guinea

Simfer is a Guinean-registered company and holder of an iron-ore mining concession called the "Simandou Project" in the Simandou mountains of Southeast Guinea. The "Simandou Project" partners include the Republic of Guinea, Rio Tinto, Aluminium Corporation of China ("Chinalco"), and the International Finance Corporation ("IFC"). The proposed Simandou Project includes: (i) an open-pit iron-ore mine in the Simandou mountain range; (ii) approximately 670 km of railway across Guinea to transport ore to the coast; (iii) a new port facility; and (iv) associated infrastructure, such as housing, roads, quarries, and power generation and distribution.

The Simandou Project has been collecting data on chimpanzees in the Pic de Fon Classified Forest since 2007 to guide the development and implementation of a mitigation plan–the Pic de Fon Management Plan and the Simandou Project Biodiversity Offsets Strategy. The study has identified an estimated 36–46 western chimpanzees (*Pan troglodytes verus*) living in the forest in

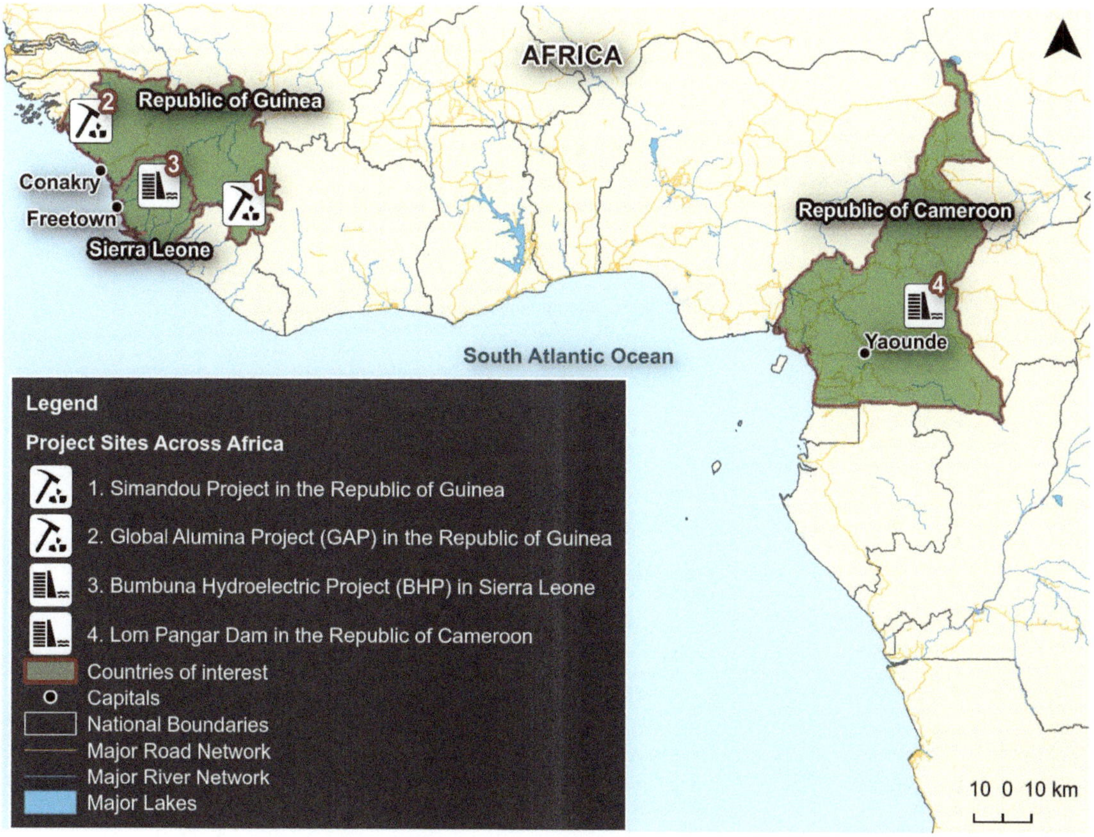

Figure 1. Sites in Africa where private sector or development projects are seeking to use offsets and compensation projects to counterbalance residual negative impacts to great apes and their habitat. Sites include (1) the Simandou Project in the Republic of Guinea, (2) the Global Alumina Project (GAP) in the Republic of Guinea, (3) the Bumbuna Hydroelectric Project (BHP) in Sierra Leone, and (4) the Lom Pangar Dam in the Republic of Cameroon.

the Simandou range dispersed across one or two communities. The number of chimpanzees along the railway is unknown, but 2,750 chimpanzee nests were recorded along the rail study area. The Social and Environmental Impact Assessment (SEIA) outlines the range of impacts on chimpanzees caused by mining activities and forecasts "best-case" and "worst-case" scenarios, dependent on the number of communities of chimpanzees and the relationships between them. Assuming that there are two separate communities, the worst-case scenario predicts a high degree of chimpanzee mortality when the communities are forced together as they lose habitat and move away from mining activities. An estimated 25% of the core of the chimpanzees' range would be permanently and irrecoverably lost to mining.

Mitigation proposed for the chimpanzees in the Simandou mountains includes controlling hunting, protecting habitat currently within the chimpanzee's range that will not be lost to mining, and creating additional habitat for chimpanzees both prior to and during mining activities. The SEIA predicts that, despite mitigation efforts, the sub-montane forest habitat where chimpanzees are living will be impacted, and the project is therefore investigating an offset site to compensate for residual damage to this unique habitat and other species living there. Simfer has formed a technical group called the Simandou Offsets Working Group with representatives from Simfer, the Environment Ministry, and the NGO Guinée Ecologie.

On 26th May 2014, the Government of Guinea, Rio Tinto, Chinalco and the IFC, signed an Investment Framework for Blocks 3 and 4 of Simandou, which now makes it the largest combined iron-ore and infrastructure project ever developed in Africa.

Global Alumina Project (GAP), Republic of Guinea

The "Guinea Alumina Project" (GAP) is a 690-km^2 bauxite-mining concession in northwest Guinea that was developed in 2008 by the Global Alumina Corporation–a joint venture of BHP Billiton, Global Alumina, Dubai Aluminum Company Ltd., and the Mubadala Development Company. The initial exploitation zone was approximately 100 km^2, to be operational for 16 years. GAP includes a mine, an alumina refinery, a steam and power plant, a port facility, additional infrastructure on the concession, and a port facility 82 km from the refinery. GAP expected to employ about 12,000 workers during the 4-year construction period and more than 2,100 employees thereafter.

GAP originally hired the company Bechtel to conduct a chimpanzee Critical Habitat study, but the IFC requested that the study be redone. A second Critical Habitat study was conducted, although only two weeks were allowed for the fieldwork. The second study estimated that a minimum of 50 chimpanzees was living in the area surveyed [48]. GAP then hired the Wild Chimpanzee Foundation (WCF) to conduct a longer-term survey of chimpanzees covering the entire concession, to assist with the design and implementation of a conservation management plan, and to look for conservation sites to offset the expected decrease in the chimpanzee population caused by the

Table 1. Publicly available documentation for each site.

Project	Country	Document type	Source
Simandou Project	Guinea	Social and Environmental Baseline Study 2010	http://ifcext.ifc.org/ifcext/spiwebsite1.nsf/0/A87B7EA570082C41852578E700569CED/$File/Vol.%20D_Biodiversity%20baseline_FINAL.pdf
		Social and Environmental Impact Assessment Chapter 12. Biodiversity	http://www.riotintosimandou.com/documents/Mine/M_Ch12_TerrBiodiv_EN.pdf
		Social and Environmental Impact Assessment Annex12.E. West African Chimpanzee - Supplementary Baseline and Impact Assessment Information	http://www.riotintosimandou.com/documents/Mine/M_An12E_Chimp_EN.pdf
		Environmental and Social Action Plan July 2013	http://ifcext.ifc.org/ifcext/spiwebsite1.nsf/0/A87B7EA570082C41852578E700569CED/$File/Simandou%20Project%20ESAP%20July%202013%20FINAL.pdf
Global Alumina Project	Guinea	Social and Environmental Assessment	http://ifcext.ifc.org/IFCExt/spiwebsite1.nsf/78e3b305216fcdba85257a8b0075079d/8a0ee1048673cb16852576ba000e2cac?opendocument
		Critical Habitat Assessment 2008	http://ifcext.ifc.org/ifcext/spiwebsite1.nsf/0/8A0EE1048673CB16852576BA000E2CAC/$File/Guinea%20Critical%20Habitat%20Assessment%20Report.pdf
		Action Plan 2008	http://ifcext.ifc.org/ifcext/spiwebsite1.nsf/0/8A0EE1048673CB16852576BA000E2CAC/$File/Post%20Comm%20Action%20Plan%20FINAL%20FINAL290808.pdf
Bumbuna	Sierra Leone	Environmental Impact Assessment 2005	http://www-wds.worldbank.org/external/default/WDSContentServer/WDSP/IB/2005/03/10/000012009_20050310135611/Rendered/PDF/E10930V.02.pdf
		2005	http://bumbuna.sl/admin/images/news/ESAP%20M4%20Draft%20Final%20Report%20for%20transmission%20JJ%2005.03.09.pdf
		Environmental and Social Advisory Panel Report 2010	http://www.bumbuna.sl/admin/images/news/ESAP%20M1%20-%20Final%20Report%2010.11.04.pdf
		ESAP mission report 2010	http://www.bumbuna.sl/admin/images/news/ESAP%20M5%20Draft%20Final%20Report%20ver%202%20JJ%2003%2011%2010.pdf
Lom Pangar	Cameroon	Social and Environmental Impact Assessment 2010	http://www.edc-cameroon.org/IMG/pdf/sde/ANNEXE%204%20PNDD%20projet%20110111.pdf
		Environmental and Social Assessments 2011	http://www-wds.worldbank.org/external/default/WDSContentServer/WDSP/IB/2012/03/07/000350881_20120307112131/Rendered/PDF/673550BR0P11400ffiicial0Use0Only090.pdf

project. Information on the predicted habitat and population decline as a result of project activities and on suggested mitigation for mining activities on chimpanzees is not publicly available.

The Global Alumina Corporation (GAC) is now a branch of the newly formed Emirates Global Aluminum (EGA) founded by Mubadala and DUBAL, and predicted to become the fifth-largest aluminum company in the world by production in 2014 [49].

The Bumbuna Hydroelectric Project (BHP), Sierra Leone

The Bumbuna Hydroelectric Project (BHP) Phase I is a 50-MW, water regulation and hydropower facility located on the Seli River near Bumbuna, Sierra Leone. The project consists of an 88-m high asphalt concrete-faced rock-fill dam, a water intake structure, two spillways with associated tunnels, an above-ground powerhouse with two 25-MW turbo-generator units and a 30-km wide, Y-shaped reservoir. The USD 91.8 million project was funded by the African Development Bank, the Government of Italy, Organization of the Petroleum Exporting Countries (OPEC), the Netherlands Clean Development Facility, the World Bank, UK DFID, and the Government of Sierra Leone.

An Environmental Impact Assessment (EIA) was completed in 2005 [50]. An Environmental Management and Mitigation Plan (EMP) was designed to include the construction phase, and the mitigation and monitoring activities associated with the first component, and includes the preparation and initiation of the Bumbuna Watershed Management Plan. In 2004, the World Bank assembled an Environmental and Social Advisory Panel (ESAP) to review these studies. In 2006, a full biodiversity assessment was conducted as a follow-up to the BHP EIA. As part of this, the size, distribution and socio-ecology of the chimpanzee population in the immediate catchment area was studied to determine the impact on chimpanzees of the filling of the reservoir and the associated loss of riparian forest [51]. The 2006 study estimated that four communities totaling 33–58 chimpanzees used the area to be flooded and that the main impacts would be (1) loss of the chimpanzees' natural resources, (2) an increase in human-wildlife

Table 2. Summary of predicted project impacts on great apes and proposed mitigation measures.

Project	Country	Type of project	Main Project components	Estimated # of apes possibly affected by project activities	Main predicted impacts on apes	Main proposed mitigation measures for ape conservation	Proposed offset site
Simandou	Guinea	Iron-ore mining	(i) An open-pit iron-ore mine in the Simandou mountain range; (ii) approximately 670 km of railway across Guinea to transport ore to the coast; (iii) a new port facility; (iv) associated infrastructure, incl. housing, roads, quarries, power generation and distribution	<50 at the mine site and unknown number along the railway	Open-pit mining will remove an area approximately 6–8 km long, 1–1.5 km wide and 300 m deep. One quarter of the core of the chimpanzees' range would be permanently and irrecoverably lost to mining	(i) Improving control of hunting; (ii) protecting habitat within the chimpanzee's current range; iii) creating additional habitat for chimpanzees both prior to and during mining activities	Undecided
Global Alumina Corporation	Guinea	Bauxite mining	(i) Mine sites in the 690-km^2 concession in northwest Guinea; (ii) an alumina refinery; (iii) a steam and power plant; (iv) a port facility 82 km from the refinery; (v) additional infrastructure on the concession	50–>100 in 247.6 km^2 of 2008 Critical Habitat survey area	Information not available	Information not available	Undecided
Bumbuna	Sierra Leone	Hydroelectric dam	(i) An 88-m high asphalt concrete-faced rock-fill dam; (ii) a water intake structure; (iii) two spillways with associated tunnels; (iv) an above-ground powerhouse with two 25-MW turbo-generator units; (v) a 30-km wide, Y-shaped reservoir	33–58	(i) loss of natural resources for the estimated four communities of about 33–58 chimpanzees using the area to be flooded; (ii) an increase in human-wildlife conflict as farmers and wildlife are forced closer together by reduction in available land; (iii) the prevention of movement of chimpanzees across the Seli River and between chimpanzee communities on each side of the river resulting in a decrease in their genetic viability	(i) Initiation of monitoring and awareness programs throughout the catchment area; (ii) incorporation of specific conservation activities into a Watershed Management Plan, including hunting controls, environmental awareness, fire control and zonation to give more protection to the most important remaining forest patches; (iii) creation of a Wildlife Conservation Area within the catchment	Loma Mountains National Park

Table 2. Cont.

Project	Country	Type of project	Main Project components	Estimated # of apes possibly affected by project activities	Main predicted impacts on apes	Main proposed mitigation measures for ape conservation	Proposed offset site
Lom Pangar	Cameroon	Hydroelectric dam	(i) A 50-m high dam located on the River Lom in Cameroon's East Region; (ii) a 610-km^2 reservoir area; (iii) a hydroelectric power plant; (iv) a transmission line; (v) a rural electrification scheme along this transmission line.	About 990 gorillas are located in the greater Deng Deng area, with over 50% of this population resident in Deng Deng NP itself; the others are located in the Forest Management Unit UFA 10–65 and in the Belabo Forest.	The key impacts identified by the project include: (i) loss of natural habitat due to flooding and infrastructure footprint; (ii) risk of reducing the viability of a distinct population of gorillas and other Red-Listed species; (iii) the risk that construction activities will induce significant loss of natural habitat;, and (iv) the risk of future habitat loss due to increased human pressures in the region.	Minimization: an adjustment to the pipeline route to avoid central Deng Deng and other dense forest areas and an offset site. Compensation: i) Extension of Deng Deng National Park by 9,000 ha, and creation of protection corridors to allow movement of gorilla populations in and out of Deng Deng National Park; ii) Conservation management programs in two Forest Management Units; iii) Creation of community forests (e.g. Belabo Forest) to support sustainable livelihoods and gorilla protection; iv) Development of mechanisms to control access to the region, especially during construction phase to prevent illegal logging; v) development of a financing mechanism to support conservation management and maintenance of ecoguards.	Forest areas around Deng Deng National Park perimeter

conflict as farmers and wildlife are forced closer together by reduction in available land, and (3) the prevention of movement of chimpanzees across the Seli River and between chimpanzee communities on each side of the river. The study predicted that these effects could result in reduced viability of the chimpanzee population in the BHP catchment over the long term due to genetic isolation [51]. Recommendations for the mitigation and offset of these impacts included initiating a monitoring and awareness program in the catchment area, and incorporating conservation activities such as hunting controls, environmental awareness, fire control and zonation to give more protection to the most important remaining forest patches in a Watershed Management Plan. The 2006 study also suggested the establishment and management of a Wildlife Conservation Area in the catchment area, to be called the Bumbuna Conservation Area (BCA), to preserve and protect biodiversity in the Bumbuna watershed and to serve as a biodiversity offset. The ESAP's view was that this BCA would not be very effective for chimpanzees, given the small size of the area and the many people living and farming within it. It therefore recommended creating an offset conservation area outside the catchment area. It also recommended that the Loma Mountain Non-Hunting Forest Reserve, an even more diverse 396-km^2 forest, 50 km from the dam with a population of approximately 1,065 chimpanzees [52], be upgraded to national park status as additional compensation.

The dam construction was completed and the reservoir area flooded in 2009. In 2012, Loma Mountains National Park (LMNP) was proclaimed, although parliamentary approval is still pending. Measures for the park have included: (a) posting and training staff; (b) providing equipment and materials; (c) conducting consultations on the park boundary, (d) resurveying and physically demarcating the boundary; (e) completing Reserve Settlement Courts Sittings; (f) completing a Process Framework, including a socio-economic baseline study; (g) developing a Management Plan; and (h) developing a provisional 5-year budget. These activities were funded by the original loan guarantee turned into a grant to support environmental and social mitigation at Bumbuna, as well as a GEF project that has been building government capacity to conserve national biodiversity. Sustainable financing for the LMNP has not yet been secured [53]. Since no long-term monitoring of chimpanzees was undertaken, the responses of the chimpanzees to the flooding remain unknown. A new Bumbuna Phase II project has now been launched that will expand the Bumbuna hydroelectric station and flood an even greater area.

Lom Pangar Hydropower Project (LPHP), Cameroon

The Lom Pangar Hydropower Project (LPHP) consists of a regulating, 50-m high dam located on the River Lom in Cameroon's East Region, a 610-km^2 reservoir area, a hydroelectric power plant, a transmission line, and a rural electrification scheme along this transmission line. Estimated costs for the LPHP are USD 393 million from the African Development Bank (AfDB), the Central African States Development Bank (BDEAC), the European Investment Bank (EIB), the French Agency for Development (AFD), the Government of Cameroon, and the World Bank.

The Environmental and Social Assessment noted that the main impact would be the flooding of natural forest, that none of the flooded habitat was critical, but that the dam site was located next to portions of the Deng Deng forest that included critical habitats and populations of the Critically Endangered (CR) Western lowland gorilla (*Gorilla gorilla gorilla*) and the Endangered (EN) Central chimpanzee (*Pan troglodytes troglodytes*). About 990 gorillas are located in the greater Deng Deng area, with over 50% of this population resident in Deng Deng National Park itself; the others are located in the Forest Management Unit UFA 10–65 and in the Belabo Forest [54]. Suggested mitigation included an adjustment to the pipeline route to avoid the central Deng Deng and other dense forest areas and control of access to the area to prevent illegal logging during the construction phase. Nonetheless, the project was predicted to have significant impacts on natural habitats that should be compensated. Studies funded by the AFD and carried out by the Wildlife Conservation Society (WCS), estimated a population size of several hundred gorillas in the greater Deng Deng area, which included a communal forest and a logging concession. As a result, a decision was made that the company would provide a high-level compensation package designed to strengthen protection of the newly designated 580-km^2 Deng Deng National Park, extend its boundary, and establish a corridor to other forest areas [55]. A third-party analysis indicated that an annual payment of USD 700,000 per year would be required to provide sufficient financing to meet the conservation management needs in the region. The activities in Lom Pangar are considered compensation rather than an offset since no efforts were made to quantify losses or gains to biodiversity and offset them to achieve no net loss.

Analysis

In the following, we analyze to what extent these field projects follow international best-practice principles for biodiversity offsets and whether these principles adequately generate conservation benefits for great apes in light of their EN and CR status, shrinking habitats, and vulnerability to disturbance. Table 3 provides a summary of our findings.

Limits to what can be offset

It is generally accepted that there are limits to what can be offset [36,38,39]. Some residual impacts on biodiversity cannot be fully compensated for by a biodiversity offset given the irreplaceability or vulnerability of the biodiversity affected [36]. Ecological criteria used to make decisions on where biodiversity offset limits should be drawn include: "levels of conservation concern", "magnitude of the estimated residual impact", and "opportunity and feasibility of offsets on the ground" [38,39].

All great apes are listed as EN or CR [13]. They are highly vulnerable to habitat disturbances due to their life history, behavior and susceptibility to human diseases [9,56]. They are important seed dispersers and play a role in shaping forest structure [57,58]. Due to the high degree of their "impactability" and low degree of their "offsetability", (i.e. their vulnerability and irreplaceability), these ecological considerations advocate in favor of an extremely high threshold for offsetting apes.

In addition to these ecological factors, offsetting great apes raises serious ethical questions. Great apes are our closest living relatives [59,60], exhibit many of the same emotions as humans [61,62], practice tool-use [63], hunt cooperatively [64,65], and show evidence of culture and traditions [62,66] as well as a capacity for language [68]. Some human communities in the region have religious, cultural and traditional taboos against hunting and eating great apes because of their close resemblance to humans [69–72]. We contend that from an ethical standpoint also, offsets for apes should require an extremely high threshold.

The IFC [44] recognizes both ecological and ethical values, and theoretically sets high standard thresholds for offsetting apes and ape habitat. IFC Guidance Note 6 divides Critical Habitat into two tiers with the likelihood of project investment in a Tier 1 habitat substantially lower than in a Tier 2 habitat. A footnote in

Table 3. Summary of project's implementation of international best practice principles for biodiversity offsets with respect to great ape conservation.

Project	Limits to what can be offset	Adherence to the mitigation hierarchy	Additional conservation outcomes	Landscape context	No net loss	Long-term outcomes
Simandou	Chimpanzees are not considered beyond the limit to what can be offset.	All possible mitigation on site and changes to the mining plan will theoretically be followed before offsets are considered for residual damage to chimpanzees and their habitat.	Unknown.	The potential offset sites being considered are priority areas for chimpanzees.	No net loss may occur in medium-term but there will probably be a short-term and potentially long-term loss of chimpanzees.	Dependent on financing mechanisms, coordination with other offset projects and whether commitment to offset site is in perpetuity. Long-term outcomes therefore unknown at this point.
Global Alumina Corporation	Chimpanzees are not considered beyond the limit to what can be offset.	Some infrastructure and original mining plan already in place before 2008 Critical Habitat study was conducted.	Unknown	Unknown	Unknown because offset site is still not certain.	Dependent on financing mechanisms, site selection, coordination with other projects. Long-term outcomes are therefore unknown at this point.
Bumbuna	Chimpanzees are not considered beyond the limit to what can be offset.	Site selection and infrastructure were already in place before the ESAP was engaged.	Yes, if assumption of baseline decline of chimpanzees in Loma Mountains is correct. No, if assumption is incorrect. Yes, if Loma Mountains protected in perpetuity, but no if not.	The Loma Mountains National Park is a priority area for chimpanzees.	Unknown. Specific calculations on losses and gains of individual chimpanzees were not made.	No
Lom Pangar	Chimpanzees and gorillas are not considered beyond the limit to what can be offset.	No. Mitigation hierarchy was not specifically applied.	Yes, if assumption of baseline decline of chimpanzees in Deng Deng is correct. No, if assumption is incorrect. Yes, if Deng Deng National Park protected in perpetuity, and the adjacent areas are effectively managed, but no if not.	Deng Deng is a priority area for gorillas and chimpanzees and studies indicated that an area larger than Deng Deng forest needed protection and management and the compensation project was designed to address that.	Unknown. Specific calculations on losses and gains of individual chimpanzees and gorillas were not made. Key concern was to maintain gorilla population in the region.	Dependent on both company compliance with financing commitments (30 year annual payments) and securing longer-term financing, and whether Deng Deng will be protected in perpetuity.

Guidance Note 6 states that "special consideration should be given to great apes given their anthropological and evolutionary significance in addition to ethical considerations. Where populations of CR and EN great apes exist, a Tier 1 habitat designation is probable" [44].

In practice, however, the presence of apes in a project area does not seem to have deterred companies, governments or funders from investing in activities that will be detrimental to great ape habitat and likely to result in their decline. Both the GAP and the Simandou project are located in areas of Critical Habitat for chimpanzees, and both have received funding from the IFC. The BHP and the Lom Pangar Dam projects were financed by the World Bank, and both are considering or have implemented offsets or compensation projects for negative impacts to great ape habitat. In summary it seems that none of these projects have considered great apes to be beyond the limits of what can be offset.

Adherence to the mitigation hierarchy

BBOP Principles [36] emphasize that biodiversity offsets are only appropriate after compliance with the mitigation hierarchy; that is, after avoidance, minimization and on-site rehabilitation measures have been exhausted [36]. Biodiversity offsets are therefore a mechanism of last resort [36,43,46]. In two of the projects profiled above, mitigation measures were not designed until they were already under way, resulting in the need for more off-site compensation than if mitigation measures had been included from the onset. For example, much of the BHP infrastructure was already completed when the ESAP was engaged, and the ESAP concluded that mitigation options for chimpanzees impacted by the dam were limited given that the dam site had already been selected [51]. Similarly in the case of the GAP, a Critical Habitat study for chimpanzees was not requested until the IFC was approached for a loan in 2008 [48], when much of the infrastructure already existed. The concept of "avoidance" was better integrated into the Simandou and Lom Pangar projects. For Simandou, the mining infrastructure development and plans for the sequence of exploitation activities were adapted to reduce impacts on chimpanzees. For Lom Pangar, costly mitigation measures were undertaken, including rerouting the pipeline to avoid the center of Deng Deng.

Additional conservation outcomes

BBOP states that a biodiversity offset should achieve additional conservation outcomes beyond results that would have occurred had the biodiversity offset not taken place [36]. The most common form of "additionality" in countries with offset policies is habitat restoration [31]. Biodiversity offset best practices also state that biodiversity offsets can achieve "additionality" by protecting areas where there is imminent or projected loss of biodiversity [36,73]. Two of the projects we examined based additionality on "averted loss" by updating the protected status of an area; the Loma Mountains National Park in Sierra Leone for BHP, and the Deng Deng National Park in Cameroon for the Lom Pangar Dam, and in the case of Lom Pangar extending areas for conservation around the park to protect great ape habitat. Both GAP and the Simandou project have provided a short list of potential offsets sites, indicating that their offsets will also consider "averted loss" as the counterbalance to actual loss on site. In these cases "no net loss" is working from an assumption of a pre-existing baseline rate of loss, assuming that habitat in the offset site is under threat or will be in the future [74]. This may be true given how many forests are threatened throughout the range of great apes. The result remains nonetheless a net loss of habitat against the extent and condition of that habitat at the time the project is implemented [73].

Maron *et al.* [31] emphasize that calculating the expected benefit of a conservation action–such as the purchase of a new reserve–requires "explicit estimation of the change in conservation value (e.g., population size of a threatened species) both with and without the action taking place, and calculation of the difference between these two scenarios. It is difficult to accurately (1) ascertain a baseline number of apes, (2) estimate the magnitude of change in ape numbers that would have occurred without project activities, (3) predict the magnitude of a population decline resulting from project activities; and (4) determine how much compensation is appropriate based on (1), (2) and (3). It is easier to estimate numbers of apes than of some other species given that their conspicuous nests can be used as indices of abundance. However, it is still extremely difficult to estimate numbers accurately [75]. Few studies have assessed the long-term impacts of extractive industries and other forms of habitat disturbance on apes [14]. As a result, it is very difficult for projects to measure losses, gains and additionality. Nevertheless, such estimations are necessary [36], and there should be consistency as to how they are generated and at what scale they are being assessed.

The projects we examined approached the challenge of measuring "additionality" in different ways, without common standards for measuring losses or gains. For the GAP, researchers from WCF proposed a mathematical formula to predict losses of chimpanzees, and gave a dollar value that companies should pay to compensate for this loss [76–78]. For BHP, it was predicted that all 33–58 chimpanzees would be impacted, but it was not specified whether "impact" would result in their death. The Loma Mountains compensation was assumed to be far greater than the loss [51]. In the case of the Lom Pangar project, the project appraisal indicated that the project would have significant and irreversible environmental impacts, including the loss of natural habitat and the risk of reducing the viability of a distinct population of gorillas and other Red-Listed species. Bolstering the protection of the national park and designating new areas for conservation management were considered additional.

Another challenge in estimating losses, gains and additionality under the current framework is that methodologies do not take into account the cumulative impacts of multiple projects, which can be far greater than the sum of the impact of individual projects. This lack of information on cumulative impacts is in part a result of offsets being funded, designed and implemented on a project-by-project basis.

Since great apes are distributed across equatorial Africa, they are likely present in or around many mining concessions. When concessions are adjacent to each other, there will be few available locations for apes to escape the mining activities. For example, the GAP concession is adjoined by a concession held by Compagnie des Bauxites de Guinée (CBG) to the east and a concession held by Russian Aluminium (RUSAL) to the north. Chimpanzees fleeing noise and other human disturbance in the GAP concession may not have accessible undisturbed habitat to move into. These projects state that they have conducted cumulative impact assessments, but these generally refer only to the direct effects on the environment from their own activities and not their impacts in combination with the activities of other companies. Tools exist to aid such an analysis. They include the Cumulative Impact Assessment (CIA), the Regional Cumulative Impact Assessment (RCIA), and the Strategic Environmental Assessment (SEA). The IFC recognizes that the "CIA should be an integral component of a good environmental and social impact assessment (ESIA) or a separate stand-alone process". They also recognize, however, that the "CIA is evolving and there is no single accepted state of global practice". In addition, the IFC Performance Standard 1 "does not expressly require, or put the sole onus on, private sector clients to undertake a CIA" [79]. Without better coordination and accounting for cumulative impacts, the risk is that offsets and compensation projects will be insufficient to offset the total cumulative loss of EN and CR species nationally or regionally over time, leading to overall species loss.

Landscape context

Several authors have encouraged biodiversity offsets to be designed in a landscape context [38,39]; however, there is little guidance regarding how this may be accomplished. The Simandou Project SEIA explicitly stated that the project would ensure that offsets were aligned with national biodiversity priorities. Both the Simandou Project and the GAP are considering sites identified in an IUCN action plan for West African chimpanzees [80]. Studies conducted in the Loma Mountains and Deng Deng forests determined both these areas to be important for apes and biodiversity in general. Thus these biodiversity offset and compensation sites appear to be located in national priority sites for these species and seem to be complying with this best practice principle. However, when offset projects are designed on a project-by-project basis without coordination or integration with other offset or compensation projects or other conservation initiatives, opportunities for aggregating sites are missed. Aggregating protection of larger areas of habitat, or connected forest patches, would have a better chance of maintaining viable populations of apes over the long term.

Operating on a project-by-project basis does not rule out placing the offset location into a larger landscape context but could still result in the protection of multiple small, isolated and vulnerable sites, impacting the 'additionality' potential of the offset project.

No net loss

One of the reasons that offset design and implementation continue to be *ad hoc* is that there are differing interpretations even as to the meaning of "no net loss" [37,38]. The IFC defines no net loss as: "the point at which project-related impacts on biodiversity are balanced by measures taken to avoid and minimize the project's impacts, to undertake on-site restoration and finally to offset significant residual impacts, if any, on an appropriate geographic scale (e.g., local, landscape-level, national, regional)". As discussed above, the challenges of accurately estimating the losses and gains of individual apes are enormous. Even if this were possible, and even if a project could result in an increase in ape numbers at a particular site, a small local increase would not necessarily contribute to the viability of the population as a whole. The worst-case scenario would be that the criterion of no net loss would merely result in many isolated offset projects protecting isolated individuals, groups or communities. While that may counterbalance losses of individual apes from the activities of individual projects, it does not necessarily contribute to protecting viable populations that would survive in the long term. This is again due to offsets being designed and implemented on a project-by-project basis.

By not coordinating with other projects, the projects we examined may have missed opportunities to aggregate offsets and create larger, more robust offset areas. Species viability in forest patches depends on many factors, including the size and shape of habitat patches and connectivity between patches. Not only does fragmentation disrupt the distribution of the species, it also affects the ecological processes that are part of the ecosystem [81]. Designing biodiversity offsets on a project-by-project basis could indeed temporarily result in a "no net loss" or "net gain", but in the long term, species viability could be eroded if offset sites exist in isolation from each other.

Long-term outcomes

BBOP addresses the need for long-term protection by emphasizing in their best practice principles that the outcomes of a biodiversity offset should last at least as long as the project's impacts, and preferably in perpetuity [36]. The ability of a biodiversity-offset project to deliver long-term outcomes depends on both biological and financial factors. Above, we have already discussed the biological factors affecting long-term outcomes. When we examined the financial sustainability of these projects, we found no consistency in the way offsets and compensation projects for apes are being funded in Sub-Saharan Africa.

In the BHP and Lom Pangar projects, the intended source of conservation funding is revenue from electricity production, which would be disbursed to specific conservation projects on an annual basis. However, disruption of either dam's operations would threaten this funding. In the case of BHP, problems with tariffs and the distribution system have curtailed profits and caused operational disruptions. In the case of Lom Pangar, the annual conservation payments were designed to last 30 years, but there are no specific payment guarantees to ensure the consistency of payments or any identified system for recourse if the project does not comply. This lack of financial security places compensation projects at risk, as well as the great ape populations whose security is dependent on effective conservation management. No information is currently available on how the GAP or Simandou offsets will be funded.

Discussion

Adherence to Biodiversity Offset Principles

Our review of these four projects that are investigating or already employing great ape offset or compensation projects demonstrates that the degree to which international best practices are applied is mixed. Despite the vulnerability of apes to disturbances, the long time it takes ape populations to recover from disturbances, their EN and CR status, and the ethical questions surrounding offsetting apes, all four projects assumed that great apes and their habitats could be at least compensated if not totally offset to achieve a no net loss. Following the mitigation hierarchy in most circumstances appears to have been challenging because project sites had already been selected using non-biodiversity criteria, and in some cases infrastructure development had already taken place, decreasing the options for mitigation and thus increasing residual impacts. Predicting losses, gains and "additionality" for apes is challenging, and most projects avoided making this calculation altogether. In all cases additionality was based on "averted losses". The projects might achieve "no net loss" or even support small local increases in numbers ("net gain") in isolated locations, but long-term outcomes of the two projects that have already been implemented are questionable given the uncertainty of the financing mechanisms and the fact that all offset or compensation projects were being designed on a case-by-case basis. Offset and compensation sites might be placed in priority locations, but without an overall offset strategy, opportunities are being missed for aggregating offsets in time and location and integrating them with species conservation objectives and other national biodiversity priorities. In summary, even though the projects we examined may result in temporary no net loss or even a net gain in species numbers, the current trajectory for great ape biodiversity offset projects is unlikely to result in no net loss over the long term. This indicates that, even if these projects adhere closely to international biodiversity offset principles, this will not ultimately generate a meaningful conservation outcome, and will not be sufficient to protect great ape populations over the long term, or contribute to species recovery.

Our assessment revealed three challenges that limit the effectiveness of efforts to compensate for impacts on great apes even if adherence to biodiversity offsets principles is improved. The first is that current great ape offset and compensation projects fail to account for cumulative development impacts at larger scales. The second is that great ape offset projects are less likely to make a meaningful contribution to great ape conservation if conducted on a project-by-project basis that results in multiple, isolated projects. This could occur even if offsets are placed in a larger, landscape context, since each location within the larger area could be disconnected from the other. The third is that biodiversity offset principles do not require offset projects to be fully operational and delivering the required biodiversity compensation before impacts from the development project occur. We suggest that a fundamental shift in the way offsets are designed and implemented is needed.

National Offset Strategies as an alternative trajectory

The most effective means of ensuring that biodiversity offset projects adhere to existing international best practice principles and contribute more effectively to great ape conservation is to develop National Offset Strategies, supported by conservation trust funds with non-wasting endowments. This new trajectory would have the following advantages.

National Offset Strategies would provide a framework for managing biodiversity offsets in a coordinated and transparent manner, consistent with national biodiversity strategies, including national protected area system plans and species recovery plans. Coordinating offsets and compensation projects through a National Offset Strategy would help ensure that sites are selected strategically, providing synergies with other conservation areas and between offset sites, and ensuring contribution to a landscape-level approach to great ape conservation. It would also help to ensure compliance with the mitigation hierarchy at the outset and that offset sites are prioritized in terms of timing of investment and location.

National Offset Strategies would maximize conservation benefits, for example, by establishing connectivity, buffering existing conservation areas, creating larger areas by aggregating offsets, or by selecting sites in different parts of the country, especially where several distinct and unconnected areas may be needed to buffer against the spread of disease such as Ebola–a major threat to humans and great apes in Africa [82].

In addition to increasing the conservation benefits, National Offset Strategies would also benefit project developers in a number of ways. They could establish a common set of rules, leveling the playing field, helping protect companies from reputational risk, and raising standards of industrial development projects. They would not limit offsets to those companies applying for funding from financial institutions with offset standards or companies with internal offset standards. National Offset Strategies would also allow companies to entrust offset management and implementation to a permanent entity, such as a conservation trust fund, with the responsibility of funding and implementing the strategy, rather than assuming the burden and liability of managing offsets in perpetuity (conservation banks in the U.S. provide a similar function). Investing in aggregated offset sites identified by National Offset Strategies could also decrease transaction costs by pooling resources for formulating offset methodologies, biological surveys, priority setting, or conservation trust fund development, which would otherwise be incurred by developers. With coordinated planning, companies could also share infrastructure or take advantage of efficiencies that lead to greater avoidance and minimization of impacts before offsets are even considered.

National Offset strategies should be developed as a result of a science-based, multi-stakeholder process and should include the following components:

A species recovery goal for EN and CR species rather than no net loss. Given the worsening global species extinction crisis [83–85], the international goal of biodiversity offsets should not just be "no net loss" but rather to make a measurable contribution towards *recovery* of EN- or CR-listed species. The idea that biodiversity offsets should contribute to species recovery has precedent in the U.S. where biodiversity offsets originated. A key aspect of offset policy in the U.S. is that the Endangered Species Act (ESA) seeks to ensure that a species listed under the ESA "recovers", that its conservation status improves to the point where it is no longer endangered (i.e. "in danger of extinction throughout all or a significant portion of its range") or "threatened" (i.e. "likely to become an endangered species within the foreseeable future throughout all or a significant portion of its range"). Thus, the United States Fish and Wildlife Service (USFWS) develops "recovery plans" for endangered species listed under the ESA to guide conservation decisions for that species. A federal permit allowing an impact on endangered species is only granted if the development activity will not "appreciably reduce" prospects for both the survival and recovery of the species [86].

Articulation of and adherence to no-go zones. Some biodiversity values are not offsetable, either because of their vulnerability and irreplaceability or because of their location. Industrial activities, for example, should not be permitted in World Heritage Sites. National Offset Strategies could play a key role in helping to determine which locations and which species should be off limits because their biodiversity value cannot be offset.

Cumulative impact assessments Sectoral EIAs. The need for projects to take into account the cumulative impacts of neighboring projects impacting EN and CR species when designing offsets is a growing concern [25,38,87,88]. We suggest that a nationwide assessment of the collective and cumulative impacts of planned and ongoing projects impacting EN or CR species should be part of the national offset design process, We also believe that assessment of cumulative impacts *before* offsets are designed should be an additional international principle as this could affect the magnitude of the offset required.

An opportunity for aggregating offsets. The challenge is to develop mechanisms to help ensure that biodiversity offset and compensation projects are implemented in the context of larger frameworks for endangered species conservation rather than on a project-by-project basis. "Conservation banking" was pioneered in the U.S. in part to address the problem of isolated biodiversity offset projects [89]. Conservation banks are lands that are conserved and permanently managed for threatened species. In exchange for permanently protecting and managing the land for the species of concern, the U.S. Fish and Wildlife Service (USFWS) agrees on a set number of habitat or species credits that bank owners may sell for that area. Developers whose projects have unavoidable adverse impacts on that species may purchase the credits from conservation bank owners to offset these impacts [90]. Conservation banks are expected to support a viable population of a species or contribute to the maintenance of a population by expanding an area managed for the species [89]. They provide an aggregated approach to offsets rather than a series of smaller, less viable projects. Conservation banks are permitted by USFWS, the same agency that also develops species recovery plans. Similar opportunities for aggregating offsets should be made available in National Offset Strategies.

Offset implementation to be complete before development occurs. The timing of biodiversity offsets can be an important factor in determining their success [38,73,91]. If development takes place before the offset is implemented, biodiversity may be lost [38], including loss of resources that are key to an EN or CR species' survival (e.g., a tree that must reach maturity to bear fruit necessary for ape survival). Allowing development to proceed before the offset is complete also creates a significant risk to biodiversity if the offset project fails. To increase the likelihood of project success, it is therefore important that offsets are in place before development occurs. Bekessy *et al.* [73] argue for a biodiversity "savings bank" approach rather than a "lending bank" in which the public "owns all the risk of failure", suggesting that "the biodiversity value of offsets should be realized before assets are liquidated", and that assets can only be traded once it has been demonstrated that they have matured (reached ecological equivalence with whatever losses they are being traded against) ([73], p.153).

There is precedence for this approach in the U.S., where conservation banks must demonstrate measurable conservation benefits before issuing credits. The conservation benefit is then far less speculative than if offset activities are concurrent with development, an approach that experts in the U.S. have found usually results in biodiversity losses [88,89,92].

Because restoration of tropical rainforests is a very lengthy process [31], and because great ape population increases would not be apparent for many years given their slow reproductive rates, some may argue that demonstrating an increase in great ape numbers before development can occur is not realistic. At a minimum, however, we suggest that areas proposed for higher protected area status, should at least have been created and already have appropriate levels of trained staff, necessary equipment and secure long-term funding before any development is allowed to occur.

Long-term funding. Industrial projects impacting EN and CR species should also ensure guaranteed, permanent funding for offset projects specifically targeting EN and CR species. We propose that Conservation Trust Funds (CTFs) with non-wasting, or permanent, endowments (which already exist in many countries and could accommodate multiple offset projects) could be the most effective mechanism. CTFs provide independent and permanent entities that can assume responsibility and liability for financing, managing and evaluating offsets and compensation projects in perpetuity. This is an important function, as conservation project management is not part of the core business of most companies. CTFs for countries where trust creation is not possible legally, or where governance is weak, can be located offshore, while maintaining in-country management and operations to ensure multi-stakeholder representation and promote civil society.

Offset insurance sites. Due to the high risk involved in biodiversity offsets, particularly in countries with weak conservation capacity and governance concerns, but also as a result of natural disasters or other unforeseen causes for project failure [93], we propose that National Offset Strategies also identify "insurance" sites. Developers would then be required to invest in both offset and "insurance" sites in case an offset fails.

Conclusion

Developing and enforcing a National Offset Strategy would help promote adherence to current best practice principles for biodiversity offsets and would be more likely than the current trajectory to contribute to species recovery objectives. Because offsets for many large-scale development projects need to be permanent, and because mechanisms for ensuring permanent protection on private lands may be lacking, the default is likely to be the creation of new protected areas, or the expansion or improvement of the management of existing protected areas. National Offset Strategies would, therefore, be likely to contribute to the establishment of well-managed national protected area systems containing a representative cross section of a country's biodiversity.

Globally, the current rate of species extinctions is 1,000 times the predicted background rate of extinction [83]. Habitat degradation and conversion remain a leading cause of biodiversity decline [21]. A new approach to their design and implementation is needed if offsets are to be a useful tool for great ape recovery.

Here we have presented a new framework for designing and implementing biodiversity offsets that we believe has a greater likelihood of protecting EN and CR species in the long term. We stress, however, that caution should be exercised in the use of offsets as a tool for ape conservation until further field-based evidence of their effectiveness is available. Biodiversity offsets remain an unproven mechanism, and the risk of failure is magnified in countries with poor governance and recent histories of civil conflict. More time is needed to gauge whether offsets are truly contributing to no-net-loss objectives and to progress towards more "evidence based" approaches to offset design [94], even in developing countries where biodiversity offsets have been in use for some time [37,72,88,91,95,96]. The international conservation community, development organizations, the private sector, and international lending banks should ensure not only a precautionary approach to the use of conservation offsets, but also that offsets are designed and implemented in the context of National Offset Strategies.

Acknowledgments

We are most grateful to Deborah Mead, John F. Oates, Chris Ransom, and Roger Fotso for reviewing the manuscript, and to anonymous reviewers for their comments and suggestions.

Author Contributions

Conceived and designed the experiments: RK CFK. Performed the experiments: RK CFK TH MSD. Analyzed the data: RK CFK TH MSD. Contributed to the writing of the manuscript: RK CFK TH AL HR RV RAM MSD AR EAW.

References

1. Mittermeier RA, Rylands AB, Wilson DE, editors (2013) Handbook of the Mammals of the World. Volume 3: Primates. Barcelona, Spain: Lynx Edicions. 951 p.
2. Dinerstein E, Varma K, Wikramanayake E, Lumpkin S (2010) Wildlife Premium Market+REDD: Creating a financial incentive for conservation and recovery of endangered species and habitats. Available: http://www.hcvnetwork.org/resources/folder.2006-09-29.6584228415/Wildlife_Premium-REDD%20Oct%2013%202010%20-2-%20-2.pdf. Accessed 2014 Jun 4.
3. Williamson EA, Maisels F, Groves CP (2013) Hominidae. In: Mittermeier RA, Rylands AB, Wilson DE, editors. Handbook of the Mammals of the World. Volume 3: Primates. Barcelona, Spain: Lynx Edicions. 792–843.
4. Tutin CEG, Williamson EA, Rogers ME, Fernandez M (1991) Gorilla dispersal of *Cola lizae* in the Lopé Reserve, Gabon. J Trop Ecol 7: 181–199.
5. Voysey BC, McDonald KE, Rogers ME, Tutin CEG, Parnell RJ (1999) Gorillas and seed dispersal in the Lopé Reserve, Gabon. I: Gorilla acquisition by trees. J Trop Ecol 15: 39–60.
6. Ancrenaz M, Lackman-Ancrenaz I, Elahan H (2006) Seed spitting and seed swallowing by wild orang-utans (*Pongo pygmaeus morio*) in Sabah, Malaysia. J Trop Biol Conserv 2: 65–70.
7. Jones CG, Lawton JH, Shachak M (1997) Positive and negative effects of organisms as physical ecosystem engineers. Ecology 78: 1946–1957.
8. Boogert NJ, Paterson DM, Laland KN (2006) The implications of niche construction and ecosystem engineering for conservation biology. BioScience 56: 570–578.
9. Plumptre AJ (1995) The effects of trampling damage by herbivores on the vegetation of the Parc National des Volcans, Rwanda. Afr J Ecol 32: 115–129.
10. Rogers ME, Voysey BC, McDonald KE, Parnell RJ, Tutin CEG (1998) Lowland gorillas and seed dispersal: the importance of nest sites. Am J Primatol 45: 45–68.
11. Weber AW (1987) Ruhengeri and Its Resources: an Environmental Profile of the Ruhengeri Prefecture. Kigali, Rwanda: ETMA/USAID. 171 p.
12. Macfie EJ, Williamson EA (2010) Best Practice Guidelines for Great Ape Tourism. Gland, Switzerland: IUCN/SSC Primate Specialist Group. 78 p.
13. IUCN (2014) IUCN Red List of Threatened Species. Version 2014.2. Available: http://www.iucnredlist.org. Accessed 2014 Sep 20.
14. Arcus Foundation (2014) State of the Apes: Extractive Industries and Ape Conservation. Cambridge, UK: Cambridge University Press. 377 p.
15. Poulsen JR, Clark CJ, Mavah G, Elkan PW (2009) Bushmeat supply and consumption in a tropical logging concession in northern Congo. Conservation Biology 23: 1597–1608.
16. Köndgen S, Kühl H, N'Goran PK, Walsh PD, Schenk S, et al. (2008) Pandemic human viruses cause decline of endangered chimpanzees. Curr Biol 18: 260–264.
17. Emery Thompson M, Kahlenberg SM, Gilby IC, Wrangham RW (2007) Core area quality is associated with variance in reproductive success among female chimpanzees at Kibale National Park. Anim Behav 73: 501–512.
18. Kahlenberg SM, Thompson ME, Muller MN, Wrangham RW (2008) Immigration costs for female chimpanzees and male protection as an immigrant counterstrategy to intrasexual aggression. Anim Behav 76: 1497–1509.
19. Mitani JC, Watts DP, Amsler SJ (2010) Lethal intergroup aggression leads to territorial expansion in wild chimpanzees. Curr Biol 20: R507–R508.

20. Ryan SJ, Walsh PD (2011) Consequences of non-intervention for infectious disease in African great apes. PLoS One 6: e29030.
21. Butler RA, Laurance WF (2008) New strategies for conserving tropical forests. Trends Ecol Evol 23: 469–472.
22. African Development Bank (2009) Oil and Gas in Africa. New York: Oxford University Press. 233 p.
23. UNDP (2013) The 2013 Human Development Report – The Rise of the South: Human Progress in a Diverse World. United Nations Development Programme.
24. Weng L, Boedhihartono A, Dirks PHGM, Dixon J, Lubis MI, et al. (2013) Mineral industries, growth corridors and agricultural development in Africa. Global Food Security 2: 195–202.
25. Edwards DP, Sloan S, Weng L, Dirks P, Sayer J, et al. (2014) Mining and the African environment. Conserv Lett 7: 302–311.
26. Morgan D, Sanz C (2007) Best Practice Guidelines for Reducing the Impact of Commercial Logging on Great Apes in Western Equatorial Africa. Gland, Switzerland: IUCN/SSC Primate Specialist Group. 48 p. Available: http://www.primate-sg.org/best_practice_logging/Accessed 2014 May 29.
27. Wich SA, Garcia-Ulloa J, Kuehl HS, Humle T, Lee JSH, et al. (2014) Will oil palm's homecoming spell doom for Africa's great apes? Curr Biol 24: 1659–1663.
28. Hockings K, Humle T (2009) Best Practice Guidelines for the Prevention and Mitigation of Conflict between Humans and Chimpanzees. Gland, Switzerland: IUCN/SSC Primate Specialist Group. 50 p.
29. Goossens B, Setchell JM, Tchidongo E, Dilambaka E, Vidal C, et al. (2005) Survival, interactions with conspecifics and reproduction in 37 chimpanzees released into the wild. Biol Conserv 123: 461–475.
30. Humle T, Colin C, Laurans M, Raballand E (2011) Group release of sanctuary chimpanzees (*Pan troglodytes*) in the Haut Niger National Park, Guinea, West Africa: ranging patterns and lessons so far. Int J Primatol 32: 456–473.
31. Maron M, Hobbs RJ, Moilanen A, Matthews JW, Christie K, et al. (2012) Faustian bargains? Restoration realities in the context of biodiversity offset policies. Biol Conserv 155: 141–148.
32. Kiesecker JM, Copeland H, Pocewicz A, Nibbelink N, McKenney B, et al. (2009) A framework for implementing biodiversity offsets: selecting sites and determining scale. BioScience 59: 77–84.
33. Madsen B, Carroll N, Moore Brands K (2010) State of biodiversity markets: offset and compensation programs worldwide. Washington, DC: Forest Trends Association. Available: http://www.forest-trends.org/publication_details.php?publicationID=2388. Accessed 2014 May 16.
34. McKenney B, Kiesecker JM (2010) Policy development for biodiversity offsets: a review of offset frameworks. Environ Manag 45: 165–176.
35. Suding KN (2011) Toward an era of restoration in ecology: successes, failures and opportunities ahead. Ann Rev Ecol Evol Syst 42: 465–487.
36. BBOP (2013) To No Net Loss and Beyond: An Overview of the Business and Biodiversity Offsets Programme (BBOP). 20 p. Available: http://www.forest-trends.org/documents/files/doc_3319.pdf. Accessed 2014 May 24.
37. Bull JW, Suttle KB, Gordon A, Singh NJ, Milner-Gulland EJ (2013) Biodiversity offsets in theory and practice. Oryx 47: 369–380.
38. Gardner TA, von Hase A, Brownlie S, Ekstrom JM, Pilgrim JD, et al. (2013) Biodiversity offsets and the challenge of achieving no net loss. Conserv Biol 27: 1254–1264.
39. Pilgrim JD, Brownlie S, Ekstrom JMM, Gardner TA, von Hase A, et al. (2013) A process for assessing the offsetability of biodiversity impacts. Conserv Lett 6: 376–384.
40. Quétier F, Regnery B, Levrel H (2014) No net loss of biodiversity or paper offsets? A critical review of the French no net loss policy. Environ Sci Pol 38: 120–131.
41. The Biodiversity Consultancy (2013) Government Policies on biodiversity offsets. http://www.thebiodiversityconsultancy.com/wp-content/uploads/2013/07/Government-policies-on-biodiversity-offsets3.pdf. Accessed 2014 May 16.
42. Rio Tinto (2008) Rio Tinto and Biodiversity: Biodiversity Offset Design. Rio Tinto plc and Rio Tinto Ltd.
43. IFC (2012a) Performance Standard 6: Biodiversity Conservation and Sustainable Management of Living Natural Resources. Washington, DC: International Finance Corporation.
44. IFC (2012b) Guidance Note 6: Biodiversity Conservation and Sustainable Management of Living Natural Resources. Washington, DC: International Finance Corporation.
45. Equator Principles (2013) Available: http://equator-Principles.com/resources/equator_Principles_III.pdf. Accessed 2014 May 16.
46. ICMM IUCN (2012) Independent report on biodiversity offsets. Prepared by The Biodiversity Consultancy. Available: http://www.icmm.com/biodiversity-offsets. Accessed 2014 May 20.
47. Kiesecker JM, Copeland H, Pocewicz A, McKenney B (2010) Development by design: blending landscape level planning with the mitigation hierarchy. Front Ecol Environ 8: 261–266.
48. Ecology, Environment Inc., Kormos R (2008) Critical Habitat Assessment Report Guinea Alumina Corporation Report. Guinea, West Africa. 182 p. Available: http://ifcext.ifc.org/ifcext/spiwebsite1.nsf/0/8A0EE1048673CB16852576BA000E2CAC/$File/Guinea%20Critical%20Habitat%20Assessment%20Report.pdf Accessed 2014 May 25.
49. Reuters (2014) CORRECTED-Emirates Global aluminium project clears Guinea hurdle. Available: http://www.reuters.com/article/2014/06/25/guinea-aluminium-emirates-idUSL6N0P62AE20140625. Accessed 2014 Sep 17.
50. World Bank (2005) Sierra Leone – Completion of the Bumbuna Hydroelectric Project – Environmental Impact Assessment and Resettlement Action Plans. Washington, DC: World Bank. Available: http://documents.worldbank.org/curated/en/2005/01/5653720/sierra-leone-completion-bumbuna-hydroelectric-project-environmental-impact-assessment-resettlement-action-plans. Accessed 2014 May 16.
51. ZSL (2006) Baseline Primate Survey and Monitoring Programme. Final Report. Annexe 8 to: Baseline Biodiversity Surveys, Completion of the Bumbuna Hydroelectric Project. Berkshire, UK: Nippon Koei UK Co. Ltd.
52. Brncic TM, Amarasekaran B, McKenna A (2010) Sierra Leone National Chimpanzee Census. Tacugama Chimpanzee Sanctuary, Freetown, Sierra Leone. 115 p. Available: http://tacugama.com/downloads/SLNCCP_Final_Report_TCS_Sep2010.pdf. Accessed 2014 May 24.
53. Haider SW (2013) Sierra Leone – SL Bumbuna Hydroelectric Environmental and Social Management Project: P086801 – Implementation Status Results Report: Sequence 19. Washington, DC: World Bank. Available: http://documents.worldbank.org/curated/en/2013/06/17822689/sierra-leone-sl-bumbuna-hydroelectric-environmental-social-management-project-p086801-implementation-status-results-report-sequence-19. Accessed 2014 May 16.
54. Maisels F, Strindberg S, Ambahe R, Ambassa E, Yara CN, et al. (2013) Deng Deng National Park and UFA 10-065, Republic of Cameroon. Wildlife and Human Impact Survey 2012. Wildlife Conservation Society. 37 p.
55. Maisels F, Ambahe R, Ambassa R, Fosso B, Puomegne J-B, et al. (2011) Gorilla population in Deng Deng National Park and a logging concession. Gorilla J 4: 18–19.
56. White A, Fa JE (2014) The bigger picture: indirect impacts of extractive industries on apes and ape habitat. In: Arcus Foundation, editor. State of the Apes 2013: Extractive Industries and Ape Conservation. Cambridge: Cambridge University Press. 197–225.
57. Plumptre AJ (1995) The effects of trampling damage by herbivores on the vegetation of the Parc National des Volcans, Rwanda. Afr J Ecol 32: 115–129.
58. Rogers ME, Voysey BC, McDonald KE, Parnell RJ, Tutin CEG (1998) Lowland gorillas and seed dispersal: the importance of nest sites. Am J Primatol 45: 45–68.
59. Chen FC, Li WH (2001) Genomic divergences between humans and other hominoids and the effective population size of the common ancestor of humans and chimpanzees. Am J Hum Genet 68: 444–456.
60. Scally A, Dutheil JY, Hillier LW, Gregory E, Jordan GE, et al. (2012) Insights into hominid evolution from the gorilla genome sequence. Nature 483: 169–175.
61. Warren Y, Williamson EA (2004) Transport of dead infant mountain gorillas by mothers and unrelated females. Zoo Biol 23: 375–378.
62. Anderson JR (2011) A primatological perspective on death. Am J Primatol 73: 410–414.
63. McGrew WC (1992) Tool-use by free-ranging chimpanzees: the extent of diversity. J Zool Lond 228: 689–694.
64. Boesch C (1994) Cooperative hunting in wild chimpanzees. Anim Behav 48: 653–667.
65. Boesch C (2002) Cooperative hunting roles among Taï chimpanzees. Hum Nature 13: 27–46.
66. Whiten A, Goodall J, McGrew WC, Nishida T, Reynolds V, et al. (1999) Cultures in chimpanzees. Nature 399: 682–685.
67. van Schaik CP, Ancrenaz M, Borgen G, Galdikas B, Knott C, et al. (2003) Orangutan cultures and the evolution of material culture. Science 299: 102–105.
68. Gardner A, Gardner BT (1980) Comparative psychology and language acquisition. In: TA Sebok, J Umiker-Sebok, editors. Speaking of Apes: A Critical Anthology of Two-way Communication with Man. New York: Plenum Press. 287–329.
69. Thompson-Handler N, Malenky RK, Reinartz GE, editors (1995) Action Plan for *Pan paniscus*: Report of Free-ranging Populations and Proposals for Their Preservation. Milwaukee (WI): Zoological Society Milwaukee County. 105 p.
70. Ham R (1998) Nationwide Chimpanzee Census and Large Mammal Survey: Republic of Guinea. Republic of Guinea: European Union. 115 p.
71. Rijksen HD, Meijaard E (1999) Our Vanishing Relative: The Status of Wild Orangutans at the Close of the Twentieth Century. Wageningen: Tropenbos Publications. 480 p.
72. Ancrenaz M, Dabek L, O'Neil S (2007) The cost of exclusion: recognizing a role for local communities in biodiversity conservation. PLOS Biol 5: 2443–2448.
73. Gibbons P, Lindenmayer DB (2007) Offsets for land clearing: no net loss or the tail wagging the dog? Ecol Manag Restor 8: 26–31.
74. Bekessy S, Wintle B, Lindenmayer DB, McCarthy M, Colyvan M, et al. (2010) The biodiversity bank cannot be a lending bank. Conserv Lett 3: 151–158.
75. Kühl H, Maisels F, Ancrenaz M, Williamson EA (2008) Best Practice Guidelines for Surveys and Monitoring of Great Ape Populations. Gland, Switzerland: IUCN/SSC Primate Specialist Group. 36 p.
76. WCF (2011) Biodiversity Management Plan for the Guinea Alumina Concession in Sangaredi, Guinea, Report 2008–2010. Germany: Wild Chimpanzee Foundation. 63 p.
77. WCF (2012a) Etat de la faune et des menaces dans les aires protégées terrestres et principals zones de forte biodiversité de Republique de Guinée. Germany: Wild Chimpanzee Foundation. 83 p.
78. WCF (2012b) Offset Strategy for the Guinea Alumina Project. Germany: Wild Chimpanzee Foundation.

79. IFC (2013) Good Practice Handbook. Cumulative Impact Assessment and Management: Guidance for the Private Sector in Emerging Markets" Washington, DC: International Finance Corporation. 82 p.
80. Kormos R, Boesch C, Bakarr MI, Butynski, TM editors (2003). West African Chimpanzees: Status Survey and Conservation Action Plan. Gland, Switzerland and Cambridge, UK: IUCN. IUCN/SSC Primate Specialist Group. 219 p.
81. Leader-Williams N, Dublin H (2000) Charismatic megafauna as 'flagship species'. In: Entwistle A, Dunstone N, editors. Priorities for the Conservation of Mammalian Diversity: Has the Panda had its Day? Cambridge, UK: Cambridge University Press. 53–81.
82. Walsh PD, Abernethy KA, Bermejo M, Beyersk R, De Wachter P, et al. (2003) Catastrophic ape decline in western equatorial Africa. Nature 422: 611–614.
83. Pimm SL, Ayres M, Balmford A, Branch G, Brandon K, et al. (2001) Can we defy nature's end? Science 293: 2207–2208.
84. Pimm SL, Jenkins CN, Abell R, Brooks TM, Gittleman JL, et al. (2014) The biodiversity of species and their rates of extinction, distribution, and protection. Science 344: 1246752 [DOI:10.1126/science.1246752].
85. Butchart SHM, Walpole M, Collen B, van Strien A, Scharlemann JPW, et al. (2010) Global biodiversity: indicators of recent declines. Science 328: 1164–1168.
86. Endangered Species Act (ESA) of 1973 As Amended through the 108th Congress Department of the Interior U.S. Washington, DC: Fish and Wildlife Service. 44 p. Available: http://www.nmfs.noaa.gov/pr/pdfs/laws/esa.pdf. Accessed 2014 May 24.
87. Brownlie S, Botha M (2009) Biodiversity offsets: Adding to the conservation estate, or "no net loss"? Impact Assessment and Project Appraisal 27: 227–231.
88. Kormos R, Kormos C (2011) Towards a strategic national plan for biodiversity offsets for mining in the Republic of Guinea, West Africa with a focus on chimpanzees. Report to the Arcus Foundation. Available: http://www.bicusa.org/wp-content/uploads/2013/07/GuineaBiodiversityOffsetKormosKormos2011.pdf. Accessed 2014 May 24. 92 p.
89. Carroll N, Fox J, Bayon R, editors (2008) Conservation and Biodiversity Banking: A Guide to Setting Up and Running Biodiversity Credit Trading Systems. London, UK and Sterling, VA: Earthscan.
90. USFWS (2003) Guidance for the Establishment, Use, and Operation of Conservation Banks. Memorandum to Regional Directors, Regions 1–7 and Manager, California Nevada operations. May 2. Washington, DC: US Fish and Wildlife Service.
91. Bendor T (2009) A dynamic analysis of the wetland mitigation process and its effects on no net loss policy. Landscape Urban Plan 89: 17–27.
92. Bean M, Kihslinger R, Wilkinson J (2008) Design of U.S. Habitat Banking Systems to Support the Conservation of Wildlife Habitat and At-Risk Species. Washington, DC: Environmental Law Institute. 978-1-58576-136-4, ELI Project No.0629-01. Available: http://www.eli.org/sites/default/files/eli-pubs/d18_02.pdf. Accessed 2014 May 24.
93. Stratus Consulting Inc. (2003) A Nationwide Survey of Conservation Banks. Prepared for the Northwest Fisheries Science Center. Seattle, WA: NOAA Fisheries.
94. Sutherland WJ, Pullin AS, Dolman PM, Knight TM (2004) The need for evidence-based conservation, Trends Ecol Evol 19: 305–308.
95. Kate K ten, Bishop J, Bayon R (2004) Biodiversity Offsets: Views, Experience, and the Business Case. Gland, Switzerland, Cambridge, UK and London, UK: IUCN and Insight Investment.
96. Fox J, Nino-Murcia A (2005) Status of species conservation banking in the United States. Conserv Biol 19: 996–1007.

20

Fast Growing, Healthy and Resident Green Turtles (*Chelonia mydas*) at Two Neritic Sites in the Central and Northern Coast of Peru: Implications for Conservation

Ximena Velez-Zuazo[1,2], Javier Quiñones[3]*, Aldo S. Pacheco[2,4], Luciana Klinge[2], Evelyn Paredes[5], Sixto Quispe[3], Shaleyla Kelez[2]*

1 Department of Biology, University of Puerto Rico-Rio Piedras, San Juan, Puerto Rico, **2** ecOceánica, Lima, Perú, **3** Laboratorio Costero de Pisco, Instituto del Mar del Perú, Ica, Perú, **4** Instituto de Ciencias Naturales Alexander von Humboldt, CENSOR Laboratory, Universidad de Antofagasta, Antofagasta, Chile, **5** Unidad de Investigaciones en Depredadores Superiores, Instituto del Mar del Perú, Callao, Perú

Abstract

In order to enhance protection and conservation strategies for endangered green turtles (*Chelonia mydas*), the identification of neritic habitats where this species aggregates is mandatory. Herein, we present new information about the population parameters and residence time of two neritic aggregations from 2010 to 2013; one in an upwelling dominated site (Paracas ~14°S) and the other in an ecotone zone from upwelling to warm equatorial conditions (El Ñuro ~4°S) in the Southeast Pacific. We predicted proportionally more adult individuals would occur in the ecotone site; whereas in the site dominated by an upwelling juvenile individuals would predominate. At El Ñuro, the population was composed by (15.3%) of juveniles, (74.9%) sub-adults, and (9.8%) adults, with an adult sex ratio of 1.16 males per female. Times of residence in the area ranged between a minimum of 121 and a maximum of 1015 days (mean 331.1 days). At Paracas the population was composed by (72%) of juveniles and (28%) sub-adults, no adults were recorded, thus supporting the development habitat hypothesis stating that throughout the neritic distribution there are sites exclusively occupied by juveniles. Residence time ranged between a minimum of 65 days and a maximum of 680 days (mean 236.1). High growth rates and body condition index values were estimated suggesting healthy individuals at both study sites. The population traits recorded at both sites suggested that conditions found in Peruvian neritic waters may contribute to the recovery of South Pacific green turtles. However, both aggregations are still at jeopardy due to pollution, bycatch and illegal catch and thus require immediate enforcing of conservation measurements.

Editor: Elliott Lee Hazen, UC Santa Cruz Department of Ecology and Evolutionary Biology, United States of America

Funding: Green turtle research at El Ñuro was supported with funds from PADI Fundation, Idea Wild, Pro Natura Japan, Rufford Small Grants Foundation, Patagonia Footwear through 1% for the Planet, OAK Foundation through Duke Minigrants and Umanotera Foundation through Krilca Gifts. The Instituto del Mar del Perú provided funds for sea turtle research at Paracas. The funders had no role in study design, data collection and analysis, decision to publish, or preparation of the manuscript.

Competing Interests: The authors have declared that no competing interests exist.

* Email: javierantonioquinones@gmail.com (JQ); shaleyla.kelez@ecoceanica.org (SK)

Introduction

Green turtles (*Chelonia mydas*) experience ontogenic habitat changes during their life cycle, switching from a juvenile oceanic epipelagic phase where they drift and disperse for several years [1–3], to a juvenile, sub-adult and adult neritic phase, where they feed and grow to reach sexual maturity [4,5]. During the neritic phase, green turtles inhabit diverse ecosystems [5] such as sea grass beds [4,6], mixed-bottoms composed of algae, sea grass and mangroves [7–12], and coral reefs, but usually nearby sea grass or algae beds for foraging [13]. All of these ecosystems are considered important development habitats of juveniles until reaching sexual maturity [3]. These areas can also be shared with adults who use them for foraging in between reproductive periods [5,14]. Given the considerable time spent in these neritic areas, studies at these habitats represent unique opportunities to understand fundamental aspects of their life history, population dynamics [15–17], feeding ecology [18,8], and functional role in the ecosystem [19]. Furthermore, the fact that green turtles are still endangered as a result of unsustainable practices (e.g., indiscriminate hunt for human consumption), habitat degradation and fisheries by-catch [20,21] studies in neritic habitats provide highly valuable information to propose conservation actions for this species such as preserving the habitat quality and the resources they feed on [22].

The Southeast Pacific holds a large aggregation of green turtles [23] recruiting from different nesting areas around the tropical-equatorial realm. In this region, this species may encounter different neritic conditions such as those imposed by the cold, nutrient-rich, upwelling coastal waters of the Humboldt Current System extending from central Chile (~35°S) to northern Peru (~6°S), and a transitional area (~6°–3°S) characterized by the convergence of northward upwelling waters with the warm, less-productive waters coming from the equator and warm tropical

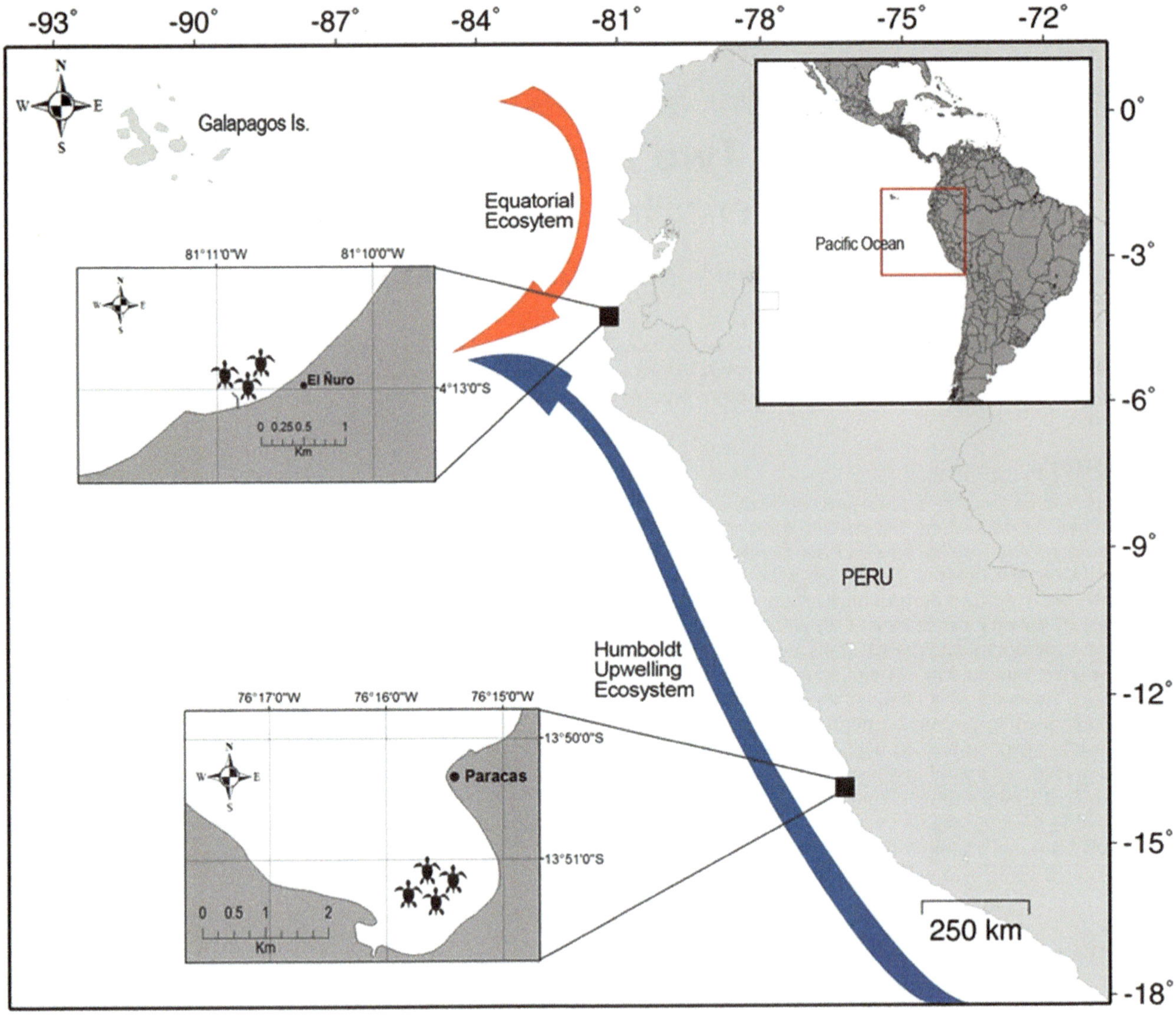

Figure 1. Green turtle (*Chelonia mydas*) neritic aggregations at two sites along the coast of Peru. (A) El Ñuro, at the northern part of the Peruvian coast, a region characterized by the convergence of the Humboldt upwelling ecosystem and the warm Equatorial ecosystem and (B) Paracas, characterized by permanent upwelling in the central coast of Peru. Map was created using Maptool available at www.seaturtle.org/maptool/.

waters that predominate in the coastal area of the northern tip of Peru towards Ecuador [24]. These conditions influence the existence and boundaries of different ecoregions [25] where green turtles may find a variety of foraging resources from distinct habitats ranging from coastal upwelling embayments to mangrove areas. Green turtles also have nesting rookeries located within the tropical realm [26,27].

Overall, the functionality of the different neritic habitats and their relationship with the distribution of each green turtle life stage within this region is still poorly understood and the information available is far from conclusive. A single study, based on the analyses of routes taken by post-nesting females from Galapagos, suggests sea surface temperature (SST) ≤25°C may act as a barrier to migration [28], but this might only be affecting adult females since juvenile and sub-adult green turtles are present in areas with lower SST. In central-south Peru, green turtles are abundant in warm neritic waters occurring during El Niño events [29]. Indeed, ca. 1000 tons of green turtles were captured and landed by artisanal fishermen during the El Niño 1987 [30]. However, landing data combines all turtle catch sites, thus precluding the assessment of habitat-life stage relationships. Green turtles can also be present in colder and temperate neritic habitats [8]. In Peru, green turtle aggregations have been registered inhabiting several areas along the coastline like Paracas bay ~14°S [29,30,31], Tambo de Mora ~13.3°S, Lobos de Afuera island ~7°S, Lobos de Tierra island ~6.5°S, Sechura bay ~5.7°S, Casitas y Punta Restin ~4.5°S and several areas along the coast of Tumbes (Punta Sal, Punta Mero, Bocapan, Puerto Pizarro) ~3.5–4°S [29,31,32]. Yet, the roles of these different habitats in the population structure need to be evaluated.

Neritic habitats off Peru might constitute important feeding grounds for this species. Results from gut content studies suggest remarkable differences in the diet composition of green turtles from northern [33] and south-central sites along the coast of Peru [29,30,34]. These differences in food resources could reflect differences in the composition of the aggregation; thus, north versus south differences in life-stage distribution could also be expected. Furthermore, given that breeding areas are located in the equatorial realm, dispersal of post-hatchling individuals occurs towards southern locations of Peru thus southern neritic benthic habitats could be dominated by recruiting juveniles (i.e., developmental habitats) [16], while a mixed group composed by juveniles, sub-adults and adults could be most common in neritic habitats at the most northern locations in the coast of Peru.

Here we provide new information of two aggregations in an upwelling and an ecotone (Humboldt-Equatorial) site. The aims of this study were: (1) to estimate and compare population parameters at each site such as size frequencies, growth rates, and sex ratios in adults, and (2) to estimate and compare time of residency at each site. We predict that there will be proportionally more adult individuals in the ecotone site compared to the upwelling site where juvenile individuals would be relatively more abundant.

Materials and Methods

Ethics statements

According to the Ley Forestal y de Fauna Silvestre del Perú N° 29763 (Forestry and Wildlife Law of Peru) and its Reglamento de la Ley Forestal y de Fauna Silvestre through a Decreto Supremo DS No 014-2001-AG (Supreme Decree for regulation) requires permissions for conducting any research activity involving CITES (Convention of International Trade in Endangered Species of Wild Fauna and Flora) enlisted species present in the Peruvian territory outside of national protected natural areas. The Servicio Nacional Forestal y de Fauna Silvestre (National Forest and Wildlife Service) grants permission upon request and after a technical opinion from a specialist indicating that the research activities will not jeopardize the conservation status of endangered species. All research activities at El Ñuro were conducted under permits RD N° 0383-2010-AG-DGFFS-DGEFFS and RD N°0606-2011-AG-DGFFS-DGEFFS. The Servicio Nacional de Areas Naturales Protegidas (National Service for Natural Protected Areas) provided the permissions for conducting sea turtle research in Paracas National Reserve (N208-2013-SERNANP-RNP/J, N105-2013- SERNANP-RNP/J, N087-2011- SERNANP-RNP/J). All permits allowed the manipulation and tagging of the animals as well as the collection of small tissue samples. All turtles were handled by trained personnel and following research and ethical protocols as in other sea turtle projects available in [35]. To minimize distress on captured turtles we kept them aboard for a maximum of 20 minutes, each individual was covered with wet towels to reduce overheating, and we released them in the same location were originally captured. In this study, no sea turtles were injured or killed during sampling manipulations.

Study locations

El Ñuro - the ecotone site. In-water surveys for green turtles were conducted in El Ñuro (4°13′S; 81°10′W, Fig. 1), a large sandy neritic area in the coast of the department of Piura, northern Peru. Green turtles were sampled in an area to the north-east of a pier in this area. As mentioned before, this area represents an ecotone between the cold upwelling waters of the Warm Temperate Southeastern Pacific Ecoregion and the warm and tropical conditions of the Tropical Eastern Pacific Ecoregion [25]. Sea surface temperature (SST) during the study period was on average 22.2°C (range: 19.3°–25.9°C). The seabed at this area is mainly composed by sandy sediment with scattered boulder reefs. The rocks and to some extent, the sediment substratum, are covered by the invasive green algae *Caulerpa* sp. (authors observations). The El Ñuro pier is used by ca. 350 artisanal fishermen mainly targeting demersal Peruvian hake (*Merluccius gayi*), large oceanic fish such as tuna (*Thunnus albacares*) and swordfish (*Xiphias glaudius*), and Humboldt squid (*Dosidiscus gigas*). Fishermen conduct day-fishing trips leaving between 3 and 4 am and returning from 11 am to 4 pm.

A nylon entanglement net (80 m×5 m, mesh size = 20.32 cm) was used to capture green turtles. Each in-water survey lasted 5 days, each day the net was deployed for around 5 hours usually starting at 8am and finishing at 1pm. The bottom part of the net was generally on top of the sea floor at ~5 m depth. In order to attract sea turtles to the net, some small pieces of Humboldt squid mantle were tied to the upper part of the net. Two to four free divers monitored the net every 30 minutes searching for sea turtles and helping to bring them on-board. A total of eight in-water surveys were conducted from the 31st of May 2010 to the 16th of August 2013, every 23.3 weeks on average (SD ±5.3).

Paracas – the upwelling site. In-water surveys were conducted at La Aguada a small inlet located in the South Eastern part of Bahia Paracas, in central Peru (13°51′S; 76°15′W, Fig. 1). These shallow waters are located within a cold temperate upwelling region, being one of the main feeding grounds in the Southeast Pacific [36]. SST during the study averaged 18.8°C (range: 15.1°–22.8°C). Green turtles were sampled within an estimated area of *ca.* 3.9 km^2. The seabed in this area is mostly sand covered by algae *Ulva* sp., *Enteromorpha* sp., *Chondracanthus* sp. [37] and *Caulerpa* sp. During the austral spring, summer and autumn, the pelagic medusae phase of the large jellyfish *Chrysaora plocamia* is very abundant [38]. This area is used by few fishermen targeting fish species such as mullets (*Mugil cephalus*) and silversides (*Odontesthes regia regia*).

A set of four entanglement nets (total length: 976 m×4 m, mesh size = 60–65 cm), traditionally used by local artisanal fisherman for catching sea turtles, was used. Each in-water survey lasted 2 days and the nets were deployed randomly within the area, usually starting at 5am and finishing at 1pm, around 8 hours on average. The net operates just below of the water column surface with no bait, as traditionally used by local artisanal fishermen in this area. When a sea turtle was captured we approached the net sailing and carefully brought it on board for evaluation. A total of 21 in-water surveys were conducted every 8 weeks on average (SD±7.9) from the 4th of March 2010 to the 12th June 2013.

Collection of biological data at both sites

At both sites the following body measurements were taken from each turtle captured: the notch to tip Curved Carapace Length and Curved Carapace Width (CCLn-t, CCWn-t) and two measurements of the tail (Total Tail Length-TTL and Post-cloacal Tail Length) to 0.1 cm with a 100 cm soft measuring tape. In El Ñuro, notch to tip Straight Carapace Length and Straight Carapace Width (SCLn-t, SCWn-t) to 0.1 cm was obtained using a 100 cm Haglöf tree caliper. In Paracas, SCLn-t was not measured but estimated using the regression formula SCLn-t = 0.919CCLn-t+1.9645 (R^2 = 0.9845) estimated from turtles at El Ñuro. Green turtles were weighted using a 100 or 200 Kg scale (1 kg precision). For the largest turtles at El Ñuro, weight was not obtained using a scale, but estimated using the regression equation between size (CCLn-t) and weight (W): W = 1.916 (CCLn-t)–87.611, R^2 = 0.91; estimated using data of smaller individuals. In Paracas all turtles were weighed. The sex of the turtles was determined based on the secondary sexual characteristics i.e. length of the tail and nails. All captured green turtles were marked with external Inconel tags. At El Ñuro, tags were always applied in

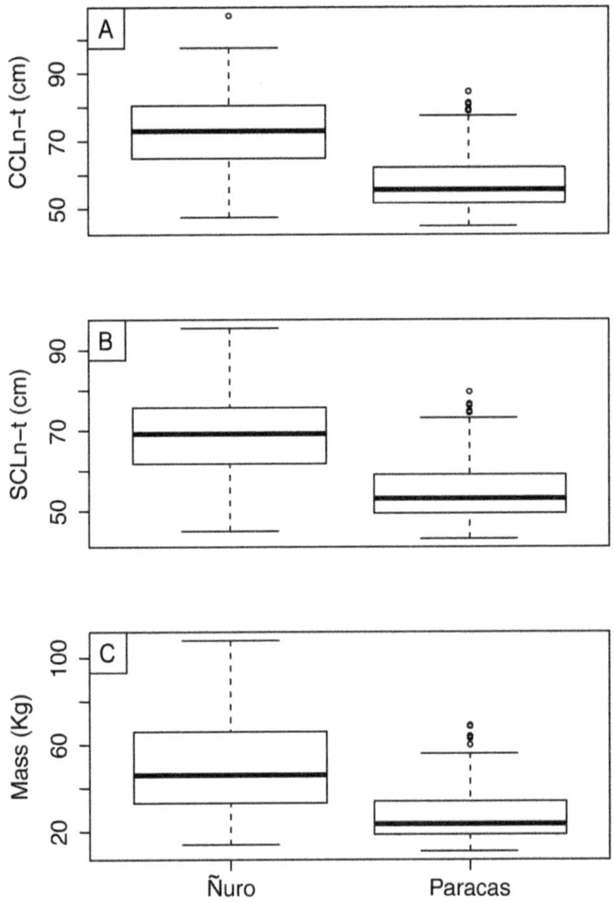

Figure 2. Box plots of (A) curve carapace length, (B) straight carapace length and (C) body weight of green turtles at both study sites.

both front flippers, except for nine individuals were tags were applied in only one of their flippers. In Paracas, Inconel tags were applied at both back flippers.

Population parameters, residency time, growth rate and body condition

Green turtles were arbitrarily classified in three different life-stages (i.e., juvenile, sub-adult and adult) using the minimum and mean CCLn-t size of nesting females in Galapagos (Patricia Zarate, personal communication). All individuals with CCLn-t, at first capture, under the minimum size of nesting females (< 60.7 cm) were categorized as juveniles, all between the minimum and the mean size (≥60.7 and ≤86.7 cm) as sub-adults, and all over the mean (>86.7 cm) as adults. Residence time was estimated as the time period elapsed between the first capture of the turtle and the last recapture and was expressed in days. Growth rates were estimated from the capture-recapture records as the increments in size (both for CCLn-t and SCLn-t) in cm per year. The body condition index (BCI) was calculated using the following formula: BCI = body mass/SCL [3]. This index is used as an indirect predictor of the nutritional status and/or health condition of the animal [39]. To calculate growth rates and BCI, only recapture intervals greater than 300 days were used, therefore reducing the effect of measurement error and seasonality effects [40,41,42].

Statistical analysis

Green turtle size (CCLn-t and SCLn-t) and weight were compared using a two- way PERMANOVA test considering site (Ñuro and Paracas) and sampling years (2010–2013) as fixed factors. PERMANOVA is a semi-parametric test based on Euclidian distance and calculates the significance using permutations. This test is robust enough in cases when data normality and variance homogeneity is not achieved such as in the case of our data set. This analysis also allows testing for significant effects of the interaction term (i.e. site×sampling years). Body condition index values were compared using PERMANCOVA test using a three-way design, considering CCLn-t as a co-variable and life stage (juveniles and subadults), site (Ñuro and Paracas), and years (2010–2013). Since adults were present only in el Ñuro, body condition index values of this life stage were not included in the statistical test. All statistical tests were performed using the PRIMER v6+PERMANOVA software [43].

Results

Number of capture and recapture individuals

At el Ñuro, a total of 228 green turtles were captured, with an average of 28.4 individuals per survey (range: 5–45). Of the total captured individuals, 154 were unique individuals while 51 were recaptures among surveys, 20 within surveys, and three captures from which data was not possible to get collected but it was used for CPUE estimates. For all subsequent analyses, within-survey recaptures were excluded. The mean number of new captures and recaptures was 19.5 new turtles per survey (range: 4–31, SD ±10.4) and 6.4 turtles (range: 0–13, SD ±4.5) respectively. One turtle, an adult female, had a tag from the Galapagos archipelago, applied in a nesting beach at Bahia Barahona, Isla Isabela, in 2010 (Macarena Parra personal communication). Also, two juvenile hawksbills (*Eretmochelys inmbricata*) were captured during the surveys. Two tagged green turtle from our project were found dead, one at El Ñuro and one at Los Organos, a beach 8 km to the north of El Ñuro. At Paracas, a total of 160 green turtles were captured, with an average of 5.3 individuals per survey (range: 1–27). Of the total captured turtles, 133 were unique individuals while 27 were recaptures among surveys. All turtles captured within surveys were excluded. The mean number of new captures and recaptures was 7.2 (range: 1–27, SD ±5.7) and 0.7 turtles (range: 0–5, SD ±1.3) respectively. One adult olive ridley sea turtle (*Lepidochelys olivacea*) was captured during surveys. In addition, one tagged turtle was found dead in the Pisco shores.

Size, weight, size classes and sex ratio

Overall, green turtles at El Ñuro were larger and heavier compared to the individuals captured at Paracas (Fig. 2, detailed body measurements are provided in Table S1). The size (CCLn-t) of captured green turtles at el Ñuro varied between 47.5 cm and 107 cm, with a mean size of 72.4 cm (SD ±10.9), while at Paracas the size ranged from 44.9 to 84.5 cm, with an average size of 57.7 cm (SD ±8.7). The PERMANOVA test indicated that differences for both CCLn-t (Pseudo-$F_{1,\ 356}$ = 196.9, $P<0.05$) and SCLn-t (Pseudo-$F_{1,\ 356}$ = 192.9, $P<0.05$) were significant. In addition, the interaction term time×year was also significant (CCLn-t; Pseudo-$F_{3,\ 356}$ = 3.1, $P<0.05$ and SCLn-t; Pseudo-$F_{3,\ 356}$ = 2.9, $P<0.05$) indicating that significant differences were persisting during all sampling years. At El Ñuro, green turtles weighed on average 50.2 kg (SD ±21.4), ranging from 14 to 108 kg while at Paracas individuals weighted on average 27.7 kg (range: 11–69 kg, n = 160, SD ±13.0). The PERMANOVA test also detected significant differences in weight (Pseudo-$F_{1,}$

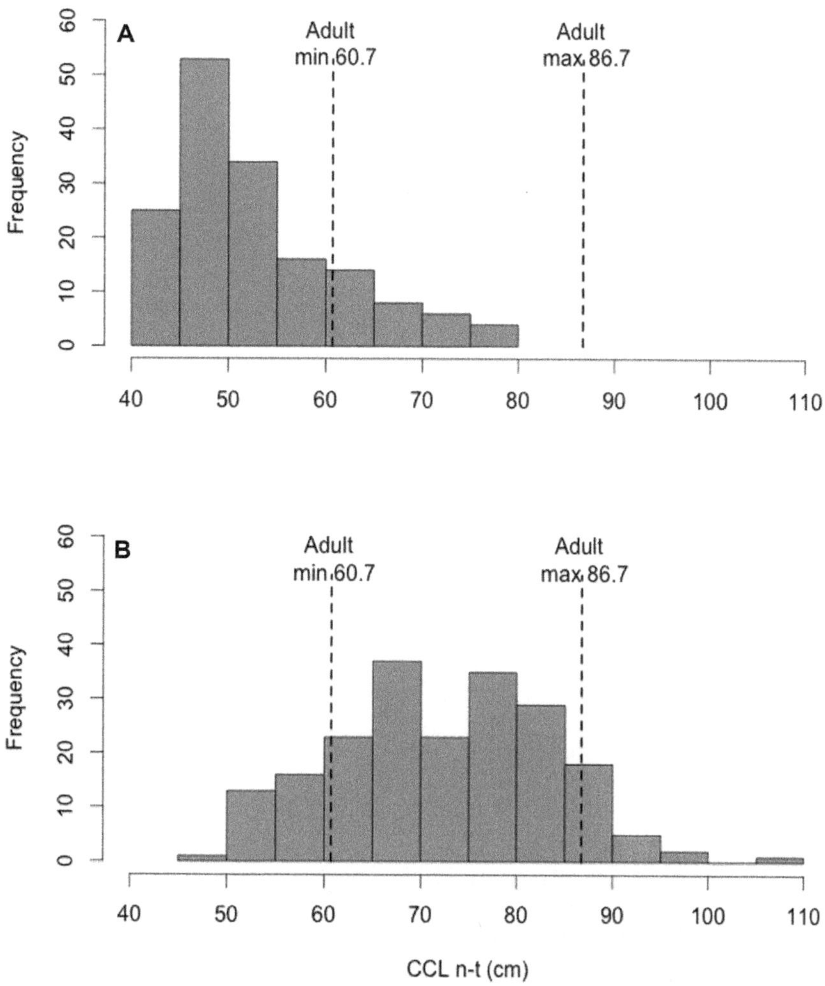

Figure 3. Size (CCLn-t) distribution of green turtles captured at (A) Paracas and (B) el Ñuro. Dotted lines are the minimum and mean size reported for adult green turtles in Galapagos Island [36], the largest green turtle rookery near Peru.

$_{356} = 146.2$, $P<0.05$) and the interaction term time×year (Pseudo-$F_{3,\ 356} = 3.3$, $P<0.05$).

At El Ñuro, the majority of individuals (81.7%; n = 166) were between 70 and 89 cm length (Figure 3) and 74.9% (n = 152) of the captured individuals were classified as sub-adults, 15.3% (n = 31) as juveniles and 9.9% (n = 20) as adults (Figure 3). The mean CCLn-t of adult individuals was 90.6 cm (range = 86.7–107, SD ±4.9). At Paracas, 73% (n = 117) individuals were classified as juveniles and 27% (n = 43) sub-adults; no adults were captured (Fig. 3). The mean CCLn-t for juveniles was 53.1 (range: 44.9–60.4 cm, SD ±3.7) and for sub-adults 69.5 (range: 61.3–84.5 cm, SD ±6.5). At El Ñuro, from all unique turtles captured, eight were identified as adult females and seven were identified as adult males. The mean TTL for adult males, including measurements obtained in subsequent recaptures, was 30.9 cm (range = 19–45 cm, SD ±7.8). These males ranged in size (CCLn-t) from 71.6 to 83.5 cm and averaged 79.4 cm. The sex ratio in captured adult sea turtles was 0.85 F: 1M. Since at Paracas no adults were recorded, the proportion of males and females of this aggregation remain unknown.

Time of residency, growth rates and body condition index

At El Ñuro, the mean recapture rate was 26%. In general, most of the individuals evaluated were captured only once (n = 115) but there were turtles captured more than once during these four years of surveys (two times: n = 31; three times: n = 15; four times: n = 3). There were more sub-adult turtles recaptured in comparison to the other life-stages (Fig. 4). The mean recapture time interval was 331.1 days (SD ±215.4), with a minimum of 121 days and a maximum recapture interval of 1015 days. At Paracas, the mean recapture rate was 12.5%. Most of the individuals evaluated were captured only once (n = 133), a few twice (n = 18) and only nine turtles were captured three times during these four years of surveys. There were more juvenile turtles recaptured in comparison to sub-adults (Fig. 4). The mean recapture interval was 277 days (SD ±236.2), with a minimum of 65 days and a maximum recapture interval of 680 days.

At El Ñuro, from all 45 recapture records (sub-adult = 44, adult = 1), we used 20 records with recapture intervals equal or higher than 300 days. The mean growth rate was estimated at 2.83±1.51 cm $year^{-1}$ (range: −0.1–5.58). At Paracas, three usable recaptures yielded a mean growth rate of 6.77±1.75 cm

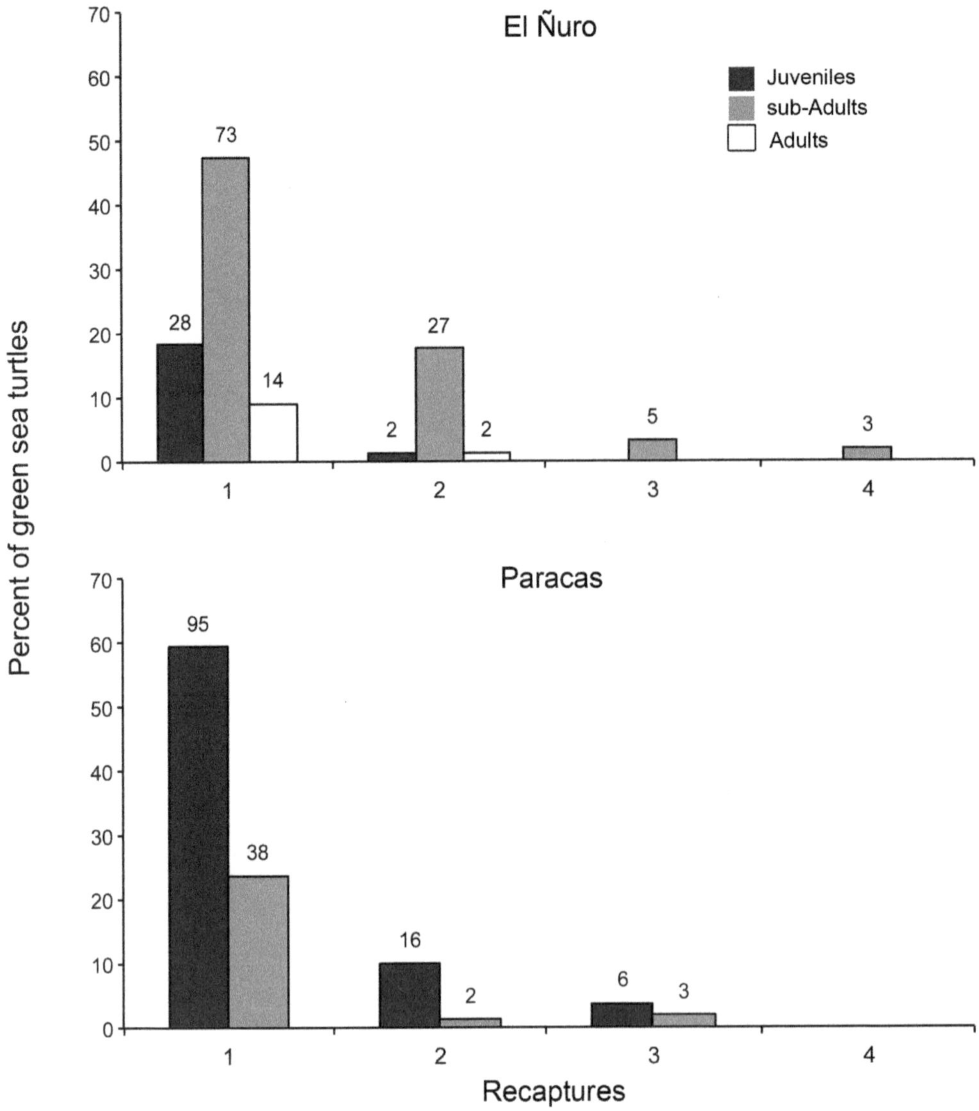

Figure 4. Number of times (%) that each turtle classified as juveniles (dark gray bar), sub-adult (light gray bar), or adult (white bar) was recaptured in El Ñuro and Paracas. The total number of turtles caught is indicated on top of each column.

year^{-1} (range: 4.77–8.07). Since we did not have equally and sufficient recaptures of individuals in all life stages, particularly from Paracas, no contrasting statistics were performed. Body condition index values at El Ñuro showed slight variation among life-stages and years. The lowest BCI exhibited per year and stage was for the juveniles in 2013 (BCI = 1.4) while adult green turtles exhibited the highest BCI in 2013 (1.6). In general, the mean BCI for each life stage (juveniles, sub-adults and adults) was 1.5. At Paracas, estimates of BCI were quite similar among life-stages and years (Fig. 5). Juvenile green turtles exhibited both, the lowest and maximum BCI (2010, BCI = 1.4 and 2013, BCI = 1.6 respectively) while adult green turtles exhibited similar values; the mean BCI for juveniles and sub-adults was the same (1.5). The PERMANCOVA test comparing BCI values of juveniles and sub-adults (excluding adults since these where only present at El Ñuro) detected marginal differences for the interaction term co-variable (CCL)×year×site (Pseudo-$F_{3, 281} = 2.6814$, $P = 0.053$).

Discussion

This is the first study comparing two neritic aggregations of green turtles in the Southeast Pacific. The presence of almost exclusively juvenile individuals in the upwelling site (La Aguada in Paracas) provides support to the development habitat hypothesis supporting the role of neritic habitats exclusively used by juveniles throughout the benthic life stage. Overall, the results of our study highlight the importance of neritic habitats for juveniles, sub-adults and adults green turtles in Peruvian waters and in the eastern Pacific in general. Preliminary growth rates and body index condition estimates suggest that green turtles at both sites are thriving in very favorable habitats exhibiting the highest values reported for foraging populations in the eastern Pacific.

The distribution of green turtles life-stages along these two sites suggests a spatial segregation pattern between sites; while the El Ñuro aggregation is composed of mainly sub-adults, with few juveniles and adults, for green turtles caught in Paracas the aggregation is dominated by juveniles and few sub-adults. Paracas

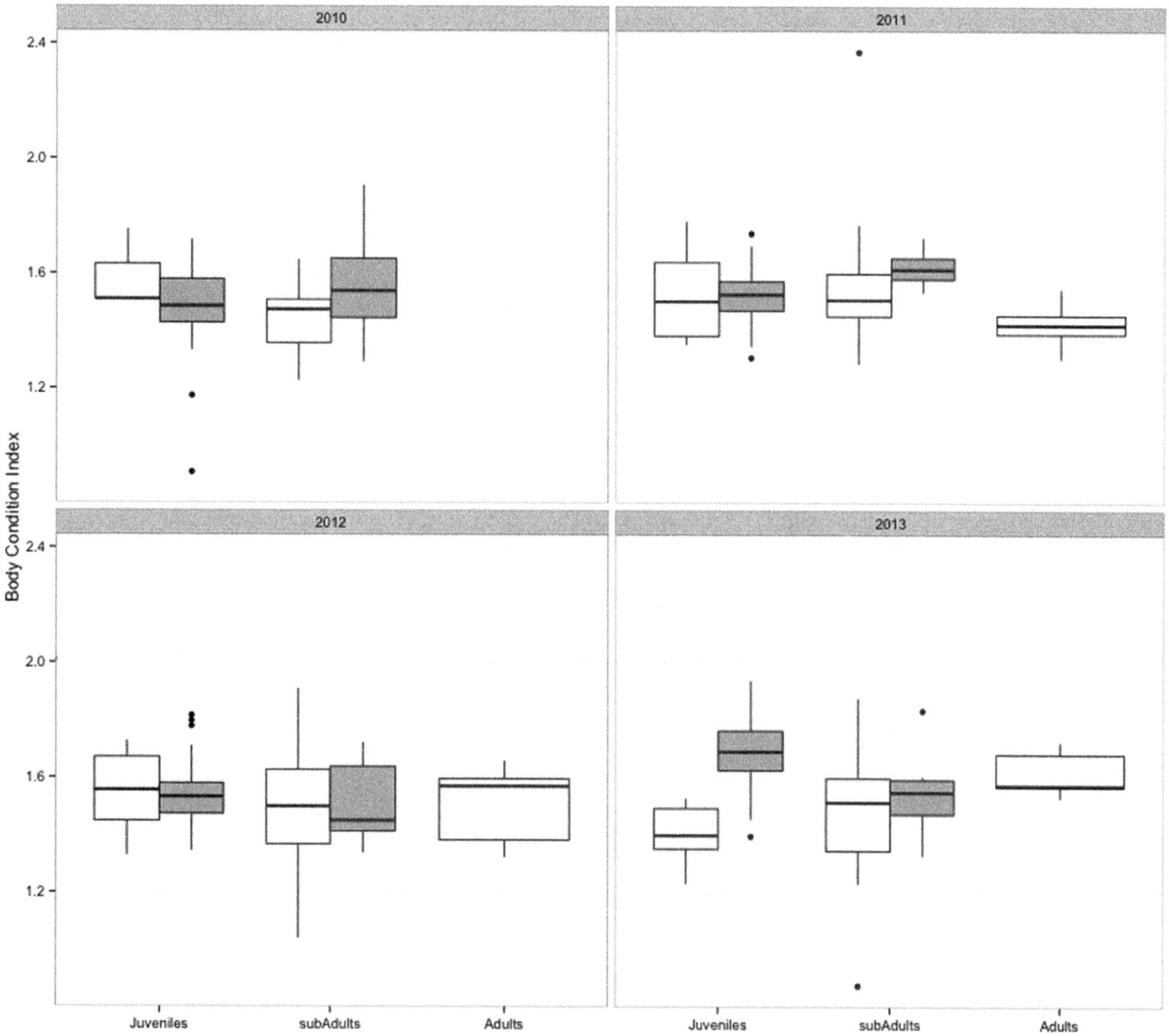

Figure 5. Box plots of body conditions values per green turtle life stages at the different sampling years at El Ñuro (grey boxes) and Paracas (white boxes).

is located within one of the most productive and permanent upwelling cells along the Peru-Chile coast [44]. In this area, early studies on the diet of this species suggested macroalgae as their principal prey item [29]. Recently, green turtles in this area have been reported preying intensively on the large jellyfish *Chrysaora plocamia* (up to 100 cm in bell diameter) [30,45,46] that can be very abundant in the water column, particularly during summer [38]. Upwelling is less intense and more seasonal in northern Peru, but still highly productive conditions exist there. At El Ñuro, the diversity and abundance of seaweeds is low; however, the algae *Caulerpa* sp. is very common and distributed in large patches along the seabed. Green turtles have been observed having pieces of this algae sticking out of their mouths (authors observations). However, we do not know whether green turtles really feed on this algae or this is unintentionally trapped in the mouth when disturbing the seabed searching for other benthic prey. Also, we cannot rule out the possibility of the influence of an external source of food (e.g., fish discard), particularly near El Ñuro pier. Diet studies complemented with isotope data are necessary to understand the trophic ecology at El Ñuro.

Our estimates of growth rates suggest that the productivity may be playing a major role in the development of juveniles and sub-adults. Growth rate estimates obtained from green turtles at El Ñuro were the second highest estimates reported so far, after estimates obtained for the green turtle aggregation in Laguna Ojo de Liebre, in Mexico [47]. The green turtle aggregation from Paracas exhibited a high growth rate as well (6.8 cm/y ±1.8), but this estimate is preliminary and should be taken with caution. The growth rate estimates for Paracas were obtained from three recapture records, after filtering the dataset to comply with the minimum recapture interval established. Yet, these first growth rate estimates, at both sites, were high and individuals, particularly at Paracas, may be growing faster than individuals at El Ñuro. This difference is somehow expected since individuals at Paracas are significantly smaller individuals (mean CCLn-t = 57.7 cm) compared to green turtles at El Ñuro (mean CCLn-t = 72.42 cm)

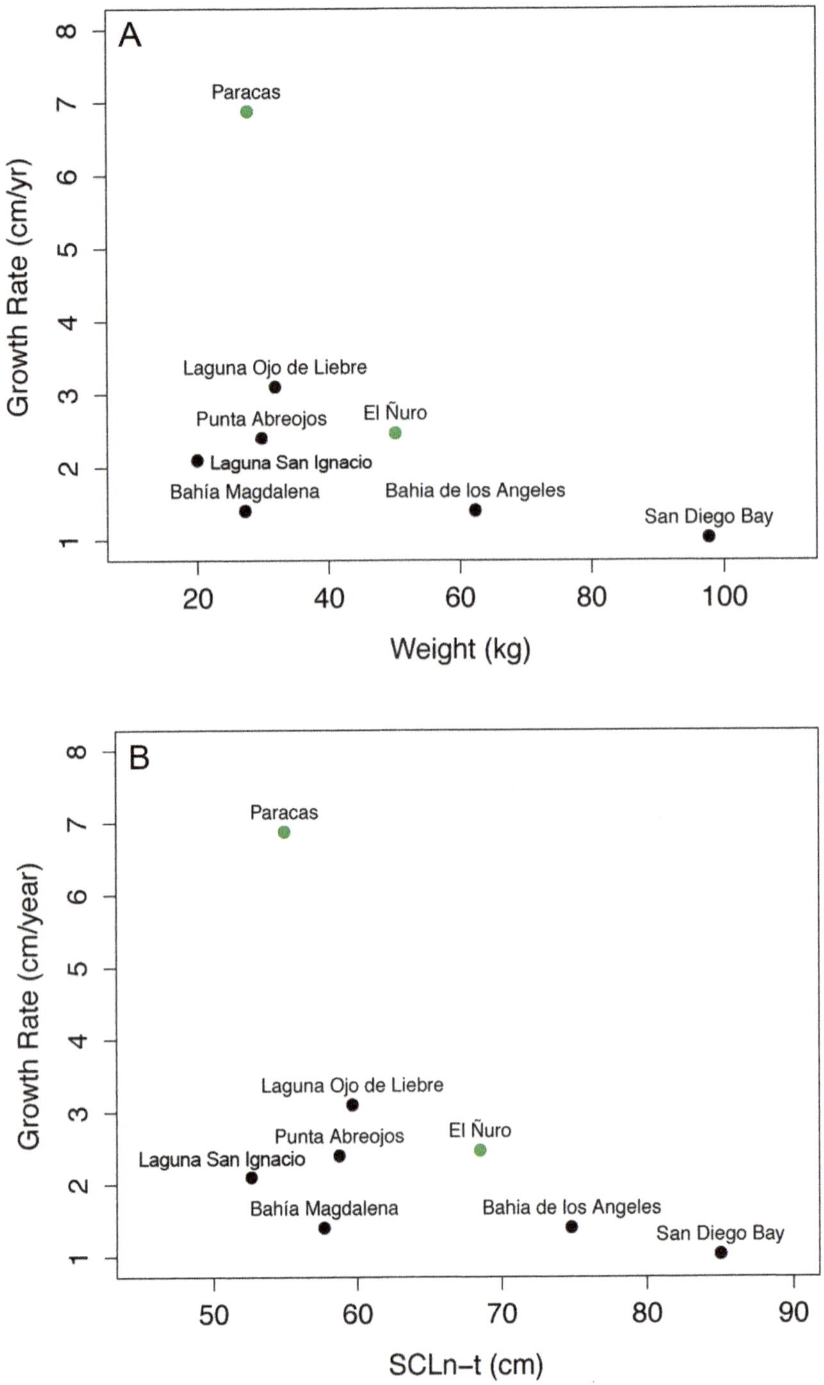

Figure 6. Relationship between weight and growth rate (A) and length (SCLn-t) and growth rate (B) for foraging aggregations of green sea turtles in the eastern and central Pacific (black dots), including our study (green dots).

and it is known that green turtles grow faster at smaller sizes e.g. [48,49,50]. Our results suggest that green turtles in Peru are growing faster than individuals at different locations in the central and northeast Pacific and this pattern is observed when comparing both weight and SCL versus growth rate (Fig. 6, Table 1). Even though the coastal region of Peru is located within tropical latitudes, the characteristics of the prevalent upwelling coastal system i.e. cold, low oxygen and nutrient-rich waters, resemble a more temperate realm extending northward close to the equator (~4°S). Green turtles inhabiting temperate feeding grounds may boost their metabolism for growing faster compared to their counterparts living in warm habitats [51]. The combination of younger stages, low temperatures and high prey availability may yield fast growing green turtles.

In line with these results, BCI estimations for green sea turtles at both sites were high in comparison to other populations in the east Pacific. For example, in Mexico, mean estimates for different sites ranged from 1.27 to 1.42 while in Peru the mean estimates were

Table 1. Size, preliminary growth rates, and body condition index of green turtles (*Chelonia mydas*) at foraging grounds in the eastern Pacific.

Study site	N	Juv:Adt (% or ind)	CCL mean (±SD)	SCL mean (±SD)	Weight mean (±SD)	Growth rate mean (±SD)	BCI mean (±SD)	Habitat	Adult criteria	Reference
Palmyra Atoll	211	80:20	69.7 (16.1)	----	44.6 (29.7)	----	----	Reef with algae cover	CCL>85 cm (Gulf of Carpentaria) [57]	[58]
San Diego Bay	210	----	----	85 (17.3)	97.7 (52.1)	1.03	----	Power plant	----	[59,49]
Bahia de los Angeles	200	112:88 ind	80.9	74.8 (SE 0.7)	62.3	1.4 (0.93)	1.42 (0.015)	Benthic community dominated by macroalgae	MNS at Michoacan (SCL>77,3 cm) [60,61]	[8,51]
Laguna Ojo de Liebre	137	96:4	----	59.66 (10.39)	31.9 (18)	3.1 (2.2)	1.32 (0.16)	(Lagoon) *Zostera marina*, benthic macroalgae	MNS at Michoacan (SCL>77,3 cm)	[47]
Laguna San Ignacio	220	99:1	----	52.66 (7.71)	20 (10.9)	2.1 cm (1.3)	1.27 (0.23)	(Lagoon) seagrass beds and mangrove swamps	MNS at Michoacan (SCL>77,3 cm)	[47]
Punta Abreojos	604	94:6	----	58.74 (10.49)	29.8 (17.9)	2.4 (1.2)	1.34 (0.17)	(Lagoon) Mangrove forest, seagrass and algae	MNS at Michoacan (SCL>77,3 cm)	[47]
El Pardito	59	85:15	----	65.95 (11.10)	41.8 (22.4)	----	1.38 (0.13)	Islet within Gulf of California	MNS at Michoacan (SCL>77,3 cm)	[47]
Bahía Magdalena	169	97:3	----	57.68 (8.83)	27.2 (12.9)	1.4 (0.7)	1.32 (0.15)	(Lagoon) seagrass beds and mangrove	MNS at Michoacan (SCL>77,3 cm)	[47]
Gorgona National Park	86	83:3 ind	----	58.4 (7.8)	28.8 (10.7)	----	----	Soft and sandy bottoms, coral and soft corals areas	ND	[62]
Isla Plata	68	----	61.56 (5.69)	----	27 (7.58)	----	----	----	----	[63]
El Ñuro	203	90:10	72.4 (10.9)	68.5 (10.1)	50.2 (21.4)	2.8 (1.5)	1.5 (0.16)	Sandy bottom with patches of algae	MNS at Galapagos Island [3]	This study
Paracas	160	100:0	57.7 (8.7)	55 (8.0)	27.7 (13)	6.8 (1.8)	1.55 (0.14)	Sandy bottom with patches of algae	MNS at Galapagos Island	This study

Information includes site name, samples size (n), proportion of juveniles to adults (Juv: Adt) reported either as a percentage or as total count of samples individuals, mean and standard deviation (SD) of curve carapace length (CCL), straight carapace length (SCL) in cm, weight in kg, growth rate in cm year^{-1}, and body condition index (BCI). We include information of the habitat and the size criteria to distinguish juveniles from adults (mean nesting size-MNS of females).

1.49 and 1.55. These suggest that green turtles from El Ñuro and Paracas are thriving in neritic waters of central and northern Peru and may have found excellent habitat conditions for fast growth and development. Juveniles and sub-adult individuals may take advantage of such conditions before migrating as adults into relatively less productive waters off the nesting areas in the equatorial region.

Coastal and oceanic waters off Peru have been described as "a major sink for marine turtles in the Pacific" due to the high number of bycatch of all sea turtles species [23]. The highly mobile and migratory nature of sea turtles and the fact that the Peruvian coast may constitute an important migratory-displacement habitat could explain the occurrence of such large by-catch. However, our study emphasizes the fact that green turtles use coastal waters in this region to spend considerable amounts of time at localized sites in the neritic realm. At El Ñuro individuals monitored from 2010 to 2013 have stayed or departed and returned at intervals of up to 1015 days (2.7 years) while at Paracas the maximum residency time observed is 680 days. Green turtles may stay at the same area because the habitat may provide enough food resources and refuge. As stated before, prey availability is high, particularly in Paracas, while in the seabed at El Ñuro turtles use several caves and crevices within rocky reefs (Aldo S. Pacheco personal observations) for resting which may facilitate the localized residence at this site. Recapture rates were higher at El Ñuro in comparison to Paracas; however, strict comparisons between sites should be taken with caution since capture strategies and efforts were different. Sampling at Paracas was done using considerable larger nets compared to the one deployed at El Ñuro. Despite these differences, some plausible explanations can be drawn. Low recapture rates at Paracas may imply that the population is rather large and individuals are constantly moving throughout the area, likely exploiting actively natural resources as in the likely case of juveniles recruiting from the oceanic habitat [3,5]. In addition, we cannot rule out the possibility that the local conditions such as shallow depth of our sampling site is preferred by juveniles and that the adults may be distributed in deeper waters within the general area. The presence of almost exclusively juveniles provides support to the development habitat hypothesis, stating that habitats occupied only by juveniles may be found during the neritic stage as stated by Meylan et al. [16] who examined an extensive data set supporting this hypothesis for Caribbean green turtles. As an example of contradictory evidence they cite the work of Hays-Brown & Brown [29] reporting the presence of a mix of adult and juveniles individuals in Pisco, a few kilometers north of Paracas. However, Hays-Brown & Brown [29] present percentages of straight carapace length based on 416 individuals landed in the port of Pisco, but no indication is given regarding the localities where green turtles were caught. This precludes any assessment of size stage segregation between habitats in this whole area. Our data from La Aguada (Paracas) suggest that only juveniles may occur in specific sites. Nevertheless, the lack of adults can be also attributable to the effect of the former high fishery in Pisco-Paracas and the current illegal catch that may have depleted the number of adults in the area. Moreover, from 1976 larger animals were targeted by the legal fishery as a consequence of a ministerial decree which banned the capture of green turtles smaller than 80 cm of "total" length (not carapace length) until the complete closure of the fishery in 1995 [32]. In contrast, at El Ñuro sub-adults and adults were more common residents, possibly with less displacement movements given the differences of both areas. These turtles may use the area as a permanent residence habitat taking advantage of an additional food supply such as fish and squid discards released into the water during landing operations in the pier. At this location illegal catch is rare, and locals take care of the green turtles. Although this study did not explicitly examine the biotic and abiotic reasons explaining the residence times in a given habitat, these results highlight the importance of the use of neritic residences along the Southeast Pacific coast.

Outlook and conservation issues

In this study, we emphasize the presence of green turtle aggregations at two neritic sites of the Peruvian coast. Unlike most of the studies in Peru on biological aspects of sea turtles whose source comes from interactions with fisheries [23,30,52–55], from poaching [29,31,34] or from stranding [56], we present results from a less disturbed situation. Nevertheless, some aspects still need research attention such as the degree of connectivity between sites and it should be assessed with further satellite tracking and genetic research. In addition, isotopic studies may reveal important aspects of the trophic ecology of this species such as the role of different upwelling regimes in food supply. Giving the high density of humans living in coastal areas, green turtles in neritic habitats will be more exposed to negative human interactions and although the green turtle is under the protection of international and Peruvian laws, there are threats at both sites. At El Ñuro the aggregation occurs very close to an artisanal fishing pier, so green turtles are constantly at risk of interacting with human debris and the potential for vessel collision is high. Furthermore, snorkeling with green turtles has become a regular tourist activity that may have undesirable consequences on the feeding and migratory behavior of this species. In Paracas, a traditional sea turtle fishery existed [29], and although now illegal, it still occurs in significant numbers (Javier Quiñones, unpublished data). Collision with vessels is a constant threat to green turtles in this site due to high amount of recreational boats (particularly Jet Ski) navigating the bay. The biological traits examined in our study (e.g., growth rate, body condition) suggest that green turtles inhabiting the coast of Peru thrive under excellent habitat conditions which lately may cause a positive impact on the recovery of the whole population. However, if the aforementioned anthropogenic impacts are not mitigated the endangered situation of this species might remain. We strongly encourage enhancing conservation and educational programs for this species especially within artisanal fishermen communities.

Supporting Information

Table S1 Lengths, weight and growth rate and body condition index measurements of all individuals studied at el Ñuro.

Acknowledgments

We deeply thank Pacifico Adventures, Maurice Epstein, Gonzalo Villegas, Cathy Craig, Arla and Daryl Sinclair, Amy White, Rocio and Miguel Santander, Paola Nalvarte, Dagnia Nolasco, Carlos Kouri, Cynthia Céspedes, Enrique Basurto and many volunteers for providing lodging and logistic support during the surveys at El Ñuro. Capt. Pasache Jr., Segundo and Carlos Pizarro were crucial to conduct the in-water surveys. We deeply thank Cesar Mejia Villa, Cristobal Meca Andrade, Luis Delgado Arosti and several students from Universidad San Luis Gonzaga for their help during surveys at La Aguada.

Author Contributions

Conceived and designed the experiments: XVZ JQ SK ASP. Performed the experiments: XVZ JQ SK ASP LK EP SQ. Analyzed the data: XVZ JQ ASP. Contributed reagents/materials/analysis tools: XVZ JQ SK ASP LK EP SQ. Wrote the paper: XVZ ASP JQ SK.

References

1. Carr A, Meylan AB (1980) Evidence of passive migration of green turtle hatchlings in sargassum. Copeia 1980: 366–368.
2. Carr A (1987) New perspectives on the pelagic stage of sea turtle development. Conserv Biol 1: 103–121.
3. Luschi P, Hays GC, Papi F (2003) A review of long-distance movements by marine turtles, and the possible role of ocean currents. Oikos 103: 293–302.
4. Bjorndal KA (1980) Nutrition and grazing behavior of the green turtle *Chelonia mydas*. Mar Biol 56: 147–154.
5. Musick JA, Limpus CJ (1997) Habitat utilization and migration in juvenile sea turtles. In: Lutz PL, Musick JA, eds. The Biology of Sea Turtles. CRC Press, Boca Raton, Florida. pp 137–164.
6. Mortimer JA (1981) The feeding ecology of the West Caribbean green turtle (*Chelonia mydas*) in Nicaragua. Biotropica 13: 49–58.
7. Lopez-Mendilaharsu M, Gardner SC, Seminoff JA, Riosmena-Rodriguez R (2005) Identifying critical foraging habitats of the green turtle (*Chelonia mydas*) along the Pacific coast of the Baja California peninsula, Mexico. Aquat Conserv Mar Freshw Ecosyst 15: 259–269.
8. Seminoff JA, Resendiz A, Nichols WJ (2002) Diet of east Pacific green turtles (*Chelonia mydas*) in the central Gulf of California, México. J Herpetol 3: 447–453.
9. Arthur KE, McMahon KM, Limpus CJ, Dennison WC (2009) Feeding ecology of green turtles (*Chelonia mydas*) from Shoalwater Bay, Australia. Mar Turtle Newsletter 123: 6–12.
10. Carrion J, Zarate P, Seminoff J (2010) Feeding ecology of the green turtle (*Chelonia mydas*) in the Galapagos Islands. J Mar Biol Assoc UK 90: 1005–1013.
11. Senko J, Koch V, Megill W, Carthy R, Templeton R, et al. (2010) Fine scale daily movements and habitat use of East Pacific green turtles at a shallow coastal lagoon in Baja California Sur, Mexico. J Exp Mar Biol Ecol 391: 92–100.
12. Taquet C, Taquet M, Dempster T, Soria M, Ciccione S, et al. (2006) Foraging rhythms of the green sea turtle (*Chelonia mydas*) on seagrass beds in N'Gouja Bay, Mayotte (Indian Ocean), determined by acoustic transmitters and listening station. Mar Ecol Prog Ser 306: 295–302.
13. Chaloupka M, Limpus C, Miller J (2004) Green turtle somatic growth dynamics in a spatially disjunct Great Barrier Reef metapopulation. Coral Reefs 23: 325–335.
14. NMFS (1998) Recovery plan for East Pacific green turtle (*Chelonia mydas*). National Marine Fisheries Service, Silver Spring, MD.
15. Bjorndal KA, Bolten AB, Chaloupka MY (2003) Survival probability estimates for immature green turtles *Chelonia mydas* in the Bahamas. Mar Ecol Prog Ser 252: 273–281.
16. Meylan PA, Meylan AB, Gray JA (2011) The ecology and migrations of sea turtles 8. Tests of the developmental habitat hypothesis. Bull Am Mus Nat Hist 357: 1–70.
17. Patricio AR, Velez-Zuazo X, Diez CE, Van Dam R, Sabat AM (2011) Survival probability of immature green turtles in two foraging grounds at Culebra, Puerto Rico. Mar Ecol Prog Ser 440: 217–227.
18. Bjorndal KA (1997) Foraging ecology and nutrition of sea turtles. In: Lutz PL, Musick JA, eds. The Biology of Sea Turtles CRC Press, Boca Raton, Florida. pp 137–164.
19. Wabnitz CC, Balazs G, Beavers S, Bjorndal KA, Bolten AB, et al. (2010). Ecosystem structure and processes at Kaloko Honokhau, focusing on the role of herbivores, including the green sea turtle *Chelonia mydas*, in reef resilience. Mar Ecol Prog Ser 420: 27–44.
20. Wallace BP, Lewison RL, McDonald SL, McDonald RK, Kot CY, et al. (2010) Global patterns of marine turtle bycatch. Conserv Lett 3: 131–142.
21. Wallace BP, Kot CY, DiMatteo AD, Lee T, Crowder LB, et al. (2013) Impacts of fisheries bycatch on marine turtle populations worldwide: toward conservation and research priorities. Ecosphere 4: art40, doi:10.1890/ES12-00388.1
22. Christianen MJA, Herman PMJ, Bouma TJ, Lamers LPM, van Katwijk MM, et al. (2014) Habitat collapse due to overgrazing threatens turtle conservation in marine protected areas. Proc R Soc B 281, 20132890, doi:10.1098/rspb.2013.2890
23. Alfaro-Shigueto J, Mangel JC, Bernedo F, Dutton PH, Seminoff JA, et al. (2011) Small-scale fisheries of Peru: a major sink for turtles in the Pacific. J Appl Ecol 48: 1432–1440.
24. Swartzman G, Bertrand A, Gutíerrez M, Bertrand S, Vasquez L (2008) The relationship of anchovy and sardine to water masses in the Peruvian Humboldt Current System from1983 to 2005. Prog Oceanogr 79: 228–237.
25. Spalding M, Fox H, Allen G, Davidson N, Ferdaña Z, et al. (2007) Marine ecoregions of the world: a bioregionalization of coastal and shelf areas. Bioscience 57: 573–583.
26. Forsberg K, Casabonne F, Castillo J (2012) First evidence of green turtle nesting in Peru. Mar Turtle Newsletter 133: 9–11.
27. Wester J (2010) Turtle encounters on the coast of northern Peru, a global analysis of conservation. MSc thesis. Leiden University, Nederlands.
28. Seminoff J, Zárate P, Coyne M, Foley DG, Parker D, et al. (2008) Post-nesting migrations of Galápagos green turtles *Chelonia mydas* in relation to oceanographic conditions: integrating satellite telemetry with remotely sense ocean data. Endang Species Res 4: 57–72.
29. Hays-Brown C, Brown WM (1982) Status of sea turtles in the southeastern Pacific: emphasis on Peru. In: Bjorndal K, ed. Biology and Conservation of Sea Turtles. Smithsonian Institution Press, Washington DC. pp 235–240.
30. Quiñones J, González Carman V, Zeballos J, Purca S, Mianzan H (2010) Effects of El Niño-driven environmental variability on black turtle migration to Peruvian foraging grounds. Hydrobiologia 645: 69–79.
31. de Paz N, Reyes JC, Echegaray M (2002) Datos sobre captura, comercio y biología de tortugas marinas en el área de Pisco - Paracas. In: Mendo J, Wolff M, eds. I Jornada Científica: Bases ecologicas y socioeconomicas para el manejo de los recursos vivos de la Reserva Nacional de Paracas. Universidad Nacional Agraria La Molina, Perú. pp 125–129.
32. Aranda C, Chandler MW (1989) Las tortugas marinas del Perú y su situación actual. Bol de Lima 62: 77–86.
33. Santillan LA (2008) Análisis de la dieta de *Chelonia mydas agassizii* "Tortuga verde del Pacifico" en la Bahía de Sechura, Piura - Perú. Master Thesis. Universidad Nacional Agraria La Molina.
34. Paredes R (1969) Introducción al estudio biológico de *Chelonia mydas agassizii* en el perfil Pisco. Undergraduate Thesis. Universidad Nacional Federico Villareal, Perú. 83 pp.
35. Ehrhart LM, Ogren LH (1999) Studies in foraging habitats: capturing and handling turtles. In: Eckert KL, Bjorndal KA, Abreu-Grobois FA, Donnelly M, eds. Research and Management Techniques for the Conservation of Sea Turtles. IUCN/SSC Marine Turtle Specialist Group Publication No. 4. pp. 61–64.
36. Marquez R (1990) Sea turtles of the world. FAO Species Catalogue 125: 1–81.
37. Gil-Kodaka P, Mendo J, Fernandez E (2002) Diversidad de macroalgas del submareal en la Reserva Nacional de Paracas y notas sobre su uso potencial. In: Mendo J, Wolff M, eds. I Jornada Científica: Bases ecológicas y socioeconómicas para el manejo de los recursos vivos de la Reserva Nacional de Paracas. Universidad Nacional Agraria La Molina, Perú. pp 154–163.
38. Quiñones J (2008) *Chrysaora plocamia* Lesson, 1830 (Cnidaria, Scyphozoa), frente a Pisco, Perú. Informe Instituto del Mar del Perú 35: 221–230.
39. Bjorndal KA, Bolten AB, Chaloupka MY (2000) Green turtle somatic growth model: evidence for density dependence. Ecol Appl 10: 269–282.
40. Krueger BH, Chaloupka MY, Leighton PA, Dunn JA, Horrocks JA (2011) Somatic growth rates for a hawksbill turtle population in coral reef habitat around Barbados. Mar Ecol Prog Ser 432: 269–276.
41. Kubis S, Chaloupka M, Ehrhard L, Bresette M (2009) Growth rates of juvenile green turtles *Chelonia mydas* from three ecologically distinct foraging habitats along the east central coast of Florida, USA. Mar Ecol Prog Ser 289: 257–269.
42. McMichael E, Seminoff JA, Carthy R (2008) Growth rates of wild green turtles, *Chelonia mydas*, at a temperate foraging habitat in the northern Gulf of Mexico: assessing short-term effects of cold-stunning on growth. J Nat Hist 42: 2793–2807.
43. Anderson M, Gorley RN, Clarke RK (2008) PERMANOVA+ for PRIMER: Guide to Software and Statistical Methods.
44. Echevin V, Aumont O, Ledesma J, Flores G (2008) The seasonal cycle of surface chlorophyll in the Peruvian upwelling system: a modelling study. Prog Oceanogr 79: 167–176.
45. Mianzan HW, Cornelius PFS (1999) Cubomedusae and Scyphomedusae. In: Boltovskoy D, ed. South Atlantic zooplankton. Backhuys Publishers, Leiden. pp 513–559.
46. Riascos JM, Villegas V, Caceres I, Gonzalez JE, Pacheco AS (2013) Patterns of a novel association between the scyphomedusa *Chrysaora plocamia* and the parasitic anemone *Peachia chilensis*. J Mar Biol Assoc UK 93: 919–923.
47. López-Castro MC, Koch V, Marical-Loza A, Nichols WJ (2010) Long-term monitoring of black turtles *Chelonia mydas* at coastal foraging areas off the Baja California Peninsula. Endang Species Res 11: 35–45.
48. Bjorndal KA, Bolten AB (1988) Growth rates of immature green turtles, *Chelonia mydas*, on feeding grounds in the southern Bahamas. Copeia 1988: 555–564.
49. Eguchi T, Seminoff JA, LeRoux RA, Prosperi D, Dutton DL, et al. (2012) Morphology and growth rates of the green sea turtle (*Chelonia mydas*) in a northern-most temperate foraging ground. Herpetologica 68: 76–87.
50. Patricio R, Diez CE, van Dam RP (2014) Spatial and temporal variability of inmature green turtle abundance and somatic growth in Puerto Rico. Endang Species Res 23: 51–62.
51. Seminoff JA, Jones TT, Resendiz A, Nichols WJ, Chaloupka MY (2003) Monitoring green turtles (*Chelonia mydas*) at a coastal foraging area in Baja California, Mexico: multiple indices describe population status. J Mar Biol Assoc UK 83: 1355–1362.
52. Kelez S, Velez-Zuazo X, Manrique C (2003) New evidence on the loggerhead sea turtle *Caretta caretta* (Linnaeus 1758) in Peru. Ecol apl 2: 141–142.
53. Castro J, de la Cruz J, Ramírez P, Quiñones J (2012). Sea turtles by-catch during El Niño 1997–1998, in northern Peru. Lat Am J Aquat Res 40: 970–979.
54. Alfaro-Shigueto J, Mangel JC, Caceres C, Seminoff JA, Gaos A, et al. (2010) Hawksbill turtles in Peruvian coastal fisheries. Mar Turtle Newsletter 129: 19–21.
55. Alfaro Shigueto J, Mangel JC, Seminoff JA, Dutton PH (2008) Demography of loggerhead turtles *Caretta caretta* in the southeastern Pacific Ocean:

fisheries-based observations and implications for management. Endang Species Res 5:129–135.
56. Rosales CA, Vera M, Llanos J (2010) Stranding and incidental catch of sea turtles in the coastal Tumbes, Peru. Rev Peru Biol 17: 293–301.
57. Hamann M, Schauble CS, Simon T, Evans S (2006) Demographic and health parameters of green sea turtles *Chelonia mydas* foraging in the Gulf of Carpentaria, Australia. Endang Species Res 2: 81–88.
58. Sterling EJ, McFadden KW, Holmes KE, Vintinner EC, Arengo F, et al. (2013) Ecology and conservation of marine turtles in a Central Pacific foraging ground. Chelonian Conserv Biol 12: 2–16.
59. Eguchi T, Seminoff JA, LeRoux RA, Dutton PH, Dutton DL (2010) Abundance and survival rates of green turtles in an urban environment: coexistence of humans and an endangered species. Mar Biol 157: 1869–1877.
60. Koch V, Brooks L, Nichols WJ (2007) Population ecology of the green/black turtle (*Chelonia mydas*) in Bahia Magdalena, Mexico. Mar Biol 153: 35–46.
61. Koch V, Nichols WJ, Peckham H, de La Toba V (2006) Estimates of sea turtle mortality from poaching and bycatch in Bahia Magdalena, Baja California Sur, Mexico. Biol Conserv 128: 327–334.
62. Amorocho D, Reina R (2007) Feeding ecology of the East Pacific green turtle (*Chelonia mydas agassizii*) at Gorgona National Park, Colombia. Endang Species Res 3: 43–51.
63. Muñoz Perez JP (2009) Identificación y estudio preliminar de los sitios críticos para anidación, forrajeo y descanso de las tortugas marinas en la costa centro y norte del Ecuador. Bachelor Thesis. Universidad San Francisco de Quito, Ecuador. 27 pp.

21

Genetic Variability and Population Structure of *Disanthus cercidifolius* subsp. *longipes* (Hamamelidaceae) Based on AFLP Analysis

Yi Yu[1], Qiang Fan[1], Rujiang Shen[1], Wei Guo[3], Jianhua Jin[1], Dafang Cui[2]*, Wenbo Liao[1]*

1 Guangdong Key Laboratory of Plant Resources and Key Laboratory of Biodiversity Dynamics and Conservation of Guangdong Higher Education Institutes, School of Life Sciences, Sun Yat-Sen University, Guangzhou, China, **2** College of Forestry, South China Agriculture University, Guangzhou, China, **3** Department of Horticulture and Landscape Architecture, Zhongkai University of Agriculture and Engineering, Guangzhou, China

Abstract

Disanthus cercidifolius subsp. *longipes* is an endangered species in China. Genetic diversity and structure analysis of this species was investigated using amplified fragments length polymorphism (AFLP) fingerprinting. Nei's gene diversity ranged from 0.1290 to 0.1394. The AMOVA indicated that 75.06% of variation was distributed within populations, while the between-group component 5.04% was smaller than the between populations-within-group component 19.90%. Significant genetic differentiation was detected between populations. Genetic and geographical distances were not correlated. PCA and genetic structure analysis showed that populations from East China were together with those of the Nanling Range. These patterns of genetic diversity and levels of genetic variation may be the result of *D. c.* subsp. *longipes* restricted to several isolated habitats and "excess flowers production, but little fruit set". It is necessary to protect all existing populations of *D. c.* subsp. *longipes* in order to preserve as much genetic variation as possible.

Editor: Ting Wang, Wuhan Botanical Garden, Chinese Academy of Sciences, Wuhan, China

Funding: Support for this study was provided by a grant by the National Natural Science Foundation of China (Project 31170202) to WL, a grant by the National Natural Science Foundation of China (Project 30670141) to WL, a grant by the National Natural Science Foundation of China (Project 31100159) to QF, a grant by the National Natural Science Foundation of China (Project 40672017) to JJ. The funders had no role in study design, data collection and analysis, decision to publish, or preparation of the manuscript.

Competing Interests: The authors have declared that no competing interests exist.

* Email: lsslwb@mail.sysu.edu.cn (WL); cuidf@scau.edu.cn (DC)

Introduction

Disanthus Maxim. (Hamamelidaceae) is a monotypic genus endemic to China and Japan [1]. *Disanthus cercidifolius* subsp. *longipes* is the only subspecies in *Disanthus* which is distributed in China (southern Zhejiang, central and northwestern Jiangxi, and southern Hunan), while its sister, *D. c.* subsp. *cercidifolius*, is endemic to Japan [2]. *D. c.* subsp. *longipes* was first reported by Cheng [3] in 1938, and then revised by Chang [4] in 1948. In a systematic study of *Disanthus* [2], the author believed that the Chinese species of *Disanthus* was a sister population of *D. c.* subsp. *cercidifolius*, which is distributed in warm and humid forests of the Cathayan Land, and also was a Tertiary relic. *D. c.* subsp. *longipes* was characterized by its morphology and preference for humid, acid soils and shady habitats. The inflorescences are paired, axillary; Capitula are 2-flowers, purple; Leaves are heart shaped, green then turning to purple, orange, and red. It is usually a small tree, 2–3 m high, occasionally reaching heights of 6–8 m in forests when growing along streams. Because of severe habitat fragmentation that caused population decline, *Disanthus* was listed in the 1992 Red List of Endangered Plant Species of China [5], the 1994 IUCN Red List of Threatened Species (www.iucnredlist.org), the Key Wild Plants under State Protection [6] and the 2004 China Species Red List as bring a species at high risk of extinction in the wild [7].

Although study of *Disanthus* has attracted many investigators [8–22], little attention has been paid to its genetic analysis and population structure, except Xiao [23]. Xiao investigated the genetic diversity of *D. c.* subsp. *longipes* based on nine allozyme loci and found a higher genetic variation within populations as well as significantly lower variation among populations. However, sampling of Xiao's study was limited to part of the distribution area of *D. c.* subsp. *longipes* only. So a comprehensive study of the populations' genetic structure at different geographic scales is still needed.

Preserving the genetic diversity of endangered species is one of the primary goals in conservation planning. Because survival and evolution of species depended on the maintenance of sufficient genetic variability within and among populations to accommodate new selection pressures caused by environmental changes [24,25]. For endemic endangered species, intraspecific variation is a prerequisite for any adaptive changes or evolution in the future, and have profound implications for species conservation [26,27]. The knowledge of the levels and patterns of genetic diversity is important for designing conservation strategies for threatened and endangered species [28,29]. So identifying variations with molecular markers has provided the abundant information concerning genetic diversity in plant species [30–36]. Amplified fragment length polymorphism (AFLP) [37] is a PCR-based technique which has been successfully applied to the identification and estimation of molecular genetic diversity and population structure [32–34,38–41]. This technique can generate information on multiple loci in a single assay without prior sequence knowledge, and [42]. Using AFLP markers, we would know (1) the degree of genetic diversity within and among populations; (2) which factors might explain genetic variation; and (3) how to apply this information to develop recommendations for management of this endangered species.

Materials and Methods

Ethics Statement

Field studies were approved by Hunan Provincial Bureau of Forestry for collection in Xinning County (1XN), the Dupanglin National Nature Reserve (2DP), the Mangshan National Nature Reserve (3MS), and approved by Guangdong Provincial Bureau of Forestry for collection in Nanling National Nature Reserve (4NL), and approved by Zhejiang Provincial Bureau of Forestry for collection in QianJiang Source National Forestry Park (5QJ) and Zhulong town, Longquan City (6ZL), and approved by Jiangxi Provincial Bureau of Forestry for collection in Guanshan National Nature Reserve (7GS), and Mount Sanqingshan National Park (8SQ).

Specimen collection

The specimens were collected from eight populations from April to June of 2008. All populations but one grew in evergreen and deciduous broad-leaved mixed forests (often lived with species such as in genus *Cyclobalanopsis* and *Sorbus*, *etc.*) at altitudes of 450–1200 m. The exception (3MS) was the one growing in bamboo forests in the Mangshan Mountains of Yizhang. For molecular analysis, 10–11 individuals per population were sampled. The locations and information of populations are provide in Table 1 and Fig. 1. Before DNA extraction, all the dried leaves were preserved in silica gel [43]. All the voucher specimens are deposited at the Herbarium of Sun Yat-sen University (SYS).

DNA extraction and AFLP reactions

Genomic DNA was extracted from silica gel-dried leaves using the cetyl trimethylammonium bromide (CTAB) method [44]. The extracted DNA was dissolved in 100 μL of Tris-hydrochloride (TE) buffer [10 mmol/l Tris-HCl (pH 8.0), 1 mmol/EDTA (pH 8.0)] and used as a template for the polymerase chain reaction (PCR).

AFLP reactions were performed following the method reported by [37] with the following modifications. The restriction digest and ligation steps were done as separate reactions. For the digestion, approximately 500 ng of genomic DNA was incubated at 37°C (EcoRI) or 65°C (MseI) for 2 h in a 20 μL volume reaction containing 10× H Buffer (TOYOBO, Shanghai) and 10 U restriction enzymes EcoRI or MseI. For the ligation, 20 μL of a ligation mix consisting of 10× T4 DNA Ligase (TOYOBO, Shanghai), 1 μL EcoRI-adapter, 1 μL MseI-adapter, and 2 U T4 DNA Ligase was added to the sample and kept at 22°C for 3 h. After ligation, the samples were diluted 10-fold with sterile deionized water (dH_2O). A pre-selective polymerase chain reaction (PCR), using PTC-200 thermocycler (MJ research, Waltham, MA) was done using primer pairs with a single selective nucleotide extension. The reaction mix (total volume 20 μL) consisted of 4 μl template DNA from the restriction/ligation step, 1 μL primer (EcoRI/MseI), and 15 μL AFLP Core Mix (13.8 μL dH_2O, 1.6 μL $MgCl_2$, 1.6 μL dNTPs (2.5 mM), 1 U Taq DNA polymerase, and 10× H buffer). After an initial incubation at 94°C for 2 min, 20 cycles at 94°C for 20 s, 56°C for 30 s, and 72°C for 2 min, with a final extension at 60°C for 30 min, were performed. The PCR products of the amplification reaction were diluted 10-fold with dH_2O and used as a template for the selective amplification using two AFLP primers, each containing three selected nucleotides.

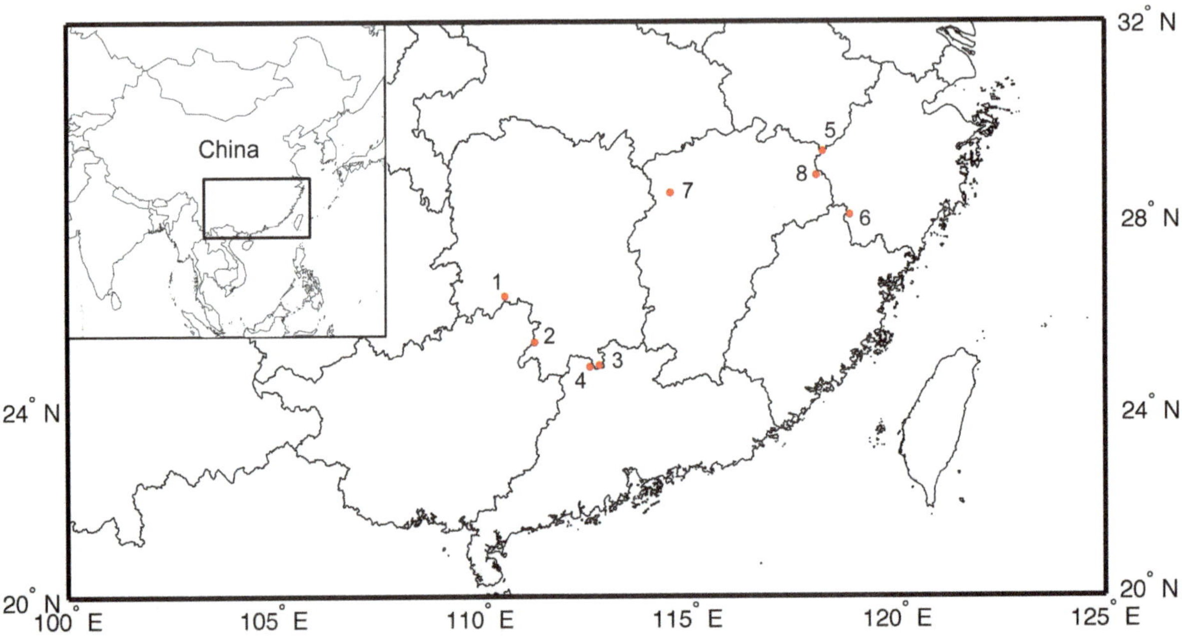

Figure 1. Location of eight populations in two groups sampled in a study of genetic diversity of *Disanthus cercidifolius* subsp. *longipes*. Populations are represented by black dots and located as Table 1. Note: 1:1XN, 2:2DP, 3:3MS, 4:4NL, 5:5QJ, 6:6ZL, 7:7GS, 8:8SQ.

Table 1. Information of location of populations.

ID	GPS co-ordinates	Location of populations	Location type	Altitude (m)	Collection voucher
1XN	26°25′08″N, 110°36′31″E	Xinning county, Hunan Province	mountain slope on the edge of forests	690–740	SHEN Rujiang and GUO Wei
2DP	25°27′43″N, 111°20′20″E	Dupanglin National Nature Reserve, Hunan Province	in bushes by stream	700–850	SHEN Rujiang and GUO Wei
3MS	24°58′12″N, 112°53′16″E	Mangshan National Nature Reserve, Hunan Province	in bamboo forests by stream	680–923	SHEN Rujiang and GUO Wei
4NL	24°56′18″N, 112°39′40″E	Nanling Naitional Nature Reserve, Guangdong Province	mountain slope	680–700	SHEN Rujiang
5QJ	29°23′59″N, 118°12′58″E	Qianjiang Source National Forestry Park, Zhejiang Province	mountain slope	650–700	SHEN Rujiang and GUO Wei
6ZL	28°06′28″N, 118°51′27″E	Zhulong town, Longquan City, Zhejiang Province	mountain slope by streams	1040	SHEN Rujiang and GUO Wei
7GS	28°33′22″N, 114°35′42″E	Guanshan National Nature Reserve, Jiangxi Province	mountain slope	560–600	CHEN Lin
8SQ	28°54′57″N, 118°03′52″E	Mount Sanqingshan National Park, Jiangxi Province	mountain slope in forests	620–1260	Observation team of Sun Yat-sen University

Nine primer combinations labeled with fluorescent 6-carboxy fluorescein (6-FAM) were probed for selective amplification, and only primer combinations with the greatest number of bands (EcoRI/MseI: AAC/CTG; ACA/CTG and ACT/CAT) were selected. Selective amplification was done using the following touchdown PCR conditions: 94°C for 2 min first, then 10 cycles at 94°C for 20 s, and 66°C for 30 s, with a 1°C decrease per cycle then extension at 72°C for 2 min; followed by 20 cycles at 94°C for 20 s, 56°C for 30 s, and 72°C for 2 min. After amplification, 3 μL of the samples were diluted 3-fold with sterile deionized water (dH_2O), and mixed with 10 μL formamide and 0.2 μL Size standard-600 (Beckman Coulter, Fullerton, CA). The mix was used for sequence analysis.

Raw data were collected on a CEQ8000 Sequencer (Beckman Coulter). AFLP products were resolved using a Beckman Coulter CEQ8000 genetic analyzer. Semi-automated fragment analysis was performed using the fragment analysis software of the CEQ8000. The chromatograms of fragment peaks were scored as present (1) or absent (0), and a binary qualitative data matrix was constructed. A total of 82 individuals were run twice with all primer combinations.

Data analysis

As a measure of population diversity, the binary data matrix was input to POPGENE version 1.32 [45], assuming Hardy–Weinberg equilibrium. The following indices were used to quantify the amount of genetic diversity within each population examined: the number of AFLP fragments ($Frag_{tot}$), the percentage of polymorphic fragments ($Frag_{poly}$), Nei's (1973) gene diversity (H), and Shannon's information index (I). In addition, the number of unique fragments ($Frag_{uni}$) and DWs, frequency-down-weighted marker values [46], were calculated as measures of divergence. For each population, the number of occurrences of each AFLP marker in that population was divided by the number of occurrences of that particular marker in the total dataset, then these values were summed up [46,47]. The value of DW was expected to be high in long-term isolated populations where rare markers should accumulate due to mutations, whereas newly established populations were expected to exhibit low values [46]. To even out the unequal sample sizes, DWs were calculated using five randomly chosen individuals. The value of DW was calculated by AFLPdat [48], which is based on R version 2.9.0 (www.r-project.org).

Nei's genetic distance [49] of *D. c.* subsp. *longipes* populations were calculated using the software POPGENE version 1.32 [45], then cluster analysis was performed based on the unweighted pair group method with arithmetic averaging (UPGMA) [50] using NTSYS pc version 2.1e [51]. In addition, based on the genetic distance index devised by [52], the UPGMA dendrogram of individuals was drawn using TREECON version 1.3b [53]. The robustness of the branches was estimated by 1000 bootstrap replicates.

ARLEQUIN version 3.0 [54] was used to perform an analysis of molecular variance (AMOVA) [55] to assess the hierarchical genetic structure among populations and within populations. The AFLPdat program [48], based on R version 2.9.0 (www.r-project.org), was used to convert the AFLP data matrix to the ARLEQUIN input format. The AMOVA was performed by partitioning genetic variation among and within populations regardless of their geographic distribution. In this study, the traditional *F*-statistics [56] cannot be used in dominant marker AFLP, therefore Φ-statistics [55] replace *F*-statistics. The significance of Φ values was tested by 1000 permutations. Φ-statistics are computed from a matrix of Euclidean squared distances between every pair of individuals [40]. Two models of the eight populations were tested to investigate regional relationships. Firstly, we treat all populations as a single group to obtain a value for Φ_{st} as an overall measure of population divergence (a two-level analysis), and then we divided the populations into two groups, the East China region and the Nanling Range (three-level hierarchical analyses).

A distance matrix of Φ_{st} between every pair of populations was calculated in ARLEQUIN as a measure of interpopulation genetic differentiation, from which 1000 bootstrapped replicate matrices were then computed, so gene flow (*Nm*) based on Φ_{st} [56] could be calculated. Isolation-by-distance was investigated by computing the correlation between geographic distance and genetic distance (Φ_{st}) between every pair of populations and applying the Mantel test, using NTSYSpc version 2.1e [51]. The Mantel *z*-statistic value was tested non-parametrically by creating a null distribution of *z* using 1000 random permutations and comparing the observed *z* value [40].

Table 2. Number of loci evaluated for each of three AFLP primer combinations utilized in assays of 82 individuals of *Disanthus cercidifolius* subsp. *longipes* from eight populations.

Primer combination	Number of loci
EcoRI-AAC/MseI-CTG	144
EcoRI-ACA/MseI-CTG	152
EcoRI-ACT/MseI-CAT	157
Total	453

The AFLP data were also subjected to a principal components analysis (PCA), which may help reveal unexpected relationships among a large number of variables, reducing them to two or three new uncorrelated variables so they retain most of the original information [57]. We chose Jaccard's similarity coefficient [50] to calculate the eigenvalue and eigenvector. The standardized data were projected onto the eigenvectors of the correlation matrix and represented in a two-dimensional scatter plot [57]. Plots of samples in relation to the first three principal components were constructed with populations designated as either populations of the East China region or populations of the Nanling Range. The data from the PCA was analyzed using the computer program NTSYSpc version 2.1e [51]. A two-dimensional representation of genetic relationships among *D. c.* subsp. *longipes* genotypes was carried out using SPSS version 16.0 [58].

STRUCTURE version 2.3.1 [59] was used to investigate structure at the individual level. In this study, we inferred that each individual of *D. c.* subsp. *longipes* comes purely from one of the populations. It was applied using a "no admixture" model, 100 000 burn-in period, and 50 000 MCMC replicates after burn-in. The approach requires that the number of clusters K be predefined, and the analysis then assigns the individuals to clusters probabilistically [40]. We performed ten runs for each value of K (1 to 10). To determine the K value, we used both the LnP(D) value and Evanno's ΔK [60]. LnP(D) is the log likelihood of the observed genotype distribution in K clusters and can be output by STRUCTURE simulation [59]. Evanno's ΔK took consideration of the variance of LnP(D) among repeated runs and usually can indicate the ideal K. The suggested $\Delta k = M(|L(k+1)-2L(k)+L(k-1)|)/S[L(k)]$, where L(k) represents the kth LnP(D), M is the mean of 10 runs, and S their standard deviation[60,61]. The output uses color coding to show the assignments of individuals in each population to the clusters.

Population structure was also investigated by HICKORY version 1.1 [62], in order to assess the importance of inbreeding in the data and the assumption of Hardy-Weinberg equilibrium. HICKORY makes it possible to evaluate departures from Hardy-Weinberg equilibrium in dominant as well as co-dominant markers [40]. The program AFLPdat [48] was used to convert the AFLP data matrix to the HICKORY input format. The deviance information criterion (*DIC*) criteria, *Dbar*, *Dhat*, and *pD*, for assessing the importance of inbreeding were computed using the default values: Burn-in 5000; sample 100 000; thin 20 [62].

Results

Population genetic diversity

In this study, three of the nine AFLP primer combinations were used (Table 2). The number of fragments for each primer combination (with percentage of polymorphisms within parenthesis), were: EcoRI-AAC/MseI-CTG: 144 (98.61%), EcoRI-ACA/MseI-CTG: 152 (99.34%), and EcoRI-ACT/MseI-CAT: 157 (96.18%). The length of the fragments varied from 64 bp to 560 bp. The three AFLP primer combinations produced a total of 453 fragments in 82 individuals, of which 444 (98.01%) were polymorphic. The total number of fragments per population ($Frag_{tot}$) varied between 198 (7GS) and 227 (5QJ). The percentage of polymorphism ($Frag_{poly}$) across the eight populations ranged from 43.71% (7GS) to 50.11% (5QJ). The number of fragments that only occur in one population ($Frag_{uni}$) varied between 6 (7GS) and 16 (8SQ). The DW ranged from 56.85 (5QJ) to 111.19 (1XN), mean 76.88, and SD = 21.11. The diversity within a population (H) ranged from 0.1290 (2DP) to 0.1394 (5QJ) (Table 3).

AMOVA

The two-level AMOVA in ARLEQUIN gave a Φ_{st} value of 0.2328 ($P<0.001$), with of 23.28% variation among populations and 76.72% within populations (Table 4). The three-level hierarchical AMOVA analyses of the two-group models shows that three-quarters of the variation (75.06%) was concentrated within populations, while the between group component (5.04%) was less than the between populations-within-group component (19.90%). All three Φ values were significant based on a 1000 permutation test (Table 4).

Pairwise genetic distance and gene flow

The pairwise genetic distance (Φst) and gene flow (*Nm*) matrix were used to establish the level of genetic divergence among the populations (Table 5). Estimates of pairwise genetic distance using AFLP date ranged from 0.1364 ($P<0.001$) for the most closely related populations (4NL and 6ZL), to 0.3428 ($P<0.001$) in the most divergent populations (3MS and 8SQ). The gene flow (*Nm*) among populations ranged from 0.4792 to 1.5830. Except for one population in Nanling National Nature Reserve (4NL), most of populations' Φst were above 0.2, and the corresponding *Nm* were below 1. These results indicate that the genetic distance of *D. c.* subsp. *longipes* is not close, yet nor dependent on geographic distance.

Relationship between geographic distance and genetic distance (The Mantel tests)

A Mantel test showed no correlation between the geographic distance and the Φst genetic distance ($r = 0.38917$, $P = 0.9920$, Fig. 2). This result implies that *D. c.* subsp. *longipes* does not demonstrate a historical pattern of isolation-by-distance.

Principal components analysis

A PCA of the AFLP-based distance data was performed to examine relationships among the populations of *D. c.* subsp. *longipes*. The PCA showed that populations from East China and the Nanling Range clearly do not cluster separately (Fig. 3). The

Table 3. Region, population (ID), sample size (N), total number of fragments per population ($Frag_{tot}$), percentage of polymorphic fragments ($Frag_{poly}$), number of fragments that only occur in one population ($Frag_{uni}$), frequency-down-weighted marker values (DW), Nei's (1973) gene diversity (H), Sannon's index (I) sampled in a study of *Disanthus cercidifolius* subsp. *longipes* (H. T. Chang) K. Y. Pan populations.

Region	ID*	N	$Frag_{tot}$	$Frag_{poly}$	$Frag_{uni}$	DW	H	I
Nanling range	1XN	10	216	47.68	12	111.19	0.1357±0.1779	0.2106±0.2570
	2DP	11	203	44.81	11	56.96	0.1290±0.1790	0.1991±0.2580
	3MS	10	204	45.03	10	93.96	0.1373±0.1839	0.2100±0.2652
	4NL	10	215	47.46	9	58.09	0.1383±0.1710	0.2136±0.2608
East China region	5QJ	11	227	50.11	10	56.85	0.1394±0.1778	0.2172±0.2564
	6ZL	10	209	46.14	12	77.45	0.1393±0.1822	0.2141±0.2631
	7GS	10	198	43.71	6	64.85	0.1353±0.1852	0.2062±0.2662
	8SQ	10	223	49.23	16	95.68	0.1373±0.1775	0.2138±0.2560
All	-	82	453	98.01	-	-	0.1781±0.1633	0.2918±0.2223

Table 4. Analysis of molecular variance (AMOVA) in *Disanthus cercidifolius* subsp. *longipes* for 82 individuals from eight populations.

Model	Source of variation	df	Sum of squares	Variance component	Percentage variance %	Φ-statistics	*P*
Two levels	among populations	7	1055.95	11.14	23.28	$\Phi_{st}=0.23283$	$P<0.001$
	within populations	74	2716.05	36.7	76.72		
	Total	81	3772	47.84			
three levels	between groups	1	237.66	2.46	5.04	$\Phi_{ct}=0.05040$	$P<0.05$
	between populations-within-groups	6	818.29	9.73	19.90	$\Phi_{sc}=0.20955$	$P<0.001$
	within populations	74	2716.05	36.70	75.06	$\Phi_{st}=0.24939$	$P<0.001$
	Total	81	3772	48.89			

Computed with ARLEQUIN version 3.0. The *P* values represent the probability of obtaining an equal or more extreme value by chance, estimated from 1000 permutations.

natural divide is separated into three clusters: cluster 1 (1XN, 2DP), cluster 2 (5QJ, 8SQ), and cluster 3, a mixture that includes some individuals from cluster 1 or cluster 2. The first and second principal component axes, PC1 and PC2, accounted for 25.46% and 25.17% of the total variation, respectively.

Hierarchical and cluster analysis

Nei's genetic distances [49] between populations showed two main clusters of populations (Fig. 4). The cluster that comprises 5QJ and 8SQ is clearly differentiated from the others. Populations 1XN and 2DP are another cluster, and population 4NL and 6ZL are grouped with 7GS, although the UPGMA dendrogram of the 82 individuals based on Nei and Li's genetic distance is not clearly grouped (Fig. 5). Although most individuals in the same population cluster together, some individuals in different populations (except 1XN and 3MS) also are clustered together.

In the genetic structure analysis (STRUCTURE), the highest estimate of the likelihood of the data, given the number of clusters chosen, was k = 10 (ten clusters). We can't get a clear knee in Lnp(D) (Fig. 6). Evanno's ΔK took consideration of the variance of LnP(D) among repeated runs and usually can indicate the best k. See the Fig. 6, we can find that when the k = 6 and we get the highestΔK value. So the best k = 6. The diagram (Fig. 7) showing assignment of individuals to the clusters revealed a clinal structure to the data. In the six-cluster model (k = 6), gene pool 1 (including 5QJ, 8SQ) and gene pool 2 (including 1XN, 2DP) are restricted with a few from another pool, only being distributed in the southwest (gene pool 1) and northeast (gene pool 2). Gene pool 3 (including 3MS, 4NL), while Gene pool 4, pool 5 and pool 6, consist of 6ZL and 7GS. These 4 pools have the most widespread pattern.

The HICKORY results showed that there is inbreeding in populations of *D. c.* subsp. *longipes* (Table 6), because the *DIC* and *Dbar* parameter were lower in "*Full model*" than in other models. This pattern of results allows the "*Full model*" to be considered best, according to the HICKORY manual [62].

Table 5. Pairwise genetic distance (*Φst*, lower diagonal, *P*<0.001) and gene flow (*Nm*, upper diagonal) between eight populations of *Disanthus cercidifolius* subsp. *longipes* based on AFLP data.

	1XN	2DP	3MS	4NL	5QJ	6ZL	7GS	8SQ
1XN	-	1.4673	0.7122	1.1283	0.7637	0.7610	0.8506	0.6140
2DP	0.1456	-	0.8055	1.1586	0.6833	0.7813	0.7398	0.5624
3MS	0.2598	0.2369	-	1.4639	0.6874	0.7389	0.8037	0.4792
4NL	0.1814	0.1775	0.1459	-	1.0727	1.5830	1.2885	0.6644
5QJ	0.2466	0.2679	0.2667	0.1890	-	1.1441	0.7897	1.3148
6ZL	0.2473	0.2424	0.2528	0.1364	0.1793	-	1.1783	0.6917
7GS	0.2272	0.2526	0.2373	0.1625	0.2405	0.1750	-	0.5466
8SQ	0.2893	0.3077	0.3428	0.2734	0.1598	0.2655	0.3138	-

Significance levels are based on 1000 iterations and indicate that the probability that random *Φst* values are higher than the observed values. *Nm* = (1 - *Fst*)/4**Fst* [56].

Discussion

Genetic diversity in endemic and endangered species

Accurate estimates of genetic diversity are useful for optimizing sampling strategies aiming at the conservation and management of genetic resources [26,63,64]. According to Hamrick and Godt (1989), there are strong associations between geographic range and genetic diversity [65]. Allozyme analyses concluded that endemic and geographically limited plant species generally possess less genetic variation, due to genetic drift and restricted gene flow [66–68]. However, in our study, the percentage of polymorphic fragments ($Frag_{poly}$) and Shannon's information index (I) are higher than those of endemic species based on allozyme [65]. This may account for the new technique can generate more genetic diversity information. Historical events have also been shown to be responsible for variations in genetic diversity [68]. Numerous allozyme studies and an increasing number of cpDNA and mtDNA studies now provide substantial evidence that putative refugee plant populations harbor higher levels of genetic diversity relative to their likely descendant populations [69]. *Disanthus cercidifolius* subsp. *longipes* is a Tertiary relic, which is distributed in warm and humid forests of the Cathayan Land in the Tertiary [2,70]. Lots of present endemics, several of which inhabit Pleistocene refugia during the Quaternary glacial period, were able to maintain higher levels of diversity because of population stability during the glacial cycle. *Disanthus cercidifolius* subsp. *longipes* may be one of them and preserved mutation during the

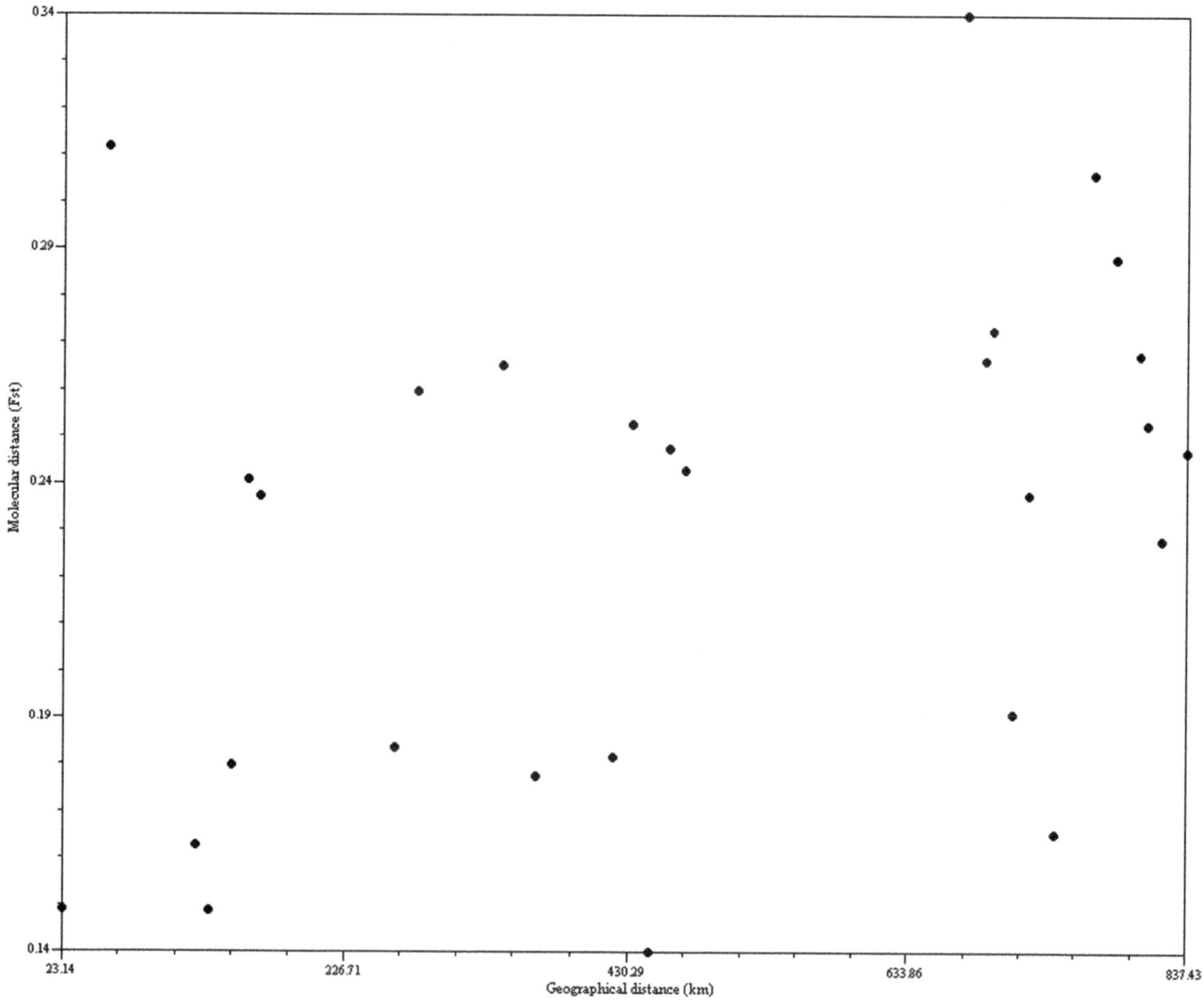

Figure 2. *Disanthus cercidifolius* subsp. *longipes*, bivariate plot showing no correlation between matrices of pair-wise geographic (km) and genetic distance (Φst) in eight populations, comprising 82 individuals. Computed with NTSYSpc ver. 2.1e.

relatively long glacial period. So it shows high genetic diversity [69,71]. Compared by AFLP, RAPD, and ISSR markers, the genetic diversity indices in this study revealed an intermediate level of genetic diversity of *D. c.* subsp. *longipes* compared to other endangered species [34,41,66,72,73].

Results of studies of genetic diversity within populations show that all populations were on the similar level, except for the Dupanglin National Nature Reserve (2DP) which is shown to be lower than the others. The lack of genetic diversity resulting from the total existing individual in population 2DP is limited (fewer than 100 individuals). In practice, a larger population often has higher genetic variation [72,74].

Genetic variation and gene flow

In this study AMOVA showed that the largest portion (76.72%) of genetic variance is contributed by genetic variation within populations (Φ_{st} value 0.23283) and only a small portion (23.28%) is due to differences among groups. Some other studies of endemic and endangered species show a similar pattern [34,41,66,72,73]. Moderate or high diversity and low population partitioning in rare plants have previously been attributed to a number of factors [66,75–77], including insufficient length of time for genetic diversity to be reduced following a natural reduction in population size and isolation; adaptation of the genetic system to small population conditions; recent fragmentation (human disturbance) of a once-continuous range, i.e., genetic system; and extensive gene flow due to the combination of bird pollination and high outcrossing rates [66,78]. The direct estimate of gene flow (*Nm*) based on Φ_{st} 0.23283 was 0.82374, which means that the number of migrants per generation is lower than one. The present distribution of *D. c.* subsp. *longipes* is restricted to several isolated habitats [5,7]. However, fossil and palynological evidences suggested that, *Disanthus* was wide spread in warm and humid forests in the Tertiary [2,70]. Accordingly, a reasonable hypothesis is that the modern range of *Disanthus* species was the result of population fragmentation and contraction after the Quaternary glacial cycles. Considering that *D. c.* subsp. *longipes* has poor success in pollination, "excess flowers production, but little fruit set" [79], and fragmented habitat, inbreeding, which generally leads to decreased fitness, or inbreeding depression, of a population, some of the above causes result in the endangered status of *D. c.* subsp. *longipes*.

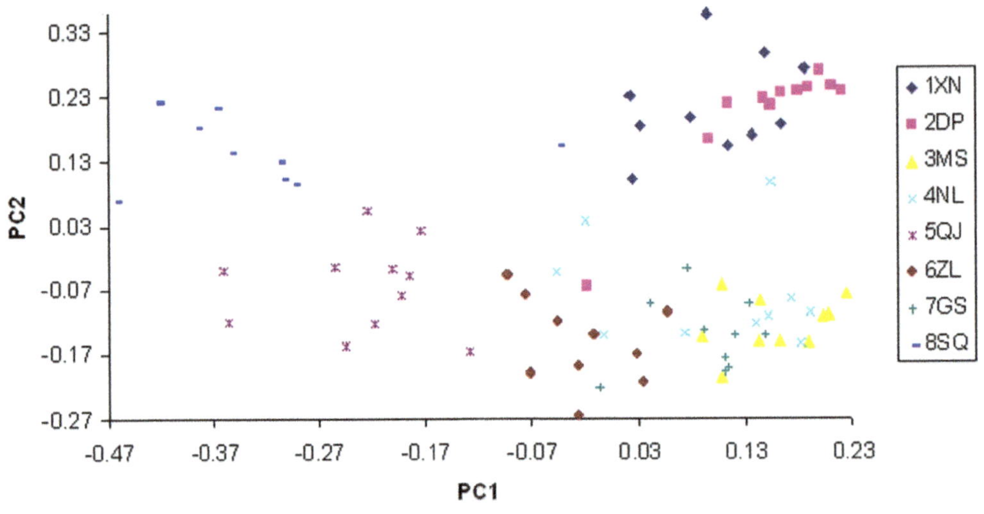

Figure 3. Principal coordinates analysis of 82 individuals from eight populations, based on dissimilarity matrix (Jaccard's coefficient). Accessions are plotted according to the values of first (x axis) and the second (y axis) components and with different symbols according to population. Principal coordinate axes shown (pc1 and pc2) represent 25.46% and 25.17% of respective variance in the dissimilarity matrix.

Genetic structure at different hierarchical levels

Although AFLP markers are dominant, they provide no information on heterozygote frequencies, and our investigation provides no direct information on the reproductive strategy of *D. c.* subsp. *longipes*. In general, outcrossing and long-lived seed plants maintain the most genetic variation within populations, while predominantly selfing, short-lived species harbor comparatively higher variation among populations [65,80]. According to data based primarily on allozyme analysis, Φ_{st} =0.2 for outcrossing species and 0.5 for inbreeders [65]. The overall Φ_{st} value 0.23283 and the gene flow value for *D. c.* subsp. *longipes* are similar to other species that outcross [81–84]. Permutation tests of the fixation index (Φ_{st} =0.23) indicated significant genetic structuring. The populations of *D. c.* subsp. *longipes* spread from east to west (Fig. 1). And in the field investigation of our study, we

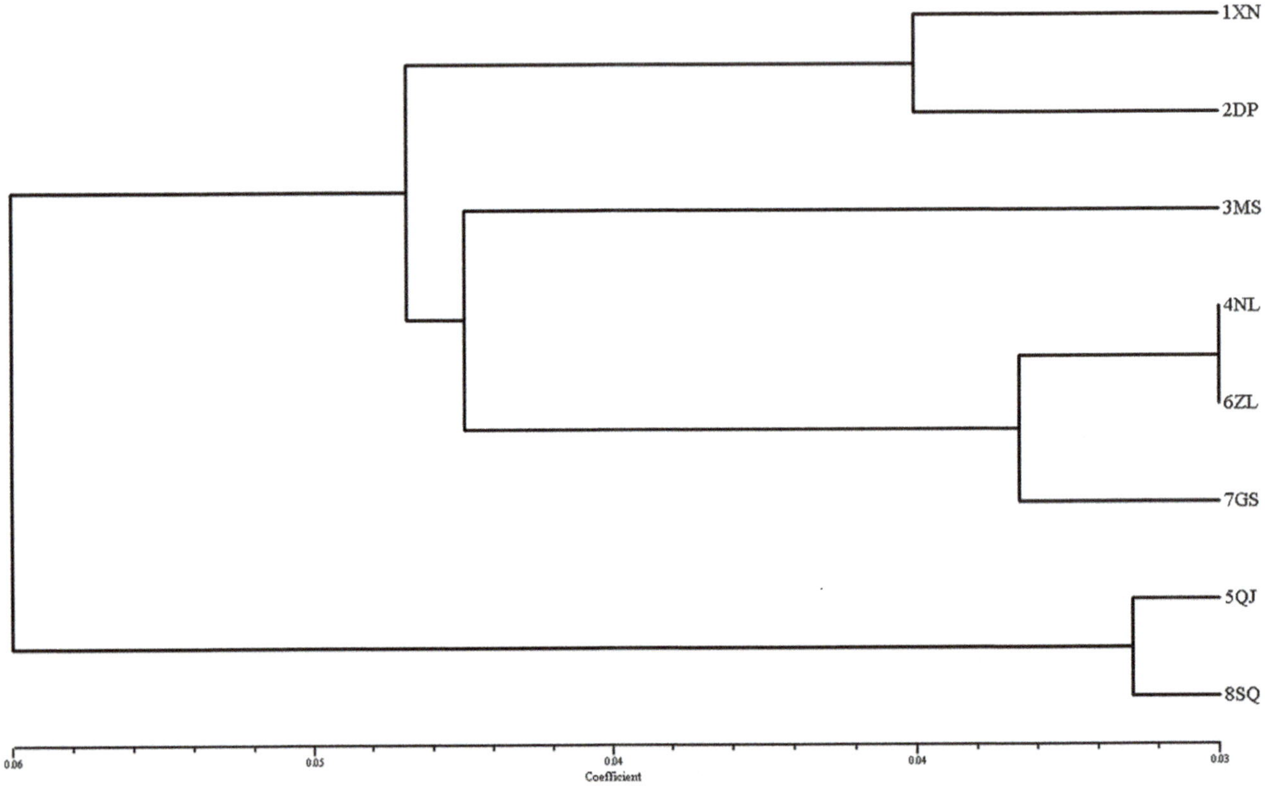

Figure 4. UPGMA dendrogram based on genetic distance. Note: A = 1XN, B = 2DP, E = 3MS, G = 4NL, H = 5QJ, I = 6ZL, J = 7GS, K = 8SQ.

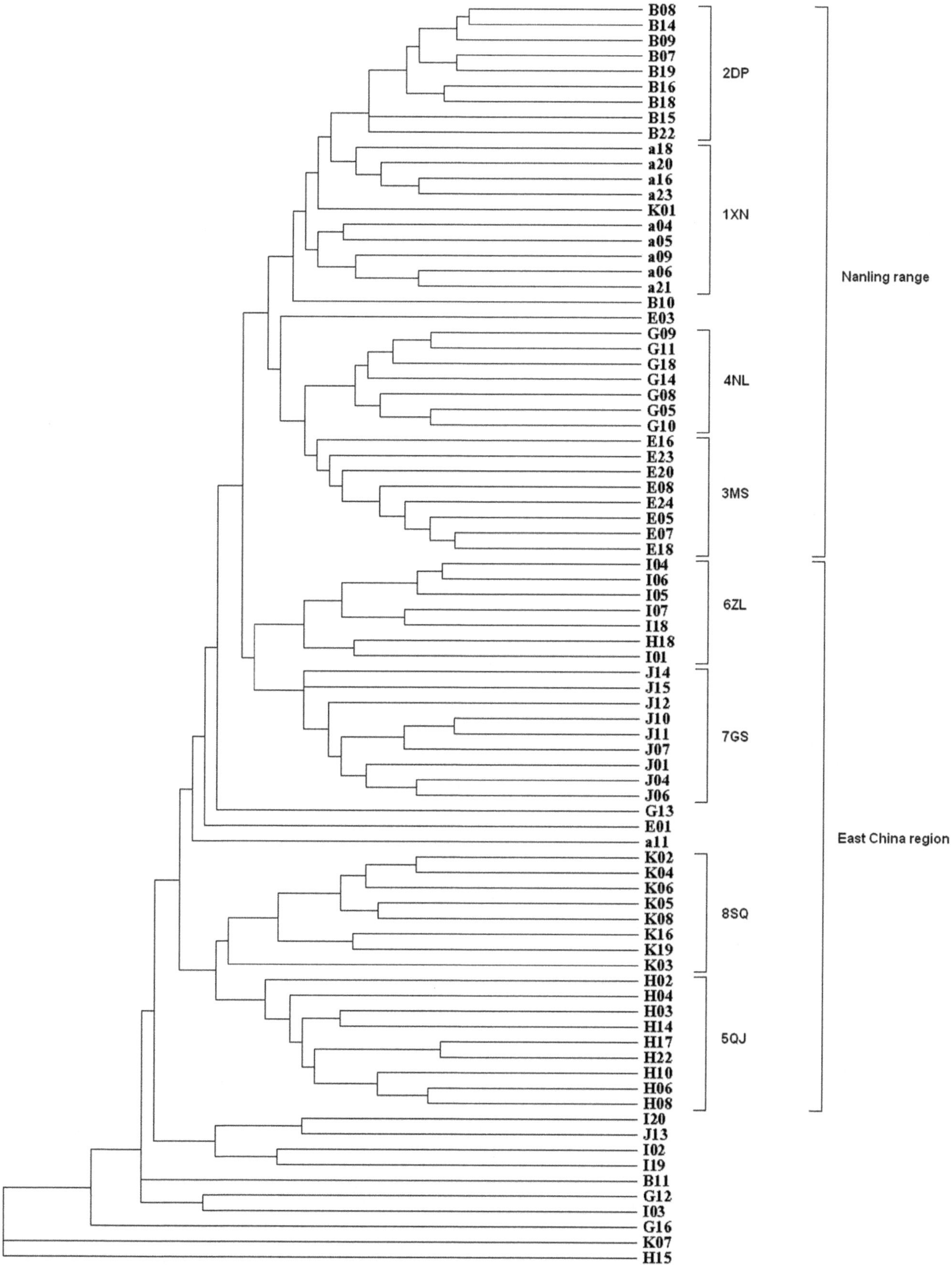

Figure 5. The UPGMA dendrogram based on Nei & Li's genetic distance.

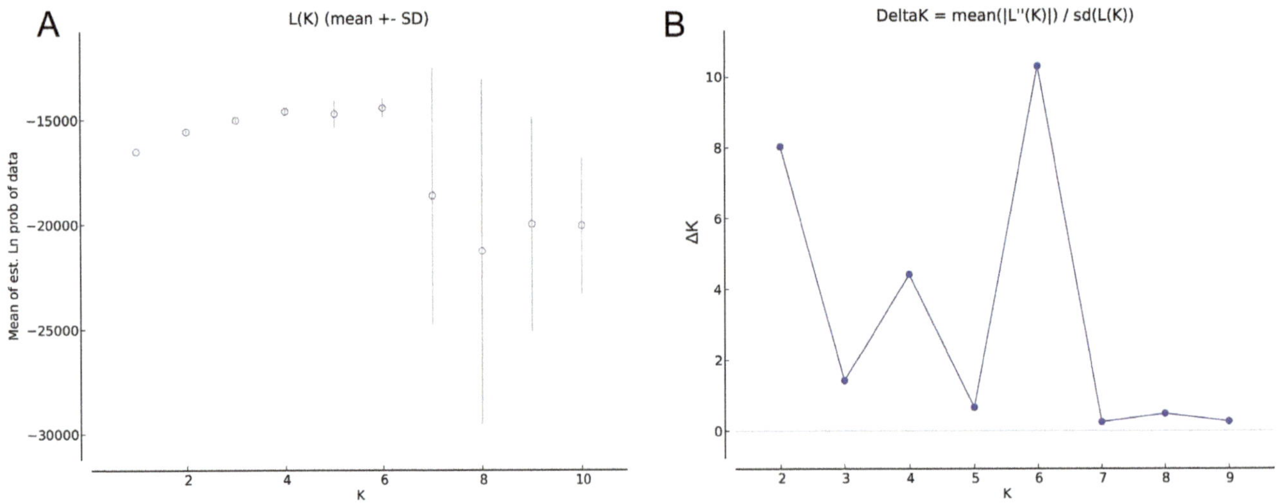

Figure 6. The mean LnP(D) and ΔK over 10 repeats of STRUCTRUE simulations. A. Mean lnP(D) value with k = 1–10. B. ΔK value with k = 2–9.

found that *D. c.* subsp. *longipes* is one of the dominant species in its habitat.

HICKORY software analysis suggests that *D. c.* subsp. *longipes* populations are not genetically differentiated. There is inbreeding in populations of *D. c.* subsp. *longipes* (Table 6), because the *DIC* and *Dbar* parameter were lower in "*Full model*" than that in other models [62]. While inbreeding generally leads to decreased fitness (inbreeding depression) of a population, it also can be advantageous, allowing plants to adapt to disadvantageous conditions [85–88]. *D. c.* subsp. *longipes* originally had an outcrossing system, but under some conditions (e.g., poor efficiency in wind or insect pollination), it would adopt inbreeding or mixed systems to reduce the risk of reproductive failure. This result is consistent with Xiao's conclusions [89] from his study of the reproductive ecology of *D. c.* subsp. *longipes*. If we use co-dominant markers, it most likely would provide an affirmative result and better understanding of the processes.

Hierarchical cluster analysis (Fig. 5) revealed that most individuals of a specific population were grouped together, despite the mixing-in of some individuals from different populations. Most geographically close populations tended to cluster together. The DW values show that the two close populations are clearly divergent, that one is long-term isolated (DW value higher) and the other is newly established (DW value lower). The DW value of 7GS is close to average, but its site is isolated. It might be that there once had been other populations between 6ZL and 7GS. In fact, there is a population of *D. c.* subsp. *longipes* in the Junfeng Mountains, Jiangxi Province, according to Pan's data [2], but we have not obtained any specimens for analysis since 2008. Moreover, although *D. c.* subsp. *longipes* in the Jinggang Mountains, Jiangxi Province, might have been transplanted from Guanshan National Nature Reserve based on the data [90], we found a large population area of *D. c.* subsp. *longipes* in the Jinggang Mountains in October 2009, evidence that the present distribution of *D. c.* subsp. *longipes* appears to be relic from a once extensive range and population.

Genetic structure analysis of the individual samples using STRUCTURE shows that six gene pools are represented in the data, and each gene pool is relatively independent except a few mixtures from other clusters. 1XN and 2DP are the first cluster separated from the other populations; 5QJ and 8SQ form the distinct cluster 2 from the others; 3MS and 4NL also cluster with each other and become distinct from rest populations. These three clusters show high correlation with geographic distribution. And the rest three clusters mixed in 6ZL and 7GS. There is a close relationship between 6ZL and 7GS. It shows that 6ZL is related to 7GS rather than to 5QJ or 8SQ, although 6ZL is close to 5QJ and 8SQ in geographic location. It seems likely that there were some populations between current populations 6ZL and 7GS, or gene flow could have been accomplished by pollen or seed dispersal.

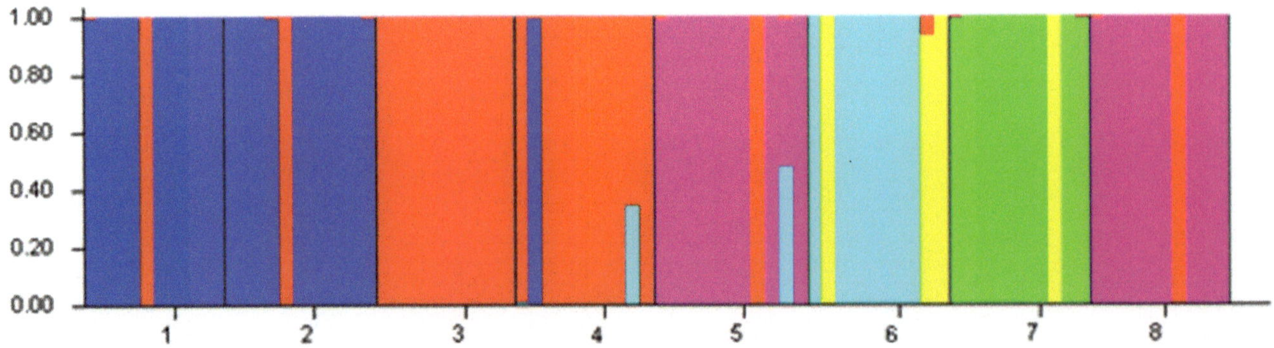

Figure 7. Diagram of results from STRUCTURE(k = 6).

Table 6. Genetic structure analysis using HICKORY ver. 1.1 [62].

	Parameter				
Model	*Dbar*	*Dhat*	*pD*	*DIC*	*f*
Full	6750.93	5395.03	1355.9	8106.82	0.980572
f=0	6780.4	5350.93	1429.47	8209.87	
Theta =0	12092.5	11716.4	376.128	12468.6	0.994715
f free	6844.98	5363.43	1481.55	8326.53	0.500214

Default values for computations were used as follows: Burn-in 5000; sample 100 000; thin 20.

The results of PCA based on AFLP data indicate a different result, which the eight populations form a triangle that has 1XN and 2DP at one angle, 5QJ and 8SQ at a second angle, but 3MS and 4NL mixed with 6ZL and 7GS in the third angle. The reasons for divergence of the East China region and the Nanling Range are not clear.

Guide for conservation measures

In the face of the species extinction generated by human beings [91,92], the urgent need for the conservation of biodiversity is global. In fact, many endemic species became endangered by the loss or fragmentation of habitats and change in natural conditions [26,29,64,93,94]. *D. c.* subsp. *longipes*, with its narrow archipelago-like distribution, habitat fragmentation, and reproductive-physiologic barriers, is typical species that can be easily affected. Although *D. c.* subsp. *longipes* is endangered in China, it has not yet been given high priority for protection. The original aim of this work is to provide some insight into the population biology of *D. c.* subsp. *longipes* in order to guide conservation measures. The guides which can be suggested by our study listed below. Firstly, Most of the genetic variation is located within populations, so we need to protect all the existing populations of *D. c.* subsp. *longipes* in order to preserve as much genetic variation as possible. Secondly, we should promote the further research and technology development to enable better reproduction of *Disanthus*. Finally, it is necessary to enforce the governmental law to forbid people stealing or purchasing wild *Disanthus*.

Conclusions

The results of the AFLP on populations of *Disanthus cercidifolius* subsp. *longipes* show a pattern of high within-population diversity and low among-population divergence. Although the distribution of *D. c.* subsp. *longipes* could be grouped as the East China region and the Nanling Range, the population divergence is clear. The estimates of genetic differentiation and gene flow suggest that the species primarily outcrosses, but can resort to mixed reproductive strategies. These patterns of genetic diversity and levels of genetic variation may be the result of *D. c.* subsp. *longipes* restricted to several isolated habitats and "excess flowers production, but little fruit set". The status of relic and their low reproductive success results in *D. c.* subsp. *longipes* being endangered.

Acknowledgments

We are grateful to the anonymous reviewers and the editor for their critical review; their comments substantially contributed to the revision and improvement of this work. We thank Dr. Sufang Chen for her valuable advice on writing.

Author Contributions

Conceived and designed the experiments: YY RS WL. Performed the experiments: YY RS. Analyzed the data: YY WG QF. Contributed reagents/materials/analysis tools: JJ DC. Wrote the paper: YY WL DC.

References

1. Mabberley DJ (1997) The Plant-book: A Portable Dictionary of the Vascular Plants. Cambridge University Press.
2. Pan K, Lu A, Wen J (1991) A systematic study on the genus *Disanthus* Maxim. (Hamamelidaceae). Cathaya 3: 1–28.
3. Cheng WC (1938) Observations on Mong-shan (Hunan). Science(Sci Soc China) 22: 400.
4. Chang HT (1948) Additions to the Hamamelidaceus flora of China. Sunyatsenia 7: 70.
5. Fu L (1992) China Plant Red Data Book: Rare and Endangered Plants. Beijing: Science Press.
6. State Forestry Administration and Ministry of Agriculture (1999) Key Wild Plants under State Protection. Beijing.
7. Wang S, Xie Y (2004) China Species Red List. Beijing: Higher Education Press.
8. Zavada M, Dilcher D (1986) Comparative pollen morphology and its relationship to phylogeny of pollen in the Hamamelidae. Ann Missouri bot Gard 73: 348–381.
9. Endress PK (1989) A suprageneric taxonomic classification of the hamamelidaceae. Taxon 38: 371–376.
10. Pan KY, Lu AM, Wen J (1991) A systematic study on the genus *Disanthus* Maxim. (Hamamelidaceae). Cathaya 3: 1–28.
11. Li J (1997) Systematics of the Hamamelidaceae based on morphological and molecular evidence. Ph.D. Thesis, University of New Hampshire.
12. Li J, Bogle AL, Donoghue MJ (1999) Phylogenetic relationships in the Hamamelidoideae inferred from sequences of trn non-coding regions of chloroplast DNA. Harvard Papers in Botany 4: 343–356.
13. Shi XH, Xu BM, Li NL, Sun YT (2002) Preliminary study on dormancy and germination of *Disanthus cercidifolius* Maxim var. *longipes* HT Chang seeds. Seed 6: 5–7.
14. Xiao YA, He P, Deng HP, Li XH (2002) Numerical analysis of population morphological differentiation of *Disanthus cercidiifolius* Maxim. var. *longipes* in Jinggangshan. Journal of Wuhan Botanical Research 20: 365–370.
15. Li K, Tang X (2003) The community characteristics and species diversity of the *Disanthus cercidifolius* var. *longipes* shrubland in the Guanshan nature reserve of Jiangxi Province. Journal of Nanjing Forestry University 27: 73–75.
16. Li X, Xiao Y, Hu W, Zeng J (2005) Isozymes analysis of genetic differentiation of the endangered plant *Disanthus cercidifolius* var. *longipes* HT Chang (Hamamelidaceae). Journal of Jianggangshan University (Natural Sciences) 26: 34–38.
17. Huang S, Fang Y, Tan X, Yan J, Fang S (2007) Effects of different concentrations of NAA on cutting regeneration of *Disanthus cercidifolius* var. *longipes*. Journal of Plant Resources and Environment 16: 74.
18. Gao P, Yang A, Yao X, Huang H (2009) Isolation and characterization of nine polymorphic microsatellite loci in the endangered shrub *Disanthus cercidifolius* var. *longipes* (Hamamelidaceae). Molecular Ecology Resources 9: 1047–1049.
19. Bogle A, Philbrick C (1980) A generic atlas of hamamelidaceous pollens. Contributions to the Gray Herbarium, Harvard University 210: 29–103.
20. Chang HT (1979) Hamaelidaceae. Flora Reipublicae Popularis Sinicae. Beijing: Science Press. pp.36–116.
21. Zhang ZY, Chang HT, Endress PK (2003) *Disanthus* Maximowicz. In: Z. Wu and P. H. Raven, editors. Flora of China. Beijing: Science Press and St. Louis: Missouri Botanical Garden Press.

22. Endress PK (1993) Hamamelidae. In: K. Kubitzki, J. G. Rohwer and V. Bittrich, editors. The Families and Fenera of Vascular Plants. Berlin: Springer-Verlag.
23. Xiao YA (2001) The study on population adaptability and genetic diversity of the endangered species *Disanthus cercidifolius* Maxim. var. *longipes* HT Chang. M.Sc. Thesis, Southwest China Normal University.
24. Soule M, Simberloff D (1986) What do genetics and ecology tell us about the design of nature reserves? Biological Conservation 35: 19–40.
25. Barrett S, Kohn J (1991) Genetic and evolutionary consequences of small population size in plants: implications for conservation. Genetics and conservation of rare plants: 3–30.
26. Ge S, Hong D, Wang H, Liu Z, Zhang C (1998) Population genetic structure and conservation of an endangered conifer, *Cathaya argyrophylla* (Pinaceae). International Journal of Plant Sciences 159: 351–357.
27. Millar C, LiBBY W (1991) Strategies for conserving clinal, ecotypic, and disjunct population diversity in widespread species. Genetics and conservation of rare plants: 149–170.
28. Francisco-Ortega J, Santos-Guerra A, Kim S, Crawford D (2000) Plant genetic diversity in the Canary Islands: a conservation perspective. American Journal of Botany 87: 909.
29. Qiu Y, Li J, Liu H, Chen Y, Fu C (2006) Population structure and genetic diversity of *Dysosma versipellis* (Berberidaceae), a rare endemic from China. Biochemical Systematics and Ecology 34: 745–752.
30. Song J, Murdoch J, Gardiner SE, Young A, Jameson PE, et al. (2008) Molecular markers and a sequence deletion in intron 2 of the putative partial homologue of LEAFY reveal geographical structure to genetic diversity in the acutely threatened legume genus *Clianthus*. Biological Conservation 141: 2041–2053.
31. Perez-Collazos E, Segarra-Moragues JG, Catalan P (2008) Two approaches for the selection of Relevant Genetic Units for Conservation in the narrow European endemic steppe plant *Boleum asperum* (Brassicaceae). Biological Journal of the Linnean Society 94: 341–354.
32. Yang J, Qian ZQ, Liu ZL, Li S, Sun GL, et al. (2007) Genetic diversity and geographical differentiation of *Dipteronia* Oliv. (Aceraceae) endemic to China as revealed by AFLP analysis. Biochemical Systematics and Ecology 35: 593–599.
33. Prohens J, Anderson GJ, Herraiz FJ, Bernardello G, Santos-Guerra A, et al. (2007) Genetic diversity and conservation of two endangered eggplant relatives (*Solanum vespertilio* Aiton and *Solanum lidii* Sunding) endemic to the Canary Islands. Genetic Resources and Crop Evolution 54: 451–464.
34. Jian SG, Zhong Y, Liu N, Gao ZZ, Wei Q, et al. (2006) Genetic variation in the endangered endemic species *Cycas fairylakea* (Cycadaceae) in China and implications for conservation. Biodiversity and Conservation 15: 1681–1694.
35. Pfosser M, Jakubowsky G, Schluter PM, Fer T, Kato H, et al. (2005) Evolution of *Dystaenia takesimana* (Apiaceae), endemic to Ullung Island, Korea. Plant Systematics and Evolution 256: 159–170.
36. Strand AE, Leebens-Mack J, Milligan BG (1997) Nuclear DNA-based markers for plant evolutionary biology. Molecular Ecology 6: 113–118.
37. Vos P, Hogers R, Bleeker M, Reijans M, van de Lee T, et al. (1995) AFLP: a new concept for DNA fingerprinting. Nucleic Acids Res 23: 4407–4414.
38. Armstrong TTJ, De Lange PJ (2005) Conservation genetics of *Hebe speciosa* (Plantaginaceae) an endangered New Zealand shrub. Botanical Journal of the Linnean Society 149: 229–239.
39. Kim SC, Lee C, Santos-Guerra A (2005) Genetic analysis and conservation of the endangered Canary Island woody sow-thistle, *Sonchus gandogeri* (Asteraceae). Journal of Plant Research 118: 147–153.
40. Andrade IM, Mayo SJ, Van den Berg C, Fay MF, Chester M, et al. (2007) A preliminary study of genetic variation in populations of *Monstera adansonii* var. *klotzschiana* (Araceae) from north-east brazil, estimated with AFLP molecular markers. Annals of Botany 100: 1143–1154.
41. Tang SQ, Dai WJ, Li MS, Zhang Y, Geng YP, et al. (2008) Genetic diversity of relictual and endangered plant *Abies ziyuanensis* (Pinaceae) revealed by AFLP and SSR markers. Genetica 133: 21–30.
42. Wolfe A, Liston A (1998) Contributions of PCR-based methods to plant systematics and evolutionary biology. Molecular systematics of plants II: 43–86.
43. Chase MW, Hills HH (1991) Silica gel: an ideal material for field preservation of leaf samples for DNA studies. Taxon 40: 215–220.
44. Doyle JJ, Doyle JL (1987) A rapid DNA isolation procedure for small quantities of fresh leaf tissue. Phytochemical bulletin 19: 11–15.
45. Yeh FC, Yang RC, Boyle TBJ, Ye ZH, Mao JX (1997) POPGENE, the user-friendly shareware for population genetic analysis. Molecular Biology and Biotechnology Centre, University of Alberta, Canada.
46. Schönswetter P, Tribsch A (2005) Vicariance and dispersal in the alpine perennial *Bupleurum stellatum* L.(Apiaceae). Taxon: 725–732.
47. Ortiz M, Tremetsberger K, Talavera S, Stuessy T, García-Castaño JL (2007) Population structure of *Hypochaeris salzmanniana* DC.(Asteraceae), an endemic species to the Atlantic coast on both sides of the Strait of Gibraltar, in relation to Quaternary sea level changes. Mol Ecol 16: 541–552.
48. Ehrich D (2006) AFLPdat: a collection of R functions for convenient handling of AFLP data. Mol Ecol Notes 6: 603–604.
49. Nei M (1978) Estimation of average heterozygosity and genetic distance from a small number of individuals. Genetics 89: 583–590.
50. Sneath PHA, Sokal RR (1973). Numerical Taxonomy, 2nd edn. San Francisco: Freeman.
51. Rohlf FJ (2000) NTSYS-pc: numerical taxonomy and multivariate analysis system, version 2.1. New York: Exeter Software.
52. Nei M, Li WH (1979) Mathematical model for studying genetic variation in terms of restriction endonucleases. Proc Natl Acad Sci USA 76: 5269–5273.
53. Van de Peer Y, De Wachter Y (1994) TREECON for Windows: a software package for the construction and drawing of phylogenetic trees for the Microsoft Windows environment. Comp Applic Biosci 10: 569–570.
54. Excoffier L, Laval G, Schneider S (2005) Arlequin (version 3.0): An integrated software package for population genetics data analysis. Evolutionary Bioinformatics 1: 47–50.
55. Excoffier L, Smouse P, Quattro J (1992) Analysis of molecular mariance inferred from metric distances among DNA haplotypes: Application to human mitochondrial DNA restriction data. Genetics 131: 479–491.
56. Wright S (1950) Genetical structure of populations. Nature 166: 247–249.
57. Nie Z-L, Wen J, SUN H (2009) AFLP Analysis of *Phryma* (Phrymaceae) Disjunct between Eastern Asia and Eastern North America. Acta Botanica Yunnanica 31: 289–295.
58. Inc S (2007) SPSS 16.0 for Windows. SPSS Inc, Chicago, IL.
59. Pritchard JK, Stephens M, Donnelly P (2000) Inference of population structure using multilocus genotype data. Genetics 155: 945–959.
60. Evanno G, Regnaut S, Goudet J (2005) Detecting the number of clusters of individuals using the software STRUCTURE: a simulation study. Mol Ecol 14: 2611–2620.
61. Zhang DL, Zhang HL, Wang MX, Sun JL, Qi YW, et al. (2009) Genetic structure and differentiation of *Oryza sativa* L. in China revealed by microsatellites. Theor Appl Genet 119: 1105–1117.
62. Holsinger K, Lewis P (2003) HICKORY: a package for analysis of population genetic data V. 1.0. University of Connecticut, Storrs, USA.
63. Cardoso M, Provan J, Powell W, Ferreira P, De Oliveira D (1998) High genetic differentiation among remnant populations of the endangered *Caesalpinia echinata* Lam.(Leguminosae-Caesalpinioideae). Molecular Ecology 7: 601–608.
64. Bouza N, Caujapé-Castells J, González-Pérez M, Batista F, Sosa P (2002) Population structure and genetic diversity of two endangered endemic species of the Canarian laurel forest: *Dorycnium spectabile* (Fabaceae) and *Isoplexis chalcantha* (Scrophulariaceae). International Journal of Plant Sciences 163: 619–630.
65. Hamrick J, Godt M (1989) Allozyme diversity in plant species. In: A. H. D. Brown, M. T. A. Clegg, L. Kahler and B. S. Weir, editors. Plant Population Genetics, Breeding and Genetic Resources. Sunderland, Mass: Sinauer. pp.43–63.
66. Ge XJ, Yu Y, Zhao NX, Chen HS, Qi WQ (2003) Genetic variation in the endangered Inner Mongolia endemic shrub *Tetraena mongolica* Maxim.(Zygophyllaceae). Biological Conservation 111: 427–434.
67. Hamrick JL, Godt MJW (1996): Conservation genetics of endemic plant species. In Conservation genetics: case histories from nature. In: Avise JC, Hamrick JL, Editors. New York: Chapman and Hall. pp.281–304.
68. Karron J (1991) Patterns of genetic variation and breeding systems in rare plant species. Genetics and Conservation of Rare Plants: 87–98.
69. Ge XJ, Yu Y, Zhao NX, Chen HS, Qi WQ (2003) Genetic variation in the endangered Inner Mongolia endemic shrub *Tetraena mongolica* Maxim. (Zygophyllaceae). Biol Conserv 111: 427–434.
70. Wu Z, Lu A, Tang Y, Chen Z, Li DZ (2003) The families and gerera of angiosperms in China. Beijing: Science Press.
71. Lewis PO, Crawford DJ (1995) Pleistocene Refugium Endemics Exhibit Greater Allozymic Diversity than Widespread Congeners in the Genus *Polygonella* (Polygonaceae). American Journal of Botany 82: 141–149.
72. Kwon J, Morden C (2002) Population genetic structure of two rare tree species (*Colubrina oppositifolia* and *Alphitonia ponderosa*, Rhamnaceae) from Hawaiian dry and mesic forests using random amplified polymorphic DNA markers. Mol Ecol 11: 991–1001.
73. Lacerda DR, Acedo MDP, Filho JPL, Lovato MB (2001) Genetic diversity and structure of natural populations of *Plathymenia reticulata* (Mimosoideae), a tropical tree from the Brazilian Cerrado. Molecular Ecology 10: 1143–1152.
74. Jian S, Zhong Y, Liu N, Gao Z, Wei Q, et al. (2006) Genetic variation in the endangered endemic species *Cycas fairylakea* (Cycadaceae) in China and implications for conservation. Biodiversity & Conservation 15: 1681–1694.
75. Schaal B, Hayworth D, Olsen K, Rauscher J, Smith W (1998) Phylogeographic studies in plants: problems and prospects. Mol Ecol 7: 465–474.
76. Slatkin M (1987) Gene flow and the geographic structure of natural populations. Science 236: 787.
77. Zawko G, Krauss S, Dixon K, Sivasithamparam K (2001) Conservation genetics of the rare and endangered *Leucopogon obtectus* (Ericaceae). Mol Ecol 10: 2389–2396.
78. Maguire T, Sedgley M (1997) Genetic diversity in *Banksia* and *Dryandra* (Proteaceae) with emphasis on Banksia cuneata, a rare and endangered species. Heredity 79: 394–401.
79. Xiao YA, Zeng JJ, Li XH, Hu WH, He P (2006) Pollen and resource limitations to lifetime seed prodution in a wild population of the endagered plant *Disanthus cercidifolius* Maxim. var. *longipes* H. T. Chang (Hamamelidaceae). Acta Ecologica Sinica 26: 496–502.
80. Huang Y, Zhang C, Li D (2009) Low genetic diversity and high genetic differentiation in the critically endangered *Omphalogramma souliei* (Primulaceae): implications for its conservation. Journal of Systematics and Evolution 47: 103–109.

81. Kothera L, Richards C, Carney S (2007) Genetic diversity and structure in the rare Colorado endemic plant *Physaria bellii* Mulligan (Brassicaceae). Conservation Genetics 8: 1043–1050.
82. Jacquemyn H, Honnay O, Galbusera P, Roldan-Ruiz I (2004) Genetic structure of the forest herb Primula elatior in a changing landscape. Molecular Ecology 13: 211–219.
83. Juan A, Crespo M, Cowan R, Lexer C, Fay M (2004) Patterns of variability and gene flow in *Medicago citrina*, an endangered endemic of islands in the western Mediterranean, as revealed by amplified fragment length polymorphism (AFLP). Mol Ecol 13: 2679–2690.
84. Morjan CL, Rieseberg LH (2004) How species evolve collectively: implications of gene flow and selection for the spread of advantageous alleles. Mol Ecol 13: 1341–1356.
85. Allard R, Jain S, Workman P (1968) The genetics of inbreeding populations. Adv Genet 14: 55–131.
86. Antonovics J (1968) Evolution in closely adjacent plant populations. V. Evolution of self-fertility. Heredity 23: 219–238.
87. Jain S (1976) The evolution of inbreeding in plants. Annual Review of Ecology and Systematics 7: 469–495.
88. Lloyd D (1980) Demographic factors and mating patterns in angiosperms.
89. Xiao YA (2005) Studies on Reproductive Ecology and photosynthetic adaptability of the Endangered Plant *Disanthus cercidifolius* var.*longipes* HT Chang. Ph.D. Thesis, Southwest China Normal University.
90. Liu R (1999) *Disanthus cercidifolius* var. *longipes*. Plants 4: 7.
91. Vitousek P, Mooney H, Lubchenco J, Melillo J (1997) Human domination of Earth's ecosystems. Science 277: 494.
92. Chapin F, Zavaleta E, Eviner V, Naylor R, Vitousek P, et al. (2000) Consequences of changing biodiversity. Nature 405: 234–242.
93. Palacios C, Gonzalez-Candelas F (1997) Lack of genetic variability in the rare and endangered *Limonium cavanillesii* (Plumbaginaceae) using RAPD markers. Molecular Ecology 6: 671–675.
94. Qiu Y, Luo Y, Comes H, Ouyang Z, Fu C (2007) Population genetic diversity and structure of *Dipteronia dyerana* (Sapindaceae), a rare endemic from Yunnan Province, China, with implications for conservation. Taxon 56: 427–437.

22

Sperm Preservation by Freeze-Drying for the Conservation of Wild Animals

Takehito Kaneko[1]*, Hideyuki Ito[2,3], Hidefusa Sakamoto[2,3], Manabu Onuma[4], Miho Inoue-Murayama[3,4]

1 Institute of Laboratory Animals, Graduate School of Medicine, Kyoto University, Kyoto, Japan, **2** Kyoto City Zoo, Kyoto, Japan, **3** Wildlife Research Center of Kyoto University, Kyoto, Japan, **4** Center for Environmental Biology and Ecosystem studies, National Institute for Environmental Studies, Ibaraki, Japan

Abstract

Sperm preservation is a useful technique for the maintenance of biological resources in experimental and domestic animals, and in wild animals. A new preservation method has been developed that enables sperm to be stored for a long time in a refrigerator at 4°C. Sperm are freeze-dried in a solution containing 10 mM Tris and 1 mM EDTA. Using this method, liquid nitrogen is not required for the storage and transportation of sperm. We demonstrate that chimpanzee, giraffe, jaguar, weasel and the long-haired rat sperm remain viable after freeze-drying. In all species, pronuclei were formed after the injection of freeze-dried sperm into the mouse oocytes. Although preliminary, these results may be useful for the future establishment of "freeze-drying zoo" to conserve wild animals.

Editor: Stefan Schlatt, University Hospital of Münster, Germany

Funding: This work was supported in part by a Grant-in-Aid for Scientific Research from JSPS (No. 30332878 and 25290082) and the Cooperation Research Program from Wildlife Research Center of Kyoto University. The funders had no role in study design, data collection and analysis, decision to publish, or preparation of the manuscript.

Competing Interests: The authors have declared that no competing interests exist.

* Email: tkaneko@anim.med.kyoto-u.ac.jp

Introduction

Many animal species are endangered or threatened with extinction [1]. However, assisted reproductive techniques can help species that are experiencing reproductive difficulties in captive and wild conditions [2]. Gamete preservation, especially sperm, is one of the effective methods in assisted reproductive techniques. Although, cryopreservation has been used as a standard method for storing sperm, there are a lot of animals for which a freezing and collecting method for sperm has not been established. Furthermore, the continuous supply of liquid nitrogen and mechanical maintenance of equipment is required for storing frozen sperm. Valuable sperm samples that are stored in liquid nitrogen might be lost if liquid nitrogen supplies cease, especially during disasters such as earthquakes and typhoons. Therefore, safe facilities and equipment are needed to store frozen genetic resources and prevent unexpected accidents [3].

Freeze-drying is the ultimate method for storing biological material. Using this method, samples can be kept for a long time in a refrigerator (4°C) or at ambient temperature [4,5]. Freeze-dried sperm can also remain viable, and successful results have been reported for various mammals [6–11]. In the mouse and rat especially, freeze-drying methods have improved, resulting in high fertility [12–16]. At present, freeze-drying methods enable mouse and rat sperm to be preserved for 3 to 5 years at 4°C in a solution containing 10 mM Tris and 1 mM ethylenediaminetetraacetic acid (EDTA) (TE buffer) [4,5]. Furthermore, freeze-dried sperm can be easily and safely transported worldwide at ambient temperature [17]. In addition, short term preservation at ambient temperature that requires neither liquid nitrogen nor dry ice is possible [18]. In this study, we freeze-dry sperm collected from chimpanzee, giraffe, jaguar, weasel and the long-haired rat, and then estimate the viability of the sperm to apply freeze-drying of sperm for conserving wild animals.

Materials and Methods

Media

TE buffer (10 mM Tris, 1 mM EDTA, pH 8.0; Applied Biosystems/Ambion, Austin, TX, USA) was used as a solution for freeze-drying the sperm [4,5,14]. The medium used for manipulation, including the collection and handling of oocytes and ICSI, was H-CZB medium, a modified CZB medium [19,20] with 20 mM Hepes-Na, 5 mM $NaHCO_3$ and 0.1 mg/ml polyvinyl alcohol (cold water soluble; M_r 30,000–70,000) added instead of bovine serum albumin [21].

Sperm collection

A chimpanzee (*Pan troglodytes*, 20 years old), reticulated giraffe (*Giraffa camelopardalis reticulata*, 13 years old), and jaguar (*Panthera onca*, 25 years old) kept in the Kyoto City Zoo were used in this study. Animal care and procedures of chimpanzee referred to the Guidelines for Association of Zoos and Aquariums (AZA). Chimpanzees were maintained in the room (36.94–43.87 m^2) with feeding (3 times/day). Environmental enrichments such as a tower (9 m), trees and artificial ant-hill were prepared outside. Samples from jaguar came from a specimen that died by circulatory failure. Samples from the Japanese weasel (*Mustela itatsi*, age-indeterminate) came from a specimen that died at the wildlife Rescue Center of Kyoto City Zoo after rescue from traffic accident. Sperm was also obtained from a road-killed Ryukyu long-haired rat (*Diplothrix legata*, age-indeterminate) kept in the National Institute for Environmental Studies. It was found on the road at Kunigami, Okinawa, Japan. In this study, animals that are specified for threatened species and can be collected sperm in zoo or field were used. In the chimpanzee and giraffe, sperm ejaculated naturally on the floor were collected in a plastic bag. Testis and epididymides of the jaguar and weasel were collected after the death of an individual at the Kyoto City Zoo. These sperm and tissues were transported to Kyoto University within 10 min. Testis and epididymides of the long-haired rat were also transported at 4°C to Kyoto University by delivery service.

Freeze-drying sperm

Freeze-drying of sperm was carried out using the same method for mouse and rat described by Kaneko [4,5,17]. In the chimpanzee and giraffe, each ejaculated sperm sample was diluted with TE buffer, and transferred to a 1.5 ml microcentrifuge tube for centrifugation at 800×G for 10 min. The supernatant was removed, and the sperm mass was re-suspended in 1 ml of fresh TE buffer. Jaguar, weasel and the long-haired rat sperm collected from cauda epididymides were suspended in 1 ml of TE buffer in a 1.5 ml microcentrifuge tube. Aliquots of 100 µl of sperm suspension were transferred into long-necked glass ampoules for freeze-drying (651506, Wheaton, Millville, NJ, USA). Ten ampoules of each animal were plunged into liquid nitrogen for 20 s and then connected to the manifold of a freeze-drying machine (Freeze-drying systems 77530, Labconco, Kansas City, MO, USA). The sperm suspension was dried for 4 h at a pressure of 0.04 hpa. All ampoules were flame-sealed and stored at 4°C (Fig. 1a). Freeze-dried sperm were stored for 1 month.

Oocyte collection

B6D2F1/Crlj female mice 8–16 weeks old (Charles River Japan Inc., Yokohama, Japan) were used as oocyte donors. Animals were housed in plastic cages in a specific pathogen-free barrier facility that was air-conditioned (temperature 24±1°C, humidity 40±10%) and light-controlled (lights on from 08:00 to 20:00). All animal care and procedures performed in this study conformed to the Guidelines for Animal Experiments of Kyoto University, and were approved by the Animal Research Committee of Kyoto University.

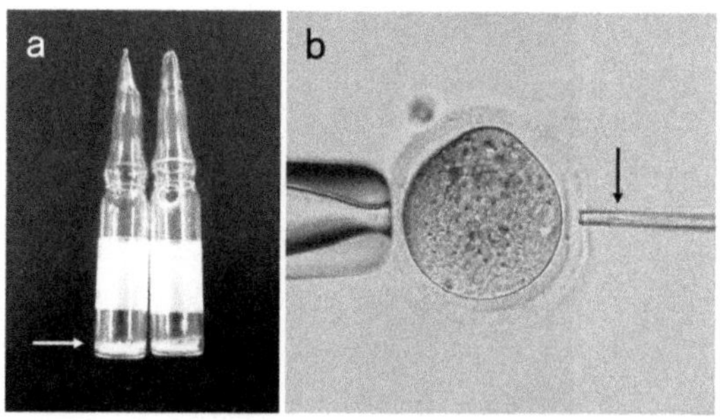

Figure 1. (a) Glass ampoules with freeze-dried giraffe sperm. Arrow shows the dried sperm on the bottom of the ampoules. (b) Sperm injection into mouse oocytes. Arrow shows giraffe sperm drawn into the injection pipette.

Females were induced to superovulate by an intraperitoneal injection of 5 IU of pregnant mare serum gonadotropin (ASKA Pharmaceutical. Co., Ltd., Tokyo, Japan) followed by an injection of 5 IU of human chorionic gonadotropin (ASKA Pharmaceutical. Co., Ltd.) 48 h later. Cumulus-oocyte complexes were collected from oviducts 13 to 15 h after the injection of human chorionic gonadotropin, and oocytes were freed from cumulus cells by treatment with 0.1% hyaluronidase in H-CZB medium for 5 min. Oocytes were rinsed in fresh H-CZB medium and kept at room temperature before sperm injection.

Intracytoplasmic sperm injection (ICSI)

ICSI was carried out using the method previously described by Kaneko [4,5,17] (Fig. 1b). One freeze-dried ampoule of each animal was used, and rehydrated sperm by adding 100 µl of sterile distilled water (Fig. 2a–e). A small volume (1–2 µl) of the sperm suspension was mixed thoroughly with a droplet of H-CZB medium containing 12% (w/v) polyvinylpyrrolidone (M_r 360,000, ICN Pharmaceuticals, Costa Mesa, CA, USA). Sperm of normal shape were selected and placed into another droplet of H-CZB medium containing 12% polyvinylpyrrolidone. A single spermatozoon was drawn by its tail into the injection pipette. Oocytes were placed in a droplet of H-CZB medium. The zona pellucida of oocytes held with a holding pipette was opened by applying a few piezo pulses. The oolemma was then opened with piezo pulses, and a sperm was introduced into the oocyte. Sham injection was carried out for control. Microinjected oocytes were cultured for 6–8 h in H-CZB medium at 37°C under 5% CO_2 and 95% air. Oocytes containing two distinct pronuclei were recorded as being fertilized.

Results

In all animal used in this study, collected sperm were already immotile. However, these showed morphologically normal shape before freeze-drying, and most of freeze-dried sperm did not separate tail, and maintained normal shape after rehydration (Fig. 2a–e). Table 1 shows the number and percentage of oocytes forming pronuclei after injection with freeze-dried chimpanzee, giraffe, jaguar, weasel and the long-haired rat sperm. No pronuclei appeared in the oocytes after sham injection. Of 29 oocytes injected with chimpanzee sperm, 14 (48%) oocytes survived, and 12 (86%) oocytes formed two distinct pronuclei (Fig. 2f). In the giraffe, all surviving oocytes (12 = 50% of 24) injected with sperm formed two pronuclei (Fig. 2g). Of 34 oocytes injected with jaguar sperm, 23 (68%) oocytes survived, and 22 (96%) formed two distinct pronuclei (Fig. 2h). All surviving (12 = 43% of 28) oocytes injected with weasel sperm formed two pronuclei (Fig. 2i). Of 19 oocytes injected with long-haired rat sperm, 11 (58%) survived and 10 (91%) formed two distinct pronuclei (Fig. 2j).

Discussion

We freeze-dried sperm collected from chimpanzee, giraffe, jaguar, weasel and the long-haired rat, and demonstrated that pronuclei were formed when these sperm were injected into the mouse oocytes. This is the first report that the viability of sperm collected from wild animals can be maintained after freeze-drying. In this study, pronuclei were formed in the mouse oocytes injected with freeze-dried sperm of different animal species (Table 1, Fig. 2f–j). Mouse oocytes have been used previously for fertility tests of sperm collected from various mammals [22,23]. Pronuclei

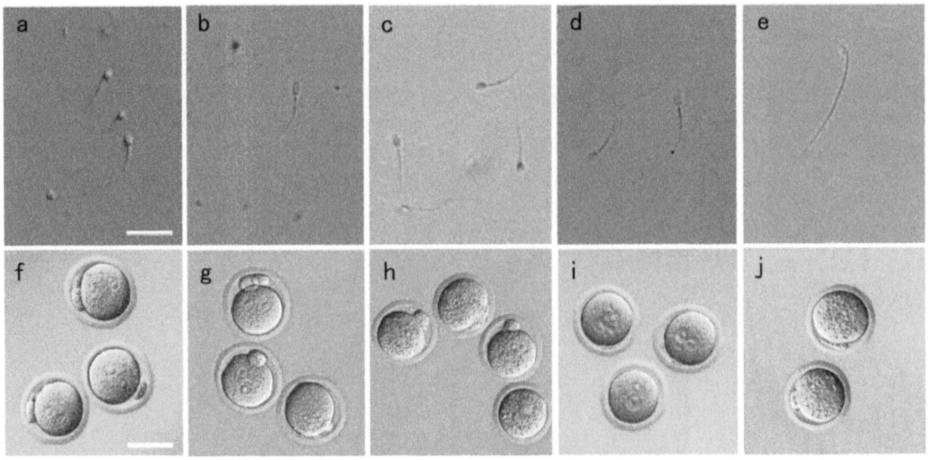

Figure 2. Freeze-dried sperm of (a) chimpanzee, (b) giraffe, (c) jaguar, (d) weasel and (e) the long-haired rat after rehydration. The common scale bar for a–e = 20 μm . Oocytes formed pronuclei after injection of the freeze-dried sperm of (f) chimpanzee, (g) giraffe, (h) jaguar, (i) weasel and (j) the long-haired rat. The common scale bar for f–j = 50 μm.

appeared in mouse oocytes when injected with sperm of various mammals by ICSI. It is well known that removal of the zona pellicida of hamster oocytes allows penetration of sperm of other mammals and pronuclei then form after penetration of sperm [24]. The results of this study further demonstrated that mouse oocytes could receive the sperm of various animals by ICSI, even wild animals such as chimpanzees, giraffes, jaguars, weasels and the long-haired rats. Although the oocytes collected from same animal species are required for detail analysis of freeze-dried sperm, mouse oocytes can be used as simple tool to estimate sperm viability in various species in which a method for collection of oocytes has not been established.

In mouse and rat, the viability of freeze-dried sperm is protected by using TE buffer, allowing sperm to be stored for a long time after freeze-drying [4,5]. Successful freeze-drying of rabbit and hamster sperm using a solution of similar composition to TE buffer has also been reported [8,10]. The epididymal sperm is tolerant of physical stress, as DNA is tightly condensed with -SS- bonds [25]. Furthermore, pH and a small volume of EDTA in the solution are helpful to protect sperm from physical stress during freeze-drying [12,14]. Although further detail studies are required, it was thought that TE buffer also protected chimpanzee, giraffe, jaguar, weasel and the long-haired rat sperm from damage during freeze-drying.

In this study, we collected chimpanzee and giraffe sperm that had been ejaculated naturally onto the floor. These sperm were already immotile before freeze-drying. Sperm also became immotile after freeze-drying even if fresh motile sperm were used. ICSI is an available technique that can fertilize oocytes with immotile sperm that have been freeze-dried or collected from dead animals [26]. The combination of assisted reproductive techniques by freeze-drying sperm and ICSI is a powerful tool for increasing animal populations [27]. Further study using sperm collected from various animals lead to efficient freeze-drying method using TE buffer as a useful tool for sperm preservation of endangered species.

Ideally, sperm and oocytes from a species should be collected for assisted reproduction of endangered species. However, the timing of oocyte collection is seasonally limited, and there are considerable difficulties in the collection of sperm and oocytes at the same time for *in vitro* fertilization. However, sperm can be preserved temporarily because sperm collection is easier than oocyte collection. Oocytes can be fertilized immediately after collection by using preserved sperm. Sperm preservation is useful for applying assisted reproduction to endangered species. Freeze-drying, especially, is relatively easy and sperm can be transported safely without using liquid nitrogen and dry ice. We believe that a "freeze-dry zoo" is the ultimate method to protect wild species

Table 1. Pronuclear formation of oocytes injected with freeze-dried sperm of various animals.

Animals	No. of oocytes injected	No. (%) of surviving oocytes [a]	No. (%) of oocytes forming pronuclei [b]
Sham injection	30	24 (80)	0 (0)
Chimpanzee	29	14 (48)	12 (86)
Giraffe	24	12 (50)	12 (100)
Jaguar	34	23 (68)	22 (96)
Weasel	28	12 (43)	12 (100)
Long-haired rat	19	11 (58)	10 (91)

The percentages were calculated from no. of [a]surviving oocytes/oocytes injected, and [b]oocytes forming pronuclei/surviving oocytes.

from extinction, not only mammals but also birds, reptiles, amphibians, and fishes.

Acknowledgments

We thank Ufugi Nature Museum of Yambaru Wildlife Conservation Center and Conservation & Animal Welfare Trust for providing Ryukyu long-haired rat.

Author Contributions

Conceived and designed the experiments: TK. Performed the experiments: TK HI MO. Analyzed the data: TK. Contributed reagents/materials/ analysis tools: TK HI HS MO MIM. Wrote the paper: TK.

References

1. Leibo SP, Songsasen N (2002) Cryopreservation of gametes and embryos of non-domestic species. Theriogenology 57: 303–326.
2. Comizzoli P, Mermillod P, Mauget R (2000) Reproductive biotechnologies for endangered mammalian species. Reprod. Nutr. Dev. 40: 493–504.
3. Dickey RP, Lu PY, Sartor BM, Dunaway HE Jr, Pyrzak R, et al. (2006) Steps taken to protect and rescue cryopreserved embryos during Hurricane Katrina. Fertil. Steril. 86: 732–734.
4. Kaneko T, Serikawa T (2012) Successful long-term preservation of rat sperm by freeze-drying. PLoS One 7: e35043. (doi: 10.1371/journal.pone.0035043).
5. Kaneko T, Serikawa T (2012) Long-term preservation of freeze-dried mouse spermatozoa. Cryobiology 64: 211–214. (doi: 10.1016/j.cryobiol.2012.01.010).
6. Keskintepe L, Pacholczyk G, Machnicka A, Norris K, Curuk MA, et al. (2001) Bovine blastocyst development from oocytes injected with freeze-dried spermatozoa. Biol. Reprod. 67: 409–415.
7. Kwon IK, Park KE, Niwa K (2004) Activation, pronuclear formation, and development in vitro of pig oocytes following intracytoplasmic injection of freeze-dried spermatozoa. Biol Reprod 71: 1430–1436.
8. Liu JL, Kusakabe H, Chang CC, Suzuki H, Schmidt DW, et al. (2004) Freeze-dried sperm fertilization leads to full-term development in rabbits. Biol. Reprod. 70: 1776–1781.
9. Sánchez-Partida LG, Simerly CR, Ramalho-Santos J (2008) Freeze-dried primate sperm retains early reproductive potential after intracytoplasmic sperm injection. Fertil. Steril. 89: 742–745.
10. Muneto T, Horiuchi T (2011) Full-term development of hamster embryos produced by injecting freeze-dried spermatozoa into oocytes. J. Mamm. Ova. Res. 28: 32–39.
11. Choi YH, Varner DD, Love CC, Hartman DL, Hinrichs K (2011) Production of live foals via intracytoplasmic injection of lyophilized sperm and sperm extract in the horse. Reproduction 142: 529–538. (doi: 10.1530/REP-11-0145).
12. Kaneko T, Whittingham DG, Yanagimachi R (2003) Effect of pH value of freeze-drying solution on the chromosome integrity and developmental ability of mouse spermatozoa. Biol. Reprod. 68: 136–139.
13. Ward MA, Kaneko T, Kusakabe H, Biggers JD, Whittingham DG, et al. (2003) Long-term preservation of mouse spermatozoa after freeze-drying and freezing without cryoprotection. Biol. Reprod. 69: 2100–2108.
14. Kaneko T, Nakagata N (2006) Improvement in the long-term stability of freeze-dried mouse spermatozoa by adding of a chelating agent. Cryobiology 53: 279–282.
15. Kaneko T, Kimura S, Nakagata N (2007) Offspring derived from oocytes injected with rat sperm, frozen or freeze-dried without cryoprotection. Theriogenology 68: 1017–1021.
16. Kaneko T, Kimura S, Nakagata N (2009) Importance of primary culture conditions for the development of rat ICSI embryos and long-term preservation of freeze-dried sperm. Cryobiology 58: 293–297. (doi: 10.1016/j.cryobiol.2009.02.004).
17. Kaneko T (2014) Mouse Molecular Embryology, Methods and Protocols, Methods in Molecular Biology 1092. In: Lewandoski M, editors. The latest improvements in the mouse sperm preservation. pp. 357–365. (doi: 10.1007/978-1-60327-292-6_21).
18. Kaneko T, Nakagata N (2005) Relation between storage temperature and fertilizing ability of freeze-dried mouse spermatozoa. Comp. Med. 55: 140–144.
19. Chatot CL, Ziomek CA, Bavister BD, Lewis JL, Torres I (1989) An improved culture medium supports development of random-bred 1-cell mouse embryos in vitro. J. Reprod. Fertil. 86: 679–688.
20. Chatot CL, Lewis JL, Torres I, Ziomek CA (1990) Development of 1-cell embryos from different strains of mice in CZB medium. Biol. Reprod. 42: 432–440.
21. Kimura Y, Yanagimachi R (1995) Intracytoplasmic sperm injection in the mouse. Biol. Reprod. 52: 709–720.
22. Lee JD, Kamiguchi Y, Yanagimachi R (1996) Analysis of chromosome constitution of human spermatozoa with normal and aberrant head morphologies after injection into mouse oocytes. Hum. Reprod. 11: 1942–1946.
23. Watanabe H, Suzuki H, Tateno H, Fukui Y (2010) A novel method for detection of chromosomal integrity in cryopreserved livestock spermatozoa using artificially fused mouse oocytes. J. Assist. Reprod. Genet. 27: 581–588. (doi: 10.1007/s10815-010-9445-0).
24. Yanagimachi R, Yanagimachi H, Rogers BJ (1976) The use of zona-free animal ova as a test-system for the assessment of the fertilizing capacity of human spermatozoa. Biol. Reprod. 15: 471–476.
25. Kosower NS, Katayose H, Yanagimachi R (1992) Thiol-disulfide status and acridine orange fluorescence of mammalian sperm nuclei. J. Androl. 13: 342–348.
26. Kishikawa H, Tateno H, Yanagimachi R (1999) Fertility of mouse spermatozoa retrieved from cadavers and maintained at 4 degrees C. J. Reprod. Fertil. 116: 217–222.
27. Li MW, Willis BJ, Griffey SM, Spearow JL, Lloyd KC (2009) Assessment of three generations of mice derived by ICSI using freeze-dried sperm. Zygote 17: 239–251. (doi: 10.1017/S0967199409005292).

23

Trends in Extinction Risk for Imperiled Species in Canada

Brett Favaro[1,2,3]*, Danielle C. Claar[1], Caroline H. Fox[4,5], Cameron Freshwater[1], Jessica J. Holden[1], Allan Roberts[6], UVic Research Derby[1,4,7]¶

1 Department of Biology, University of Victoria, Victoria, Canada, 2 Centre for Sustainable Aquatic Resources, Fisheries and Marine Institute of Memorial University of Newfoundland, St John's, Canada, 3 Department of Ocean Sciences, Memorial University of Newfoundland, St. John's, Canada, 4 Department of Geography, University of Victoria, Victoria, Canada, 5 Raincoast Conservation Foundation, Sidney, Canada, 6 Bamfield Marine Sciences Centre, Bamfield East, Canada, 7 School of Environmental Studies, University of Victoria, Victoria, Canada

Abstract

Protecting and promoting recovery of species at risk of extinction is a critical component of biodiversity conservation. In Canada, the Committee on the Status of Endangered Wildlife in Canada (COSEWIC) determines whether species are at risk of extinction or extirpation, and has conducted these assessments since 1977. We examined trends in COSEWIC assessments to identify whether at-risk species that have been assessed more than once tended to improve, remain constant, or deteriorate in status, as a way of assessing the effectiveness of biodiversity conservation in Canada. Of 369 species that met our criteria for examination, 115 deteriorated, 202 remained unchanged, and 52 improved in status. Only 20 species (5.4%) improved to the point where they were 'not at risk', and five of those were due to increased sampling efforts rather than an increase in population size. Species outcomes were also dependent on the severity of their initial assessment; for example, 47% of species that were initially listed as special concern deteriorated between assessments. After receiving an at-risk assessment by COSEWIC, a species is considered for listing under the federal Species at Risk Act (SARA), which is the primary national tool that mandates protection for at-risk species. We examined whether SARA-listing was associated with improved COSEWIC assessment outcomes relative to unlisted species. Of 305 species that had multiple assessments and were SARA-listed, 221 were listed at a level that required identification and protection of critical habitat; however, critical habitat was fully identified for only 56 of these species. We suggest that the Canadian government should formally identify and protect critical habitat, as is required by existing legislation. In addition, our finding that at-risk species in Canada rarely recover leads us to recommend that every effort be made to actively prevent species from becoming at-risk in the first place.

Editor: Clinton N. Jenkins, Instituto de Pesquisas Ecológicas, Brazil

Funding: BF was supported by a Liber Ero postdoctoral fellowship (http://liberero.ca/). CHF was supported by the Raincoast Conservation Foundation (http://www.raincoast.org/) and the National Sciences and Engineering Research Council of Canada (http://www.nserc-crsng.gc.ca/). DCC and CF were supported by University of Victoria Graduate Fellowships. Catering for the UVIC Research Derby event was provided by the University of Victoria Environmental Science and Biology departments. The funders had no role in study design, data collection and analysis, decision to publish, or preparation of the manuscript.

Competing Interests: The authors have declared that no competing interests exist.

* Email: brett.favaro@mi.mun.ca

¶ Membership of the UVic Research Derby is provided in the Acknowledgments.

Introduction

Unsustainable exploitation, climate change, ocean acidification and other anthropogenic impacts have resulted in a global extinction rate that is as much as 1000 times the historic background rate [1–3]. Given the irreversibility of extinctions, preventing or reversing the continuing decline of at-risk species is a major focus of conservation [4]. Preserving global biodiversity is also considered essential for human well-being and the maintenance of ecosystem processes [3,5–7].

Many countries have legislation that explicitly protects species at risk of extinction. In general, such legislation is designed to identify vulnerable taxa, establish recovery plans, prevent further declines, and promote recovery [8]. Recognizing that habitat loss is the leading cause of extinction [9–12], the identification and preservation of habitat is often required, contributing to the stabilization and recovery of threatened species [13]. Despite the implementation of laws and conservation programs, global biodiversity continues to decline [3,5,14].

In Canada, species at risk are identified and protected in a multi-step process. The process begins with the Committee on the Status of Endangered Wildlife in Canada (COSEWIC), an independent scientific body formed in 1977. COSEWIC assesses the status of candidate species that are potentially at risk of extinction or extirpation [15]. This body only considers scientific evidence relevant to a species' recovery potential, and ignores socioeconomic costs or benefits of protection [15]. Their assessments are based on the Canadian extent of the species' range, even if the species occurs in the United States or elsewhere [15]. Species assessed as at-risk by COSEWIC do not automatically secure legal protection from the committee's decision. Rather, protection is issued under the Species at Risk Act (SARA), which was passed in 2003, and which formalized the use of COSEWIC assessments as the scientific basis for listing decisions [16]. Upon receipt of a COSEWIC assessment, the Minister of the

Environment must issue a response statement indicating whether the species will be listed at the status recommended by the assessment, or whether more information is required [17]. The ultimate decision whether to list may also incorporate the socio-economic impacts of listing. If listed, it becomes illegal to kill or harm individuals of that species and the species' critical habitat must be identified and protected to the extent possible [18]. Protections derived from SARA automatically apply to federal lands (including oceans), but for these measures to apply outside of these areas (e.g. in provincial or private land) complementary protections must be implemented that applies to the species and its habitat [19]. If the Minister of the Environment decides that provincial or territorial law is insufficient to effectively protect a listed species, the federal government has the power to apply SARA's protections to provincial or territorial lands [20]. Recovery strategies and action plans (required by SARA) are mandated to be released within a specified timeframe, the length of which depends on the species' designation and when it was listed [21,22]. Furthermore, SARA requires that management plans be produced for species that are not at immediate risk of extinction or extirpation, but are nevertheless of 'special concern' since they may become at-risk in the future [23]. Species of special concern do not receive the full protections offered to species listed under SARA as threatened, endangered, or extirpated.

Previous studies have identified biases, limitations, and a general lack of implementation associated with at-risk species legislation in Canada (Table 1). For instance, harvested species and those found in northern territories are less likely to be listed under SARA [24,26], as are species threatened by biological resource use, including those that are unintentionally harvested [27]. In addition, recovery strategies and action plans are often not completed within deadlines established by SARA [22], leading to concerns that species are not receiving timely protections, despite being listed. This finding was confirmed in a recent federal court decision [28]. Moreover, COSEWIC itself has been criticized for its assessment criteria and apparent biases [31–32]. For example, species with low information availability tend to receive less severe assessments than would be suggested under the precautionary principle [30].

In this study, we examine trends in the status of species repeatedly assessed by COSEWIC, used here as a proxy for Canada's effectiveness in species conservation. Using records obtained from COSEWIC [33] and the SARA Public Registry [34], we analyze trends in the designations of species with two or more assessments to determine whether species are, on average, improving, deteriorating, or remaining stable in status. In addition, we examine outcome differences across taxonomic groups, and whether the basic obligation to identify critical habitat for listed species has been met. Finally, we assess whether species listed for longer have better outcomes. While this study builds on previous work reviewing the process of listing species under SARA (Table 1), it is the first to assess the overall trends in status of at-risk species that have been assessed more than once by COSEWIC.

Methods

COSEWIC Wildlife Species Database

Species, subspecies, and populations (hereafter 'species') that are assessed by COSEWIC are placed into one of five categories on a scale of increasing risk of extinction or extirpation [32]: not at risk, special concern, threatened, endangered, or extinct or extirpated (hereafter extirpated). A sixth category, data deficient, is used when there are insufficient data to classify a species. The criteria for categorization depends on a variety of factors including changes in total numbers of mature individuals, whether a species has a small and declining population size, and the changes in size of its range [35]. COSEWIC's assessment criteria were updated in 2001 [35]. Although the same assessment categories were employed after the revision in 2001, species required a more severe decline in total population to qualify as either endangered or threatened. For example, prior to 2001 the criteria for an Endangered assessment (Criteria A1: [35]) required a reduction of ≥50% in the total number of mature individuals over the last 10 years or 3 generations. As of 2001, this same category required a ≥70% decline [36].

The definition of full recovery that we employ here is a change in COSEWIC assessment status to 'not at risk.' Although this may not equate to a full ecological recovery (e.g. [37,38]) it allows us to broadly assess trends with a consistent metric.

Data collection

We identified all species that COSEWIC has assessed more than once using the Wildlife Species Search Engine on the COSEWIC website [33]. Assessments occurred between 1977 (COSEWIC's establishment) and December 2013 (when we conducted our search). We recognize that our analysis may be limited by biases associated with which species have been assessed multiple times and that COSEWIC listing criteria have changed since 2001. However, because the 2001 revision tightened criteria for more-severe listings, the expected bias introduced by the criteria change would be a greater proportion of less severe species listings. We used the 'change in status' tab on the website to include any species 'in a higher risk category' (N = 81); 'in a lower risk category' (N = 36); 'no longer at risk' (N = 21); 'changed' (N = 20); 'reassigned' (N = 69); and 'no change' (N = 272); populating our database with a total of 499 species. In our analysis, we excluded any species that had ever been assessed as 'data deficient' by COSEWIC (N = 30 species), or that had experienced some form of reassignment (e.g. a species split into multiple designatable units, N = 106 species), because changes in status for these species cannot be interpreted as a true change in extinction risk. We excluded six species that were both data deficient and reclassified, leaving a total of 369 species in our analysis.

We collected the history of COSEWIC assessments for each species on its respective species summary page. We recorded the date and status designation of each COSEWIC assessment for each species, and the taxonomic group to which that species belongs (due to small sample sizes, we combined molluscs and arthropods into a single 'invertebrates' category, and combined lichens and mosses). We then used the SARA public registry [34] to record whether species were SARA listed, and to access COSEWIC status reports for each species. To account for differences between species' life histories, COSEWIC and the IUCN place rates of decline into a biological context by scaling assessments by the generation time (GT) of each species [35,39]. Therefore we also recorded each species' generation time, if available. If the species was SARA-listed as extirpated, endangered, or threatened, we recorded whether a recovery strategy had been completed. If the recovery strategy was completed, we recorded whether the strategy indicated that critical habitat had been fully, partially, or not identified. If a recovery strategy was not complete for the species, we recorded that critical habitat had not been identified. For SARA-listed species, Environment Canada provided us with the dates when species were listed.

For some species, the COSEWIC summary explained that an apparent improvement in status was due to the discovery of new populations through increased sampling, and not because of an

Table 1. Summary of review papers related to endangered species assessment and legislation in Canada.

Publication	Reference #	Primary findings
Vanderzwaag and Hutchings (2005)	[51]	Review of SARA implementation related to marine fish. Paper advocates for biodiversity preservation by implementing marine protected areas and modernizing Fisheries Act.
Mooers et al. (2007)	[52]	First identification of taxonomic and regional biases in SARA listing. Northern species and marine fish and terrestrial mammals unlikely to receive SARA-listing.
Findlay et al. (2009)	[24]	Commercially harvested species, species managed by DFO, and species that occur entirely within Canada are less likely to receive listing.
Lukey et al. (2009)	[29]	Changes in assessment status from 'endangered' to 'threatened' often occur without sufficient justification for the change. Assessment criteria are not always applied consistently.
Lukey et al. (2010)	[30]	COSEWIC assessments do not follow the precautionary principle – lack of information is associated with assessments of species to lower risk categories.
Mooers et al. (2010)	[25]	Most SARA-listed species lack recovery plans. Scientific advice is insufficiently reflected in conservation policy for at-risk species. First review of changes in COSEWIC status across assessments.
Powles (2011)	[31]	General overview of marine fish assessed by COSEWIC.
Taylor and Pinkus (2013)	[53]	Only 17% of recovery strategies led by DFO included critical habitat, as opposed to 63% for those led by Environment Canada. Recovery strategies written after court judgments related to SARA were more likely to identify critical habitat.
Waples et al. (2013)	[8]	Comparison of United States' Endangered Species Act (ESA) with SARA. ESA should adopt a single national scientific body, while SARA should adopt strict deadlines for listing action. The emphasis on socioeconomic factors should also be reduced.
McCune et al. (2013)	[27]	Human disturbance, invasive species, residential development, and ultimately loss of habitat are major threats to the majority of SARA-listed species. Threats differ by taxonomic grouping.
Schultz et al. (2013)	[26]	Imperiled marine fish are unlikely to receive SARA listing if the forecasted cost of listing exceeds $90,000 per decade. The threshold for freshwater fish is $5,000,000. Rationale used in the decision to list was inconsistent between freshwater and marine fish.
Auditor General of Canada (2013)	[22]	All three government agencies involved with species at risk (Environment Canada, Fisheries and Oceans Canada, and Parks Canada) are not meeting obligations to complete recovery strategies, action plans, or management plans. Only seven of 97 required action plans were complete.

actual recovery in population. We made note of these cases (N = 20) so as to distinguish them from improvements due to conservation action.

Data analysis

We assigned each COSEWIC status a numerical value in descending order of severity (5 = not at risk, 4 = special concern, 3 = threatened, 2 = endangered, 1 = extirpated). We then calculated the overall change for each species, between the first and most recent COSEWIC assessment. For example, a species that was initially classified as special concern, and then deteriorated to endangered in its most recent assessment would receive a score of −2. For species requiring critical habitat designation (extirpated, endangered and threatened), we performed a Pearson's chi-squared test to determine whether critical habitat identifications (full, partial, or not identified) across the ten taxonomic groups differed from the proportions across all species. We used the statistical software R for computations and data plots [40].

We identified species whose most recent COSEWIC assessment occurred after the species had been listed under SARA for at least three of that species' GT, to identify species for which some recovery may be biologically possible. We then employed a cumulative link mixed model using the "ordinal" package in R [41] to test whether the number of GTs since SARA-listing was associated with a change in final COSEWIC status across species. Cumulative link mixed models test the influence of one or more independent variables (herein: number of generation times since listing) on an ordinal dependent variable (change in COSEWIC status), and are analogous to generalized linear mixed effects models in that they allow for the incorporation of a random effect, or 'grouping' factor (taxonomic order) [41–43].

Results

There were a total of 369 species across ten taxonomic groups that met our criteria for inclusion (Table S1, Fig. 1). For all taxonomic groups (except marine fish), the majority of species declined or remained the same in status (Table S1, Fig. 2). However, species trajectories varied substantially based on their initial assessment (Fig. 3). For species initially classified as not at risk, endangered, or extirpated, the most common outcome was to remain at their initially assessed status. However, for species initially classified as special concern or threatened, deterioration in status was the most common outcome. For example, 47% of species listed as special concern deteriorated in status (Fig. 3). Only 20 species (5.4%) received a 'not at risk' assessment after previously being listed in an at-risk category (three terrestrial mammals, nine birds, four freshwater fish, one marine fish, two vascular plants, and one lichen). Five of these cases were not due to conservation action, but were instead due to increased sampling effort. A total of 221 species are listed under SARA as threatened, endangered, or extirpated (Fig. 4), and therefore their critical habitat should be identified. Overall, and for all taxonomic groups, critical habitat has not been fully identified for more than half of SARA-listed species (Fig. 4). Further, critical habitat identifications (full, partial, or not identified) significantly varied across the ten taxonomic groups ($\div^2 = 51.90$, df = 18, $p < 0.001$).

There were 163 species that met our criteria for inclusion in the Generation Time analysis (listed as extirpated, endangered, or

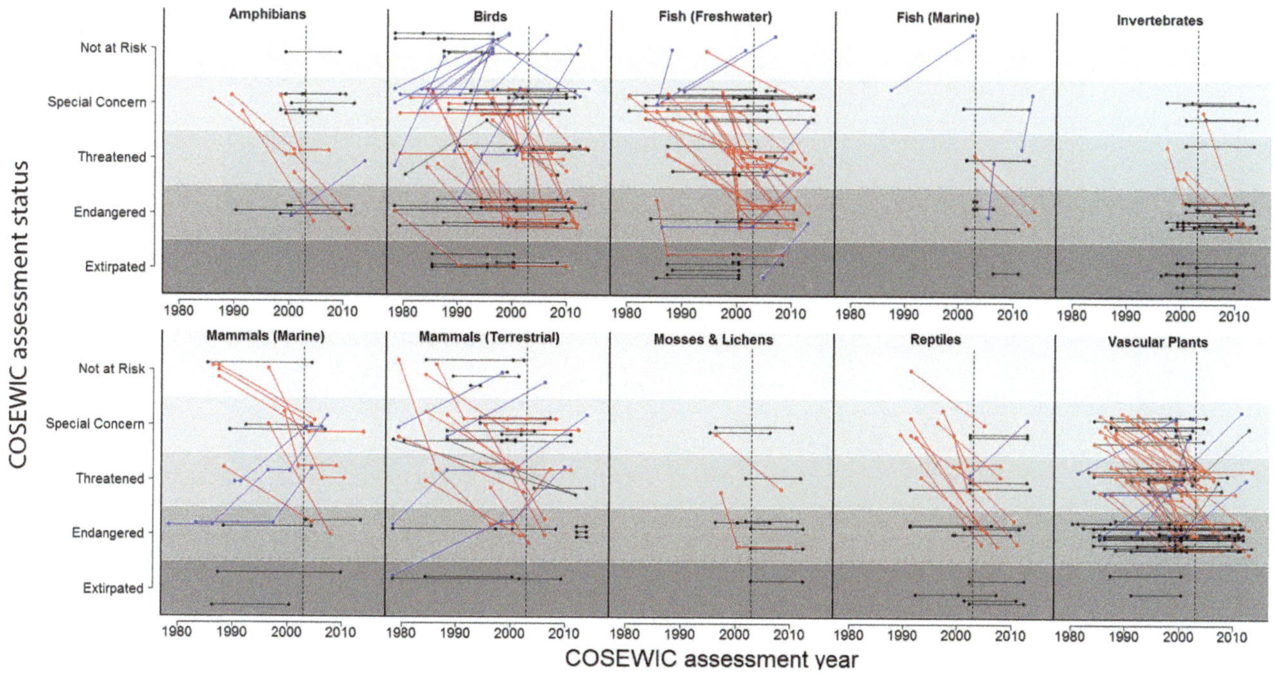

Figure 1. Overview of COSEWIC assessment statuses for all species that have been assessed more than once, and that have never been data deficient or taxonomically reassigned. Each species is represented by points (assessment date and outcome) connected by lines. Species that deteriorated in status from their first to final COSEWIC assessments are red, species that have improved are blue, and species that have remained constant are black. The vertical dotted line indicates 2003, or the passing of SARA. Species whose apparent recovery was due to increased sampling effort, and not biological recovery (N = 20) are not shown.

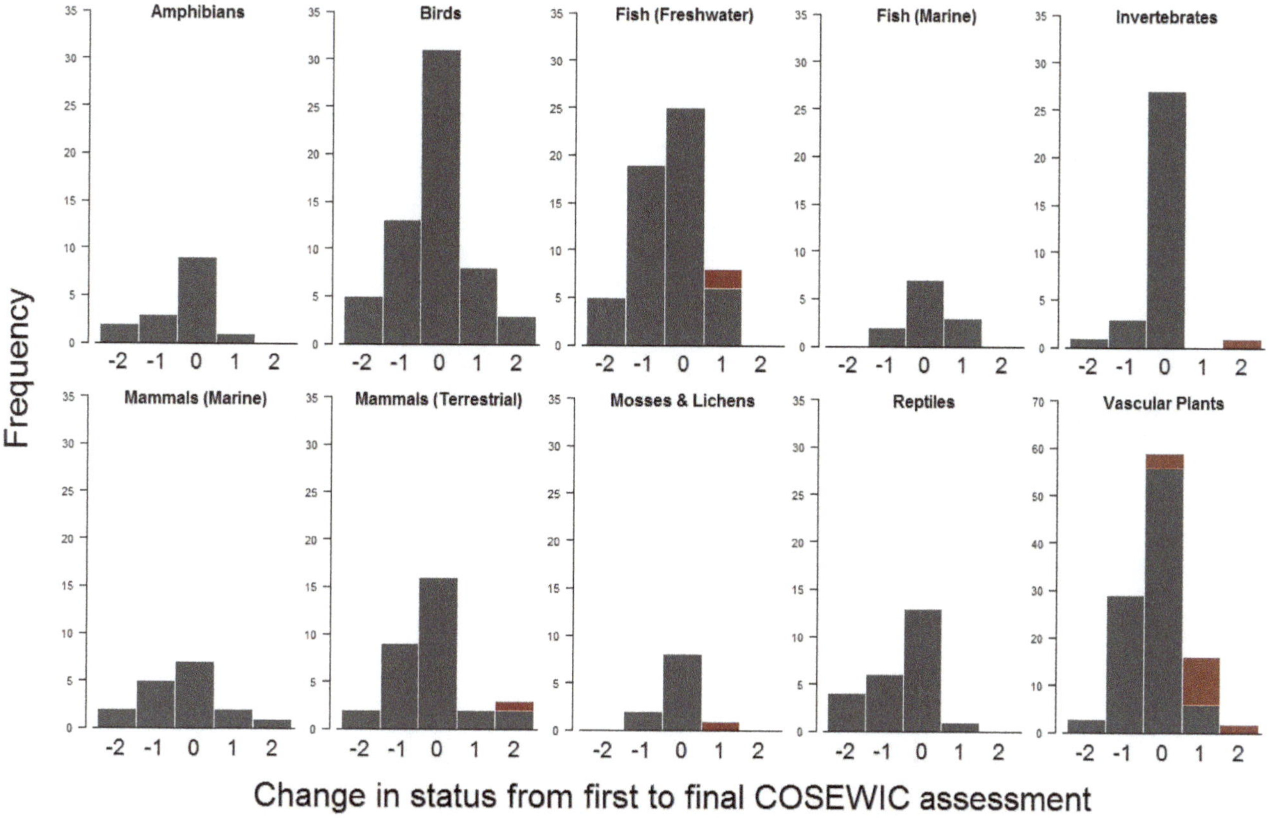

Figure 2. Frequencies of occurrence of change in COSEWIC assessed status for each taxonomic group. Positive numbers indicate improvement (e.g. a transition from endangered to threatened would be +1, while endangered to special concern would be +2), negative numbers indicate deterioration, and zero indicates no change across assessments. Red bars indicate apparent recoveries due to increased sampling efforts. Note that the y-axis for vascular plants is scaled differently from other taxonomic groups.

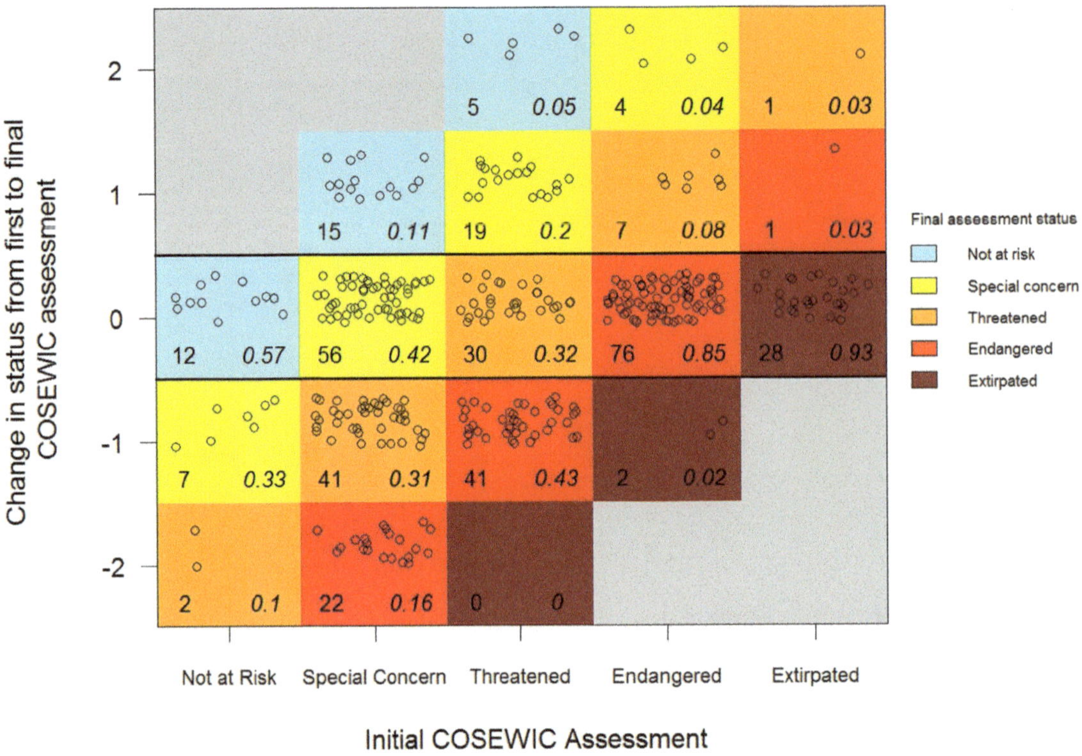

Figure 3. Trends in COSEWIC statuses for each species, grouped by initial assessments. Proportions are grouped by initial assessment status (column). Change from first to final assessment is indicated on the Y axis, and colours indicate the final assessment status. For example, of species that were initially assessed as threatened (third column), 19 of them (or a proportion of 0.2) improved by one status level (+1 on Y-axis), ultimately placing them into the special concern category (yellow).

threatened; GT reported; schedule 1), and 69 that have been listed for three or more GT. The number of GT since listing ranged from 0.11 to 35.8 (mean ±1 S.D., 3.7±4.3, Figure 5). If a species has been adequately protected, the probability of a species improving in status should increase with the number of generations since listing. However, there was an estimated decline in the probability of a species improving as the number of generations since listing increased ($\hat{\beta} = -0.05$, $\{\hat{\theta}_j\} = \{-2.40, -0.18, 2.44, 4.31\}$, 95% C.I. = −0.124 to 0.008), but it was not significant (Wald $p = 0.084$). This model had a small absolute gradient ($7.4*10^{-6}$) and a condition number of the Hessian of 720, indicating that the model was able to converge and was well-defined, respectively [42]. The cumulative odds ratio was between 0.88 and 1.01 (95% C.I.), meaning that for each unit increase in GT since listing, the odds of a species improving in status ranged from increasing by 1% to decreasing by 12%.

Discussion

For species that have been assessed more than once by COSEWIC, improvement was rare, and recovery to a 'not at risk' status occurred in only 5.7% of cases. Moreover, one quarter of those were apparent recoveries due to increased sampling, rather than increases in population size driven by conservation action. Contrary to the intent of endangered species legislation in Canada, the probability of a species improving in status did not increase with the number of generations since initial listing under SARA. Moreover, species had a greater probability of deteriorating in status with the number of generations since listing, although this relationship was marginally statistically non-significant (Wald $p = 0.084$). In contrast, species recovery in the United States was strongly correlated with the number of years of protection under the Endangered Species Act [44], indicating that endangered species legislation can be effective. These results suggest a potential failure of Canadian legislation, its subsequent implementation, or both. While COSEWIC's assessment criteria were modified in 2001 [35], the revisions made it harder to qualify for more-severe categories. Therefore we have no evidence to suggest that observed declines in status were an artifact of changes in criteria.

The lack of observed recovery for SARA-listed species may be due to the lack of implementation of the law. For example, for those species without identified critical habitat, the habitat protection provisions of SARA cannot be fully implemented. Overall, the proportion of species with critical habitat identified was higher than what has been reported in a previous study [25], although our study focused on a subset of species that have been assessed multiple times, and that have never been data deficient. These are the species for which it should have been most feasible to identify critical habitat, as they have been scrutinized for a relatively long time and have sufficient data to complete an accurate assessment. The fact remains that despite the importance of critical habitat identification, the proportion of species whose critical habitat was fully identified remained low, and varied significantly across the ten taxonomic groups. This pattern suggests that there are considerable differences between the protections that species should receive under SARA, and what is actually achieved.

Since our data demonstrate that it is rare for at-risk species to recover in Canada, it is essential that substantial efforts be made to prevent species from becoming at-risk in the first place. Given that

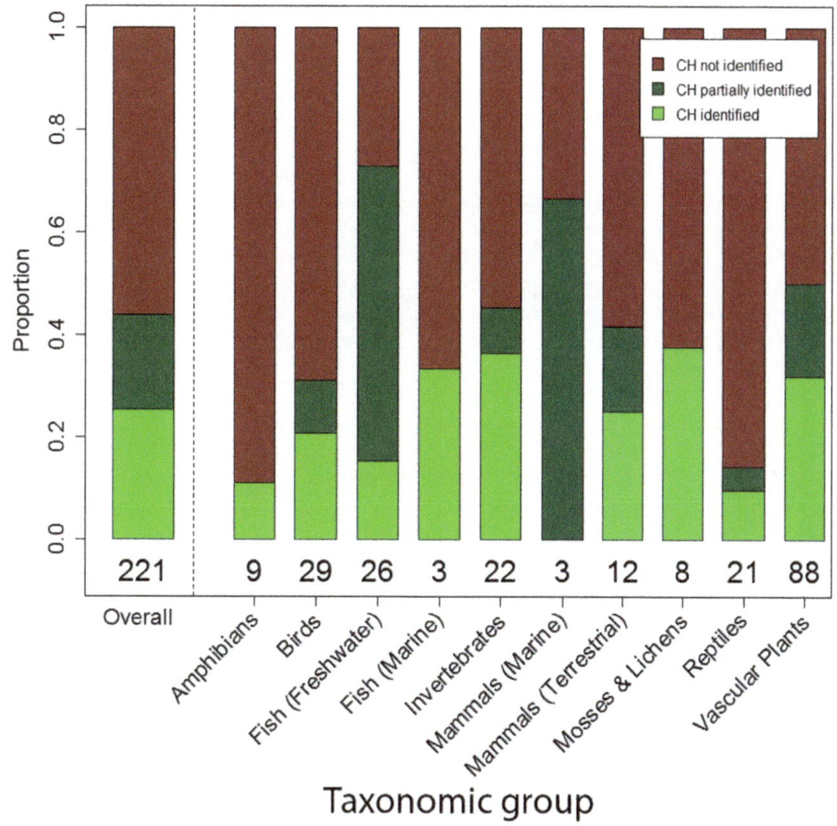

Figure 4. Proportion of SARA-listed species (listed as extirpated, endangered, or threatened) that have critical habitat (CH) fully identified (light green), partially identified (dark green), or not at all identified (dark red), overall and by taxonomic group. Values under bars indicate the number of species in each taxonomic category.

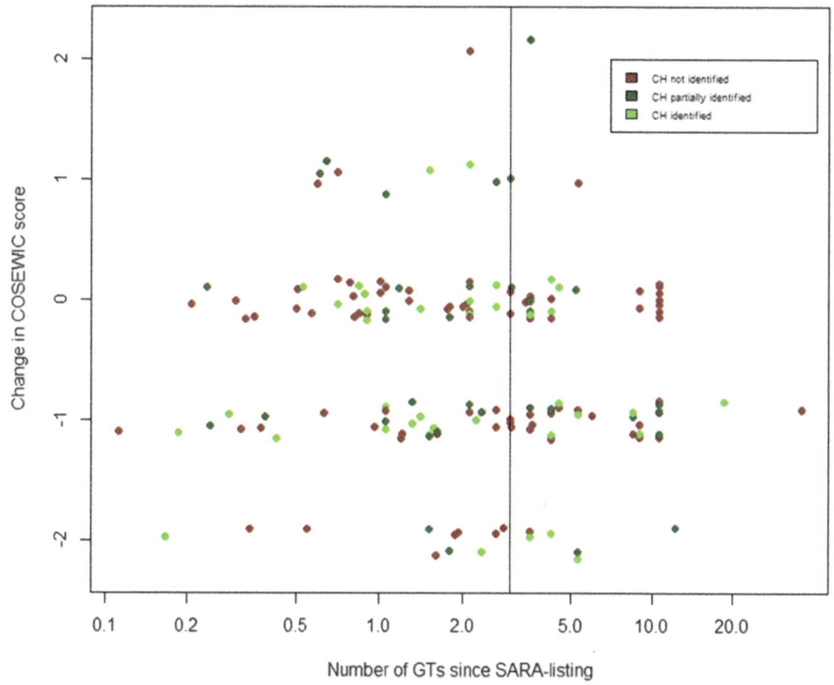

Figure 5. Change in COSEWIC assessment statuses versus the number of GT that have passed since initial SARA listing for all schedule 1 species listed as threatened, endangered, or extirpated. Dot colour indicates whether CH has been fully identified (green), partially identified (dark green), or not identified (red). The black vertical line indicates three GT.

imperiled species are usually threatened by loss of habitat [45], recent weakening of federal laws that protect habitat in Canada [25,46,47,48] will be unhelpful in the long term, as it may result in additional species declining to the point where they receive an at-risk designation. Habitat protection should not be limited to critical habitat – it should be managed appropriately for all habitat such that protection does not become critical. The experience in the United States demonstrates that protection of critical habitat is associated with improved conservation outcomes relative to species without such protection [44,47]. In addition, since we found that the most common outcome for a species assessed as special concern was to deteriorate, an assessment of special concern does not currently result in sufficient protection to promote recovery.

The single most common outcome across taxonomic groups was for species to remain at the same status across assessments, and the number of species that declined outnumbered those that recovered by a ratio of approximately 2:1. These findings are alarming for three reasons. First, by definition, a species that has been assessed as anything other than 'not at risk' is at elevated risk of extinction or extirpation given current conditions, and therefore maintaining a species at a threatened status should not be interpreted as a conservation victory. Second, it takes a substantial decline in population size or range size to trigger a change in assessment status, so real declines (or increases) could still occur within species held at the same COSEWIC threat level across assessments. Third, a 'not at risk' designation only means that the species is not at elevated risk of extinction or extirpation – it does not imply that the population has recovered to historical levels. Even species that are classified 'not at risk' can be heavily depleted and unable to serve their historic roles in ecosystem structure or function. This has implications for managers in the United States as well as Canada, because many at-risk species have ranges that extend into both countries. At the very least, a successful species at risk program should demonstrate species recovering to a point where they are not at risk of extinction or extirpation given current conditions. Currently, this goal is not being achieved in Canada for the overwhelming majority of species.

Recommendations

Our results lead us to make three core recommendations for at-risk species in Canada. First, given that it was much more common for species of special concern to deteriorate than to improve, we should recognize that a special concern listing warns of a coming deterioration, and we therefore suggest that the protections associated with this listing should be strengthened. Second, given the poor outcomes of at-risk species in Canada, it should be a policy priority to prevent species from becoming at-risk in the first place. The importance of critical habitat indicates that future legislation should be underpinned by a strong mandate to conserve habitat and we recommend that any legislative changes that may reduce habitat protection (e.g. the Fisheries Act [48]) should be reconsidered. Third, to experience conservation benefits from SARA, this law must be fully implemented. Implementation requires that critical habitat be fully identified and subsequently protected for SARA-listed species. The federal government should also be prepared to enact its 'safety net' provision, in the event that species in these regions are not receiving adequate protection to enable recovery (as it did with greater sage-grouse, *Centrocercus urophasianus* [49]).

Finally, even if these recommendations were accepted and put into effect, recovery takes time. Effective management requires that conservation measures be sustained over the long term, even if positive outcomes are not immediately observed.

Supporting Information

Table S1 Summary counts of species across taxonomic groups, for all species assessed more than once by COSEWIC. Values for species listed for >3 generation times (GT) refer to the subset of species for which GTs were reported.

Data S1 Raw data for the present study in both wide and long formats.

Acknowledgments

We thank Environment Canada for providing us with dates that species were listed under SARA. The idea for this paper originated within a Research Derby conducted at the University of Victoria [50]. The 2013 UVic Research Derby author group consisted of the authors of this paper, as well as A.F. Martin, K. Pawluk, D. Roberts, and J. Robinson, and can be contacted through B. Favaro (brett.favaro@mi.mun.ca). We thank S. Pinkus and three anonymous reviewers for providing useful comments.

Author Contributions

Conceived and designed the experiments: BF DCC CHF CF JJH. Performed the experiments: BF DCC CHF CF JJH. Analyzed the data: BF DCC CHF CF JJH AR. Wrote the paper: BF DCC CHF CF JJH AR.

References

1. Pimm SL, Russell GJ, Gittleman JL, Brooks TM (1995) The future of biodiversity. Science 269: 347–350.
2. Chapin III FS, Zavaleta ES, Eviner VT, Naylor RL, Vitousek PM, et al. (2000) Consequences of changing biodiversity. Nature 6783: 234–242.
3. Rands MRW, Adams WM, Bennun L, Butchart SHM, Clements A, et al. (2010) Biodiversity conservation: Challenges beyond 2010. Science 329: 1298–1303.
4. Arponen A (2012) Prioritizing species for conservation planning. Biodivers Conserv 21: 875–893.
5. Hassan RM, Scholes R, Ash N, editors. (2005) Ecosystems and human well-being: Current state and trends. Findings of the condition and trends working group millennium ecosystem assessment series. Washington DC: Island Press.
6. Hooper DU, Chapin III FS, Ewel JJ, Hector A, Inchausti P, et al. (2005) Effects of biodiversity on ecosystem functioning: A consensus of current knowledge. Ecol Monogr 75: 3–35.
7. Barton J, Pretty J (2010) What is the best dose of nature and green exercise for improving mental health? A multi-study analysis. Environ Sci Tech 44: 3947–3955.
8. Waples RS, Nammack M, Cochrane JF, Hutchings JA (2013) A tale of two acts: Endangered species listing practices in Canada and the United States. Bioscience 63: 723–734.
9. Brooks TM, Mittermeier RA, Mittermeier CG, Da Fonseca GAB, Rylands AB, et al. (2002) Habitat loss and extinction in the hotspots of biodiversity. Conserv Biol 16: 909–923.
10. Kareiva P, Wennergren U (1995) Connecting landscape patterns to ecosystem and population processes. Nature 373: 299–302.
11. Wilcove DS, Rothstein D, Jason Dubow, Phillips A, Losos E (1998) Quantifying threats to imperiled species in the United States. Bioscience 48: 607–615.
12. Homan RN, Windmiller BS, Reed JM (2004) Critical thresholds associated with habitat loss for two vernal pool-breeding amphibians. Ecol App 14: 1547–1553.
13. Taylor MFJ, Sattler PS, Evans M, Fuller RA, Watson JEM, et al. (2011) What works for threatened species recovery? An empirical evaluation for Australia. Biodivers Conserv 20: 767–777.
14. "Secretariat of the Convention on Biological Diversity." (2010) Global biodiversity outlook 3. Montreal. 94 p.
15. COSEWIC (2013) Terms of reference: Committee on the status of endangered wildlife in Canada (COSEWIC), revised and approved by COSEWIC November 2011.
16. *SARA (Species at Risk Act)* (2002) An act respecting the protection of wildlife species at risk in Canada. SC 2002, c 29, s 15. Available: http://Laws-lois.justice.gc.ca/PDF/S-15.3.pdf. Accessed 2014 August 11.

17. *SARA (Species at Risk Act)* (2002) An act respecting the protection of wildlife species at risk in Canada. SC 2002, c 29, s 27. Available: http://Laws-lois.justice.gc.ca/PDF/S-15.3.pdf. Accessed 2014 August 11.
18. *SARA (Species at Risk Act)* (2002) An act respecting the protection of wildlife species at risk in Canada. SC 2002, c 29, ss 32–36. Available: http://Laws-lois.justice.gc.ca/PDF/S-15.3.pdf. Accessed 2014 August 11.
19. *SARA (Species at Risk Act)* (2002) An act respecting the protection of wildlife species at risk in Canada. SC 2002, c 29, ss 34(1)-35(1,2). Available: http://Laws-lois.justice.gc.ca/PDF/S-15.3.pdf. Accessed 2014 August 11.
20. *SARA (Species at Risk Act)* (2002) An act respecting the protection of wildlife species at risk in Canada. SC 2002, c 29, ss 3435. Available: http://Laws-lois.justice.gc.ca/PDF/S-15.3.pdf. Accessed 2014 August 11.
21. *SARA (Species at Risk Act)* (2002) An act respecting the protection of wildlife species at risk in Canada. SC 2002, c 29, ss 41–45. Available: http://Laws-lois.justice.gc.ca/PDF/S-15.3.pdf. Accessed 2014 August 11.
22. Auditor General of Canada (2013) Chapter 6: Recovery planning for species at risk. In: Anonymous Report of the commissioner of the environment and sustainable development. Office of the Auditor General of Canada. pp. 1–22. Available: http://www.oag-bvg.gc.ca/internet/English/parl_cesd_201311_06_e_38676.html. Accessed 2014 August 11.
23. *SARA (Species at Risk Act)* (2002) An act respecting the protection of wildlife species at risk in Canada. SC 2002, c 29, ss 65–72. Available: http://Laws-lois.justice.gc.ca/PDF/S-15.3.pdf. Accessed 2014 August 11.
24. Findlay C, Elgie S, Giles B, Burr L (2009) Species listing under Canada's Species at Risk Act. Conserv Biol 23: 1609–1617.
25. Mooers AO, Doak DF, Findlay CS, Green DM, Grouios C (2010) Science, policy, and species at risk in Canada. Bioscience 60: 843–849.
26. Schultz JA, Darling ES, Côté IM (2013) What is an endangered species worth? Threshold costs for protecting imperilled fishes in Canada. Mar Policy 42: 125–132. http://dx.doi.org/10.1016/j.marpol.2013.01.021.
27. McCune JL, Harrower WL, Avery-Gomm S, Brogan JM, Csergő A, et al. (2013) Threats to Canadian species at risk: An analysis of finalized recovery strategies. Biol Conserv 166: 254–265. http://dx.doi.org/10.1016/j.biocon.2013.07.006.
28. *Western Canada Wilderness Committee, David Suzuki Foundation, Greenpeace Canada, Sierra Club of British Columbia Foundation, and Wildsite v. Minister of Fisheries and Oceans and Minister of the Environment* (2014) Canada Federal Court, 2014 FC 148, T-1777-12. Available: http://cas-ncr-nter03.cas-satj.gc.ca/rss/T-1777-12%20SARA%20decision%2014-02-2014%20ENG.pdf. Accessed 2014 August 11.
29. Lukey JR, Crawford SS (2009) Consistency of COSEWIC species at risk designations: Freshwater fishes as a case study. Can J Fish Aquat Sci 66: 959–971.
30. Lukey JR, Crawford SS, Gillis D (2010) Effect of information availability on assessment and designation of species at risk. Conserv Biol 24: 1398–1406.
31. Powles H (2011) Assessing risk of extinction of marine fishes in Canada - the COSEWIC experience. Fisheries 36: 231–246.
32. Kowalchuk KA, Kuhn RG (2012) Mammal distribution in Nunavut: Inuit harvest data and COSEWIC's species at risk assessment process. Ecol Soc 17: 4.
33. COSEWIC (2013) Database of wildlife species assessed by COSEWIC. Available: http://www.cosewic.gc.ca/eng/sct1/searchform_e.cfm. Accessed 2014 August 11.
34. Government of Canada (2014) Species at risk act public registry. Available: http://www.sararegistry.gc.ca/default_e.cfm. Accessed 2014 August 11.
35. COSEWIC (2011) COSEWIC's Assessment Process and Criteria. Available: http://www.cosewic.gc.ca/pdf/Assessment_process_and_criteria_e.pdf. Accessed 2014 August 11.
36. COSEWIC (2005) Original criteria and definitions used in the status assessment of species from October 1999 to May 2001. Available: http://www.cosewic.gc.ca/eng/sct0/original_criteria_e.cfm. Accessed 2014 August 11.
37. Redford KH, Amato G, Baillie J, Beldomenico P, Bennet EL, et al. (2011) What does It mean to successfully conserve a (vertebrate) species? BioScience 61:39–48.
38. Westwood A, Reuchlin-Hugenholtz E, Keith DM (2014) Re-defining recovery: A generalized framework for assessing species recovery. Biological Conservation 172:155–162.
39. IUCN Standards and Petitions Subcommittee (2014) Guidelines for using the IUCN Red List categories and criteria. Version 11. Available: http://www.iucnredlist.org/documents/RedListGuidelines.pdf. Accessed 2014 August 11.
40. R Core Team (2013) R: A language and environment for statistical computing. R Foundation for Statistical Computing, Vienne, Austria. http://www.R-project.org.
41. Christensen RHB (2013) Ordinal – Regression models for Ordinal Data. R package version 2013.9-30. Available: http://www.cran.r-project.org/package=ordinal/.
42. Tutz G, Hennevogl W (1996) Random effects in ordinal regression models. Comput Stat Data An 22: 537–557.
43. Christensen RHB (2013) A tutorial on fitting Cumulative Link Models with the ordinal package. Available: ftp://ftp.stat.math.ethz.ch/R-CRAN/web/packages/ordinal/vignettes/clm_tutorial.pdf.
44. Taylor MFJ, Suckling KF, Rachlinski JJ (2005) The effectiveness of the endangered species act: A quantitative analysis. Bioscience 55: 360–367.
45. Venter O, Brodeur NN, Nemiroff L, Belland B, Dolinsek IJ, et al. (2006) Threats to endangered species in Canada. Bioscience 56: 903–910.
46. Favaro B, Reynolds JD, Côté IM (2012) Canada's weakening aquatic protection. Science 337: 154.
47. Kirchhoff D, Tsuji LJS (2014) Reading between the lines of the 'Responsible resource development' rhetoric: The use of omnibus bills to 'streamline' Canadian environmental legislation. Impact Assessment and Project Appraisal: 1–13. DOI: 10.1080/14615517.2014.894673.
48. Hutchings JA, Post JR (2013) Gutting Canada's fisheries act: No fishery, no fish habitat protection. Fisheries 38: 497–501.
49. Canada Gazette (2013) Emergency Order for the Protection of the Greater Sage-Grouse. Species At Risk Act. Government of Canada. P.C. 2013–1245. Available: http://canadagazette.gc.ca/rp-pr/p2/2013/2013-12-04/html/sup-eng.php Accessed 2014 July 22.
50. Favaro B, Braun DC, Earth2Ocean Research Derby (2013) The 'research derby': a pressure cooker for creative and collaborative science. Ideas Ecol Evol 6:40–66.
51. Vanderzwaag DL, Hutchings JA (2005) Canada's marine species at risk: Science and law at the helm, but a sea of uncertainties. Ocean Dev Int Law 36: 219–259.
52. Mooers AO, Prugh LR, Festa-Bianchet M, Hutchings JA (2007) Biases in legal listing under Canadian endangered species legislation. Conserv Biol 21: 572–575.
53. Taylor EB, Pinkus S (2013) The effects of lead agency, nongovernmental organizations, and recovery team membership on the identification of critical habitat for species at risk: Insights from the Canadian experience. Environ Rev 21: 93–102.

Permissions

All chapters in this book were first published in PLOS ONE, by The Public Library of Science; hereby published with permission under the Creative Commons Attribution License or equivalent. Every chapter published in this book has been scrutinized by our experts. Their significance has been extensively debated. The topics covered herein carry significant findings which will fuel the growth of the discipline. They may even be implemented as practical applications or may be referred to as a beginning point for another development.

The contributors of this book come from diverse backgrounds, making this book a truly international effort. This book will bring forth new frontiers with its revolutionizing research information and detailed analysis of the nascent developments around the world.

We would like to thank all the contributing authors for lending their expertise to make the book truly unique. They have played a crucial role in the development of this book. Without their invaluable contributions this book wouldn't have been possible. They have made vital efforts to compile up to date information on the varied aspects of this subject to make this book a valuable addition to the collection of many professionals and students.

This book was conceptualized with the vision of imparting up-to-date information and advanced data in this field. To ensure the same, a matchless editorial board was set up. Every individual on the board went through rigorous rounds of assessment to prove their worth. After which they invested a large part of their time researching and compiling the most relevant data for our readers.

The editorial board has been involved in producing this book since its inception. They have spent rigorous hours researching and exploring the diverse topics which have resulted in the successful publishing of this book. They have passed on their knowledge of decades through this book. To expedite this challenging task, the publisher supported the team at every step. A small team of assistant editors was also appointed to further simplify the editing procedure and attain best results for the readers.

Apart from the editorial board, the designing team has also invested a significant amount of their time in understanding the subject and creating the most relevant covers. They scrutinized every image to scout for the most suitable representation of the subject and create an appropriate cover for the book.

The publishing team has been an ardent support to the editorial, designing and production team. Their endless efforts to recruit the best for this project, has resulted in the accomplishment of this book. They are a veteran in the field of academics and their pool of knowledge is as vast as their experience in printing. Their expertise and guidance has proved useful at every step. Their uncompromising quality standards have made this book an exceptional effort. Their encouragement from time to time has been an inspiration for everyone.

The publisher and the editorial board hope that this book will prove to be a valuable piece of knowledge for researchers, students, practitioners and scholars across the globe.

List of Contributors

Nathan F. Bendik
Watershed Protection Department, City of Austin, Austin, Texas, United States of America

Thomas A. Morrison
Wyoming Cooperative Fish and Wildlife Research Unit, Department of Zoology and Physiology, University of Wyoming, Laramie, Wyoming, United States of America

Andrew G. Gluesenkamp
Texas Parks and Wildlife Department, Austin, Texas, United States of America

Mark S. Sanders and Lisa J. O'Donnell
Balcones Canyonlands Preserve, City of Austin, Austin, Texas, United States of America

Gregory R. Moyer
United States Fish and Wildlife Service, Conservation Genetics Laboratory, Warm Springs, Georgia, United States of America

Edgardo Díaz-Ferguson
United States Fish and Wildlife Service, Conservation Genetics Laboratory, Warm Springs, Georgia, United States of America
Department of Fisheries and Allied Aquacultures, Auburn University, Auburn, Alabama, United States of America

Jeffrey E. Hill
Program in Fisheries and Aquatic Sciences Tropical Aquaculture Laboratory, University of Florida, Ruskin, Florida, United States of America

Colin Shea
Department of Biology, Tennessee Technological University, Cookeville, Tennessee, United States of America

Allison M. Benscoter, James I. Watling and Frank J. Mazzotti
University of Florida, Fort Lauderdale Research and Education Center, Davie, Florida, United States of America

Joshua S. Reece
Valdosta State University, Department of Biology,Valdosta, Georgia, United States of America

Reed F. Noss
University of Central Florida, Department of Biology, Orlando, Florida, United States of America

Laura A. Brandt
United States Fish and Wildlife Service, Davie, Florida, United States of America

Stephanie S. Romañach
United States Geological Survey, Southeast Ecological Science Center, Davie, Florida, United States of America

Howard Peters, Bethan C. O'Leary, Julie P. Hawkins and Callum M. Roberts
Environment Department, University of York, York, United Kingdom

Kent E. Carpenter
International Union for Conservation of Nature, Global Marine Species Assessment, Biological Sciences, Old Dominion University, Norfolk, Virginia, United States of America

Miklos Blaho, Tamas Herczeg, Andras Barta, Gabor Horvath, Adam Egri, Denes Szaz and Alexandra Farkas
Environmental Optics Laboratory, Department of Biological Physics, Physical Institute, Eötvös University, Budapest, Hungary

Gyorgy Kriska
Group for Methodology in Biology Teaching, Biological Institute, Eötvös University, Budapest, Hungary

Danube Research Institute, Centre for Ecological Research, Hungarian Academy of Sciences, Budapest, Hungary

Nikolett Tarjanyi
Danube Research Institute, Centre for Ecological Research, Hungarian Academy of Sciences, Budapest, Hungary

Laszlo Czinke
Environmental Optics Laboratory, Department of Biological Physics, Physical Institute, Eötvös University, Budapest, Hungary

Group for Methodology in Biology Teaching, Biological Institute, Eötvös University, Budapest, Hungary, Budapest, Hungary

Rochelle Steven, J. Guy Castley and Ralf Buckley
International Centre for Ecotourism Research, Griffith University, Gold Coast, Queensland, Australia

Ruidong Wu
Institute of International Rivers and Eco-security, Yunnan University, Kunming, Yunnan, China
Yunnan Key Laboratory of International Rivers and Transboundary Ecosecurity, Yunnan University, Kunming, Yunnan, China

Yongcheng Long
The Nature Conservancy China Program, Kunming, Yunnan, China

George P. Malanson
Department of Geography, The University of Iowa, Iowa City, Iowa, United States of America

Paul A. Garber
Department of Anthropology, University of Illinois, Urbana, Illinois, United States of America

Shuang Zhang, Peng Zhao and Longzhu Wang
The Nature Conservancy China Program, Kunming, Yunnan, China

Diqiang Li
Institute of Forest Ecology, Environment, and Protection, Chinese Academy of Forestry, Beijing, China

Hairui Duo
School of Nature Reserve, Beijing Forestry University, Beijing, China

Mark W. Schwartz
John Muir Institute of the Environment, University of California Davis, Davis, California, United States of America
Department of Environmental Science & Policy, University of California Davis, Davis, California, United States of America

Lacy M. Smith
Department of Environmental Science & Policy, University of California Davis, Davis, California, United States of America

Zachary L. Steel
Graduate Group in Ecology, University of California Davis, Davis, California, United States of America

Mehregan Ebrahimi
School of Biological Sciences, Flinders University, Adelaide, South Australia, Australia
Department of Biology, Shiraz University, Shiraz, Fars Province, Iran

C. Michael Bull
School of Biological Sciences, Flinders University, Adelaide, South Australia, Australia

Tara M. Cornelisse, Michelle K. Bennett and Deborah K. Letourneau
Environmental Studies Department, University of California Santa Cruz, Santa Cruz, California, United States of America

Amanda J. Chunco
Department of Environmental Studies, Elon University, Elon, North Carolina, United States of America

Somphouthone Phimmachak
Department of Biology, National University of Laos, Vientiane, Lao PDR
Department of Zoology, Kasetsart University, Bangkok, Thailand

Niane Sivongxay
Department of Biology, National University of Laos, Vientiane, Lao PDR

Bryan L. Stuart
North Carolina Museum of Natural Sciences, Raleigh, North Carolina, United States of America

Kathrin A. Witzenberger
Zoo Hoyerswerda, Hoyerswerda, Germany
Trier University, Department of Biogeography, Trier, Germany

Axel Hochkirch
Trier University, Department of Biogeography, Trier, Germany

Damon P. Little
Lewis B. and Dorothy Cullman Program for Molecular Systematics, The New York Botanical Garden, Bronx, New York, United States of America

Patrick Knopf and Christian Schulz
Lehrstuhl für Evolution und Biodiversität der Pflanzen, Ruhr-Universität Bochum, Bochum, Nordrhein-Westfalen, Bundesrepublik Deutschland

Daniel Frynta, Olga Š imková, Silvie Lišková and Eva Landová
Department of Zoology, Faculty of Sciences, Charles University in Prague, Prague, Czech Republic

Maria João Santos
Center for Spatial and Textual Analysis, Spatial History Project, and Bill Lane Center for the American West, Stanford University, Stanford, California, United States of America

Terry Watt
Terry Watt Planning Associates, San Francisco, California, United States of America

Stephanie Pincetl
California Center for Sustainable Communities, Institute of the Environment University of California Los Angeles, Los Angeles, California, United States of America

Jared G. Underwood
Pacific Reefs National Wildlife Refuge Complex, U.S. Fish and Wildlife Service, Honolulu, Hawai'i, United States of America

Mike Silbernagle
Oahu National Wildlife Refuge Complex, U.S. Fish and Wildlife Service, Haleiwa, Hawai'i, United States of America

Mike Nishimoto
Maui National Wildlife Refuge Complex, U.S. Fish and Wildlife Service, Kihei, Hawai'i, United States of America

Kim Uyehara
Kauai National Wildlife Refuge Complex, U.S. Fish and Wildlife Service, Kilauea, Hawai'i, United States of America

Diego Lirman, Stephanie Schopmeyer, Crawford Drury and Andrew C. Baker
Rosenstiel School of Marine and Atmospheric Science, University of Miami, Miami, Florida, United States of America

Victor Galvan
Punta Cana Ecological Foundation, Punta Cana, Dominican Republic

Iliana B. Baums
Pennsylvania State University, State College, Pennsylvania, United States of America

John M. Maniscalco
Department of Science, Alaska SeaLife Center, Seward, Alaska, United States of America
Institute of Marine Science, University of Alaska Fairbanks, Fairbanks, Alaska, United States of America

Alan M. Springer and Milo D. Adkison
Institute of Marine Science, University of Alaska Fairbanks, Fairbanks, Alaska, United States of America

Pamela Parker
Department of Science, Alaska SeaLife Center, Seward, Alaska, United States of America

Wei Liu, Dongxue Yin, Jianjun Liu and Na Li
College of Forestry, Northwest A & F University, Yangling, China

Rebecca Kormos
Department of Integrative Biology, University of California, Berkeley, California, United States of America

Cyril F. Kormos
The WILD Foundation, Berkeley, California, United States of America

Tatyana Humle
Durrell Institute of Conservation and Ecology, School of Anthropology and Conservation, University of Kent, Canterbury, United Kingdom

Helga Rainer
Conservation Program, The Arcus Foundation, Cambridge, United Kingdom

Annette Lanjouw
Strategic Initiatives and Great Apes Program, The Arcus Foundation, Cambridge, United Kingdom

Ray Victurine
Business Conservation Initiative and Conservation Finance, Wildlife Conservation Society, Bronx, New York, United States of America

Russell A. Mittermeier
Conservation International, Arlington, Virginia, United States of America

Mamadou S. Diallo
Guinée Écologie, Conakry, Republic of Guinea

Anthony B. Rylands
Conservation International, Arlington, Virginia, United States of America

Elizabeth A. Williamson
Scottish Primate Research Group, School of Natural Sciences, University of Stirling, Scotland, United Kingdom

Ximena Velez-Zuazo
Department of Biology, University of Puerto Rico-Rio Piedras, San Juan, Puerto Rico ecOceánica, Lima, Perú

Javier Quiñones and Sixto Quispe
Laboratorio Costero de Pisco, Instituto del Mar del Perú, Ica, Perú

Aldo S. Pacheco
ecOceánica, Lima, Perú
Instituto de Ciencias Naturales Alexander von Humboldt, CENSOR Laboratory, Universidad de Antofagasta, Antofagasta, Chile

Luciana Klinge and Shaleyla Kelez
ecOceánica, Lima, Perú

Evelyn Paredes
Unidad de Investigaciones en Depredadores Superiores, Instituto del Mar del Perú, Callao, Peru

Yi Yu, Qiang Fan, Rujiang Shen, Wenbo Liao and Jianhua Jin
Guangdong Key Laboratory of Plant Resources and Key Laboratory of Biodiversity Dynamics and Conservation of Guangdong Higher Education Institutes, School of Life Sciences, Sun Yat-Sen University, Guangzhou, China

Dafang Cui
College of Forestry, South China Agriculture University, Guangzhou, China

Wei Guo
Department of Horticulture and Landscape Architecture, Zhongkai University of Agriculture and Engineering, Guangzhou, China

Takehito Kaneko
Institute of Laboratory Animals, Graduate School of Medicine, Kyoto University, Kyoto, Japan

Hideyuki Ito and Hidefusa Sakamoto
Kyoto City Zoo, Kyoto, Japan
Wildlife Research Center of Kyoto University, Kyoto, Japan

Manabu Onuma
Center for Environmental Biology and Ecosystem studies, National Institute for Environmental Studies, Ibaraki, Japan

Miho Inoue-Murayama
Wildlife Research Center of Kyoto University, Kyoto, Japan
Center for Environmental Biology and Ecosystem studies, National Institute for Environmental Studies, Ibaraki, Japan

Brett Favaro
Department of Biology, University of Victoria, Victoria, Canada
Centre for Sustainable Aquatic Resources, Fisheries and Marine Institute of Memorial University of Newfoundland, St John's, Canada
Department of Ocean Sciences, Memorial University of Newfoundland, St. John's, Canada

Danielle C. Claar, Cameron Freshwater and Jessica J. Holden
Department of Biology, University of Victoria, Victoria, Canada

Caroline H. Fox
Department of Geography, University of Victoria, Victoria, Canada
Raincoast Conservation Foundation, Sidney, Canada, 6 Bamfield Marine Sciences Centre, Bamfield East, Canada

Allan Roberts
Bamfield Marine Sciences Centre, Bamfield East, Canada

UVic Research Derby
Department of Biology, University of Victoria, Victoria, Canada
Department of Geography, University of Victoria, Victoria, Canada
School of Environmental Studies, University of Victoria, Victoria, Canada

Index

A
African Jewelfish, 9-17

Altitude, 48-49, 57, 83-85, 89, 93-94, 168, 170, 173, 178-179, 209

Amphibian, 1, 7-8, 17, 83, 94-95, 106, 119

Animal Welfare, 3, 70, 223

Anthropogenic Disturbance, 28, 75

Artificial Ponds, 9-10

B
Biodiversity, 8, 16, 18-19, 24, 26-29, 35, 38-40, 43, 46-50, 53-59, 66-67, 81, 83, 95-96, 105, 109, 118-119, 127-128, 131, 139, 149, 168, 181-182, 184, 187-194, 207, 217-219, 224, 226, 230

Biogeographical Region, 28

Bird, 26, 40-44, 46-47, 49, 69, 74, 141, 147, 166, 213

Birdwatching, 40, 43, 47

Body Weight, 119, 121, 198

Breeding, 38-39, 41, 74, 77, 94, 96-99, 103-107, 119-120, 128-129, 141, 159, 164, 166, 175, 179, 197, 218, 230

Burrow, 29, 68-70, 72-73, 75-76

C
Cat, 96-97, 99, 101, 103, 105-106, 209-210

Cicindela Ohlone, 75-76, 81-82

Climate Change, 18, 26-27, 38, 94, 131, 136-139, 147, 157, 164, 168, 224

Coastal Development, 25, 28-30, 33-35

Cone Snails, 28-29, 36-37, 39

Conservation, 1, 7-9, 13, 16-22, 24-29, 31, 36, 38-43, 46-61, 64-69, 72-76, 80-84, 89, 94-97, 104-109, 116-117, 119-120, 123, 127-141, 145, 147, 149, 151, 156-157, 165, 167-168, 175, 178-183, 185-195, 197, 204-207, 212, 217-220, 223-226, 228, 230-231

Conus, 28-39

D
Data Deficiency, 28, 37

Densities, 9-10, 12-13, 15-16, 60-61, 63-64, 66, 120, 141, 164-165

Density Dependence, 77, 80, 205

Dispersal, 7, 18-21, 23-28, 30, 36, 68-70, 72-74, 81, 94, 168, 173-175, 179, 181, 192-193, 197, 216, 218

Dna Barcode, 108-109, 118

Dna Detection, 9, 11, 15

E
Ecological Traits, 18-21, 23-25

Edna, 9-17

European Wildcat, 96-97, 99, 101, 103-106

Eurycea Tonkawae, 1-2, 4, 7-8

Ex Situ, 96-99, 101, 103-107, 128, 157, 167, 178, 180

Expenditure, 43, 59, 61, 65-66

Experimental Design, 10, 156

Extinction, 1, 7, 18-19, 24-34, 36-41, 59, 66-67, 74-76, 78-81, 83, 94, 101, 106-107, 119, 128, 139, 165, 191-192, 194, 207, 217, 220, 223-225, 230-231

Extirpated Patches, 75-76, 81

F
Felis S. Catus, 96

Felis S. Silvestris, 96-97

Foraging, 70, 72, 75-76, 79, 128, 159, 162, 164, 195-196, 200, 202-203, 205-206

Fragmentation, 18-19, 21-27, 29-31, 67, 105, 135, 149-150, 153-154, 157, 167, 173, 179, 190, 213, 217

Funds, 40-41, 43, 46, 59-60, 120, 131, 137-138, 191-192, 195

G
Genetic Diversity, 95-96, 98-99, 104-107, 119, 167-169, 172-175, 178-180, 207-210, 212-213, 217-219

Genetic Integrity, 96, 101, 105-106

Genotypes, 97, 102, 108, 111, 114, 149-156, 179, 210

Geographic Distribution, 57, 61, 67, 83-84, 171, 209, 216

Global Threats, 28, 31

Grasses, 75

H
Habitat, 3, 8, 16, 18-19, 21-31, 33-34, 36-40, 49-50, 53, 59-61, 63-70, 72-76, 78-79, 81-91, 93-95, 108, 119-120, 136-140, 144, 146-147, 149-150, 152, 155, 167-168, 173, 175, 178, 181-196, 200, 203-205, 207, 213, 216-217, 224-226, 228-231

Habitat Loss, 24, 26, 28-30, 37, 39, 64, 66, 75, 139, 181, 186, 224, 230

Habitat Management, 75, 81

Human Disturbance, 48-51, 53-54, 57, 166, 173, 175, 189, 213, 226

Human Disturbance Index, 48, 50

I
In Situ, 38, 96-97, 99, 101, 105, 129, 149, 151, 156, 167, 175, 178
Investment, 46, 57, 59-60, 66, 183, 187, 191, 194

L
Land Use Policy, 131, 133, 139
Land-use Planning, 48-49, 57
Lao Newt, 83, 95
Laotriton Laoensis, 83-84, 88, 90, 94-95
Logistic Regression Models, 9, 11-15

M
Mammal, 8, 26-27, 42, 46, 49, 57, 124, 128-129, 165-166, 193, 231
Management, 7, 10, 13, 17-18, 26, 39-40, 43, 46, 48, 55, 57, 66, 68-69, 72, 74-76, 78-82, 86, 104, 107, 109, 117, 119, 128, 130, 132, 137, 140-142, 145-147, 156, 166, 168, 175, 182-194, 205-206, 208, 212, 225-226, 230
Mapping, 49-50, 106
Mating, 68, 72-76, 79, 82, 173, 179, 219
Matrix Modeling, 78, 80
Metapopulation, 74-76, 79-82, 205
Microsatellites, 96-98, 105-106, 118, 149, 218
Migration, 5, 21, 34, 69, 72, 75-76, 79-81, 165, 175, 179, 196, 205
Mtdna Sequencing, 96

N
Natural Land Resources, 131
Niche Modeling, 83-84, 90, 95

O
Ohlone Tiger Beetle, 75-77, 80, 82
Organism, 2, 9, 16
Overharvesting, 83
Oviposition, 75-76, 82

P
Photoid, 1-7
Pigmentation, 1-4, 6
Plant, 19, 27, 49, 54, 56-57, 59-67, 116-118, 131-132, 140, 167-168, 173, 175, 178-180, 183, 185-187, 203, 207, 212, 217-219
Podocarpaceae, 108-109, 111-112, 114-118
Protected Area, 40-41, 43, 46, 58, 81, 83-84, 90, 93-94, 131-132, 139, 191-192
Pygmy Bluetongue Lizard, 68, 72, 74

S
Salamander, 1-3, 5-8, 95
Sampling, 2, 5, 9-10, 13-16, 39, 62, 85, 94, 97, 105, 111, 116, 118, 129, 142, 155-156, 159, 161-162, 168-170, 178-179, 197-198, 201, 204, 207, 212, 224-228
Sea-level Rise, 18, 26-27
Sequencing, 9, 37, 96-98, 105, 109, 115, 118
Socioeconomic Proxies, 48, 50
Species, 1-3, 6-13, 15-21, 24-43, 46-50, 53-69, 72-76, 80-85, 88-90, 93-97, 104-109, 111-132, 136-143, 145-147, 149-151, 153-156, 158, 164-169, 171-175, 178-179, 181-183, 186, 189-192, 194-197, 201, 204-208, 212-214, 216-231
Specimens, 31, 34, 70, 108-109, 114, 118, 208, 216
Subspecies, 18-26, 60, 75, 107, 207, 225

T
Taxa, 7, 9, 18-20, 23-25, 27-31, 34, 36-38, 40, 56, 83, 120-121, 123, 127, 168, 181-182, 224
Tourism, 28, 31, 33-37, 40-47, 181, 192

U
Urban Pollution, 28

V
Visible Implant Elastomer, 1, 8

W
Water Samples, 9-16
Water Stratum, 9-11
Waterbirds, 140-142, 145-147
Wild-id, 1, 3, 5, 7
Wildlife, 1, 3, 7-9, 16-20, 26, 46-47, 49, 57, 65-68, 74, 82, 88, 101, 105, 119, 127-128, 130, 138, 140-142, 144, 146-147, 151, 159, 164-166, 181, 184-185, 187, 191-194, 197, 220, 223-225, 230-231

Z
Zoo, 39, 69-71, 74, 96-97, 99, 101, 104-107, 119-129, 193, 220-222